Progress in Botany/Fortschritte der Botanik 56

56 Progress in Botany

Structural Botany Physiology
Genetics Taxonomy Geobotany

Fortschritte der Botanik
Struktur Physiologie Genetik
Systematik Geobotanik

Editors
H.-Dietmar Behnke, Heidelberg
Ulrich Lüttge, Darmstadt
Karl Esser, Bochum
Joachim W. Kadereit, Mainz
Michael Runge, Göttingen

Springer-Verlag
Berlin Heidelberg New York
London Paris Tokyo Hong Kong
Barcelona Budapest

With 57 Figures

ISBN 3-540-58407-2 Springer-Verlag Berlin Heidelberg New York
ISBN 0-387-58407-2 Springer-Verlag New York Berlin Heidelberg

Printed in Germany

Library of Congress Catalog Card Number 33-15850

Typesetting: Macmillan India Ltd., Bangalore 25

SPIN: 10467351 31/3130/SPS-5 4 3 2 1 0 – Printed on acid-free paper

Contents

Review

A. Structural Botany

B. Physiology

C. Genetics

D. Taxonomy

E. Geobotany

F. Special Topics

List of Editors

Section A:

Professor Dr. H.-D. BEHNKE, Zellenlehre, Universität Heidelberg, Im Neuenheimer Feld 230, 69120 Heidelberg, Germany

Section B:

Professor Dr. U. LÜTTGE, TH Darmstadt, Institut für Botanik, FB Biologie (10), Schnittspahnstraße 3–5, 64287 Darmstadt, Germany

Section C:

Professor Dr. Dr. h.c. mult. K. ESSER, Lehrstuhl für Allgemeine Botanik, Ruhr Universität, Postfach 10 21 48, 44780 Bochum 1, Germany

Section D:

Professor Dr. J.W. KADEREIT, Institut für Spezielle Botanik und Botanischer Garten, Universität Mainz, Saarstraße 21, 55099 Mainz, Germany

Sections E and F:

Professor Dr. M. RUNGE, Lehrstuhl für Geobotanik, Systematisch-Geobotanisches Institut der Universität, Untere Karspüle 2, 37073 Göttingen, Germany

Professor Dr. Dr. h.c. mult. Hubert Ziegler

Review

Stable Isotopes in Plant Physiology and Ecology

By Hubert Ziegler

Every botanist is aware of the great progress achieved in the fields of biochemistry and plant physiology using radioactive isotopes with their easily and sensitively detectable radiation. The most striking example is the exploration of the carbon pathway in photosynthesis. Recently, besides these radioactive, instable isotopes, the non-radioactive, stable isotopes have also gained increased interest in biochemistry, plant physiology and ecology. This is not only true for oxygen (^{18}O) and nitrogen (^{15}N), where there is no radioactive isotope available with a suitable half-life for laboratory research, but also for hydrogen, carbon, sulfur and others, where besides the well-known radioactive isotopes, stable isotopes are also used.

In this review some successful or promising applications of stable isotope analysis are described, with main consideration of the pairs $^{12}C/^{13}C$ and $^{1}H/^{2}H$. No completeness of the very comprehensive data can be expected.

2. Abundance of Stable Isotopes

Many chemical elements of biological importance have two or more stable isotopes. One, normally the lightest (exception: iron and strontium), is present in far greater abundance than the other(s) (Table 1). Very little information is available about the biological fate of the stable isotopes of magnesium, silicon, chlorine, potassium, calcium, iron, copper, zinc, and strontium. Strontium isotopes are extensively used in geochemistry, e.g. to determine the ages of rocks (cf. review by Graustein 1988).

Progress in Botany, Vol. 56

Table 1. Average terrestrial abundances of the stable isotopes of major elements of interest in ecological studies. (Ehleringer und Rundel 1988)

Element	Isotope	Abundance (%)
Hydrogen	^{1}H	99.985
	^{2}H	0.015
Carbon	^{12}C	98.89
	^{13}C	1.11
Nitrogen	^{14}N	99.63
	^{15}N	0.37
Oxygen	^{16}O	99.759
	^{17}O	0.037
	^{18}O	0.204
Magnesium	^{24}Mg	78.70
	^{25}Mg	10.13
	^{26}Mg	11.17
Silicon	^{28}Si	92.21
	^{29}Si	4.70
	^{30}Si	3.09
Sulfur	^{32}S	95.00
	^{33}S	0.76
	^{34}S	4.22
	^{36}S	0.014
Chlorine	^{35}Cl	75.53
	^{37}Cl	24.47
Potassium	^{39}K	93.10
	^{40}K	0.0118
	^{41}K	6.88
Calcium	^{40}Ca	96.97
	^{42}Ca	0.64
	^{43}Ca	0.145
	^{44}Ca	2.06
	^{46}Ca	0.0033
	^{48}Ca	0.18
Iron	^{54}Fe	5.82
	^{56}Fe	91.66
	^{57}Fe	2.19
	^{58}Fe	0.33
Copper	^{63}Cu	69.09
	^{65}Cu	30.91
Zinc	^{64}Zn	48.89
	^{66}Zn	27.81
	^{67}Zn	4.11
	^{68}Zn	18.57
	^{70}Zn	0.62
Strontium	^{84}Sr	0.56
	^{86}Sr	9.86
	^{87}Sr	7.02
	^{88}Sr	82.56

3. Units of Isotope Composition, Standards, and Isotope Fractionation

Isotope composition is measured by means of a specially designed mass spectrometer as the ratio of the two stable isotopes of interest present in the sample. It has been found that the analysis of the difference between a defined standard and a sample gives more reliable and more precise results than the absolute isotope ratio in the sample (Mc Kinney et al. 1950).

The isotope composition of a sample is most commonly described by the differential notation:

$$\delta X_{std}\,(‰) = (R_{sample} / R_{std} - 1) \times 1000\ ,$$

where δX_{std} is the isotope ratio in δ units relative to a standard, and R_{sample} and R_{std} are the absolute isotope ratios of the sample and standard respectively.

To give an example with the relative $^{13}C/^{12}C$ ratio[1]:

$$\delta^{13}C\,[‰] = \left[\frac{R_{sample}}{R_{std}} - 1\right] \times 1000\,.$$

If R_{sample} and R_{std} are identical, $\delta^{13}C$ becomes 0, when the relative amount of ^{13}C in the sample is higher, the $\delta^{13}C$ value is positive, and when the sample is reduced in ^{13}C, $\delta^{13}C$ will be negative. A $\delta^{13}C$ value of −28‰, e.g., means that the sample contains 28‰ less ^{13}C than the standard. In a similar way, the relative amount of other stable isotopes is expressed in the literature.

The four accepted isotopic standards for the five principal elements of biological importance are carbonate from the fossil *Bellemnitella americana* from the PeeDee formation in South Carolina (PeeDee Belemnite: PDB) for carbon; standard mean ocean water (SMOW) for hydrogen and oxygen; atmospheric air for nitrogen; and the Canyon Diablo Meteorite (CD) for sulfur. Absolute ratios of these standards are given in Table 2.

There are two kinds of isotope fractionation: thermodynamic and kinetic (cf. Vogel 1980; O'Leary 1981; Simon 1983). Thermodynamic fractionations occur in processes which are at equilibrium. Kinetic fractionations occur because different isotopes are transformed at different rates. In both cases, fractionation is proportional to differences in the masses of the isotopes (Broecker and Oversley 1976).

---

[1] Recently, instead of $\delta^{13}C$ values, a value Δ for carbon isotope discrimination has also been used (cf. Evans et al. 1986; Farquhar et al. 1989; Mágúas et al. 1994), according to the notation:

$$\Delta = \frac{\delta_a - \delta_p}{1 + \delta_p}\ ,$$

where $\delta_a = \delta^{13}C$ of the air source and $\delta_p = \delta^{13}C$ of plant material. This is especially useful for instantaneous discrimination measurements, where the $\delta^{13}C$ of the CO_2 entering and leaving the cuvette with the plant material is analyzed.

Table 2. Isotope compositions of primary standards. (Hayes 1983)

Primary standard	Isotope ratio	Accepted value ($\times 10^6$) (with 95% confidence interval)
Standard mean ocean water (SMOW)	$^{2}H/^{1}H$	155.76 ± 0.10
	$^{18}O/^{16}O$	2005.20 ± 0.43
	$^{17}O/^{16}O$	373 ± 15
PeeDee Belemnite (PDB)	$^{13}C/^{12}C$	11237.2 ± 9.0
	$^{18}O/^{16}O$	2067.1 ± 2.1
	$^{17}O/^{16}O$	379 ± 15
Air	$^{15}N/^{14}N$	3676.5 ± 8.1

In the following, a short survey is given of some fundamental points in the interpretation of isotope data and some results of more general interest for the elements of main biological interest.

4. Isotope Composition of Natural Materials

a) $^{13}C/^{12}C$ Ratio

α) Basic Results

The first report about "variations in the relative abundance of the carbon isotopes" in natural materials (e.g. limestones, atmospheric CO_2, marine and terrestrial plants) was published by Nier and Gulbransen (1939). Further data on the carbon isotope composition of plant material were given by Wickman (1952), Craig (1953, 1954), and Baertschi (1953). Craig (1953) showed that his terrestrial plants were about 27‰ diminished in ^{13}C content, relative to ^{12}C ($\delta^{13}C$ = −27‰); compared with his PDB standard, his marine plants averaged 12‰ ($\delta^{13}C$ = −12‰). Craig (1954) supposed that the depletion of ^{13}C in the biomass of terrestrial plants proceeds during the photosynthetic fixation of atmospheric CO_2 ($\delta^{13}C$ = −7‰, with respect to PDB; Keeling 1958).[2]

This was confirmed and extended by Park and Epstein (1960), who found that the key enzyme of photosynthetic CO_2 fixation, the ribulose bisphosphate

[2] In industrial areas, the $\delta^{13}C$ value of atmospheric CO_2 may be more negative, due to combustion of oil and coal. The value decreased (became more negative) by ca. 0.6‰ between 1956 and 1978 (Keeling et al. 1979). In 1994, mostly a value of −8‰ is assumed (and measured) for atmospheric CO_2. CO_2 in greenhouses or dense vegetation can have a $\delta^{13}C$ value much more negative than −8‰. Leaves near the ground are several ‰ more negative than "normal" C_3 plants (see below) or leaves from higher elevation in the same stand (Vogel 1978; Medina and Minchin 1980).

carboxylase/oxidase (Rubisco), strongly discriminates the $^{13}CO_2$; the products have on average a $\delta^{13}C$ of $-28‰$.

Remarkably, all these pioneers in plant carbon isotope fractionation research were geochemists, not botanists.

β) Carbon Isotope Fractionation in Plants Performing Different Types of Photosynthetic CO_2 Fixation

An American archaeologist, Margret Bender, was responsible for the next step. She found (1968, 1971) that – according to radiocarbon determination – kernels and cobs of *Zea mays* were seemingly 160 to 240 years younger than wood samples from the same archaeological site. This did not agree with the archaeological conclusions. She could also show that maize (and related grasses from the subfamily Panicoideae) was richer in ^{14}C content (and ^{13}C!) than most other plants. This systematically higher ^{14}C content of maize compared with the wood samples was sufficient to explain the seemingly different radiocarbon age.

At that time, it was discovered that there was a group of plants with prefixation of CO_2 during photosynthesis, producing the first products with four C-atoms (Kortschak et al. 1965; Hatch and Slack 1970). *Zea mays* also undergoes this C_4 type of photosynthesis. Later, it was found that this different $^{13}CO_2$ fractionation is a general characteristic of all C_4 plants. As far as the carbon isotopes are concerned, the C_4 plants are heavier than the C_3 plants (Fig. 1); in the latter, the first and single CO_2-fixating enzyme is the strongly discriminating ($^{13}CO_2$ against $^{12}CO_2$) Rubisco. The difference in isotope composition is the most reliable and most commonly used method to distinguish C_3 and C_4 plants, even in herbarium material.[3]

The prefixation of CO_2 in C_4 plants occurs – as is well known – via phosphoenolpyruvate (PEP) carboxylase, spatially separated (in the leaf mesophyll) from the final fixating Rubisco (in the bundle sheaths of the leaf). The PEP carboxylase does not discriminate to a degree worth mentioning relative to HCO_3^- (which is the substrate for PEP carboxylase), but enriches considerably the ^{13}C relative to atmospheric CO_2 (see Table 3). The reason is the enrichment of ^{13}C during CO_2 hydration.

The average $\delta^{13}C$ value for the several hundreds of C_4 plants analyzed is $-14‰$ (PDB). This is surprisingly positive, when we bear in mind that the prefixed CO_2 has to be finally used (after decarboxylation of the primary products) by Rubisco with its strong discrimination of $^{13}CO_2$. The reason is the kinetic isotope effect in the CO_2 fixation by Rubisco. When this enzyme completely and exclusively fixes the CO_2 prefixed by PEP carboxylase and delivered to Rubisco

[3] One example of results with herbarium material: 10 of 65 genera of Poaceae in Germany (15%) have C_4 species, for the most part not autochthonous ones, while e.g. in Namibia, 86 of 105 genera include C_4 species (82%) (Ziegler and Schulze, unpubl.).

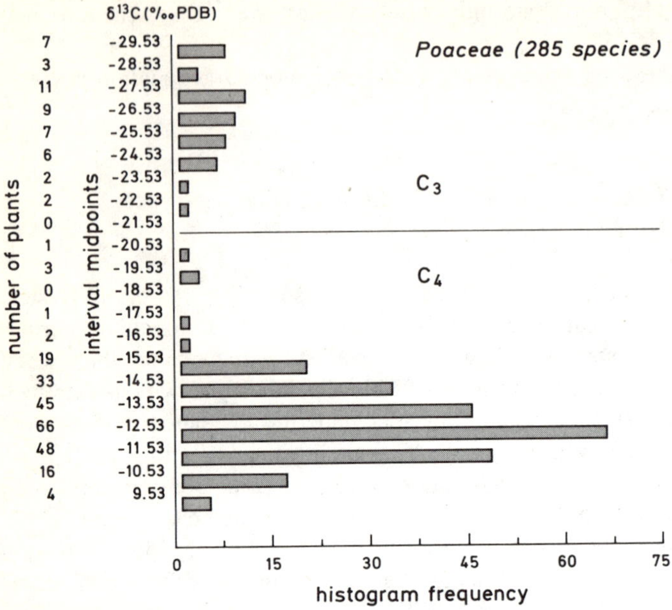

Fig. 1. $\delta^{13}C$ values of Poaceae from Namibia. Clear distinction between C_3 and C_4 species. (Ziegler and Schulze, unpubl. data)

Table 3. Expected values of isotope fractionations in various components of the CO_2 fixation process. (O'Leary 1981)

Step	Isotope discrimination ‰
Gas-phase diffusion of CO_2	4.4
Dissolution of CO_2[a]	−0.9
Liquid-phase diffusion of CO_2 or HCO_3^-	0.0
CO_2 hydration[a]	−7.0
Carboxylation of phosphoenolpyruvate	
Relative to HCO_3^-	2.0
Relative to CO_2	−5.0
Carboxylation of ribulose bisphosphate[b]	30
Respiratory decarboxylation	0.20

[a] Equilibrium isotope effect.

[b] Exact magnitude uncertain, at present 27 is mainly assumed.

by decarboxylation of malate or aspartate, the $\delta^{13}C$ values of the final photosynthetic products reflect those of the PEP carboxylase products. An increasing amount of atmospheric CO_2, passing through the mesophyll (and by-passing PEP carboxylase) shifts the $\delta^{13}C$ of the photosynthetic products to an increasingly more negative value.

There are three different types of C_4 plants, characterized by different decarboxylating enzymes in the bundle sheath, by the main transport substance from mesophyll to bundle sheath, and by some cytological characteristics of bundle sheath cells (cf. Ray and Black 1979). Hattersley (1982) could show that grass species grown under identical conditions, with these three types of C_4 photosynthesis (the NADP-ME type, the PCK type, and the NAD-ME type), have statistically significant differences in the ^{13}C values of their leaves. Hattersley interpreted these differences as differences in the rates of CO_2 and/or HCO_3^- leakage from the bundle sheath cells.

A third group of terrestrial plants separates prefixation (via PEP carboxylase) and final fixation (via Rubisco) not spatially, but in time: crassulacean acid metabolism (CAM) plants (cf. Kluge and Ting 1978; Osmond 1978; Osmond et al. 1982). They fix in the dark period via PEP carboxylase, and store the produced malic acid in the vacuoles. In the following light period, the malic acid is decarboxylated and the free CO_2 is fixed by Rubisco.

Besides this dark-prefixed, relatively ^{13}C-rich CO_2, the Rubisco also fixes atmospheric CO_2 (with great discrimination of $^{13}CO_2$) to a different degree during the light period. Since the relative contribution of dark prefixation and direct light fixation varies according to the environmental conditions (dark fixation relatively increased by water stress), the $\delta^{13}C$ values of CAM plants vary considerably over nearly the whole range of C_3 and C_4 plants (Fig. 2), and they become significantly more positive as the plants shift from predominantly light to predominantly dark fixation (Bender et al. 1973; Osmond et al. 1973; Lerman and Queiroz 1974; Medina and Troughton 1974). From the $\delta^{13}C$ values of the biomass it can be determined, therefore, whether a CAM species was grown on its natural site under more or less water stress.

Two examples can illustrate this statement:

Kluge et al. (1993) demonstrated that all species of *Kalanchoe* in Madagascar are capable of CAM. The ^{13}C values showed a clear correlation with the climate of the habitats from where the samples were derived (Fig. 3).

Within the family Orchidaceae the terrestrial species are sufficiently supplied with water and belong in general to the C_3 type of photosynthetic CO_2 fixation; this is true for all the species growing in Germany. The epiphytic (or winding) species normally have more problems with the water supply and show CAM (cf. Earnshaw et al. 1987; Fig. 4). This is why the natural (expensive!) spice "vanilla" from *Vanilla fragrans*, the only orchid of direct economic importance, is relatively rich in ^{13}C and can be distinguished by its $\delta^{13}C$ value (-16.7 to -20.8‰; Hoffman and Salb 1979) from artificial vanillin, made from coniferin or eugenol, both from C_3 sources. It was the natural spice, of course, with which Montezuma flavoured the chocolate he gave Cortés in 1520 (cf. review by Winkler and Schmidt 1980). Even individuals within one species show different $\delta^{13}C$ values when they grow terrestrially or epiphytically.

As mentioned above, aquatic plants often have significantly less negative (richer in ^{13}C) $\delta^{13}C$ values than terrestrial plants, besides the fact that they belong in most cases to the C_3 type of photosynthesis. This may have two different reasons:

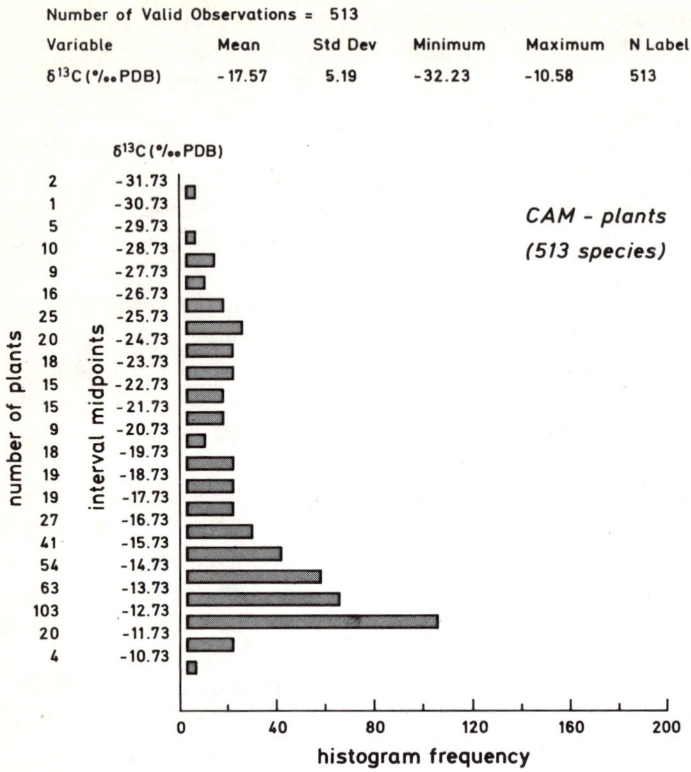

Fig. 2. $\delta^{13}C$ values of CAM species, analyzed to date by the author. They cover the whole range of C_3 and C_4 plants, with two maxima between −24 to −25‰ (C_3) and between −11 and −12‰ (C_4)

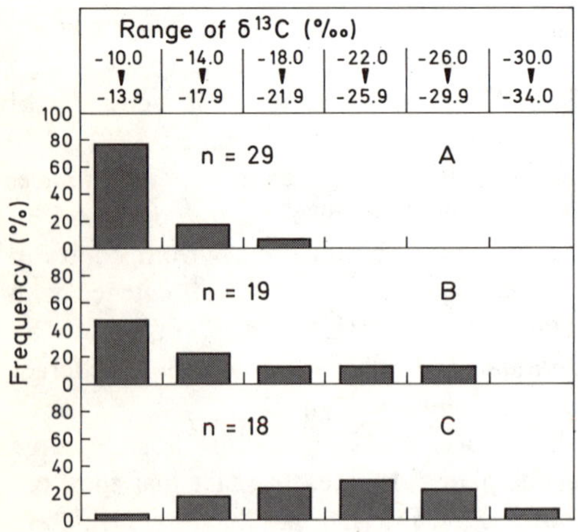

Fig. 3. Quantitative evaluation of the $\delta^{13}C$ distribution over the vegetation zones of Madagascar. The frequency is expressed as percent of the total number of samples (n) analyzed for the given zone. **A** Xerophilous bush; **B** evergreen dry forest and savanna; **C** evergreen rain forest and evergreen montane forest. (Kluge et al. 1993)

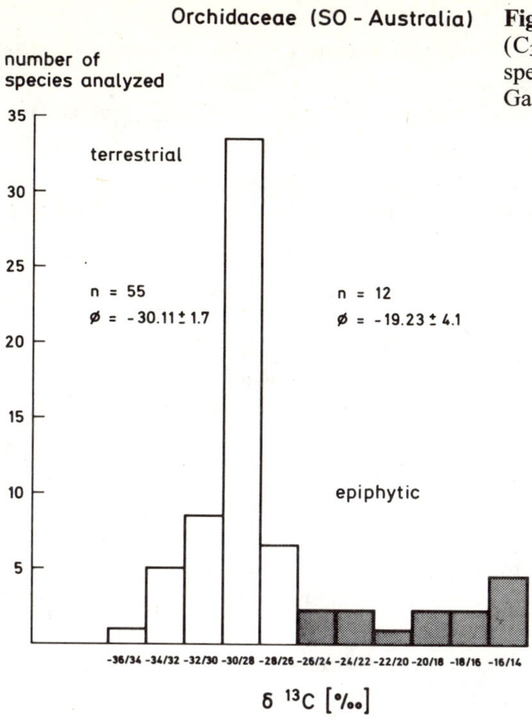

Fig. 4. $\delta^{13}C$ values from terrestrial (C_3) and epiphytic (CAM) Orchidaceae species from SO Australia. (Ziegler and Gaff, unpubl. data)

In some cases, the diffusion of CO_2 through water may be the limiting step of photosynthetic CO_2 fixation and not the action of carboxylating enzymes. Since this diffusion does not discriminate $^{13}CO_2$ (Table 3), the relatively high ^{13}C content of the biomass is understandable.

The other possibility is the action of a bicarbonate pump in the plasmalemma of cells in multicellular or – quite often – unicellular water plants. As mentioned above, ^{13}C is relatively enriched in HCO_3^- relative to free CO_2.

These two possibilities of ^{13}C enrichment could possibly explain the different $\delta^{13}C$ values of (terrestrial) lichens with green algae or cyanobacteria, respectively, as photobionts (Lange and Ziegler 1986; Lange et al. 1988; Badger et al. 1993; Máguas et al. 1994).

γ) Different Fractionation of Carbon Isotopes Due to Variable Environmental Conditions

The most important factor influencing the carbon isotope fractionation in photosynthetic organs of terrestrial plants with C_3 photosynthesis is the intercellular CO_2 concentration (C_i). Farquhar et al. (1982) gave the following formula:

$$\delta^{13}C_{leaf} = \delta^{13}C_{air} - a - (b-a)c_i/c_a \ ,$$

where a is the fractionation caused by the slower gas-phase diffusion of $^{13}CO_2$ relative to $^{12}CO_2$ (4.4‰), b is the fractionation caused by discrimination of Rubisco against $^{13}CO_2$ (27‰), and c_a is the atmospheric CO_2 concentration. There are many reports confirming a close relationship between $\delta^{13}C_{leaf}$ and c_i (e.g. Farquhar et al. 1982; Downton et al. 1985; Ehleringer et al. 1985; Seemann and Critchley 1985; cf. Fig. 5).

Also the leaf water-use efficiency (molar ratio of photosynthesis to transpiration) is correlated with c_i, according to the equation:

$$A/E = (c_a - c_i)/(1.6\,\Delta_w)\ ,$$

where A is the net photosynthetic rate, E the transpiration rate, and Δ_w is the leaf to air water vapour concentration gradient. High water-use efficiencies are related to less negative (more positive) values of $\delta^{13}C$ (cf. Farquhar and Richards 1984; Ehleringer et al. 1985; Ehleringer 1988; Farquhar et al. 1988). With this approach integrated, long-term water-use efficiency by a plant can be checked by the leaf carbon isotope ratio (Fig. 5).

Striking examples for these correlations are mistletoes and their hosts. Mistletoes are normally hemiparasites, feeding on the xylem of their hosts. To obtain sufficient water, they have to lower their water potential below that of their hosts. This is achieved by higher and longer leaf conductivity, resulting in increased water loss by transpiration, and higher c_i, resulting in more negative $\delta^{13}C$ of leaf biomass, and a lower water-use efficiency (WUE), compared with the hosts (cf. Ziegler 1986; Ehleringer et al. 1985; Ehleringer 1988).

The effect of temperature is less clear than the effect of CO_2 concentration on the carbon isotope fractionation. Nevertheless, many attempts were made to correlate $\delta^{13}C$ values with growth temperature, especially with the view that plants

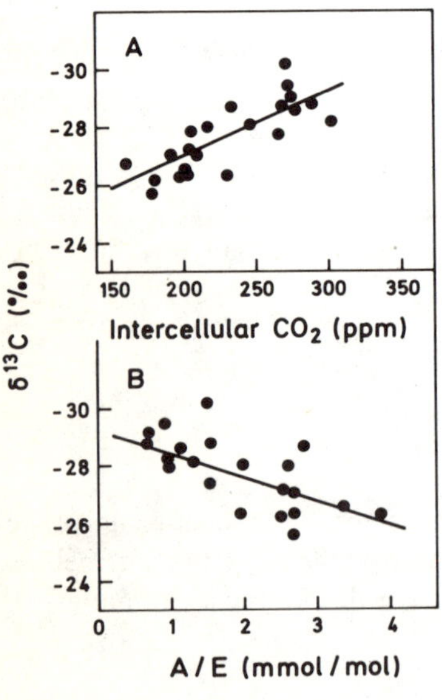

Fig. 5. A Relation between carbon isotope ratio and average daytime intercellular CO_2 concentration $[CO_2]_i$ for mistletoes and host plants in central Australia. The regression line is $\delta^{13}C = -22.54 -0.0226\ [CO_2]_i$ ($P \leq 0.01$). Standard error of the slope is 0.0042. **B** Relation between carbon isotope ratio and daily water-use efficiency (see text) for mistletoes and host plants in central Australia. The regression line is $\delta^{13}C = -29.46 + 0.863$ (A/E) ($P \leq 0.05$). Standard error of the slope is 0.227. (Ehleringer et al. 1985)

might be useful indicators for past climate (cf. literature in O'Leary 1981). Lipp et al. (1991), e.g., analyzed the cellulose in late wood of dendrochronologically dated tree rings of *Abies alba* and found significant correlations between temperature, humidity and precipitation and the $\delta^{13}C$ values (but not the $\delta^{2}H$ values) for the period 1959 to 1980. Based on this experience, they reconstructed a palaeoclimatic record for the past millenium (Fig. 6), which seems to be in good agreement with historical data (cf. Lamb 1968). It is probable that the effect of temperature does not directly influence the metabolic carbon isotope fractionation, but is mediated by the change in leaf conductivity for gas exchange and as a consequence by the change in c_i. This is especially clear, when CAM plants are considered; here, a higher temperature is mostly combined with increased water stress and a switch in the primary CO_2 fixation (see above).

δ) $\delta^{13}C$ Values in Trophic and Edaphic Chains

There are numerous reports on the use of $\delta^{13}C$ values to clarify trophic or edaphic chains. Several examples are treated, e.g. in "Stable Isotopes in Ecological Research", edited by Rundel, Ehleringer, and Nagy (1988), especially in the chapters by Ehleringer (Physiological Processes in Aridland Plants), Tieszen and Boutton (Stable Carbon Isotopes in Terrestrial Ecosystem Research), Fry and Sherr ($\delta^{13}C$ Measurements as Indicators of Carbon Flow in Marine and Freshwater Ecosystems), Schell and Ziemann (Natural Carbon Isotope Tracers in Arctic Aquatic Food Webs). In the following, a few, mainly our own, results will be discussed.

The $\delta^{13}C$ value of soil organic matter (SOM) is closely related to that of the vegetation with which it was in equilibrium for a longer period. C_3 vegetation (e.g. a forest) results in value of SOM different to C_4 grass vegetation. Also, changes from one vegetation type to another are mirrored in the SOM at differing soil depths (review in Tieszen and Archer 1990). The respiratory CO_2 from a soil cropped only with C_3 plants for at least 7 years showed a $\delta^{13}C$ value very similar to C_3 plants. Correspondingly, a soil bearing the C_4 plant maize for at least 6 years had a $\delta^{13}C$ value for the respiratory CO_2 quite near a typical C_4

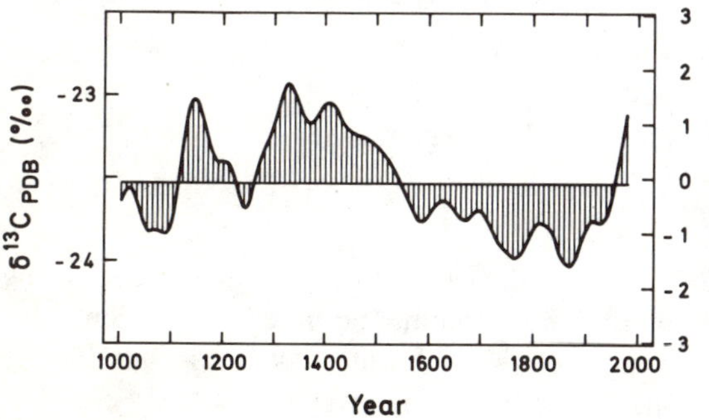

Fig. 6. $\delta^{13}C$ values (smoothed by Gaussian low-pass filter over 100 years) of cellulose extracted from late wood of firs (*Abies alba*) in the Black Forest during the period from 1004 to 1980 A.D. Values from 1850 to 1980 are corrected for input of anthropogenic fossil fuel CO_2. Temperatures are reported as deviations from their mean value. (Lipp et al. 1991)

plant. A soil cropped with maize in the year of study only showed intermediate values, with a tendency towards C_4 (Schönwitz et al. 1986).

The results of the carbon isotope determinations in mistletoes (of the C_3 families Viscaceae and Loranthaceae) growing on CAM hosts were instructive. The same parasite species had a higher ^{13}C content (less negative $\delta^{13}C$ values) on a CAM than on a C_3 host (cf. Ziegler 1994). This indicates that mistletoes cannot only be water, but also assimilate parasites (cf. Ehleringer et al. 1985). There is a highly specialized Loranthaceae species, *Tristerix aphyllus*, which was called by Mauseth et al. (1984) "one of the most highly reduced seed plant known". As we have found recently, it has traces of chlorophyll and only very few (functioning?) stomata and is obviously a full parasite: *Tristerix aphyllus* had a $\delta^{13}C$ value of $-14.56‰$ on the CAM host *Echinops chilensis* ($-12.79‰$), while the "normal", leaf-bearing species of the same genus, *Tristerix penduliflorus*, had a $\delta^{13}C$ value of $-29.08‰$ and the (C_3) host *Polylepis* sp. $-26.46‰$ (Kraus and Ziegler, unpubl.).

By measuring the $\delta^{13}C$ value in the tissues or faeces of herbivorous animals, it is possible to determine the relative proportion of C_3 or C_4 (or CAM) plants in their diets. Since in the lowland tropics most grass species are C_4, most shrubs and trees and nearly all forbs are C_3 (Vogel 1978; Tieszen et al. 1979), the feeding pattern of different animals can be checked: pure browse types (e.g. kudu antelope, giraffe), mixed feeders (goats and sheep), and pure grazers (cattle) can be distinguished (Vogel 1978). By analyzing human fossils even the main diet of ancient hunters can be identified.

De Niro and Epstein (1978) showed that the $\delta^{13}C$ values of the tissues gave a sequence hair < brain muscle < liver < fat, but only fat showed a larger difference (about 3‰ more negative than the diet). This fractionation of ^{13}C during the synthesis of fatty acids (and other substances derived from acetyl-CoA) is due to the discrimination of ^{13}C-pyruvate against ^{12}C-pyruvate by pyruvate dehydrogenase (De Niro and Epstein 1977).

Using the advantageous situation at an Australian butterfly farm, i.e. the food plants of four species of caterpillars were all C_3 plants, while the artificial nectar for the imagines was made with sucrose from sugarcane (C_4 species), it could be shown by analyzing the food plants, the biomass of the caterpillars, the nectar for the imagines and the biomass of the adult butterflies that ca. 90% of the biomass of the imagines is derived from the caterpillars (Ziegler and Wright 1993).

b) $^2H/^1H$ Ratio (cf. reviews by Schiegl 1970; Ziegler 1988; Smith and Ziegler 1990)

The basic source for the deuterium in the biomass of an autotrophic plant (using water as hydrogen source) is the water of the medium, and in terrestrial plants this means soil water. The deuterium content of the soil water again depends mainly on its concentration in precipitation. Condensate from water vapour is enriched in deuterium relative to the vapour with the result that the remaining vapour will be increasingly depleted in deuterium as moisture is removed from the air. Since the absolute water content of the atmosphere is temperature-dependent, with decreasing temperature an increasing depletion of deuterium in precipitation will result. This is called the "climatic effect" (cf. Schiegl 1970). The variations in δD (δ^2H) of natural waters caused by condensation and evaporation lie between +100‰ and −400‰ (SMOW as standard).

As a consequence of these basic relations, Schiegl (1970) distinguished the following effects on the deuterium content of the precipitation and subsequently on the soil water:

Latitude Effect. With increasing geographic latitude, the deuterium content of precipitation decreases, following the temperature gradient between the equator and the poles. The effect increases rapidly with latitude (Dansgaard 1961, 1964).

Altitude Effect. With increasing altitude, the deuterium content of precipitation will gradually decrease: at low and mid-latitudes, this effect is about 1.6‰/100 m, and at high latitudes, 4.8‰/100 m difference in altitude (Dansgaard 1964).

Seasonal Effect. The deuterium content of precipitation at higher latitudes (> 30°) varies seasonally, with a maximum in summer and a minimum in winter.

Continental Effect. Precipitation will contain less deuterium the further the air mass moves from the coast into the interior of a continent.

Total Precipitation. With an increasing amount of precipitation, the deuterium content of the precipitation decreases. This is especially important in tropical regions, where the seasonal variations are small.

As early as 1934, Washburn and Smith found that the leaf tissue water of plants was heavier than the soil water. It was found later that the leaf tissue water is enriched in deuterium as well as in oxygen 18 (see later).

This enrichment of deuterium in leaf tissue water is not due to discrimination of the H_2O against HDO during uptake into the roots (Table 4); this is reasonable,

Table 4. δD values of tissue water in different parts of plants grown under controlled conditions[a,b]

Species	δD (Tissue water)‰ Root	Shoot
Spinacia oleracea (C_3)[c]	−71.0	−33
Pisum sativum (C_3)	−75.0	−41
Amaranthus edulis (C_4)	−68.0	−42
Sedum praealtum (CAM)	−78.0	−32
Mean	−73.0	−37

[a] From Ziegler et al. (1976)
[b] δD of irrigation water = −78.2‰.
[c] δD of transpiration water = −152.1‰.

since the influx of water into plants is governed by viscous flow rather than molecular flow. Also, during the transport of water within the xylem, there is little enrichment of HDO. This was shown by comparing the δD value of the medium water with that of root pressure exudate (Fig. 7). A strong discrimination of HDO against H_2O results during transpiration in the leaves (cf. Table 4) and is the main reason for deuterium enrichment in leaf tissue water (Fig. 7; cf. historical outline by Bricout 1978). Since the export of assimilates from photosynthesizing leaves starts with leaf water (according to the mass-stream theory), it was assumed at first that the content of the sieve tubes should mirror the δD of the leaf tissue water. The analysis of sieve-tube sap of *Robinia pseudoacacia* (ca. 30 years old), obtained about 1.5 m above ground and at about the same distance from the crown basis, yielded a value very near that of the soil water, but very different from that of leaf tissue water (Fig. 7). This can be explained by an effective exchange of water between xylem and phloem. Moreover, this may also explain the small increase in HDO in xylem water versus soil water. If this assumption is valid, the deuterium content in the sieve-tube sap should become progressively higher, the closer one gets to the transpiring leaf (the shorter the time for exchange with xylem water). This was confirmed in an experiment with *Cucurbita pepo* (Fig. 7).

Bokhoven and Theeuwen (1956) were the first to describe a strong decrease in deuterium content in the organic plant material, in comparison to that of tissue water. As shown by Estep and Hoering (1981) with microalgae, the deuterium content of the organic material of a plant depends on the deuterium content of the tissue (or medium) water only when plants are growing photoautotrophically. If the cells grow heterotrophically, the δD value of the organic material depends largely on the deuterium content of the food material. This is also true for heterotrophic tissues or organs in an autotrophic higher plant, e.g. for roots. In photoautotrophic cells, the δD value of the organic material is much more negative (in the algae, about 100%) than that of the medium water (= cell water). This is also true for photoautotrophic Eubacteria (Quandt et al. 1977), which do not use H_2O as a hydrogen source in photosynthesis. This indicates that in photoautotrophic cells, not only HDO is discriminated against H_2O during the light-driven reduction of $NADP^+$ (in organisms with water as hydrogen donor; cf. Estep and Hoering 1981), but in a similar way there is also discrimination of other reducing substances (H_2S, H_2, etc.) containing deuterium against those with protium alone in the course of light-induced reduction of NAD^+ (in Eubacteria).

This discrimination is again considerably increased when acetyl-CoA is the substrate for organic syntheses, e.g. in fatty acids and terpenoids (Smith and Epstein 1970; Bricout 1978). It may be supposed that, as in the case of ^{13}C, the pyruvate dehydrogenase complex discriminates strongly against the heavy (deuterium-rich) pyruvate.

When we consider that the deuterium content in the organic substances of a plant depends primarily on that of the tissue water of photosynthesizing organs and, further, that apparently the main discrimination of the heavy isotope occurs during the light-mediated pyridine nucleotide reduction, it is not surprising that plants with different types of photosynthetic CO_2-fixation pathways differ in the

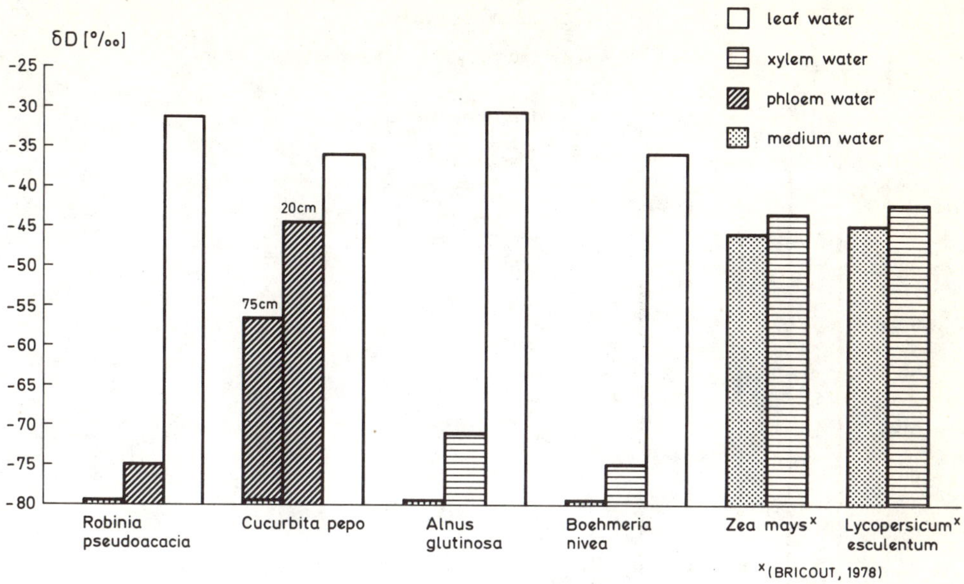

Fig. 7. δD values in medium water, leaf water (root pressure exudate), and phloem water (sieve-tube sap) of different species. The sieve-tube sap in the axis of *Cucurbita pepo* was obtained 20 and 75 cm from the leaf basis. Note the different δD values of the medium water in Munich ($-78.2‰$) and in Verrière (south France) (ca. $-46.0‰$) in the experiments of Bricout (1978). (Ziegler 1988)

deuterium content of their organic material, since these different types have a different water-use efficiency (see review by Osmond et al. 1982).

Plants of the Crassulacean acid type (CAM) of photosynthetic CO_2 fixation generally show a much higher deuterium content in the organic material than C_3 or C_4 plants from the same biotope or under the same conditions of cultivation (Fig. 8; cf. Ziegler et al. 1976; Sternberg et al. 1984). This difference can be used to distinguish among CAM plants and C_4 plants on dry sites, when the $\delta^{13}C$ values do not provide a clear distinction.

It can be shown that CAM plants in their natural habitats also have a high deuterium content in their tissue water. This was reported already by Schiegl (1970): *Opuntia* (CAM) tissue water had on average a 20‰ less negative δD value than *Populus* (C_3). At that time, there was no explanation for this observed difference. The accumulation of heavy water in the tissues of CAM plants in dry environments is easily explained: succulent CAM plants are adapted to habitats where they can occasionally fill up their large water-storing vacuoles after rare heavy rains and where long dry periods occurring in between hinder potential non-CAM competitors. During these drought periods, these CAM plants transpire slowly, but continuously (mainly at night), discriminating and therefore accumulating HDO in the remaining tissue water. Since metabolism in CAM plants

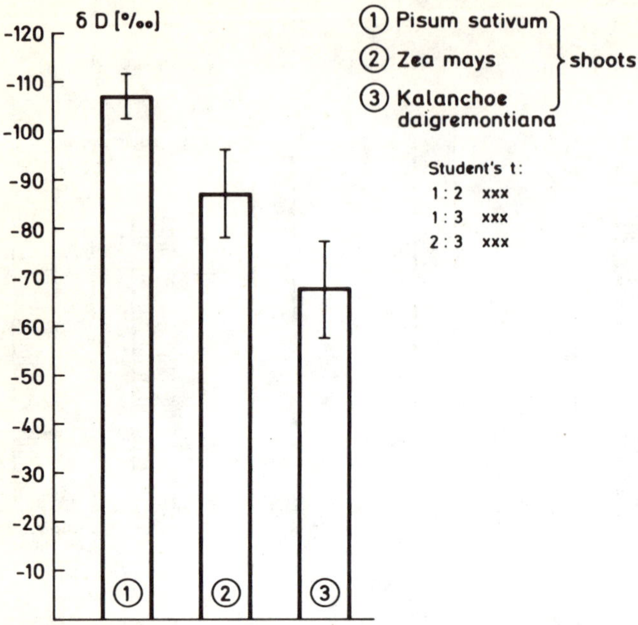

Fig. 8. δD values in the organic material of the shoots of the C_3 plant *Pisum sativum*, the C_4 plant *Zea mays*, and the CAM plant *Kalanchoe daigremontiana*, cultivated under identical conditions. All differences are statistically highly significant. (Lenhart 1979)

involving cyclic synthesis and degradation of malate continues even after long-lasting water stress (Szarek et al. 1973; Osmond 1975) and since this metabolism is again dependent on tissue water, the deuterium content in organic material of CAM plants becomes progressively higher, the stronger and longer-lasting the water-stress period (Ziegler et al. 1976). It is not completely clear, however, whether the enrichment of deuterium in the tissue water can fully explain the relatively high deuterium content in the organic material of CAM plants. Considering the situation in C_4 plants (see below), this explanation seems rather doubtful.

Are there characteristic differences in the δD values of the organic material between C_3 and C_4 plants? Smith and Epstein (1970) observed no characteristic differences between C_3 and C_4 plants in a salt marsh. The question can be answered only by cultivating plants with different photosynthetic pathways under controlled conditions, especially ensuring that the water supply has a uniform δD value. This was first done by Ziegler et al. (1976); the results are summarized below:

When irrigated with a water source with a δD value of $-78.2‰$, the C_3 plants showed mean organic material δD values of $-132‰$ for shoots and $-117‰$ for roots; C_4 plants showed δD values of $-91‰$ for shoots and $-77‰$ for roots, and CAM plants a δD value of $-75‰$ for shoots and roots.

These differences between the δD values of the different types of photosynthetic pathway are statistically significant, as shown for the organic material of the shoots in Fig. 8.

As Sternberg and De Niro (1983) and Sternberg et al. (1984) pointed out, differences in the deuterium content of the organic material can also be due in part to differences in the chemical composition or to the exchange of hydrogen bound to organic compounds (mainly in -OH and -NH_2 groups; cf. Bonhoeffer 1934) with tissue water. These workers analyzed, therefore, the δD values of cellulose nitrate from C_3, C_4, and CAM plants, growing within a small area in Val Verde County, Texas. They confirmed that CAM plants had distinctly higher D/H ratios than C_3 or C_4 plants, while there was no clear-cut difference between the C_3 and C_4 plants.

Lenhart (1979; cf. Ziegler 1988) analyzed different chemical fractions of a C_3 plant (*Pisum sativum*), a C_4 plant (*Zea mays*), and a CAM plant (*Kalanchoe daigremontiana*), which were grown under identical conditions, with irrigation water having a uniform δD value. The chemical fractions isolated from the shoot were: total organic material, H_2O extract, lipids, protein, insoluble material, and cellulose nitrate. At least in cellulose nitrate, not only the differences in the D/H ratio between C_3 and CAM plants or between C_4 and CAM plants were statistically highly significant, but also the difference between C_3 and C_4 plants (Fig. 9).

The difference in hydrogen isotope composition of dry matter in C_3 and C_4 plants could be due again to a difference in the deuterium content of the tissue water that serves as the source for photosynthetically used hydrogen. To answer

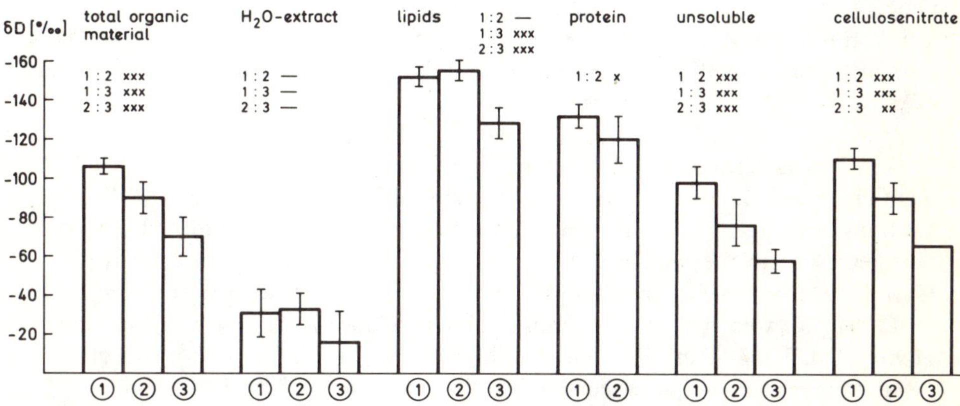

Fig. 9. δD values of different chemical fractions, including cellulose nitrate (unable to exchange hydrogen with medium water), in the leaves of *Pisum sativum* (C_3), *Zea mays* (C_4) and *Kalanchoe daigremontiana* (CAM) with statistical evaluation (t-test). The plants were cultivated under identical conditions. (Lenhart 1979). 1, Pisum sativum; 2, Zea mays, 3, Kalanchoe daigremontiana

this question, Leaney et al. (1985) cultivated again C_3 and C_4 plants under identical conditions with irrigation water having a uniform δD value. The δD value of leaf tissue water showed a marked diurnal variation; the greatest enrichment of deuterium was observed about midday. When δD values of organic material were referenced to the mean δD values of tissue water during the period of active photosynthesis, the discrimination against deuterium during photosynthetic metabolism was greater in C_3 plants (−117 to −121‰) than in C_4 plants (−86 to −109‰).

From the results it was concluded that the different water-use efficiencies of C_3 and C_4 plants are responsible for the measured difference in the deuterium composition of leaf water. However, it seemed unlikely that these physical progresses account fully for the differences in the hydrogen isotope composition of the C_3 and C_4 photosynthetic assimilates. This conclusion is in accordance with the assumptions of Sternberg and De Niro (1983).

Estep and Hoering (1981) consider that a discrete pool of organically bound hydrogen exists in photosynthetic organisms that does not undergo rapid isotopic exchange with tissue water. This pool, probably NADPH, could have a special isotope composition due to photoreduction. The differences in δD values in organic material between C_3 and C_4 plants may thus mirror differences in the kinetic properties of the thylakoid NADP reductase enzymes (Ziegler et al. 1976), or they may reflect differences in the compartmentalization of reductants during C_4 photosynthesis (cf. Leaney et al. 1985).

It seemed to be of interest in this connection to check the δD value of the organic material in plants with different types of C_4 photosynthesis, since they show characteristic differences in compartmentalization, as mentioned above for their $\delta^{13}C$ values. Together with Paul Hattersley (Canberra), we checked the δD values in the organic material of grass species with different types of C_4 photosynthesis, grown under identical conditions and irrigated with water having a uniform deuterium content. The δD values of the species with the NADP-ME type of C_4 photosynthesis were significantly different from those of the PCK and NAD-ME types. The latter two types did not differ statistically from each other.

How can this exceptional situation of the NADP-ME plants be explained? A main characteristic of this type is the formation of malate in the mesophyll cells, the transfer of bound CO_2 and reduction equivalents (in the form of malate) from the mesophyll to the bundle sheath, and the reuse of the reduction equivalents taken over from malate by $NADP^+$ immediately in the Calvin cycle. Smith et al. (1991) showed that the deuterium content of the tissue water in the leaves of the NADP-ME plant *Zea mays* is higher (δD is less negative), the greater the distance from the water-conducting tissues (xylem of bundles), since transpiration discrimination could operate over a longer distance (Table 5). Hydrogen imported in organically bound form from the mesophyll to the bundle sheath carries, therefore, a relatively high deuterium content, resulting finally in ^2H-rich photoassimilates.

Table 5. Hydrogen and Carbon isotopes from different tissues of *Zea mays*. (Smith et al. 1991)

Tissue	δD_{water} [‰ vs. SMOW]	$\delta D_{organic\ matter}$ [‰ vs. SMOW]	$\delta C_{organic\ matter}$ [‰ vs. PDB]
Intercostal cells(mesophyll)	+ 7.88	−55.18	−14.74
Veins(bundle sheath)	−19.67	−52.80	−15.08
Roots	−90.20	−52.00	−17.00

c) $^{18}O/^{16}O$ Ratio (cf. review by Sternberg 1988)

There are three stable isotopes of oxygen: ^{16}O (99.759%), ^{17}O (0.037%), and ^{18}O (0.204%). Only the $^{18}O/^{16}O$ ratios are normally determined.

The $\delta^{18}O$ varies in nature from +40‰ (heavy sedimentary carbonates) to −50‰ (light meteoric waters). Meteoric waters range from −50 to +6‰ in their $\delta^{18}O$ values (Craig 1961). Plant organic material varies in a similar way. Geologically, $\delta^{18}O$ determinations became very important in understanding thermodynamic processes and as indicators of palaeotemperatures (Urey et al. 1951; Epstein 1959; Taylor 1974; see review, Ehleringer and Rundel 1988).

Similar to deuterium, fractionation of $H_2{}^{18}O$ against $H_2{}^{16}O$ does not occur during uptake of soil water by the root or during transport of water in the xylem (Gonfiantini et al. 1965; Zundel et al. 1978). In the transpiring leaf, however, $H_2{}^{18}O$ is relatively enriched (Gonfiantini et al. 1965; Lesaint et al. 1974; Ferhi and Letolle 1977; cf. Sternberg 1988). It was found further that $\delta^{18}O$ in the cellulose of plant leaves, independently of the different photosynthetic modes, is always 27 ± 3‰ higher than for the water at the site of cellulose synthesis (Epstein et al. 1977; De Niro and Epstein 1981, Sternberg and De Niro 1983; Sternberg 1988). There is good evidence that this relation is due to an exchange of oxygen atoms between synthesized cellulose and leaf water (in a carbonyl hydration reaction) (De Niro and Epstein 1981; Sternberg 1988).

Consequently, e.g., in the field, cellulose from C_4 plants has higher $\delta^{18}O$ values than cellulose from C_3 plants (Sternberg et al. 1984).

d) $^{15}N/^{14}N$ Ratio (cf. review by Shearer and Kohl 1988; Virginia et al. 1988)

There are two stable isotopes of nitrogen in nature: ^{15}N (0.3663% in air) and ^{14}N (Nier 1950; Junk and Svec 1958; Sweeny et al. 1978). Since there is no variation in ^{15}N abundance of atmospheric N_2 (Mariotti 1983), atmospheric N_2 is used as standard for $\delta^{15}N$ analysis (see above).

Schönheimer and Rittenberg (1939) were the first to find variations in $\delta^{15}N$ in biological samples; they range from −5 to +10‰. Normally, soil N contains more ^{15}N than atmospheric N (cf. Shearer et al. 1978; Steele and Wilson 1981; Karamanos et al. 1981; Mariotti 1982; Shearer and Kohl 1988), perhaps due to discrimination of the heavier isotope by microbes during decomposition. Plants dependent on soil-derived N, therefore, should be more abundant in ^{15}N than N_2-fixing plants which obtain N_2 from the atmosphere as well as bound nitrogen from the soil. Thus, by determining $\delta^{15}N$ in plant material, a semi-quantitative analysis of N_2 fixation under field conditions is possible and quite frequently performed.

^{15}N is quite homogeneously distributed in non-nodulated plants, but quite often enriched in nodules (in the bacteroids and cortex) of N_2-fixing species (cf. Shearer and Kohl 1988); the reason for this observation is unknown. Animal tissues are mostly richer in ^{15}N than plants; the enrichment increases along subsequent trophic levels (Miyake and Wada 1967; Minegawa and Wada 1984; Schoeninger and De Niro 1984) due to the elimination of ^{14}N in the catabolic pathways (Gaebler et al. 1966).

A different field of research, where ^{15}N is used, is the application of additional ^{15}N-labelled compounds (e.g. NO_x or NH_3) from the atmosphere and an analysis of their fate in the plant (e.g. Bruckner et al. 1993; Koppisch et al. 1993; Schulze et al. 1994). $^{15}N_2$ fixation was also used for final proof that root nodules or rhizothamnia from suspected, previously undescribed, N_2-fixing species are really fixing N_2 (cf. Ziegler and Hüser 1963: *Comptonia peregrina*; Heisey et al. 1980: *Chamaebatia foliolosa*).

The many other applications of ^{15}N in biochemical studies will not be discussed here.

e) $^{34}S/^{32}S$ Ratio (cf. reviews by Fry 1988; Krouse 1988; Winner et al. 1988)

There are four stable sulfur isotopes in nature: ^{32}S (95.00%), ^{33}S (0.76%), ^{34}S (4.22%), and ^{36}S (0.014%). Thode et al. (1949) were the first to report on variations in isotope composition of S compounds in nature.

$\delta^{34}S$ values in plants and soils vary between −30 and +30‰, but much less in unpolluted areas (Krouse 1980). The variation in $\delta^{34}S$ values between sulfur in pollutants and sulfur in plants and soils is the basis for the use of sulfur isotope ratios in pollutant transfer studies.

Relatively uncomplicated is the situation in marine food webs. Since in oceanic SO_4^{2-} $\delta^{34}S$ is uniform (around +21‰) and since the isotope discrimination during SO_4^{2-} assimilation is small, members of this food web usually show $\delta^{34}S$ values around $+17 \pm 3$‰; this also holds true for coasts and islands dominated by the import of marine sulfur. An exception are biotopes around ocean floor thermal vents or marshes, where thermogenic or biogenic sulfide, normally very depleted in ^{34}S, plays a major role (cf. Krouse 1988).

Also the emission of H_2S and other reduced sulfur compounds by leaves of terrestrial plants having a high sulfur supply is preferentially the ^{32}S (Winner et al.

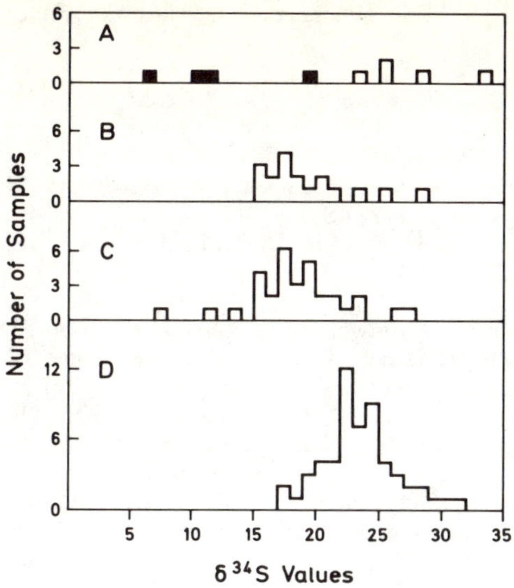

Fig. 10. $\delta^{34}S$ values of soils (**A**, *solid bars*) and air (**A**, *open bars*), *Abies balsamea* needles (**B**), *Picea glauca* needles (**C**), and terrestrial mosses (**D**), collected at the Fox Creek study site. (Winner et al. 1978)

1981) so that the remaining S content in the plant is predominantly the ^{34}S. When the $\delta^{34}S$ values of atmospheric, gaseous sulfur compounds (e.g. SO_2 or H_2S) differ from those of soil SO_4^{2-}, the relative amount of S taken up and incorporated by terrestrial plants can be determined (see Fig. 10). Epiphytic lichens and some mosses give a cumulative value for the isotope composition of atmospheric sulfur.

If the $\delta^{34}S$ in the sources of anthropogenic gaseous sulfur compounds (e.g. ores or coals) are different, even the emission region of the pollutant can be identified.

It should be mentioned, however, that the interpretation of sulfur isotope data in ecosystems is quite complex. Details can be found in the reviews mentioned above.

Acknowledgement. The isotope analyses of the author and his coworkers were performed in cooperation with W. Stichler, P. Trimborn and J. Lipp, Insitut für Hydrologie des GSF-Forschungszentrums für Umwelt und Gesundheit, Oberschleißheim, and continuously supported by the Deutsche Forschungsgemeinschaft.

References

Badger MR, Pfanz H, Büdel B, Heber U, Lange OL (1993) Planta 191:59–72. **Baertschi P** (1953) Helv Chim Acta 36:773–781. **Bender MM** (1968) Radiocarbon 10:468–472. **Bender MM** (1971) Phytochemistry 10:1239–1244. **Bender MM**, Rouhani I, Vines HM, Black CC Jr (1973) Plant Physiol 52:427–430. **Bokhoven C**, Theeuwen HHJ (1956) Ned Akad Wet B 59:78–83. **Bonhoeffer KA** (1934) Z Elektrochem 40:469–474. **Bricout J** (1978) Rev Cytol Biol Végét-Bot 1:133–209. **Broecker WS**, Oversley VM (1976) Chemical equilibria in the Earth. McGraw-Hill, New York. **Bruckner G**, Gebauer G, Schulze ED (1993) Isotopenpraxis 29:71–76.

Craig H (1953) Geochim Cosmochim Acta 3:53–92. **Craig H** (1954) J Geol 62:115–149. **Craig H** (1961) Science 133:1702–1703.

Dansgaard W (1961) Meddelelsor om Grönland 165 (2), Copenhagen. **Dansgaard W** (1964) Tellus 16:436–468. **De Niro MJ**, Epstein S (1977) Science 197:261–263. **De Niro MJ**, Epstein S (1978) Geochim Cosmochim Acta 42:495–506. **Downton WJS**, Grant JR, Robinson SP (1985) Plant Physiol 78:85–88.

Earnshaw MJ, Winter K, Ziegler H, Stichler W, Cruttwell NEC, Kerenga K, Cribb PJ, Wood J, Croft JR, Carvor KA, Gunn ZC (1987) Oecologia 73:566–572. **Ehleringer JR** (1988) In: Rundel PW, Ehleringer JR, Nagy KA (eds) Stable isotopes in ecological research. Springer, Berlin Heidelberg New York, pp 41–54. **Ehleringer JR**, Rundel PW (1988) In: Rundel PW, Ehleringer JR, Nagy KA (eds) Stable isotopes in ecological research. Springer, Berlin Heidelberg New York, pp 1–15. **Ehleringer JR**, Schulze ED, Ziegler H, Lange OL, Farquhar GD, Cowan IR (1985) Science 227:1479–1481. **Epstein S** (1959) In: Abelson PH (ed) Research in geochemistry. John Wiley and Sons, New York, pp 217–240. **Epstein S**, Thompson P, Yapp CJ (1977) Science 198:1209–1215. **Estep MF**, Hoering TC (1981) Plant Physiol 67:474–477. **Evans JR**, Sharkey TD, Berry JA, Farquhar GD (1986) Aust J Plant Physiol 13:281–292.

Farquhar GD, Richards RA (1984) Aust J Plant Physiol 11:539–552. **Farquhar GD**, O'Leary MH, Berry JA (1982) Aust J Plant Physiol 9:121–137. **Farquhar GD**, Hubick KT, Concon AG, Richards RA (1989) In: Rundel PW, Ehleringer JR, Nagy KA (eds) Stable isotopes in ecological research. Springer, Berlin Heidelberg New York, pp 21–40. **Ferhi A**, Letolle R (1977) Physiol Végét 15:363–370. **Fry B** (1988) In: Rundel PW, Ehleringer JR, Nagy KA (eds) Stable isotopes in ecological research. Springer, Berlin Heidelberg New York, pp 445–453. **Fry B**, Sherr EB (1988) In: Rundel PW, Ehleringer JR, Nagy KA (eds) Stable isotopes in ecological research. Springer, Berlin Heidelberg New York, pp 196–229.

Gaebler OH, Vitti TC, Vumirovich R (1966) J Biochem Phys 44:1245–1257. **Gonfiantini R**, Gratziu S, Tongiori E (1965) In: Isotopes and radiation in soil. Plant Nutrition Studies. International Atomic Energy Agency, Vienna, p 405. **Graustein WC** (1988) In: Rundel PW, Ehleringer JR, Nagy KA (eds) Stable isotopes in ecological research. Springer, Berlin Heidelberg New York, pp 491–512.

Hatch MD, Slack CR (1970) In: Reinhold L, Liwschitz Y (eds) Progress in phytochemistry. Wiley-Interscience, New York. **Hattersley PW** (1982) Aust J Plant Physiol 9:139–154. **Hayes JM** (1983) In: Meinschein WG (ed) Organic geochemistry of contemperaneous and ancient sediments. Society of Economic Paleontologists and Mineralogists, Bloomington, pp 5–31. **Heisey RM**, Delwich CC, Virginia RA, Bryan BA (1980) Am J Bot 67:429–431. **Hoffmann PG**, Salb M (1979) J Agric Food Chem 27:352.

Junk G, Svec HV (1958) Geochim Cosmochim Acta 14:234–243.

Karamonos RE, Voroney RP, Rennie DA (1981) Soil Sci Soc Am J 45:826–828. **Keeling CD** (1958) Geochim Cosmochim Acta 13:322–324. **Keeling CD**, Mook WG, Tans PP (1979) Nature 277:121. **Kluge M**, Ting IP (1978) Crassulacean acid metabolism. Ecological Studies 30. Springer, Berlin Heidelberg New York. **Kluge M**, Brulfert J, Lipp J, Ravelomanana D, Ziegler H (1993) Bot Acta 106:320–324. **Koppisch D**, Schulze ED, Gebauer G (1993) Isotopenpraxis 29:77–84. **Kortschak HP**, Hartt CE, Burr GO (1965)

Plant Physiol 40:209–213. **Krouse HR** (1980) In: Fritz P, Fontes JC (eds) Handbook of environmental isotope geochemistry. Elsevier, New York, pp 435–471. **Krouse HR** (1988) In: Rundel PW, Ehleringer JR, Nagy KA (eds) Stable isotopes in ecological research. Springer, Berlin Heidelberg New York, pp 424–444.

Lamb HH (1968) The Changing Climate. Section 7, Methuen, London. **Lange OL**, Ziegler H (1986) In: Marcelle R, Clijsters H, Van Poucke M (eds) Biological control of photosynthesis. Nijhoff, Dordrecht, pp 147–161. **Lange OL**, Green TGA, Ziegler H (1988) Oecologia 75:494–501. **Leaney FW**, Osmond CB, Allison GB, Ziegler H (1985) Planta 164:215–220. **Lenhart B** (1979) PhD Thesis, Techn Univ Munich. **Lerman JC**, Queiroz O (1974) Science 183:1207–1210. **Lesaint C**, Merlivat L, Bricout J, Fontes JC, Gautheret R (1974) C R Acid Sci Paris Ser D 278:2925–2930. **Lipp J**, Trimborn P, Fritz P, Moser H, Becker B, Frenzel B (1991) Tellus 43B:322–330.

Mágúas C, Griffiths H, Broadmeadow MSJ (1994) (in press). **Mariotti A** (1982) Mémoires des Sciences de la Terre no 81–13. Univ P et M Curie, Paris. **Mariotti A** (1983) Nature 303:685–687. **Mauseth JD**, Montenegro G, Walckowiak AM (1984) Can J Bot 62:847–857. **Mc Kinney CR**, Mc Crea JM, Epstein S, Allen HA, Urey HC (1950) Rev Sci Instrum 21:724–730. **Medina E**, Minchin P (1980) Oecologia 45:377–378. **Medina E**, Troughton JH (1974) Plant Sci Lett 2:357. **Minegawa M**, Wada E (1984) Geochim Cosmochim Acta 48:1135–1140. **Miyake Y**, Wada E (1967) Rec Oceangr Works Japan 9:37–53.

Nier AO (1950) Phys Rev 77:789–793. **Nier AO**, Gulbransen EA (1939) J Am Chem Soc 61:697–698.

O'Leary MH (1981) Phytochemistry 20:553–557. **Osmond CB** (1975) In: Marcelle R (ed) Environmental and biological control of photosynthesis. Junk, The Hague, pp 299–309. **Osmond CB** (1978) Annu Rev Plant Physiol 29:379–414. **Osmond CB**, Allaway WC, Sutton BG, Troughton JH, Queiroz O, Lüttge U, Winter K (1973) Nature 246: 41–42. **Osmond CB**, Winter K, Ziegler H (1982) In: Lange OL, Nobel PS, Osmond CB, Ziegler H (eds) Physiological plant ecology II. Encyclopedia Plant Physiol, New Series, vol 12B, Springer, Berlin Heidelberg New York, pp 479–547.

Park R, Epstein S (1960) Geochim Cosmochim Acta 21:110–126.

Quandt L, Gottschalk G, Ziegler H, Stichler W (1977) FEMS Microbiol Lett 1:125–128.

Ray TB, Black CC (1979) In: Gibbs M, Latzko E (eds) Photosynthesis II. Encyclopedia Plant Physiol vol 6. Springer, Berlin Heidelberg New York, pp 77–101. **Rundel PW**, Ehleringer JR, Nagy KA (1988) Stable isotopes in ecological research. Springer, Berlin Heidelberg New York.

Schell DM, Ziemann PJ (1988) In: Rundel PW, Ehleringer JR, Nagy KA (eds) Stable isotopes in ecological research. Springer, Berlin Heidelberg New York, pp 230–251. **Schiegl W-G** (1970) PhD Thesis, Univ of South Africa, Pretoria. **Schoenheimer R**, Rittenberg D (1939) J Biol Chem 127:285–290. **Schoeninger MJ**, De Niro MJ (1984) Geochim Cosmochim Acta 48:625–639. **Schönwitz R**, Stichler W, Ziegler H (1986) Oecologia 69:305–308. **Schulze ED**, Chapin FS III, Gebauer G (1994) Oecologia (in press). **Seemann JR**, Critchley C (1985) Planta 164:151–162. **Shearer G**, Kohl DH (1988) In: Rundel PW, Ehleringer JR, Nagy KA (eds) Stable isotopes in ecological research. Springer, Berlin Heidelberg New York, pp 342–374. **Shearer G**, Kohl DH, Chien SH (1978) Soil Sci Soc Am J 42:899–902. **Simon H** (1983) In: Hoppe W, Lohmann W, Markl H, Ziegler H (eds) Biophysics. Springer, Berlin Heidelberg New York, pp 301–315. **Smith BN**, Epstein

S (1970) Plant Physiol 46:738–742. **Smith BN**, Ziegler H (1990) Bot Acta 103:335–342. **Smith BN**, Ziegler H, Lipp J (1991) Naturwissenschaften 78:358–359. **Steele KW**, Wilson AT (1981) NZ J Agric Res 24:167–170. **Sternberg LSL** (1988) In: Rundel PW, Ehleringer JR, Nagy KA (eds) Stable isotopes in ecological research. Springer, Berlin Heidelberg New York, pp 124–141. **Sternberg L**, De Niro MJ (1983) Science 220:947–948. **Sternberg L**, De Niro MJ, Ajie H (1984a) Phytochemistry 23:2475–2477. **Sternberg LO**, De Niro MJ, Johnson HB (1984b) Plant Physiol 74:557–561. **Sternberg LSL**, De Niro MJ, Ajie H (1986) Planta 169:320–324. **Sweeney RE**, Liu KK, Kaplan IR (1978) In: Robinson BW (ed) Stable isotopes in the Earth. Science Div of Scientific and Industrial Res Bull 220. **Szarek SR**, Johnson HB, Ting IP (1973) Plant Physiol 52:539–541.

Taylor HP (1974) Econ Geol 69:843–883. **Thode HG**, Macnamara J, Collins CB (1949) Can J Res 27:361–373. **Tieszen LL**, Archer St (1990) In: Osmond CB, Pitelka LF, Hidy GM (eds) Plant biology of the basin and range. pp 293–321. **Tieszen LL**, Boutton TW (1988) In: Rundel PW, Ehleringer JR, Nagy KA (eds) Stable isotopes in ecological research. Springer, Berlin Heidelberg New York, pp 167–195. **Tieszen LL**, Senyimba MM, Imbamba SK, Troughton JH (1979) Oecologia 37:337–350.

Urey HC, Lowenstam HA, Epstein S, McKinney CR (1951) Bull Geol Soc Am 62: 399–416.

Virginia RA, Jarrell WM, Rundel PW, Shearer G, Kohl DH (1988) In: Rundel PW, Ehleringer JR, Nagy KA (eds) Stable isotopes in ecological research. Springer, Berlin Heidelberg New York, pp 375–394. **Vogel JC** (1978) Ecol Plant 13:89–94. **Vogel JC** (1980) Sitzungsber Heidelberger Akad Wiss, Math-Naturwiss Klasse. Springer, Berlin Heidelberg New York, pp 111–135.

Washburn EW, Smith ER (1934) J Res Natl Bur Stand 12:305–311. **Wickmann FE** (1952) Geochim Cosmochim Acta 2:243–254. **Winkler FJ**, Schmidt H-L (1980) Z Lebensm Unters Forschg 171:85–94. **Winner WE**, Bewley JD, Krouse HR, Brown HM (1978) Oecologia 35:351–361. **Winner WE**, Smith CL, Koch GW, Mooney HA, Bewley JE, Krouse HR (1981) Nature 289:672–673. **Winner WE**, Berg VS, Langston-Unkefer PJ (1988) In: Rundel PW, Ehleringer JR, Nagy KA (eds) Stable isotopes in ecological research. Springer, Berlin Heidelberg New York, pp 454–470.

Ziegler H (1986) In: Marcelle R, Clijsters H, Van Poucke M (eds) Biological control of photosynthesis. Nijhoff, Dordrecht, pp 171–185. **Ziegler H** (1988) In: Rundel PW, Ehleringer JR, Nagy KA (eds) Stable isotopes in ecological research. Springer, Berlin Heidelberg New York, pp 105–123. **Ziegler H** (1994) Carbon and hydrogen isotope discrimination in Crassulacean acid metabolism. Ecological Studies (in press). **Ziegler H**, Hüser R (1963) Nature 199:508. **Ziegler H**, Wright P (1993) Isotopenpraxis 29:45. **Ziegler H**, Osmond CB, Stichler W, Trimborn P (1976) Planta 128:85–92. **Zundel G**, Miekeley W, Grisi BM, Förstel II (1978) Rad Environ Biophys 15:203–212.

Prof. Dr. Dr. h.c. mult. Hubert Ziegler
Institut für Botanik und Mikrobiologie
Technische Universität München
Arcisstrasse 21
80333 München, Germany

A. Structural Botany

I. Cytology and Morphogenesis of the Prokaryotic Cell

By Frank Mayer and Michael Hoppert

1. Structural Organization and Biogenesis of Bacterial Flagella

Studies on the structural organization of the flagellar filament and base and on the biogenesis of these components revealed new insight into this complex system (Macnab 1992).

Together with the sensory apparatus that controls its operation, the flagellum represents a unique subsystem in the bacterial cell (Iino 1969; Emerson et al. 1970; DePamphilis and Adler 1971a,b; Silverman and Simon 1974; Manson et al. 1977; Homma and Iino 1985; Trachtenberg and DeRosier 1987; Macnab and DeRosier 1988; Blair 1990; Driks and DeRosier 1990; Homma et al. 1990; Khan 1990; Macnab 1990; Jones and Aizawa 1991; Khan et al. 1991). Likewise complex are the genetics of flagella and associated functions. Based on knowledge of the many genes that are involved, their chromosomal organization, their operon structure, the control of their expression, and the function of the gene products, processes such as flagellar biogenesis, energetics, rotation and switching, and chemotaxis can now be described in more detail (Silverman and Simon 1972, 1973; Iino 1974; Komeda et al. 1980; Ikeda et al. 1987; Macnab 1987; Stewart and Dahlquist 1987; Iino et al. 1988; Khan et al. 1988; Kihara et al. 1989; Jones and Macnab 1990; Jones et al. 1990; Gillen and Hughes 1991; Stock et al. 1991; Khan et al. 1992; Kubori et al. 1992; Müller et al. 1992).

A recent model (Macnab 1992) describing the structural organization of the flagellar filament and base of *Escherichia coli* indicates that older interpretations of electron micrographs have to be corrected as far as the inner rings are concerned. The entities, formerly called M-ring and L-ring, could be shown not to be separate structures. Rather, they are components of only one ring, now collectively termed MS-ring (Macnab 1992). This ring, together with the rod, the hook, and the filament, is postulated to rotate relative to a ring formed by Mot Proteins which is situated in the plane of the cytoplasmic membrane. Structural features described recently (Khan et al. 1991) include a switch and an export apparatus responsible for transport of the axial flagellar proteins to the outside of the cell. The biogenesis of the flagellar apparatus was analyzed by considering the morphology of the successive structures that occur during assembly, and the products of their respective genes (Macnab 1992).

Progress in Botany, Vol. 56

The initiating event of flagellar biogenesis is the insertion of the MS-ring protein into the cytoplasmic membrane and its self-assembly into the ring structure. The switch turned out to be a peripheral structure at the base of the MS-ring, located at the cytoplasmic face of the MS-ring. This finding suggests that assembly of the switch follows directly on that of the MS-ring. Experimental evidence confirmed this view (Suzuki and Komeda 1981; Jones and Macnab 1990; Kubori et al. 1992). The next step is attachment of a "rivet" (Kubori et al. 1992; Müller et al. 1992). It contains the proximal and the distal rod elements and extends from the cell membrane to just beyond the outer membrane. The export apparatus (Vogler et al. 1991; Macnab 1992) responsible for transport of the axial flagellar proteins from the inside to the outside of the cell obviously must assemble prior to the export process, and so prior to assembly of structures such as rod, hook, and filament. The gene products which comprise the P- and L-rings are the only flagellar components that have cleaved signal peptides and so are presumed to be exported by the primary cellular pathway (Jones et al. 1989; Wickner et al. 1991).

By analysis of mutants, it could be shown that hook assembly can be initiated without the P- and L-rings present, but it proceeds only to a limited extent (Kubori et al. 1992). This was assumed to be due to constraint given by the fact that in this mutant further growth of the hook is prevented by the barrier presented by the outer membrane. In wild-type cells, this constraint does not exist because the L-ring allows penetration of the outer membrane by the growing hook. Assembly of the hook proceeds to a well-defined length in wild-type cells (DePamphilis and Adler 1971). Overproduction of hook protein causes the hook-length control mechanism to fail (Jones et al. 1987), and it could be assumed that limitation of synthesis of hook protein in wild-type cells might control growth of the hook. However, this question is not yet settled. At the point where hook elongation is completed, hook-filament junction proteins and capping proteins are exported, followed by export of flagellin molecules. The junction proteins are laid down in a specific order (Homma and Iino 1985) and are believed to be necessary for mating of hook and filament. The capping proteins (Homma et al. 1985; Ikeda et al. 1987), located between junction proteins (terminating the hook) and growing filament, are assumed to bias the kinetics in favor of assembly of exported flagellin molecules into the defined filament structure. In contrast to rod and hook, there is no control of the extent of elongation of the filament. However, the rate of elongation (Iino 1974) decreases with length of the filament, presumably because the frictional coefficient for passage of a column of monomers down the hollow core within the flagellar filament increases with length. In addition, long filaments are fragile and so breakage resets the length.

The Mot proteins (Khan et al. 1988) are not needed for the assembly of the rest of the flagellar apparatus. The point at which the Mot proteins assemble themselves around the basal body is not known. It is likely that they could assemble quite early in the pathway of flagellar biogenesis (Macnab 1992).

In summary, flagellar operons (Kutsukake et al. 1988) form a coordinated system of expression, or regulon (Yamaguchi et al. 1986; Nishimura and Hirota 1989; Kutsukake et al. 1990). They are arranged in a hierarchy in which expression of an operon in a given class is necessary for expression of operons in lower classes. In the case of flagella, four such classes have been defined (1, 2, 3a, 3b) (Iino 1985; Macnab 1990). Class 1 contains only the master operon, whose expression is needed for the expression of all other operons. Class 2 operons contain many of the structural genes for components that are assembled in the early and middle stages of flagellar synthesis. These operons are under direct control of the master operon. Class 3a operons are expressed under dual control, in part by the master operon and in part by a class 2 (fliA; flagellum-specific sigma factor; Helmann and Chamberlin 1987; Ohnishi et al. 1990; Helmann 1991) operon. Expression of class 3b operons appears to be solely under the control of the fliA operon.

2. Techniques to Study the Structure of Biofilms

In a recent communication, Sutton et al. (1994) compared conventional scanning electron microscopic (SEM) techniques, low temperature SEM, and application of the Electroscan wet scanning electron microscope to study the structure of a biofilm.

Biofilms formed by bacteria are consortia of cells and extracellular polymers (EPS). The EPS enclose the cells, attach them to the substratum, and hold them in place (Marshall et al. 1971; Fletcher 1977; Zvyagintsev et al. 1977; Geesey et al. 1978; Wardell et al. 1980; Costerton et al. 1981; Geesey 1982; Hermansson et al. 1982; Eighmy et al. 1983; Nelson et al. 1985; Fletcher 1986; Allison and Sutherland 1987; Costerton et al. 1987; Hamilton 1987; Roszak and Colwell 1987; Van Loosdrecht et al. 1987; Marshall 1988; Beech and Gaylarde 1989; Siebel and Characklis 1990; Geesey et al. 1992; Marshall 1992).

Biofilms are highly hydrated, open structures, composed of 73–98% extracellular material and space (Lawrence et al. 1991).

Details of the ultrastructure of fully hydrated biofilms have not been revealed, as the routine SEM preparation techniques employed so far produce varying degrees of shrinkage artifacts.

Recently, a new SEM, the Electroscan, has become available. In this instrument, untreated living wet samples and even liquids (Baumgarten 1989) can be observed. The high vacuum requirement of conventional SEM is eliminated. It comprises a differential pumping system along the column, whereby a series of chambers, linked by small apertures, are maintained at increasing pressures along the beam path, and a detector in a gaseous environment. These properties indicate that the Electroscan should have great potential for studying the surface morphology of bacterial biofilms as there are no preparation artifacts to damage or shrink the matrix of EPS. Sutton et al. (1994) demonstrate that this expectation is fulfilled. They evaluated critical point drying, freeze-drying, low temperature SEM and Electroscan imaging for biofilm preservation. Their sample was an in vitro biofilm of the dental plaque organism *Streptococcus crista* CR3, generated on hyroxyapatite discs. Application of the Electroscan gave the best image of the biofilm. The true extent of the apparently undamaged EPS was revealed and the bacterial cells were completely buried underneath a smooth uninterrupted cover of EPS.

3. A Consortium of Sulfate-Reducing Bacteria with the Capacity to Reduce Cr(VI) to Cr(III)

At sites of industrial plants engaged in metal finishing and electroplating, severe contamination of the soil and the wastewater are frequently observed. Cr(VI) (CR^{6+}, CrO_4^{2-}, and $Cr_2O_7^{2-}$) found in these soils and waters is toxic to plants, animals, and humans. Due to a high tendency to bind to oxygen, Cr(VI) exists as an anion of high solubility. Cr(III) is much less mobile, and it precipitates as hydroxide at a pH above 5.5 (Alloway 1990). Cr(VI) is readily reduced to Cr(III) by Mn(II), Fe(II), S^{2-}, and some organic compounds (Richard and Bourg 1991). Therefore, Cr(VI) should be removed from wastewater originating from contaminated soil. Usually, this requires chemical reduction, ion exchange, or electrodepositing. The use of microbial biomass, as a Cr absorbent, offers an inexpensive alternative (Fude and Shigui 1993). It was suggested that the use of sulfate-reducing bacteria may be possible to ensure Cr(VI) reduction and precipitation, thereby decontaminating the liquid waste and providing the opportunity for Cr recovery. Fude et al. (1994) recently characterized a consortium of bacteria with tolerance to high concentrations of Cr(VI) and other toxic heavy metals. They showed that this consortium consists of a range of Gram-positive and Gram-negative rods, and that it has the capacity to reduce Cr(VI) to Cr(III) as amorphous precipitates which are associated with the bacterial surface. Over 80 to 95% of Cr could be removed from concentrations ranging from 50 to 2000 ppm. The sorption was not entirely due to a passive interaction with reactive sites on the bacterial surface as killed cells could not immobilize as much metal. When U or Zn was added with the Cr, it was also removed and could even increase the total amount of Cr immobilized. The authors suggest that the reduction occurs via H_2S, and they speculate that the consortium may be useful for the bioremediation of fluid metal-refining wastes. Pilot treatment ponds are now in operation.

4. Establishment of Biofilms in the Laboratory, and Studies of Inherent Dynamics and Interactions

The in situ degradation of organic xenobiotica present in soil and water is one of the major tasks of applied microbiology (Jain and Sayler 1987). Various approaches can be used. The autochthonous microflora could be optimized, or mixed populations of bacteria cultured in the laboratory could be brought into the soil or water. For this purpose, GEMs constructed according to the specific need could replace the autochthonous microflora. However, in the latter case, restrictions are obvious. Not only legal aspects are of concern. Questions may arise regarding the efficiency of degradation of xenobiotica in the situation during application as compared to the laboratory experiment, and the possibility should be considered that the ecological balance could be lost.

Experiments aimed at a simulation of the in situ situation are promising when populations of microorganisms are used that are capable of forming biofilm consortia. A number of questions can be addressed during the study of respective systems (Aquifer columns, microcosms, simplified biofilms):

- What are the consequences of competition between added and autochthonous microflora?
- Can added bacteria grow or, at least, survive?
- Is there horizontal gene transmission in a diversified habitate?

Winkler (1994) analyzed defined systems. He established binary biofilms consisting of *Pseudomonas putida* (pWWO) and *Pseudomonas fluorescens*. *P. putida* contained the TOL plasmid (Worsey and Williams 1975), a broad host-range conjugative plasmid carrying the structural genes for toluol degradation including that for catechol 2,3-dioxygenase. The transfer of this plasmid into the donor, *P. fluorescens*, was measured by the expression of this enzyme in the recipient. This was done by application of monoclonal antibodies raised against the enzyme. Donor and recipient cells were identified and discriminated by application of monoclonal antibodies directed against the O-specific side chains of their capsules. Colony blots, immunofluorescence microscopy, and immunoelectron microscopy were used. The data indicate that the TOL plasmid transfer rate in this binary biofilm was significantly higher than in a chemostat liquid culture.

In a second experiment, the survival rate of added *Pseudomonas cepacia* G4 PR1 was measured in oligotrophic subsoil water, both natural and sterilized prior to addition of *P. cepacia*. In the untreated sample, *P. cepacia* could survive for only 10 days; this was caused by "grazing" and by the limited capability of *P. cepacia* to compete with the autochthonous microflora present in the untreated, natural oligotrophic sample. Further analysis revealed that in Aquifer columns, *P. cepacia* could be found both in free state and adherent to particles, with the consequence that part of the population was washed out.

In a third experiment carried out with confocal laser scanning immunofluorescence microscopy, *P. cepacia* was added to a preexisting complex biofilm which had been obtained from a natural Aquifer sample during a growth period of 1 month. After 12 h of incubation, only very few *P. cepacia* cells could be seen attached to the pre-existing biofilm.

In summary, it can be stated that many more studies will have to be performed for a better understanding of the basic dynamics and structural and physiological interactions within biofilms. Without respective data, surprising findings might be expected which not always may be in favor of the solution of a problem of bioremediation.

References

Allison DG, Sutherland IW (1987) J Gen Microbiol 133:1319–1327. **Alloway BF** (1990) Heavy metals in soils. Wiley and Sons, New York.

Baumgarten N (1989) Nature, London 341:81–82. **Beech IB**, Gaylarde CC (1989) J Appl Bacteriol 67:202–207. **Blair DF** (1990) Sem Cell Biol 1:75–85.

Costerton JW, Cheng KJ, Geesey GG, Ladd TI, Nickel JC, Dasgupta M, Marrie TJ (1987) Annu/Rev Microbiol 41:435–464. **Costerton JW**, Irvin RT, Cheng KJ (1981) Annu/Rev Microbiol 35:299–324.

DePamphilis ML, Adler J (1971a) J Bacteriol 105:384–395. **DePamphilis ML**, Adler J (1971b) J Bacteriol 105:396–407. **Driks A**, DeRosier DJ (1990) J Mol Biol 211:669–672.

Eighmy TT, Maratea D, Bishop PP (1983) Appl Environ Microbiol 45:1921–1931. **Emerson SU**, Tokuyasu K, Simon MI (1970) Science 169:190–192.

Fletcher M (1977) Can J Microbiol 23:1–6. **Fletcher M** (1986) Appl Environ Microbiol 52:672–676. **Fude L**, Harris B, Urrutia MM, Beveridge TJ (1994) Appl Environ Microbiol 60:1525–1531. **Fude L**, Shigui L (1993) J Sichuan Univ Nat Sci Ed 2:266–273.

Geesey GG (1982) ASM News 48:9–14. **Geesey GG**, Mutch R, Costerton JW, Green RB (1978) Limnol Oceanogr 23:1214–1223. **Geesey GG**, Stupy MW, Bremer PJ (1992) Intern Biodeterior Biodegrad 30:135–154. **Gillen KL**, Hughes KT (1991) J Bacteriol 173:6453–6459.

Hamilton WA (1987) Symp Soc Gen Microbiol 41:361–385. **Helmann JD** (1991) Mol Microbiol 5:2875–2882. **Helmann JD**, Chamberlin MJ (1987) Proc Natl Acad Sci USA 84:6422–6424. **Hermansson M**, Kjelleberg S, Korhonen TK, Stenström TA (1982) Arch Microbiol 131:308–312. **Homma M**, DeRosier DJ, Macnab RM (1990) J Mol Biol 213:819–832. **Homma M**, Iino T (1985) J Bacteriol 162:183–189. **Homma M**, Kutsukake K, Iino T (1985) J Bacteriol 163:464–471.

Iino T (1969) J Gen Microbiol 56:227–239. **Iino T** (1974) J Supramol Struct 2:372–384. **Iino T** (1985) Genetic control of flagellar morphogenesis in *Salmonella*. In: Eisenbach M, Balaban M (eds) Sensing and response in microorganisms. Elsevier, Amsterdam pp 83–92. **Iino T**, Komeda Y, Kutsukake K, Macnab RM, Matsumara P, Parkinson JS, Simon MI, Yamaguchi S (1988) Microbiol Rev 52:533–535. **Ikeda T**, Homma M, Iino T, Asakura S, Kamiya R (1987) J Bacteriol 169:1168–1173.

Jain RK, Sayler GS (1987) Microbiol Sci 4:59–63. **Jones CJ**, Aizawa SI (1991) Adv Microb Physiol 32:109–172. **Jones CJ**, Homma M, Macnab RM (1987) J Bacteriol 169:1489–1492. **Jones CJ**, Homma M, Macnab RM (1989) J Bacteriol 171:3890–3900. **Jones CJ**, Macnab RM (1990) J Bacteriol 172:1327–1339. **Jones CJ**, Macnab RM, Okino H, Aizawa SI (1990) J Mol Biol 212:377–387.

Khan S (1990) Bacteria 12:301–343. **Khan S**, Dapice M, Reese TS (1988) J Mol Biol 202:575–584. **Khan S**, Khan IH, Reese TS (1991) J Bacteriol 173:2888–2896. **Khan IH**, Reese TS, Khan S (1992) Proc Natl Acad Sci USA 89:5956–5960. **Kihara M**, Homma M, Kutsukake K, Macnab RM (1989) J Bacteriol 171:3247–3257. **Komeda Y**, Kutsukake K, Iino T (1980) Genetics 94:277–290. **Kubori T**, Shimamoto N, Yamaguchi S, Namba K, Aizawa SI (1992) J Mol Biol 226:433–446. **Kutsukake K**, Ohya Y, Iino T (1990) J Bacteriol 172:741–747. **Kutsukake K**, Ohya Y, Yamaguchi S, Iino T (1988) Mol Gen Genet 214:11–15.

Lawrence JR, Korber DR, Hoyle BD, Costerton JW, Cauldwell DE (1991) J Bacteriol 173:6558–6567.

Macnab RM (1987) In: Neidhardt FC, Ingraham J, Low KB, Magasanik B, Schaechter M, Umbarger HE (eds) *Escherichia coli* and *Salmonella typhimurium*: cellular and molecular biology. Am Soc Microbiol, Washington DC, 1:732–759. **Macnab RM** (1990) Symp Soc Gen Microbiol 46:77–106. **Macnab RM** (1992) Annu Rev Genet 26:131–158. **Macnab RM**, DeRosier DJ (1988) Can J Microbiol 34:442–451. **Manson MD**, Tedesco P, Berg HC, Harold FM, van der Drift C (1977) Proc Natl Acad Sci USA 74:3060–3064. **Marshall KC** (1988) Can J Microbiol 34:503–506. **Marshall KC** (1992) ASM News 58:202–208. **Marshall KC**, Stout R, Mitchell R (1971) Can J Microbiol 17:1413–1416. **Müller V**, Jones CJ, Kawagishi I, Aizawa SI, Macnab RM (1992) J Bacteriol 174:2298–2304.

Namba K, Yamashita I, Vonderviszt F (1989) Annu Rev Microbiol 44:689–719. **Nelson CH**, Robinson JA, Characklis WG (1985) Biotech Bioengin 27:1662–1667. **Nishimura A**, Hirota Y (1989) Mol Gen Genet 216:340–346.

Ohnishi K, Kutsukake K, Suzuki H, Iino T (1990) Mol Gen Genet 216:139–147.

Richard FC, Bourg ACM (1991) Water Res 25:807–816. **Roszak DB**, Colwell RR (1987) Microbiol Rev 51:365–379.

Siebel MA, Characklis WG (1990) Biotech Bioengin 37:778–789. **Silverman MR**, Simon MI (1972) J Bacteriol 112:986–993. **Silverman MR**, Simon MI (1973) J Bacteriol 113:105–113. **Silverman M**, Simon MI (1974) Nature 249:73–74. **Stewart RC**, Dahlquist FW (1987) Chem Rev 87:997–1025. **Stock JB**, Lukat GS, Stock AM (1991) Annu Rev Biophys Biophys Chem 20:109–136. **Sutton NA**, Hughes N, Handley PS (1994) J Appl Bacteriol 76:448–454. **Suzuki T**, Komeda Y (1981) J Bacteriol 145:1036–1041.

Trachtenberg S, DeRosier DJ (1987) J Mol Biol 195:581–601.

Van Loosdrecht MCM, Lyklema J, Norde W, Schraa G, Zehnder AJB (1987) Appl Environ Microbiol 53:1898–1901. **Volger AP**, Homma M, Irikura VM, Macnab RM (1991) J Bacteriol 173:3564–3572.

Wardell JN, Brown CM, Ellwood DC (1980) In: Berkeley RCW (ed) Microbial adhesion to surfaces. Ellis Horwood, Chichester pp 221–230. **Wickner W**, Driessen AJM, Hartl FU (1991) Annu Rev Biochem 60:101–124. **Winkler J** (1994) Immuncytochemische Charakterisierung von Pseudomonaden und Lokalisierung der Brenzkatechin 2,3-Dioxygenase in *Pseudomonas putida* (pWWO) sowie Untersuchungen zum TOL-Plasmid-Transfer zwischen *Pseudomonas putida* (pWWO) und *Pseudomonas fluorescens* 55 in binären Biofilmen. Doct Thesis, University of Göttingen, Germany. **Worsey MJ**, Williams PA (1975) J Bacteriol 124:7–13.

Yamaguchi S, Fujita H, Ishihara A, Aizawa SI, Macnab RM (1986) J Bacteriol 166:187–193

Zvyagintsev DG, Guzev VS, Guzeva IS (1977) Microbiologiya 46:245–249.

Prof. Dr. Frank Mayer
Dr. Michael Hoppert
Institut für Mikrobiologie
Georg-August-Universität Göttingen
Grisebachstraße 8
37077 Göttingen, Germany

II. Cytology and Morphogenesis of Pollen and Spores

By Michael Hesse

1. Delimitation of Topics; Monographs, Conference Proceedings, etc.

This review on the development and structure of spores/pollen grains, on pollen germination, and pollen tube formation covers the period from 1990/1991 to the beginning of 1994; earlier papers are only referred to if necessary. It is a follow-up to my review in Prog. Bot. 52. Papers dealing with the structure and the developmental events in isolated pollen protoplasts, with allergenic pollen components, embryogenic pollen, or gene expression in pollen, etc. are not or only marginally referred to, as these topics are part of the applied research in fields like allergology, plant genetics/breeding, etc. It is, furthermore, not intended to give a full list of pollen/spore developmental papers published since 1990. I strongly refer the reader to Vol. 140 of Int. Rev. Cytol. (1992), where reviews and pertinent dates on many interesting topics of pollen development (cytoplasma organization, generative cell, sperm cells, cytoskeleton, male germ unit, etc.), covering literature up to 1991, are presented. In several cases I will refer to this volume, but only in a very abbreviated way, to avoid redundancies.

First of all, I welcome a concise glossary of pollen and spore terminology by Punt et al. (1994). In Prog. Bot. 52, I had asked the "rethorical" question: "Who will provide us with a new encyclopedic vocabulary?" – it is here, at least as a text, but with instructive sketches.

Excellent introductions to features of sexual reproduction in Spermatophyta, especially on the EM level, are given by Cresti and Tiezzi (1992) and Cresti et al. (1992). Students, teachers, and scientists working in the fields of biotechnology and plant breeding should consider these books as "vade mecum", because they give a good overview of sexual plant reproduction. The treatise *The Angiosperm Pollen: Structure and Function*, reprinted some years ago (Shivanna and Johri 1989), contains a lot of valuable information; unfortunately, however, the micrographs are printed on paper of poor quality. A proceedings volume, covering 74 contributions dealing with morphology, ecology, molecular biology and biotechnology of pollen grains (and ovules), is also an indispensable contribution to pollen/spore research (Ottaviano et al. 1992). Many aspects of tapetum development and biology are edited by Hesse et al. (1993). Papers on tapetum development, tapetal products, and also on male sterility are reviewed by Hesse

Progress in Botany, Vol. 56

and Hess (1993). Excellent overviews (and reference lists) on the male gametophyte development (especially McCormick 1993, but also many other articles in this Special Review issue of Plant Cell vol. 5, number 10), and on male gamete development in angiosperms (Tanaka 1993) should not be overlooked.

2. Preparation Methods and the Preservation of Structures

Progress in our field predominantly results from sophisticated light microscope methods, but also from advanced EM techniques. Immunocytochemistry (at the LM and EM level) and confocal microscopy, but also high resolution video recording of living pollen grains or pollen tubes are powerful tools in determining the features and fates of cytoskeletal components. Unfortunately, the problems of biological specimen preparation have not been approached with the same intensity as the improvement of resolution in transmission and scanning electron microscopy. The aspect of good fixation and staining quality for transmission electron microscopy (TEM) should not be overlooked if developmental aspects of metabolically active cells are discussed. The criteria for good fixations include among others: intact membranes; mitochondria are neither swollen nor shrunk; a finely granular cytoplasmic ground substance showing no empty spaces; no cell shrinkage, etc. These criteria are not only valid for chemical fixations, but also for physical fixation methods (e.g., cryoimmobilization, see below). In the past years, advanced fixation methods have considerably improved ultrastructure preservation (e.g., Noguchi 1990; Noguchi and Ueda 1990; Theunis et al. 1991; Del Casino et al. 1992; Lancelle and Hepler 1992; Hess 1993a; Hesse 1993a), especially concerning metabolically active cells/tissues such as anther tapetum and developing microspores/pollen grains. In the light of these findings, some earlier interpretations concerning structure and development of organelles/or products of cellular metabolism cannot be upheld: some observations need reinterpretation due to an increased knowledge based on better techniques!

This statement is valid also for delicate sporoderm strata: the valuable papers on *Vigna* by Stainier and Horvat (e.g. 1978a,b) focus on exine configuration in mature, acetolyzed pollen, while – one and a half decades later – Pérez-Muñoz et al. (1993a,b) show that the sporoderm in *Vigna* is far more complex and also includes intine and Zwischenkörper formerly lost by acetolysis.

In transmission electron microscopy the "perfect" structure preservation is, however, not enough. It has to be followed by "perfect", highly selective addition of (mostly) heavy metal atoms to the interesting biological structure, i.e., by perfectly balanced staining protocols: only on the basis of known chemical reactions of the cyto- or immunochemical stain can the TEM micrographs be interpreted accurately. Hayat (1993) recently pointed out that, apart from apt fixation protocols, optimized staining methods of TEM specimens are absolutely necessary to avoid artifacts. The author regrets the scanty description of important details of preparation techniques that accompany the result and discussion paragraphs of most papers. Progress in biological specimen preparation has, no doubt, not kept pace with recent improvements in instrumentation (TEM and SEM). Some examples: the advent of High Resolution Field Emission SEMs (HRFESEM) provided us with so far unsurpassed resolution. Due to rather coarse preparation techniques in SEM this high resolution cannot, however, be taken advantage of (cf. Takahashi 1993, p. 197: 15 nm gold sputtering obscures the surface details with a diameter of, say, 10 or less nm, visible under HRFESEM). Spatial resolution in TEM, researching the most ultrathin sections, is no better than ca. 1.5–2 nm, the instrument itself allows ca. 0.2 nm! Especially for high resolution work, excessive staining often limits resolution as it can increase the probe size unless the staining reaction is controlled. High resolution also requires the immobilization of molecules to prevent their movement during imaging, a hard claim! One should, however, bear in mind, that: "As long as unavoidable artifacts are known and interpretable, they can be accepted" (Hayat 1993, p. 7). The following sections will frequently address the "artifact" question.

Various cryotechniques have come of age and today are already used as routine methods. This is especially interesting for TEM applications. Freeze-fracture, for instance, is mainly employed for the study of biomembranes. The greatest problem in this technique, applied to spores/pollen grains, is the cleaning of the replicas by removing the pollen wall (especially the exine). Of course, the study of ultrathin sections should always accompany the study of freeze-fracture replicas. Freeze-fixation (i.e., rapid freezing = RF and high pressure freezing = HPF), followed by freeze-substitution of relatively large, inhomogeneous specimens (not only pollen tubes, cf. e.g., Lancelle and Hepler 1992, but also intact anthers, cf. Hess 1993a) gives us the possibility to compare whole, complex, metabolically highly active tissues. Naturally, also cryomethods have their limitations, as critically pointed out by Robards (1991) and Southworth (1992).

3. Some Aspects of the Early Stages in Normal and Abnormal Spore Pollen Grain Development

Even after several decades of studying pollen developments by LM and EM, all aspects of microsporogenesis and -gametogenesis have not yet been sufficiently fathomed. Important details are still not completly understood, and nearly every author highlights a new, often surprising facette.

a) Oganization of Metabolically Highly Active Cells

In the past, a lot of work has focused on structure and development of angiosperm pollen, while the developing microspores/pollen grains have been interpreted merely as absorbing cells. The "transfer cell concept", in contrast, gives rise to a more modern interpretation. Transfer cells facilitate rapid short-distance movements of solubles or compouds; they are by no means restricted to the vascular tissue; their occurrence in various plant organs and ultrastructure is referred to by Joshi et al. (1993). Charzynska et al. (1990) reinterpret the developing spores/pollen grains, discussing the role of some sporoderm features and strata (i.e., the Zwischenkörper and especially the plasmatubules – or "plasmotubules" – penetrating the intine) in the light of the "transfer cell concept". Van Aelst and Van Went (1991) report that the amount of significant organelles in the vegetative cell of maturing pollen may vary considerably between different taxa: e.g., *Nicotiana* and *Linaria* show many ER profiles, *Impatiens* only very few, while the *Papaver* type is intermediate. Van Aelst et al. (1993) compare excellently fixed cytoplasm and organelles in dehydrating (= closed anther) and rehydrating pollen grains.

Cryotechniques in TEM have revealed several previously unreported details, especially in pollen tubes (Lancelle and Hepler 1992: e.g., ultrastructure of the axial cytoplasmic "slow" and "fast" lanes, each with distinct features and the hexagonal arrangement of the cortical ER), but also in anther tapetal cells (for review see Hesse and Hess 1993), microspores (Hess 1991, 1993a), and mature pollen grains (Hess 1993b): The membranes, for instance, appear stretched and not folded, especially in vacuoles (the obvious absence of microfilaments in *Ledebouria* pollen is perhaps due to their poor contrast in freeze-substituted and resin-embedded samples). N.B.: An excellent report on membrane structure in male reproductive cells is given by Southworth (1992). Using freeze-fracturing, the author presents some results which differ significantly from those obtained by ultrathin sectioning.

b) Mosses and Ferns

The developmental and morphological studies in Bryophyta and Pteridophyta lag far behing those in the Spermatophyta. This is due in part to the fact that palynologists are much more interested in the formation and development of (angiosperm) pollen grains with their – seemingly – larger diversification. It has only recently become evident that also in mosses and ferns there is an unexpected diversity in structural and developmental features (see the impressive volume by Tryon and Lugardon 1991 on fern spores!).

Gambardella and co-workers present an impressive series of papers on the sporogenous lineage in the moss *Timiella*. Not only are studies in moss palynonology rare, the above-mentioned papers also cover very early developmental stages (Gambardella and Alfano 1991a,b,c). This makes these papers especially valuable, since our knowledge in both points is rather poor. The sporoderm development (Gambardella et al. 1993b) shows some aberrant features; interestingly, the intine seems to be restricted to the proximal spore hemisphere, which contrasts with former observations in *Timiella* (if really mature spores had been originally studied). Generally speaking, the cytoplasmic events can be demonstrated as being either less complex in preprophase sporogones, e.g., PPBs are lacking as in other bryophytes, in contrast of fern allies etc., or more complex: up to prometaphase (Gambardella and Alfano 1990). *Timiella* differs from other mosses in the origin of metaphase I spindle: while one half of the nascent spindle derives from three of the four tetrahedral microtubule arrays seen in prophase I, the second half of the spindle derives from only one, the fourth array (Gambardella et al. 1993a).

A single paper on ferns can be reported: Rudramuniyappa and Suvarnkhandi (1992) analyze aspects of sporogenesis in *Ophioglossum*; unlike in (most) Spermatophyta no (additional) callose is synthesized before meiosis – nevertheless normal haploid spores are produced (cf. passages on callose in angiosperms, below in Sect. 7).

c) Gymnosperms

As in bryophytes and pteridophytes, only very few papers can be referred to. Especially in the *Cycadophytina*, publications on the microgametogenesis and the microsporogenesis are extremely rare, and it would be very welcome if this gap were filled. Norstog (1990) presents an excellent and detailed ultrastructural study on *Microcycas* spermatozoids; remarkably, the spermatozoids show a gradient in motility and complexity: the first-formed ones are well organized and actively motile, while the last-formed are poorly organized and less motile. Wang (1990) briefly reviews our knowledge of *Ginkgo* microsporogenesis.

In the last few years, the development of the male and female gametophyte in various *Coniferophytina* taxa has received some attention. Rowley and Walles (1993) continue their long-term work on cell differentiation in *Pinus* microsporangia, now reporting on the interval between diakinesis and tetrad formation; one of the several interesting results is the confirmation that both plasmalemma and glycocalyx coating (of the tapetal cells) give rise to the specific "exine" form of the orbicules in *Pinus*. "Pericytoplasmic embayments" (as the authors call "empty spaces" between the callose and the MMC plasmalemma) and also "endomaizes" (= interlocking multiple cylinders of dilated rER, cf. Rowley and Walles 1985) seem to be no artifacts. The (male and female) embryology of Cephalotaxaceae and Taxaceae (Chen and Wang 1990; Chen et al. 1991) and the male gametophyte development in *Pinus banksiana* (Chen et al. 1990) are studied with respect to their systematic significance. One interesting finding are remnant sacci in young development stages of *Torreya* (Chen et al. 1991), which mature pollen grains lack. This result once more underlines the importance of studying young developmental stages for a better understanding of the phylogenetic and functional significance of the sporoderm stratification (see below the pertinent passage in Sect. 7).

d) Angiosperms

Prior to meiosis, the cell wall of PMCs is simple, with a middle lamella. Before the onset of callose deposition (late prophase) this wall structure changes twice dramatically (Polowick and Sawhney 1992); the authors also debate some aspects of artifacts concerning cytomixis/cytoplasmic connections, or nuclear vacuoles during prophase I, concluding that these vacuoles are indeed formed by invagination of the (inner) nuclear membrane. That these authors found no drastic reduction in the ribosome density during meiotic prophase, as formerly reported (e.g., Dickinson 1982), may be due either to phylogenetic differences, or to the improved structure preservation since the 1970s. Polowick and Sawhney (1993a,b) continue their work on the development of *Lycopersicon* pollen, all stages are now documented in detail; numerous lipid droplets – an important energy store – in mature pollen grains are surrounded by rER, which may play a role in lipid mobilization.

Microspore/pollen development is closely connected with the tapetum development. In most cases of abnormal pollen development, the anther tapetum plays a crucial role. The pathways of the tapetal secretory events and of the primary/secondary metabolism were reviewed recently (Hesse and Hess 1993). I refer, therefore, only to some new or especially interesting aspects and publications. Our knowledge of genetics, especially of gene expression, of tapetal cells and developing microspores, of individual gene products, which participate in processes unique to the tapetum, is so far rather poor. One of the first papers in this field was published by Koltunow et al. (1990). Some papers followed (e.g., Aarts et al. 1993; and especially McCormick 1993 for review), but recently advances were achieved in identifying proteins which are involved indeed in the production of pollen wall components, especially in sporopollenin formation and/or deposition (Staiger and Apel 1993; Staiger et al. 1994), or which are part of the enzymes callase (Hird et al. 1993).

Cytoplasmic male sterility in *Zea mays*, *Brassica*, *Petunia*, and *Vicia faba* is reviewed – especially with respect to genetics – by Braun et al. (1992). Usually abnormal features in microsporogenesis appear long before pollen grain abortion takes place. Majewska-Sawka et al. (1993), however, found no aberrant cytological features in the sterile microspores/pollen grains before abortion. In few cases the tapetum appears not to be involved in male sterility, as in hybrids between maize and *Tripsacum*: Kindiger (1993) addresses cytokinesis failure of first and second meiotic divisions, or the forming of aberrant spindles in the microspores. Recently, the influence of physical/chemical agents on pollen development has received the attention of basic and applied research (the latter one of course of enormous commercial interest). Physical agents (temperature), but especially chemical agents, were reviewed by Cross and Ladyman (1991): Chemicals may inhibit pollen development completely, or, applied at suitable dosages and at critical stages, may induce male-sterile mutants, i.e., the development of pollen does in fact not stop completely. Schulz et al. (1993) devoted a detailed study to how two inhibiting phenylcinnoline-carboxylates influence pollen development in

Triticum: as in earlier reports, the authors could not confirm that these chemicals influence the course of meiosis. Besides this, the authors briefly review the literature on chemically influenced pollen development published since Cross and Ladyman (1991).

The phases in accumulation/dissolution of metabolically active storage substances can often be followed by LM, there is generally no need for EM. This is demonstrated by the elegant paper by Noher de Halac et al. (1992, their results are summarized in their instructive Fig. 11). The LM level is sufficient also for studies of principal features in sterile anthers, as shown by Dawson et al. (1993) for six male-sterile *Arabidopsis thaliana* mutants (the authors announce a genetic mapping of the *ms* loci). Chaudhury et al. (1994) characterize more male sterile mutants of *A. thaliana* and identify some genes, which play an important role in male fertility/sterility. Of course, only TEM studies can reveal ultrastructural details of accumulation/dissolution processes. The fine paper by Evans et al. (1992) gives a quantitative analysis of lipid accumulation during *Brassica* pollen development; very interestingly it seems that the tapetum provides the basic substrate for the rapid and enormous lipid synthesis during the vacuolate pollen stage. Enormous masses of electron-dense deposits appear in the vacuoles of *Lycopersicon* tapetum cells, and disappear with the formation of the orbicules; presumably the orbicules act as transporting vehicle to the exine (Polowick and Sawhney 1993c). Lipids, especially fatty acids were identified and quantified in mature pollen of *Carya illinoinensis* cultivars: quantities vary greatly, but percent composition is very stable (Thompson et al. 1993). Clément et al. (1994) have studied in detail amylogenesis/amylolysis in *Lilium* pollen development. Another pertinent paper was published by Wetzel and Jensen (1992) on the accumulation of storage products; but does it really make sense to use a high voltage EM without publishing stereo-pairs of thick sections? Only such stereo-pairs can overcome the depth limitations of "ordinary" TEM micrographs.

New TEM-preparation techniques, such as freeze-fixation and freeze substitution give rise to new insights into structure and function. Tapetal secretory activities are more manifold as thought before. Fitzgerald et al. (1993b) provided evidence for protein secretion of tapetum cells during the tetrad stages. Hess and Hesse (1994) showed intensive polysaccharide exocytosis from the tapetum into the anther locule; the intitial steps of a cytoplasmic, i.e., nonplastidial, pollenkitt fraction were also elucidated due to the benefits of cryomethods. A further low temperature preparation method, observing freeze-fractured material in the SEM, provides us with attractive micrographs clearly demonstrating the structural differences between the two main tapetal types (Barnes and Blackmore 1992).

4. First and Second Pollen Mitosis

In an important, very impressive publication series, Brown and Lemmon (1991b,c,d,e, 1992b, 1994) minutely review and describe the cytokinetic apparatus and the control of cell division (unequal first mitosis) in *Phalaenopsis* and

Cypripedium (Orchidaceae); their concept of a cytoplasmic domain can serve as a good model for spatial control of cytokinesis and provides us with fascinating insights (Brown and Lemmon 1992c). They also discuss, for example, the question of the presence or absence of phragmoplasts in monocots and dicots. The authors make successful use of advanced LM techniques, using TEM merely as a completion to clarify ultrastructural aspects at higher resolution, and illustrate analogue features and aspects of the meiotic cytokinesis (spindle ontogeny) in such unrelated taxa as hornworts (Brown and Lemmon 1990), *Selaginella* (its plastid polarity is different from other taxa with monoplastidic meiosis, Brown and Lemmon 1991a) and *Magnolia* (Brown and Lemmon 1992a). The authors compare in *Magnolia* the development of normal and drug-induced microsporocytes. All these papers have, no doubt, broadened our horizon.

Investigations of generative cell (GC) development in families other than in Brassicaceae are very rare. In *Cyrtandra pendula* (Gesneriaceae), the mature GC is encaged by a framework formed from ER tubules. The abundant ER cisternae are the only organelle closely associated with the GC, therefore perhaps responsible for the undulating GC wall (Luegmayr 1993). Two sorts of membrane coatings on the GC surface of freeze-substituted monocotyledon pollen were described by Hess (1993b): the cytoplasmic face of the plasma membrane bordering the GC either bears a diffuse fibrous, "fluffy" coat (according to Noguchi and Ueda 1990) or strip-shaped projections (according to Van Went and Gori 1989). Tanaka et al. (1989) demonstrate wall material around isolated GCs in *Lilium longiflorum*. In *Pyrus communis* some new features of GC structure was shown by Tiwari (1994) using low-temperature preparation methods and TEM operating at 400 KeV.

Details of the detachment mode of the generative cell were shown recently in the orchid *Polystachia pubescens* (Schlag and Hesse 1992) and in the Gesneriaceae *Monophyllaea horsfieldii* (Luegmayr 1994). The congruent features in these two genera and the similarity to former investigations point to the possibility that the observed detachment mode of the generative cell could be a ubiquitous feature. Very unfortunately, the driving force behind the detachment process is poorly known.

In contrast to other reports (cf. Schlag and Hesse 1992 for review) that in orchids the GC is normally formed at the distal pole, in *Pterostylis concinna* it is formed at the proximal pole (Pandolfi et al. 1993). The latter authors furthermore stress that different tetrad types are formed in the same anther of this *Pterostylis* (also a common feature in other highly derived monocots, e.g., in *Typha*). Zhu et al. (1990) clarify the ultrastructure and composition of GC in *Medicago* using the "old-fashioned", but successful $KMnO_4$-fixation technique, and using also advanced computer-assisted 3-D reconstruction techniques. Actin-like filament/microfilament arrays seem to be responsible for the changing of GC shape from rounded to elongated (Hause et al. 1992; Polowick and Sawhney 1993b). The organization of the GC mitotic apparatus and the GC division mechanism itself varies slightly between plant taxa. The mitosis of the GC has been subjected to many investigations, and the respective results and – especially – the interpretations are often contradictory (cf. reviews by Palevitz and Tiezzi 1992; Tiezzi 1991; Tanaka 1993). As in other specialized topics, the reader is referred to such review articles. In an original contribution, Palevitz (1993b) highlights differences and similarities regarding the GC division in *Nicotiana* compared with *Tradescantia*; but the reasons for differing features remain speculative so far. In contrast to *Phalaenopsis* (Orchidaceae, Brown and Lemmon 1991d), Hause et al. (1992) could not observe a microtubular preprophase band in *Brassica*.

The detailed study of 1st (microspore) and 2nd (pollen) mitosis can provide us with new or renewed information, e.g.:

a) The presence of ER-like cisternae around and partly "in" the prometaphase chromosomes of *Ornithogalum* (Charzynska and Cresti 1993), this could have some connection with the regulation of the Ca^{2+}-concentration, cf. Tirlapur and Willemse (1992).
b) After studying the effects of high temperature (microspores cultured either at 32.5 °C for 24 h or at 20 °C for 24 h) on the first pollen mitosis in *Brassica* Telmer et al. (1993) list ultrastructural characters typically for potentially embryogenic microspores, and they discuss some contradicting results of former observations (Zaki and Dickinson 1990, 1991).
c) Which substances (or "events") do stimulate the 2nd pollen mitosis? Read et al. (1993) have tried to answer this only seemingly simple question, once more using (cultured) pollen tubes of *Nicotiana*; interestingly, no signals from stigma/styles are necessary for pollen tubes growth after 2nd mitosis; further development depends only on the nutritional state of the pollen tube.

Hagemann (1992) presents a detailed review on the uniparental/biparental plastid inheritance in gymno- and angiosperms. The question of male plastid inheritance (and other organelles) in some genotypes of alfalfa was studied by Zhu et al. (1990, 1991, 1992): the authors give quantitative data from GCs and sperm cells: the structure of the generative cells cannot be used to predict male plastid transmission in a particular genotype, and the authors failed to find differences in the male plastid inheritance pattern of different genotypes. Tanaka (1991) clearly demonstrates the action of the microtubules in movement and distribution of plastids during microsporogenesis in *Lilium longiflorum*. Pacini et al. (1992) focus on the development and distribution of plastids in the tapetum, on microspore/pollen development, sperm cells, etc.: the plastids of the generative cell degenerate within autophagic vacuoles, so that the sperm cells lack plastids. In *Pelargonium zonale*, the plastids persist in the mature GC, but return to a proplastid status. Plastid DNA and also mitochondrial DNA are present in *Pelargonium zonale* sperm cells (Kuroiwa et al. 1993). Occasional biparental plastid inheritance is documented excellently by Yu et al. (1994).

Dinis and Mesquita (1994) report on polymorphic and aggregated microbodies (peroxisomes) in mature *Magnolia* GC; their function is unclear, possible the microbodies are paternally transmitted.

5. Structural Connections Between VN, GC, and Sperm Cells (MGU Concept)

It is not known if a structural association (connection) between the GC and the vegetative nucleus (VN) exists; reports so far deny it (Van Aelst et al. 1989). When and how could a close association between the VN and the "nascent" sperm cells be formed? Also the discussion about the possibility of a sperm cell interconnection and/or a connection to the vegetative nucleus continues. In some taxa with tricellular pollen a MGU (male germination unit) seems to be ± the rule (e.g., Brassicaceae), while in others (e.g., Poaceae) the presence of a MGU could not be confirmed. Recently, a MGU was reported by Charzynska and Lewandowska (1990) for tricellular *Sambucus* (Caprifoliaceae) pollen.

Reviews on the MGU concept, but also new findings, were given by Mogensen (1992), on GC- and sperm cell structure by Theunis et al. (1991), Chaboud and Perez (1992), Yu et al. (1992), and Yu and Russel (1993, 1994). In brief, these well-organized, informative papers deal with: structural characters of GCs and sperm cells in situ and after their isolation, further on prevalence and implications of the MGU, its functional significance and, interestingly, some considerations over possible exceptions to the "rule", i.e., if an MGU does not exist in certain taxa, all with extensive background information. Details of sperm cell/vegetative cell ultrastucture and computer-assisted 3-D reconstruction of the male gamete organization are presented for *Nicotiana* (Yu et al. 1992) and – by spectacular, colored stereo-pairs – for *Cymbidium* (Orchidaceae) by Yu and Russel (1992); in both papers, the questions of MGU organization and MGU concept are discussed, especially with respect to bicelluar pollen (*Cymbidium*) from the standpoint of a close association between the two sperm cells and between the GC and the VN. In the *Cymbidium* paper, the more important question of sperm dimorphism is discussed: the authors claim that – statistically – sperm cells are isomorphic, but they do not exclude some, so far hidden, dimorphism. Many (controversial) findings have been published, besides others, e.g., the important question of putative differences in organelle equipment between both sperm cells: Yu et al. (1992) and also Rougier et al. (1991) strictly deny any differences, but at least in some taxa a different number of mitochondria has been reported (Charzynska and Lewandowska 1990, and references therein!).

In vegetative cells of *Arabidopsis thaliana*, the plasma membrane bordering the intine shows different features compared with the plasma membrane bordering the sperm cells, and arabino-galactan glycoproteins are present in the intine as well as at the sperm cell surface (Van Aelst and Van Went 1992); the authors consider possible relationships between the different membrane characters and the gamete function. Pectins and glycoproteins are not only (or even?) associated with the plasmalemma but especially within the zone of the cell "walls" surrounding the sperms of *Arabidopsis* (Van Aelst and Van Went 1992). This is a further step in the young history of immunolabeling of pollen grains using antibodies to characterize the vegetative plasma membranes bordering the sperm cell. This membrane has a different glycoprotein spectrum than the plasmamembrane bordering the intine, i.e., the plasmalemma. Unfortunately, to date this very promising technique "requires a compromise between preservation and visualization of the ultrastructure" (Van Aelst and Van Went 1992, p. 15ff). The authors also present a theory on the very poor significant labelling of freeze-fixed/freeze-substituted material. The present state-of-the-art in all these techniques is shown by the informative book by Newman and Hobot (1993).

6. Pollen Germination and the Early Stages of Pollen Tube Growth

The ultrastructural changes during de- and rehydration of pollen grains immediately before pollen tube germination were recorded for the first time by Van Aelst et al. (1993). Interestingly, the local uptake of exogenous solutes seems to be unnecessary to start pollen germination, to form polar growth, or to constitute polarity in the vegetative cell; only uptake of atmospherical water (vapor phase) is enough for pollen activation (Heslop-Harrison and Heslop-Harrison 1993). Fernandez and Rodriguez-Garcia (1990) and Fernandez et al. (1992) discuss aspects of aperture structure, development, and function in younger stages, but especially shortly before germination.

Pollen tubes, but also germinating pollen grains, are apt objects for studying the cytoskeleton. The cytoskeleton consists of several structural components that incite cellular and intracellular movements, create shapes and organize organelles. It plays a crucial role in cytokinesis, mitosis and meiosis, and cell-to-environment interactions, e.g., pollen tube/stylus interactions during pollination, and fertilization. Our knowledge of the cytoskeleton apparatus and its role in the dividing pollen grain and in the pollen tube development is making tempestuous progress at present. An astonishingly high number of qualitatively outstanding publications (reviews and original contributions) deals with this intriguing and repeatedly discussed problem: Tiezzi (1991), Liu and Palevitz (1992), Mogensen (1992), Palevitz (1993a,b,c), Palevitz and Liu (1992), Palevitz and Tiezzi (1992), Pierson and Cresti (1992), and Southworth (1992) present most interesting news and give excellent overviews on many general aspects of cytoskeletal components, but also on researching methods. To reproduce their manifold results and considerations here would be pure redundancy; the reader is strongly referred to at least one or two of these papers. The enormous dynamics of the cytoskeleton in activated *Hyacinthus* pollen (generative cell and sperm cells within the pollen grain and the pollen tube) is presented brilliantly by Del Casino et al. (1992, herein further outstanding literature); the authors failed to detect actin filaments. The only so far positive report in *Rhododendron* (Taylor et al. 1989) should be carefully reinvestigated. Palevitz and Liu (1992) were unable to detect actin-containing microfilaments in GC and sperms of three typical, unrelated angiosperms, and the authors conclude after debating pro's and con's – and critically evaluating the literature – that GC and sperm do not contain microfilaments.

The events of cytoskeleton evolution, and the effects of cytoskeleton inhibitors during pollen activation and pollen tube emergence, with special reference to the "monocolpate" pollen grains of *Narcissus* are referred by Heslop-Harrison and Heslop-Harrison (1992a,b). Tanaka and Wakabayashi (1992) studied the influence of colchicine and Cytochalasin B on the cytoskeleton organization in *Lilium longiflorum*, assuming that both cytoskeleton components play essential roles in pollen germination. Malhó and Pais (1992) report on the dynamics of in-vitro-grown pollen tubes in *Agapanthus* (the reader is referred especially to their Fig. 12, a very graphic diagram, showing the complex volume percentages of the pollen grain compartments "three-dimensionally"); they discuss different features of in-vivo- and in-vitro-grown tubes, but conclude that pollen germination follows a common pattern in both breeding systems.

For the first time, the MGU behavior was studied in interspecific incompatible systems by Rougier et al. (1992, *Populus deltoides* × *alba*): the MGU components and their spatial relationships in incompatible *P. alba* pollen tubes are similar to those found in compatible *P. deltoides* pollen tubes. Speranza and Calzoni (1992), Ciampolini et al. (1993), and Calzoni et al. (1993) tested inhibitory effects of monensin, a carboxylic ionophore, and its influence on ultrastructural details of pollen tube growth and protein synthesis during pollen germination in vitro: tube growth depends on a functioning Golgi apparatus; a severe inhibition of the secretory activity of the dictyosomes even at low concentration levels of monensin can therefore (almost) completely and quickly arrest pollen tube activities. The addition of monensin leads to a swelling of the dictyosome cisternae, which is especially pronounced at the tip of the pollen tube. This, in turn, results in the lack of new membranes, membrane-associated proteins, and other proteins. It is also proved that monensin induces complex changes in synthesis and composition of the pollen tube walls.

Membrane-associated calcium and calmodulin are involved in many developmental plant processes, of course also in pollen development, germination and pollen tube growth (cf. the review by Palevitz and Tiezzi 1992), and the following references are closely related to the cytoskeleton paragraph in some aspects. Using computer-assisted video analysis, Tirlapur and Cresti (1992) visualize the pattern of membrane-associated Ca^{2+} and calmodulin during pollen hydration and germination. Tirlapur and Willemse (1992) demonstrate a tip-to-base membrane-calcium gradient; in actively synthesizing tapetal cells we can principally expect a similar feature (Willemse 1993).

7. Sporoderm Development and Composition

a) Fossil Material, Ferns and Gymnosperms

Nearly a gift from the gods are well-preserved fossil pollen grains in different stages of development: Osborn and Taylor (1993) found varying ontogenetic stages of pollen sacs in Triassic material. Kurmann (1991) compares developmental stages in exine formation of a Jurassic member of Taxodiaceae with an extant one, adding new information to her review on exine ontogeny in conifers (Kurmann 1990). Some well-documented cases of microgametophyte development in fossil gymnosperms, but also in ferns and in angiosperms, are reported by Taylor (1990): of course aspects of sporoderm formation and composition dominate.

In *Lycopodium clavatum*, Uehara and Kurita (1991) failed to detect (by TEM) a perispore wall stratum, supporting Lugardon (1990), who has pointed out that in *L. clavatum* – in contrast to many other Lycopodiaceae – a perispore is lacking. That the lack of a perispore in some taxa does not depend on an artifact caused by preparation methods, is demonstated by the presence of perispore in *Isoëtes* by Uehara et al. (1991), using the same techniques as Uehara and Kurita (1991). In a series of papers, Nina Gabarayeva and Michael Zavada describe aspects of sporoderm ontogeny in primitive angiosperms and discuss their findings in relation to gymnosperm and angiosperm phylogeny (Gabarayeva 1990, 1991, 1992, 1993; Zavada and Gabarayeva 1991). They recognize three types of sexine development in gymnosperms and (primitive) angiosperms and debate the significance of a "basal layer", consisting of a foot layer and (always?) of an endexine. A lamellated exine stratum is found predominantly – and sometimes exclusively – in young developmental stages and becomes often compressed (as endexine) in mature stages; this stratum is interpreted as endexine. This is further proof that the angiosperms did not "invent" the endexine a second time: there is evolutionary continuity; the homology was shown already by Blackmore and Barnes (1987).

b) Excursus on the Phylogenetic Significance of Spore/Pollen Wall and Aperture Organization and Development

The position of apertures with respect to the pollen grain/spore polarity is of considerable interest in actuo- and paleopalynology. Although this issue thematically does not belong to this review, some interesting publications and views dealing with developmental aspects will be referred to.

The textbook view that all sulci are formed exclusively at or around the distal pollen pole is most probably an oversimplification. Some older findings report about simultaneously occurring "apertures" proximally and distally. There is much speculation if and how distal apertures could divide and shift into the aequator plane, e.g., the (mono-)sulcus or the trichotomosulcus get some sort of operculum at the "polar cap", and the remaining slits elongate towards the equator plane thereby becoming colpi. A fragmentation of a distal sulcus is more likely than an inaperturate intermediate between (mono-)sulcate and tricolpate pollen. According to Doyle and Hotton (1991), tricolpate pollen arose twice or even three times within the so-called eudicots. The (independent ?) evolution of triporate pollen grains in the eudicots (sensu Doyle and Hotton 1991) may be an adaptation facilitating the growth of the pollen tube into the stigma (Stewart and Rothwell 1993). Even though this seems very speculative, some recent reports make this hypothesis appear more probable. The variability and trends of the respective aperture condition of mature pollen grains include shifting towards the equatorial plane – this is reported for some genera within such unrelated families as Annonaceae (Le Thomas and Thanikaimoni 1987) and Iridaceae (Goldblatt et al. 1991; Goldblatt and Le Thomas 1992); but a second point should be considered in this circumstance.

As pointed out in my review in Prog. Bot. 52 (p. 27), early steps of development can provide us with surprising insights into old problems. A fine example hereof is given by Gabarayeva (1993): she found in *Asimina* (Annonaceae) that pollen tubes emerge from the proximal pole of the microspores, united as permanent tetrads. This proximal area was called in the past a secondarily developed pollen tube exit, but does a "primary" pollen tube exit, i.e., an "aperture", indeed exist in the distal pollen area? The enormous variation of the aperture condition within the Palmae includes the perhaps isolated occurrence of the triporate condition in the subtribe Sclerospermatinae; this demonstrates that the development of such microspores/pollen grains should be studied (Ferguson and Harley 1993). Guzzo et al. (1994) describe an unusual mode of aperture (sulcus) formation in *Liriodendrou* (Magnoliaceae), starting already in tetrad stage as an exine folding! Another example for the probability of this hypothesis are some highly specialized taxa with highly derived pollination mode "allowed" to have pollen grains with leptomata (Martinsson 1993). Are the leptomata situated in the Equator plane? If not, where else? These examples demonstrate a need for investigating early developmental stages, and for avoiding calling pollen grains lacking an aperture "inaperturate". Investigations of hydrated, non-acetolyzed pollen grains reveal an unexpected variety of sulcus "types", at least in monocots (Halbritter and Hesse 1993). Evolutionary-oriented botanists (palynologists, etc.) are challenged to reconsider their concepts. Rowley and Dunbar (1990) pinpointed a further common mistake: palynologists usually consider exines as highly static in form; this conviction is based upon mature pollen, while during pollen development often elongation, retraction and/or reconstruction of sporoderm elements takes place.

c) Exine Substructure Organization

The *exine substructure organization* is far from being elucidated. Three – incompatible ? – types of substructure organization are mentioned in the literature: exine tuft units (Dahl and Rowley 1991), fibrous threads associated with granules (Takahashi 1993), penrose pattern (Kedves 1991, N.B.: his model is not only valid for the exine!). Claugher and Rowley (1990) found ca. 70–100-nm-wide tufts on young *Fagus* microspores, and ca. 200–250-nm-wide tufts in mature exines – the authors cannot decide between an increase in tuft diameter during development or a large, second tuft generation; the "units" are furthermore subdivided into "subunits": at present it seems very likely that a hierarchy exists. The "sister" product of the (sporopolleninous) exine, the orbicules, was checked by Clément and Audran (1993) with respect to its cytochemistry and composition; they also claim the compatibility of their architecture model with those by Southworth, Kedves, and even Rowley. A widely overlooked problem is the absence of exine substructures in the mature (!) endexine. Blackmore and Claugher (1987) try to answer this question, but their hypothesis has not been tested yet. Rowley (1988) proposed another model: structures ("tuft units") are radially orientated of the endexine. The endexine within the complex aperture area in the Onagraceae can be divided into three zones, all with different numbers of endexine units (Horvat 1992). A detailed analysis of wall development in unrelated gymnosperm/angiosperm taxa reveals distinctly different patterns of growth – even the sporopollenin "receptor" structures differ in their mode of arrangement (Rowley 1991, Rowley and Skvarla 1993). The report of deposition of globular sporopollenin elements on the ectexine surface of *Felicia* (Jordaan and Kruger 1993) should be proved.

The *exine biochemistry* (chemical composition, changes in structure and biochemistry during pollen development) is brilliantly reviewed by Southworth (1991); besides this, she presents an interesting, extended definition of exine which includes sporopollenin, primexine matrix remnants and tapetum material deposits. The (exine) sporopollenin biochemistry (note: do not confuse sporopollenin with exine, see Southworth's definition) is reviewed by Wiermann and Gubatz (1992). The working group around Rolf Wiermann goes ahead step for step in their work on sporopollenin biosynthesis, and the "classical" view that sporopollenin is generated exclusively by carotenoid (esters) should be replaced definitely (Gubatz and Wiermann 1992a,b). Wilmesmeier et al. (1993) investigated spectroscopically sporopollenin from various bryophytes, pteridophytes, and spermatophyta; the analyzed sporopollenins show significant differences, and the authors conclude that sporopollenin is no uniform macromolecule. Chaloner et al. (1993) edited a Symposium Volume on the composition and biogenesis of sporopollenin. We are, however, still far away from structure formula(s) for the various(?) sporopollenin, and, moreover, nothing is known of the origin of sporopollenin or its precursor(s), i.e., in or at which organelle they are synthesized. Gene expression research may give some hints on sporopollenin production: Foster et al. (1992) suggest that the so-called E 2 gene plays a role in sporopollenin deposition. It is uncertain if "sporopollenin precursors" exist at all, and if yes, whether the "sporopollenin receptors" sensu Rowley et al. are identical with these precursors? The level of chemical degradation resistance probably depends on the specific packing density of exine (sub-)units rather than to the (so far hypothetical) different formulas of the sporopollenin(s), see also Rowley and Claugher (1991). The mineral content of the exine, a rather overlooked aspect, varies between populations growing on the same soil type (Dajoz and Haut 1993).

d) Young Stages: Role of Callose, Monad Versus Polyad Development

The role of the callosic special wall is not yet quite clear. Callose is more permeable than initially assumed, at least for some substances: this is confirmed by Rodriguez-Garcia and Majewska-Sawka (1992). It seems that the presence of callose is not absolutely necessary for exine formation and/or pollen fertility. In monocots, successive cytokinesis often occurs resulting in nontetrahedral microspore tetrads. In *Pandanus odoratissimus*, callose is totally absent and no tetrad, but isolated microspores are formed; nevertheless, a fully developed exine and full fertility are reported by Periasamy and Amalathas (1991). This is very remarkable, since in other *Pandanus* species (same source), as well as in (most) spermatophyta, the presence of callose around PMZ and young microspores is a prerequisite for a normal pollen development: missing callose leads to misformed exines or even pollen sterility. This is especially evident in orchids with compound pollen where exine size is directly correlated with the preceding amount of callose. (See *Peristylus spiranthes*, Zee and Siu 1990, with three exine types, according to the orientation within the pollinium).

Rowley (1991, also for review of pertinent literature) and Rowley and Flynn (1991) stress that the development of the exine within the seed plants is extremely diverse in crucial points. Even seemingly exactly timed processes, such as meiotic stages, or initiation and further development of exine, may be conspicuously asynchronous, as shown for *Pinus* (Rowley and Walles 1993), for *Tambourissa* (Rowley and Flynn 1991) or for *Eucommia* pollen (Rowley et al. 1992). The latter authors also correct former reports on the *Eucommia* apertures allegedly having unequal length: detailed ontogenetic investigations reveal no evidence for unequal apertures. Pérez-Muñoz et al. (1993a,b) describe an uncommon spongy structure of the primexine ridges in *Vigna*. Commonly, the primexine pattern is organized into probacula and protectum. These studies once more show the importance of detailed developmental investigations in as many taxa as possible.

A propos "detailed developmental investigation": many authors describing stages of development have difficulties with terms and/or with descriptions. I therefore strongly recommend the article by Blackmore et al. (1987), listing and describing nearly all stages of the male haplophase, the (changing) character states (e.g., primexine continuous – primexine restricted to small areas – primexine missing). Unfortunately, however, this article seems to be largely unknown, the only exception is Fitzgerald et al. (1993a): the development of compound pollen in *Acacia paradoxa* shows some unique features: cytoplasmic channels only present until callose dissolution, microspores do not undergo a vacuolation period, and wall bridges form by and unusual mechanism. The pathways used in compound pollen ontogeny are evidently diverse: cf. Mimosaceae – Asclepiadaceae – Orchidaceae, which are so far the best documented examples. Regarding the Asclepiadaceae, a very valuable study was published by Dannenbaum and Schill (1991): they describe minutely the developmental stages in some Asclepiadaceae pollen tetrads/polyads; they find some features similar to other taxa with polyads (e.g., the lack of callose in many orchids, as also in the Cynanchoideae: thus the current opinion that the presence of callose is – absolutely? – necessary for sporopollenin deposition is very questionable, cf. above in Section 3 – the lack of callose in *Ophioglossum*, Rudramuniyappa and Suvarnkhandi 1992).

The majority of orchids with compound pollen show the typical sporoderm development including the formation of primexine and a subsequent formation of exine and intine. In *Polystachia pubescens*, however, no primexine is formed, and the outer intine layer is simultaneously formed with the exine (Schlag and Hesse 1993). This fits very well with some other unusual features during early and late stages of sporoderm formation which indicate the possibility that in strongly derived taxa, as, e.g., in orchids, the sporoderm formation could have been modified during evolution.

e) Exine Pattern

Exine initiation, pattern formation, and exine substructure in young and mature pollen of various taxa are studied by M. Takahashi and his collaborators, using in part the high resolution of field emission scanning electron microscopes (HRFSEM). In contrast to some other views, the plasma membrane in *Bougainvillea* plays indeed a crucial role in exine pattern formation, and invaginations or (slight) withdrawals of the plasma membrane should not be interpreted as artifacts (Takahashi and Skvarla 1991a). Interestingly, Gabarayeva (1992) depicts in very young stages of *Asimina* exine formation enormous periplasmic spaces between the plasma membrane and the callose wall; in her micrographs, these spaces are partly filled in by a "cobweb-like prematrix". The control of pattern formation remains, no doubt, an enigma. A universally valid model would have to explain the small, but significant differences between the respective patterns in closely related taxa. At present, the plasma membrane model (Takahashi 1989 and 1993, for *Caesalpinia*) is favored. Uehara and Kurita (1991) provided support for this model with their findings in Lycopodiaceae. In contrast, Pérez-Muñoz et al. (1993a,b) believe that the pollen wall pattern is determined by a patterning of the spore protoplast, and that the ER is involved in pollen wall pattern formation.

The concept of plasma membrane-derived exine sculpture pattern doubtless has its merits; but not all exine elements can be determined by the plasma membrane, especially those formed towards the end of pollen development, e.g., the long supratectal spines of *Nuphar* pollen (Takahashi, (1992). In *Ipomopsis*, two developmental phases of exine formation can be distinguished: the first (punctate) exine pattern corresponding with plasma membrane invaginations is laid down in young developmental stages and the final (striate) pattern, typical for the mature pollen ornamentation, is due to differential exine thickening during the free microspore stage (Takahashi and Skvarla 1991b).

A further, peculiar exine element not governed by this concept are the viscin threads typical for most Onagraceae, some Ericaceae, and for a few Caesalpiniaceae: is the callose (completely?) resolved before viscin thread formation starts, or are the threads formed by "holes" in the callose wall? No plasma membrane fills the space between neighboring pollen grains/tetrads, only locular fluid and tapetum remnants – but most Onagraceae viscin threads are specifically sculptured! The origin of the genuine viscin thread is still unclear (Takahashi and Skvarla 1990). The initial protectum in *Caesalpinia* consists of "fibrous threads", as also revealed by high resolution field emission electron microscopy (HRFESEM). These threads become accumulated on distinct parts of the invaginated plasma membrane (Takahashi 1993). Evidently, these threads cannot be compared with young stages of viscin threads.

It seems that the outermost ectexine elements ("sculpture"), like the *Nuphar* supratectal spines (not to be confused with the "normal" exinous spines in *Nuphar*, Takahashi 1993, or *Ottelia*, Takahashi 1994), and the Onagraceae/

Ericaceae viscin threads are formed in a developmental stage "too late" for the plasma-membrane-pattern model. Viscin threads (containing sporopollenin) are so far known from Onagraceae; Ericaceae (pollen tetrads of *Rhododendron*, connected by viscin threads, germinate on stigma surface, Palser et al. 1992), and few Caesalpiniaceae (more precisely called exinal connections), but Ambwani and Kumar (1991) present a first report on such sporopolleninous pollen-connecting agents for *Daemonorops* (Arecaceae)! It should be checked whether those threads really are acetolysis-resistant. Nothing seems to be known about the pollination ecology of this palm. Nonsporopolleninous, viscous substances forming long pollen-connecting threads, are a well-known feature in various angiosperm families; a futher report on Annonaceae is given by Morawetz and Waha (1991): the authors list four functions of such threads in pollination biology. A mucilage, analogous to viscin threads and pollenkitt in terms of floral ecology, forms nonfibrous pollen-associated material, which may play a role in pollen germination (De Frey et al. 1992).

f) Exine Coating: Composition and Function

Evidently, the pollen coat does not only serve to stick pollen grains together or to protect pollen grains against UV or excessive dehydration, the widely overlooked role in pollen stigma interactions is discussed by Elleman and Dickinson (1990), who recognize in *Brassica* four respective functions. Pollenkitt is thought to contain many volatile substances which are attractive to pollinators (Henning and Teuber 1992). The nature and distribution of exine-held material in mature pollen grains is shown by Weber (1992b) for *Apium*: she distinguishes two sorts of substances and reviews aspects of preservation using different fixatives. Weber (1992a) beautifully shows the transfer of pollenkitt to the exine of mature *Smyrnium* pollen. Malachite Green as fixative additive reveals a membrane-like structure around the exine-held pollenkitt material in *Felicia* (Asteraceae) (Jordaan and Kruger 1993). Only in Brassicaceae is the outer pollen surface covered with tryphine, a complex protein/lipid mixture from the former tapetum (Murgia et al. 1991). According to Evans et al. (1992), it consists of a mixture of active enzymes, proteins, organelles, and membrane remnants in a lipid matrix. In contrast to pollenkitt, tryphine changes its volume and condition markedly during pollen hydration (Van Aelst et al. 1993).

Apart from pollenkitt and tryphine, it seems that, at least in some angiosperms, other exine-held material of uncertain nature and origin exists: some reports on Brassicaceae depict micrographs of conspicuous exine-held material without giving an interpretation (e.g., Van Aelst et al. 1993; Charzynska et al. 1989) or only a tentative one (Polowick and Sawhney 1993c: "membranous fragments"). In my view, this phenomenon is no artifact; I found correlating objects in *Tilia*, which has no tryphine, but pollenkitt, and called it "white boxes" (Hesse 1993b).

g) Intine and Other Sporoderm Strata

Only very few papers deal with nonexinous sporoderm strata, most probably because only the exine with all its spectacular features survives acetolysis and fossilization. Our knowledge of the chemical composition of exine-/intine-held proteins has made some progress: the distribution of acid phosphatase and glycoproteins was tested by Suarez-Cervera et al. (1992) in the light of their respective functions. Harris et al. (1994) analyze the histochemical composition in seagrass pollen walls. Heslop-Harrison and Heslop-Harrison (1991) give a highly instructive overview on form and function of the various intine strata including "oncus" and "Zwischenkörper", and on the diversication and variation in these only seemingly simple sporoderm strata. Rowley and Vasanthy (1993) point in *Cinnamomum* (Lauraceae) to the differences between onciform zone (o.z.) and intine (the o.z. was often misinterpreted as "outer intine"!).

New, unexpected ideas and insights are very welcome, and highlight the routine work of a professional reviewer: surprisingly, cryopreparation methods (HPF) demonstrate that in *Ledebouria* the innermost part of the exine becomes infiltrated with intine matrix substances, and is therefore "transformed" to an intine layer (Hess 1993a). Gabarayeva (1993) points out a similar process in *Asimia*: endexine lamellae become intermingled with material of the outer intine, and sporopollenin is simultaneously accumulated on the endexine lamellae. Igersheim and Weber (1993) and Weber and Igersheim (1994) reinvestigated the formation of "pollen buds" (intine protrusions) in *Ophiorrhiza* (Rubiaceae) and collected former reports that these pollen buds should serve to eliminate the vegetative nucleus (!); instead, large vacuoles were misinterpreted as being the vegetative nuclei. Reference to more such "uncommon" results should stimulate a check-up!

References

Aarts MGM, Dirkse WG, Stiekema WJ, Pereira A (1993) Nature 363:715–717. **Ambwani K**, Kumar M (1991) Phytomorphology 41:95–97.

Barnes SH, Blackmore S (1992) Arch Histol Cytol 55 (Suppl):217–224. **Blackmore S**, Barnes SH (1987) Cladistics 3:185–195. **Blackmore S**, Claugher D (1987) Rev Paleo- bot Palynol 53:175–184. **Blackmore S**, McConchie CA, Knox RB (1987) Cladistics 3:333–347. **Braun CJ**, Brown GG, Lerings III CS (1992) Cytoplasmic male sterility. In: Herrmann RG (ed) Cell Organelles. Plant Gene Research. Springer, Vienna New York, pp 219–245. **Brown RC**, Lemmon BE (1990) Am J Bot 77:1482–1490. **Brown RC**, Lemmon BE (1991a) Protoplasma 161:168–180. **Brown RC**, Lemmon BE (1991b) Protoplasma 163:9–18. **Brown RC**, Lemmon BE (1991c) Protoplasma 165:155–165. **Brown RC**, Lemmon BE (1991d) J Cell Sci 99:273–381. **Brown RC**, Lemmon BE (1991e) J Cell Sci 100:559–565. **Brown RC**, Lemmon BE (1992a) J Cell Sci 103:1031–1038. **Brown RC**, Lemmon BE (1992b) Protoplasma 167:183–192. **Brown RC**, Lemmon BE (1992c) EMSA Bull 22:48–53. **Brown RC**, Lemmon BE (1994) Sex Plant Reprod 7:87–94.

Calzoni GL, Speranza A, Li YQ, Ciampolini F, Cresti M (1993) Acta Bot Neerl 42:473–480. **Chaloner WG**, El-Ghazaly GE, Hemsley A (1993) Grana Suppl 1:1–48. **Chaboud A**, Perez R (1992) Int Rev Cytol 140:205–232. **Charzynska M**, Cresti M (1993) Protoplasma 172:77–83. **Charzynska M**, Lewandowska E (1990) Ann Bot 65:685–689. **Charzynska M**, Murgia M, Milanesi C, Cresti M (1989) Protoplasma 149:1–4. **Charzynska M**, Murgia M, Cresti M (1990) Protoplasma 158:26–32. **Chaudhury AM**, Lavithis M, Taylor PE, Craig S, Singh MB, Signer ER, Knox RB (1994) Sex Plant Reprod 7:17–28. **Chen ZK**, Wang FH (1990) Cathaya 2:41–52. **Chen ZK**, Wang FH, Zhou F (1990) Cathaya 2:113–126. **Chen ZK**, Wang FH, Tang ZX (1991) Cathaya 3:53–64. **Ciampolini F**, Li YQ, Cresti M, Calzoni GL, Speranza A (1993) Acta Bot Neerl 42: 349–355. **Claugher D**, Rowley JR (1990) Can J Bot 68:2195–2200. **Clément C**, Audran JC (1993) Grana 32:348–353. **Clément C**, Chavant L, Burrus M, Audran JC (1994) Sex Plant Reprod 7:347–356. **Cresti M**, Tiezzi A (1992) Sexual plant reproduction. Springer, Berlin Heidelberg New York, 247 pp. **Cresti M**, Blackmore S, Van Went JL (1992) Atlas of Sexual Reproduction in Flowering Plants, Springer Berlin Heidelberg New York, 249 pp. **Cross JW**, Ladyman JAR (1991) Sex Plant Reprod 4:235–243.

Dahl AO, Rowley JR (1991) Ann Sci Nat Bot 13e sér 1990–1991, 11:155–176. **Dajoz I**, Haut C (1993) Grana 32:176–183. **Dannenbaum C**, Schill R (1991) Bibl Bot 141:1–138. **Dawson J**, Wilson ZA, Aarts MGM, Braithwaite AF, Briarty LG, Mulligan BJ (1993) Can J Bot 71:629–638. **De Frey HM**, Coetzer LA, Robbertse PJ (1992) Sex Plant Reprod 5:298–303. **Del Casino C**, Tiezzi A, Wagner VT, Cresti M (1992) Protoplasma 168:41–50. **Dickinson HG** (1982) Rev Cytol Biol Vég Bot 5:5–19. **Dinis AM**, Mesquita JF (1994) Ann Bot 73:83–90. **Doyle JA**, Hotton CL (1991) Diversification of early angiosperm pollen in a cladistic context. In: Blackmore S, Barnes SH (eds) Pollen and Spores: Patterns of Diversification, Syst Assoc Spec Vol 44. Clarendon Press, Oxford, pp 169–195.

Elleman S, Dickinson HG (1990) New Phytol 114:511–518. **Evans DE**, Taylor PE, Singh MB, Knox RB (1992) Planta 186:343–354.

Ferguson IK, Harley MM (1993) Ken Bull 48:205–243. **Fernandez MC**, Rodriguez-Garcia MI (1990) Plant Cell Rep 8:667–671. **Fernandez MC**, Romero-Garcia AT, Rodriguez-Garcia MI (1992) Rev Palaeobot Palynol 72:41–48. **Fitzgerald MA**, Calder DM, Knox RB (1993b) Plant Syst Evol Suppl 7:53–62. **Foster GD**, Robinson SW, Blundell RP, Roberts MR, Hodge R, Draper J, Scott RJ (1992) Plant Sci 84:187–192.

Gabarayeva N (1990) Bot Zh 75:783–791. **Gabarayeva N** (1991) Bot Zh 76:10–19. **Gabarayeva N** (1992) Grana 31:213–222. **Gabarayeva N** (1993) Grana 32:210–220. **Gambardella R**, Alfano F (1990) Protoplasma 156:29–38. **Gambardella R**, Alfano F (1991a) Giorn Bot Ital 125:970–972. **Gambardella R**, Alfano F (1991b) Giorn Bot Ital 125:976–978. **Gambardella R**, Alfano F (1991c) Giorn Bot Ital 125:983–985. **Gambardella R**, Duckett JG, Alfano F, Gargiulo M, Squillacioti C (1993a) Bot Acta 106:350–355. **Gambardella R**, Alfano F, Gargiulo M, Squillacioti C (1993b) Bot Acta 106:356–363. **Goldblatt P**, Le Thomas A (1992) Rev Palaeobot Palynol 75:301–315. **Goldblatt P**, Manning JC, Bari A (1991) Ann Mo Bot Gard 78:950–961. **Gubatz S**, Wiermann R (1992a) Z Naturforsch 48c:10–15. **Gubatz S**, Wiermann R (1992b) Bot Acta 105:407–413. **Guzzo F**, Baldan B, Bracco F, Mariani P (1994) Can J Bot 72:352–358.

Hagemann R, (1992) Plastid genetics in higher plants. In: Herrmann RG (ed) Cell Organelles. Plant Gene Research. Springer, Vienna New York, pp 65–96. **Halbritter H**, Hesse M (1993) Grana 32:87–99. **Harris PJ**, Webster J, Weinhandl JA, Stone BA (1994) Sex Plant Reprod 7:101–106. **Hause G**, Hause B, Van Lammeren AAM (1992) Can J Bot

70:1369–1376. **Hayat MA** (1993) Stains and Cytochemical Methods. Plenum Press, New York, XVII + 455 pp. **Henning JA**, Teuber LR (1992) Crop Sci 32:653–656. **Heslop-Harrison J**, Heslop-Harrison Y (1991) Structural and functional variation in pollen intines. In: Blackmore S, Barnes SH (eds) Pollen and Spores: Patterns of Diversification, Syst Assoc Spec Vol 44. Clarendon Press, Oxford, pp 331–343. **Heslop-Harrison Y**, Heslop-Harrison J (1992a) Ann Bot 69:385–394. **Heslop-Harrison J**, Heslop-Harrison Y (1992b) Ann Bot 69:395–403. **Heslop-Harrison J**, Heslop-Harrison Y (1993) Bot Acta 106:331–337. **Hess MW** (1991) Plant Syst Evol 176:63–74. **Hess MW** (1993a) Planta 189:139–149. **Hess MW**, (1993b) Protoplasma 176:84–88. **Hess MW**, Hesse M (1994) Planta 192:421–430. **Hesse M**, (1993a) Plant Syst Evol Suppl 7:39–52. **Hesse M**, (1993b) Eur J Cell Biol Suppl 39 (Vol 61):125. **Hesse M**, Hess MW (1993) Plant Syst Evol Suppl 7:127–145. **Hesse M**, Pacini E, Willemse MTM (eds) (1993) The Tapetum, Plant Syst Evol Suppl 7:1–152. **Hird DL**, Worrall D, Hodge R, Smartt S, Paul W, Scott R (1993) Plant J 4:1023–1033. **Horvat F** (1992) Belg J Bot 125:80–88.

Igersheim A, Weber M (1993) Opera Bot Belg 6:51–59.

Jordaan A, Kruger H (1993) Ann Bot 71:97–105. **Joshi PA**, Caetano-Anollés G, Graham ET, Gresshoff PM (1993) Protoplasma 172:64–76.

Kedves M (1991) Three dimensional modelling of the biopolymer structure of the plant cell wall I. In: Kedves M (ed) Plant cell biology and development 2. Szeged, pp 63–74. **Kindiger B** (1993) Genome 36:987–997. **Koltunow AM**, Truettner J, Cox KH, Wallroth M, Goldberg RB (1990) Plant Cell 2:1201–1224. **Kurmann ME** (1990) Exine ontogeny in conifers. In: Blackmore S, Knox RB (eds) Microspores: Evolution and Ontogeny. Academic Press, London, pp 157–172. **Kurmann ME** (1991) Rev Paleobot Palynol 69:29–298. **Kuroiwa T**, Kawazu T, Uchida H, Ohta T, Kuroiwa H (1993) Eur J Cell Biol 62:307–313.

Lancelle S, Hepler P (1992) Protoplasma 167:215–230. **Le Thomas A**, Thanikaimoni G (1987) Mém Trav EPHE Inst Montpellier 17:333–353. **Liu B**, Palevitz BA (1992) Protoplasma 166:122–133. **Luegmayr E** (1993) Protoplasma 177:73–81. **Luegmayr E** (1994) Pollenmorphologische und pollenontogenetische Untersuchungen an ausgewählten Gesneriaceae. Dissertation, Formal- und naturwissenschaftliche Fakultät, Universität Wien. **Lugardon B** (1990) Pteridophyte sporogenesis. In: Blackmore S, Knox BR (eds) Microspores: Evolution and Ontogeny. Academic Press, London, pp 95–120.

Majewska-Sawka A, Rodriguez-Garcia MI, Nakashima H, Jassem B (1993) Sex Plant Reprod 6:22–32. **Malhó R**, Pais MSS (1992) Sex Plant Reprod 5:163–168. **Martinsson K** (1993) Grana 32:198–209. **McCormick S** (1993) Plant Cell 5:1265–1275. **Mogensen HL** (1992) Int Rev Cytol 140:129–147. **Morawetz W**, Waha M (1991) Beitr Biol Pflanz 66:145–154. **Murgia M**, Detchepare S, Van Went JL, Cresti M (1991) Sex Plant Reprod 4:176–181.

Newman GR, Hobot JA (1993) Resin microscopy and on-section immunocytochemistry. Springer, Berlin Heidelberg New York, 221 pp. **Noguchi T** (1990) Protoplasma 156:19–28. **Noguchi T**, Ueda K (1990) Cell Struct Funct 15:379–384. **Noher de Halac I**, Fama G, Cismondi IA (1992) Sex Plant Reprod 5:110–116. **Norstog KJ** (1990) Bot Gaz 151:275–284.

Osborn JM, Taylor TN (1993) Rev Palaebot Palynol 79:205–219. **Ottaviano E**, Mulcahy DL, Sari Gorla M, Bergamini Mulcahy G (1992) Angiosperm pollen and ovules. Springer, Berlin Heidelberg New York, 465 pp.

Pacini E, Taylor PE, Singh MB, Knox RB (1992) Ann Bot 70:179–188. **Palevitz BA** (1993a) Sex Plant Reprod 6:1–10. **Palevitz BA** (1993b) Protoplasma 174:25–35. **Palevitz BA** (1993c) Plant Cell 5:1001–1009. **Palevitz BA**, Liu B (1992) Sex Plant Reprod 5: 89–100. **Palevitz BA**, Tiezzi A (1992) Int Rev Cytol 140:149–185. **Palser BF**, Rouse JL, Williams EG (1992) Can J Bot 70:1039–1060. **Pandolfi T**, Pacini E, Calder DM (1993) Plant Syst Evol 186:175–185. **Pérez-Muñoz CA**, Jernstedt JA, Webster BD (1993a) Am J Bot 80:1183–1192. **Pérez-Muñoz CA**, Jernstedt JA, Webster BD (1993b) Am J Bot 80:1193–1202. **Periasami K**, Amalathas J (1991) Ann Bot 67:29–33. **Pierson ES**, Cresti M (1992) Int Rev Cytol 140:73–125. **Polowick PL**, Sawhney VK (1992) Protoplasma 169:139–147. **Polowick PL**, Sawhney VK (1993a) Can J Bot 71:1039–1047. **Polowick PL**, Sawhney VK (1993b) Can J Bot 71:1048–1055. **Polowick PL**, Sawhney VK (1993c) Ann Bot 72:595–605. **Punt W**, Blackmore S, Nilsson S, Le Thomas A (1994) Glossary of pollen and spore terminology. LPP Contributions Series No 1, LPP Foundation, Utrecht, 71 pp.

Read SM, Clarke AE, Bacic A (1993) Protoplasma 174:101–115. **Robards AW** (1991) Rapid feezing methods and their applications. In: Hall JL, Hawes C (eds) Electron microscopy of plant cells. Academic Press, London, pp 257–312. **Rodriguez-Garcia MI**, Majewska-Sawka A (1992) J Exp Bot 43:1659–1663. **Rougier M**, Saïd C, Russel D, Dumas C (1991) Protoplasma 162:140–150. **Rougier M**, Jnoud N, Saïd C, Russel D, Dumas C (1992) Protoplasma 168:107–112. **Rowley JR** (1988) J Palynol 23/24:29–42. **Rowley JR** (1991) J Palynol 27:323–344. **Rowley JR**, Claugher D (1991) Bot Acta 104:316–323. **Rowley JR**, Dunbar A (1990) Bot Acta 103:355–359. **Rowley JR**, Flynn JJ (1991) Ann Sci Nat Bot Paris 13eSé 1990–1991, 11:125–147. **Rowley JR**, Walles B (1985) Nord J Bot 5:241–254. **Rowley JR**, Walles B (1993) Nord J Bot 13:67–82. **Rowley JR**, Skvarla JJ (1993) Grana Suppl 2:21–25. **Rowley JR**, Skvarla JJ, Pettitt JM (1992) Rev Paleobot Palynol 70:297–323. **Rowley JR**, Vasanthy G (1993) Grana Suppl 2:49–53. **Rudramuniyappa CK**, Suvarnkhandi SS (1992) Phytomorphology 42:267–274.

Schlag M, Hesse M (1992) Sex Plant Reprod 5:131–137. **Schlag M**, Hesse M (1993) Grana 32:22–28. **Schulz PJ**, Cross JW, Almeida E (1993) Sex Plant Reprod 6:108–122. **Shivanna KR**, Johri BM (1989) The Angiosperm Pollen: Structure and Function, 2nd edn. Wiley Eastern New Delhi, 374 pp. **Sodmergen**, Luo YY, Kuroiwa T, Hu HS (1994) Sex Plant Reprod 7:51–56. **Southworth D** (1991) Exine biochemistry. In: Blackmore S, Knox RB (eds) Microspores: evolution and ontogeny. Academic Press, London, pp 193–212. **Southworth D** (1992) Int Rev Cytol 140:187–204. **Speranza A**, Calzoni GL (1992) Sex Plant Reprod 5:232–238. **Staiger D**, Apel K (1993) Plant J 4:697–703. **Staiger D**, Kappeler S, Müller M, Apel K (1994) Planta 192:221–231. **Stainier F**, Horvat F (1978a) Pollen Spores 20:195–214. **Stainier F**, Horvat F (1978b) Pollen Spores 20:341–349. **Stewart WN**, Rothwell GW (1993) Paleobotany and the evolution of plants, 2nd edn, Cambridge University Press, Cambridge. **Suarez-Cervera M**, Seoane-Camba J, Lobreau-Callen D (1992) Plant Syst Evol 183:67–81.

Takahashi M (1989) Am J Bot 76:1615–1626. **Takahashi M** (1992) Rev Palaeobot Palynol 75:317–322. **Takahashi M** (1993) Am J Bot 80:192–197. **Takahashi M** (1994) J Plant Res 107:161–164. **Takahashi M**, Skvarla JJ (1990) Am J Bot 77:1142–1148. **Takahashi M**, Skvarla JJ (1991a) Am J Bot 78:1063–1069. **Takahashi M**, Skvarla JJ (1991b) Am J Bot 78:1724–1731. **Tanaka I** (1991) J Cell Sci 99:21–31. **Tanaka I** (1993) J Plant Res 106:55–63. **Tanaka I**, Wakabayashi T (1992) Planta 186:473–482. **Tanaka I**, Nakamura S, Miki-Hirosige H (1989) Gamete Res 24:361–374. **Taylor P**, Kenrick J, Li Y, Kaul V, Gunning BES, Knox RB (1989) Sex Plant Reprod 2:254–264. **Taylor TN** (1990) Microsporogenesis in fossil plants. In: Blackmore S, Knox RB (eds) Microspores: Evolution and Ontogeny. Academic Press, London, pp 121–145. **Telmer CA**, Newcomb

W, Simmonds DH (1993) Protoplasma 172:154–165. **Theunis CH**, Cresti M, Milanesi M (1991) Bot Acta 104:324–329. **Thompson TE**, Senter SD, Grauke LJ (1993) Hort Sci 28:1191–1193. **Tiezzi A** (1991) Electron Microscopy Reviews 4:205–219. **Tirlapur UK**, Cresti M (1992) Ann Bot 69:503–508. **Tirlapur UK**, Willemse MTM (1992) Sex Plant Reprod 5:214–233. **Tiwari SC** (1994) Sex Plant Reprod 7:177–186. **Tryon AF**, Lugardon B (1991) Spores of the Pteridophyta. Springer, Berlin Heidelberg New York, 648 pp.

Uehara K, Kurita S (1991) Am J Bot 78:24–36. **Uehara K**, Kurita S, Sahashi N, Ohmoto T (1991) Am J Bot 78:1182–1190.

Van Aelst AC, Van Went J (1991) Acta Bot Neerl 40:319–328. **Van Aelst AC**, Van Went J (1992) Protoplasma 168:14–19. **Van Aelst AC**, Müller T, Dügellin M, Guggenheim R (1989) Acta Bot Neerl 38:25–30. **Van Aelst AC**, Pierson ES, Van Went JL, Cresti M (1993) Zygote 1:173–179. **Van Went J**, Gori P (1989) Submicrosc Cytol Pathol 21: 149–156.

Wang FH (1990) Microsporogenesis in cycads and *Ginkgo*. In: Blackmore S, Knox RB (eds) Microspores: Evolution and Ontogeny. Academic Press, London, pp 147–155. **Weber M** (1992a) Ann Bot 70:573–577. **Weber M** (1992b) Grana 31:17–24. **Weber M**, Igersheim A (1994) Bot Acta 107:257–262. **Wetzel CLR**, Jensen WA (1992) Sex Plant Reprod 5:117–127. **Wiermann R**, Gubatz S (1992) Int Rev Cytol 140:35–72. **Willemse MTM** (1993) Plant Syst Evol Suppl 7:107–116. **Wilmesmeier S**, Steuernagel S, Wiermann R (1993) Z Naturforsch 48c:697–701.

Yu HS, Russell SD (1992) Sex Plant Reprod 5:169–181. **Yu HS**, Russell SD (1993) Eur J Cell Biol 61:338–348. **Yu HS**, Russell SD (1994) Sex Plant Reprod 7:324–332. **Yu HS**, Hu SY, Russell SD (1992) Protoplasma 168:172–183. **Yu HS**, Huang BQ, Russell SD (1994) Sex Plant Reprod 7:313–323.

Zaki MAM, Dickinson HG (1990) Protoplasma 156:149–162. **Zaki MAM**, Dickinson HG (1991) Sex Plant Reprod 4:48–55. **Zavada M**, Gabarayeva N (1991) Bull Torrey Bot Club 118:292–302. **Zee SY**, Siu IHP (1990) Rev Paleobot Palynol 64:159–164. **Zhu T**, Mogensen HL, Smith SE (1990) Protoplasma 158:66–72. **Zhu T**, Mogensen HL, Smith SE (1991) Theor Appl Genet 81:21–26. **Zhu T**, Mogensen HL, Smith SE (1992) Eur J Cell Biol 59:211–218.

Prof. Dr. Michael Hesse
Institut für Botanik
und Botanischer Garten
der Universität Wien
A-1030 Vienna, Austria

B. Physiology

I. Plant Water Relations

By Rainer Lösch

More than 2400 papers on plant water relations were again published during the last 2 years. As in previous reviews, approximately one fifth of them will be mentioned here. Books covering various aspects of the theme are also mentioned in the following chapters.

1. Water Relations of Cells and Tissues

A reasonable amount of water in plant cells cannot move freely. It is bound by forces of different strength. Osmotic forces which essentially control water exchange between the cell and its surroundings are much weaker than those forces which keep the structurally bound water by electrostatic interactions at the surfaces of macromolecules and cell organelles.

Hydration of membrane structures is receiving increasing research interest for its importance for transmembrane exchange processes. It can be measured, e.g., by ^{1}H-NMR (Bacic et al. 1992; Fleischer and Werner 1992). Ho and Stubbs (1992) found by fluorescence spectroscopic approaches a higher hydration localized at the protein-lipid interface of membranes. A reduced hydration at these sites effects protein conformation, thereby modifying its functioning. Osmotic shrinkage, for instance, will decrease the anion channel permeability of tonoplast membrane vesicles (Andreev 1992). In a gramicidin A channel the hydration shell around the protein behaves as an integral part of the macromolecule and plays an important role in ion selectivity (Wu 1992; Poxleitner et al. 1993). Such a channel lined by water dipoles is not adequately described by analogy either to solid or to liquid or to liquid crystal states; Chiu et al. (1993) emphasize that it must be dealt with as a particular type of condensed matter. Membrane hydration changes will influence the lateral and transversal mobility of membrane components (Eriksson and Lindblom 1993) as well as the membrane folding properties. This occurs by hydration pressures due to the reorientation of the water dipoles by the bilayer. This pressure is most efficient at a distance of one to two water molecules from the bilayer surfaces (Gawrisch et al. 1992; McIntosh and Simon 1993). As a result of these bilayer-water dipole interactions, membrane phase transitions are strongly controlled by the hydration states (Mellier 1989; Webb et al. 1993). The transversal permeability through membranes of the water molecules themselves

Progress in Botany, Vol. 56

was determined to be 2.44 × 10^{-5} m s^{-1} by Snaar and van As (1992) using ^{1}H-NMR. Replacement of water molecules at the phosphate moieties by alcohols destabilizes the membranes (Chiou et al. 1992). Drought stress alters chemical composition of plasma membranes, increasing diacylglycerols and free sterols and decreasing, among others, total lipids, phospho-, and glycolipids (Drory et al. 1992; Ha and Thompson 1992; Navari-Izzo et al. 1993); but dehydration tolerance with aspect to a higher stability of membrane structures can be induced by repeated water stress treatments (Norberg et al. 1992). The particular membrane composition strongly determines the degree of desiccation tolerance of poikilohydrous organisms or tissues, e.g., pollen grains (Hoekstra et al. 1992). When dehydrated, the prevailing temperature can exert influence upon the restitution of membrane integrity (studies with yeast cell membranes: Pena et al. 1992). The osmoelastic coupling, i.e., the interactions between membrane deformations, pressures, and hydration changes, and the osmotically driven solvent and solute transport across membranes is treated from a thermodynamical point of view by Ito et al. (1989).

Cell and tissue water relations as influenced by osmotic forces were intensively studied during the past 3 years by pressure-volume (pV) analysis. Such comprehensive estimations of the plant water potential (ψ) components are widely adopted now as the state of art in water relations studies. They render increasing information about seasonal trends of turgor (ψ_P), the bulk tissue elasticity modulus ε, and osmoregulation (osmotic potential: ψ_π). Many papers, not all being itemized here, quantify the respective parameters for many species, provenances and cultivars of trees, crops, and, less often, wild-growing herbs. Sometimes tissue water relations equilibration over salt solutions is used (Livingston et al. 1992), more often pressure bomb measurements and psychrometric techniques are applied. As in previous years, some methodical improvements and caveats are reported (e.g., a comparative study of psychrometrically determined osmotic potentials from press sap and freeze-thawed leaf discs: Kikuta and Richter 1992; combination of capillary microcryoscopy in conjunction with pV analysis: Gross and Koch 1991b; consideration of the influence of tissue rehydration upon pV data: Kubiske and Abrams 1991; Schulte and Henry 1992). Psychrometric determinations can meanwhile be reliably applied by continuous in situ measurements (Dixon and Tyree 1984; Dixon 1987). Attempts are underway to replace after proper calibration the time-consuming and discontinuous psychrometric determinations of pV data by electrical impedance measurements (Colombo and Blumwald 1992).

Adaptations of plant water relations to adverse conditions were measured with respect to salinity (Pezeshki et al. 1993), cold (Goldstein and Nobel 1991; Koster and Lynch 1992), and drought stress. They occur as short-term processes diurnally (Saliendra and Meinzer 1991; George and Nissen 1992) or over longer periods.

Very often, osmoregulation occurs by net increases of solutes. These may come from changed balances of photosynthate production, consumption, and translocation under the influence of stress conditions (Zrenner and Stitt 1991; Quick et al. 1992; Li et al. 1993b). In yeast, genes were identified encoding signal kinases which trigger the production of osmotically active substances upon changes of the osmolarity gradient (Brewster et al. 1993); a plasma membrane ATPase is activated when the salt-tolerant yeast *Zygosaccharomyces rouxii* accumulates glycerol upon NaCl stress (Nishi and Yagi 1992). In higher plants, often the monosaccharides fructose and glucose play a most prominent role (Tan et al. 1992; Wang and Stutte 1992; Zhang and Archbold 1993); but also substantial increases of polyols (Wang and Stutte 1992), organic acids (Lecoeur et al. 1992), amino acids and other osmotically active substances are reported (Koster and Lynch 1992; Meier et al. 1992). Fan et al. (1993) combine high resolution ^{1}H-NMR and gas chromatography as a quick method for identification of these osmoregulatory organic substances in crude tissue extracts. The relative contributions of various solutes to total cell sap osmolality can be different in leaves of different age or different insertion levels along a shoot (Premachandra and Joly 1992). Li et al. (1992) and Premachandra et al. (1992) determined potassium ions as important osmoregulatory substances in wheat and sorghum, respectively.

Osmotic adjustment by solute accumulation keeps the symplast volume constant (Saliendra and Meinzer 1991) while the apoplastic water fraction becomes reduced (Rodrigues et al. 1993). Alternatively, a concentration of unaltered amounts of solutes occurs by lowered water contents when the symplast volume changes (Gross and Koch 1991a; Evans et al. 1992). Then the apoplastic water content increases and tissue elasticity becomes changed. This attains reasonable importance during water loss from symplast to the apoplast under freeze-dehydration stress (Zhu and Beck 1991). During several stress cycles or over an extended time of stress, it was observed (Kwon and Pallardy 1989; Wan et al. 1993) that first osmotic potentials become lowered by accumulation of solutes. Thereafter osmotic potential continues decreasing as a result of volume changes of the symplast combined with an increased tissue elasticity. The structural background of this latter development might be a persisting hydrolysis of cell wall linkages and a reduced production of polysaccharides which stiffen the cell walls (Talbott and Ray 1992; Munoz et al. 1993a,b). Enzymes with wall-loosening properties controlling such processes have been isolated by McQueen-Mason et al. (1992) and Li et al. (1993a). As a result, even over prolonged stress periods, positive ψ_P values are maintained although not at the high initial levels (Jensen et al. 1992).

A wall-loosening by xyloglucan turnover as a prerequisite for growth occurs also in roots, particularly in the tips, less intensive in proximal parts. This brings about a localized apical growth which continues under mild water stress, whereas ca. 1 cm behind the tip tightened walls prevent further elongation growth even after full turgor restoration (Pritchard et al. 1993). Pressure-probe measurements indicated that the water status of the root tip growing zone is mainly uncoupled from changes of the xylem water potential in the rest of the plant (Frensch and Hsiao 1993). Roots and leaves osmoregulate with different intensity (Koppenaal et al. 1991), and ontogenetic processes lead to pronounced osmoregulatory differences between different developmental stages and different organs (Shackel et al. 1991; Ito and Gyokusen 1993). Thus seasonally higher or lower osmoregulation has been found, e.g., by Colombo and Teng (1992) in *Picea glauca* shoots, by Linnenbrink et al. (1992) in deciduous hedgerow shrubs, by Ruiz-Sanchez et al. (1993) in almond leaves, by Grossnickle (1993) in *Tsuga heterophylla* and *Thuja plicata* seedlings.

In the grass family a difference seems to exist between C_3 and C_4 species in their ability for osmotic adaptation. Osmoregulatory solute changes are higher in C_4 grasses, whereas C_3 species have more elastic cell walls (Barker et al. 1993). It is important that the meristematic zones have sufficient osmoregulatory abilities. Indeed, grass leaf bases undergo a higher osmotic adjustment than the rest of the lamina (Thomas and Evans 1991; Walker and Hsiao 1993). An increased plasmamembrane ATPase activity possibly intensifies such a local gain of osmotically active substances (Hu et al. 1993) – see also findings with *Dunaliella* under hyperosmotic stress (Chen et al. 1991) and with yeast cell walls (Nishi and Yagi 1992). Auxin effects can be involved in such a maintenance of turgor for sustained growth (Liu et al. 1992). Matyssek et al. (1991a,b) emphasize the possibility of plant internal water redistribution from nongrowing to growing tissues, keeping in this way ψ_P sufficiently high for further growth. Indirect support of this view is that under mild water stress maize leaf elongation ceases during daytime hours, but not during night when evaporation is strongly reduced (Ephrath and Hesketh 1991). Using a pressure probe and a nanoliter osmometer Nonami and Boyer (1993) measured directly such growth-induced water potential gradients in soybean seedlings in a water-saturated atmosphere.

Normal expansion growth relies upon both import of solutes into the expanding cell (Kutschera and Froehlich 1992) and wall surface enlargement. A curvilinear relationship exists between growth rate (G) and ψ_P, which is influenced in shape by the action of auxin (IAA) and absisic acid (ABA): Hohl and Schopfer (1992); Bleiss and Ehwald (1993). IAA action reduces, and steep increases of turgor raise, cell wall xyloglucan levels which increase wall rigidity (Talbot and Ray 1992). The threshold turgor for growth, Y, is not influenced by phytohormones. The Lockhart equation for growth, $G = m \cdot (\psi_P - Y)$, thus would describe only an extreme situation, if m, the wall extensibility coefficient, were constant. Sudden changes in ψ_P, however are offset in their effects upon G by changes in wall-yielding properties (Serpe and Matthews 1992). Passioura and Fry (1992) describe a molecular model of the growing cell which accounts for homeostasis of both, m and Y, if ψ_P varies. Based on measurements with *Saprolegnia*, Kaminskyj et al. (1992) emphasized that turgor is necessary to explain extension rates of hyphal apices, but not sufficient. The cytoskeletal regulation will be involved as well in the growth process. Also Zhu and Boyer (1992) measuring *Chara* cell enlargement with a turgor clamp, postulate, additionally to the ψ_P affects on growth, the control by metabolic reactions causing synthesis and/or extension of wall polymers. Such processes have been documented by several studies concerning both metabolizing enzymes and structural carbohydrates (Talbot and Ray 1992; McQueen-Mason et al. 1992; Munoz et al. 1993a,b; Li et al. 1993a). A short, but comprehensive review of the present knowledge about water relations and growth at the level of both organs and tissues and at the level of individual cells is given by Boyer (1992).

2. Root Water Uptake and Water Movement Through the Plant

During the course of evolution, plant roots were adapted morphologically and anatomically to the various soil environments in order to acquire sufficient water and nutrients (Peterson 1992). Such structural differences related to function become obvious also in interspecies comparisons (Wilder and Johansen 1992).

The soil environment is not at all homogenous with respect to water and nutrients. A root system efficiently exploiting these resources often will not be homogenously structured, but rather adapted to the local conditions. Meyer and Barrs (1991) give an overview about destructive and nondestructive techniques to quantify the extend and shape of root systems. Using such methods, several investigations made comparative studies of the root systems of crop species or varieties under various influences. Some taxa show a very high plasticity in rooting depth, others are not as able to elongate their roots into deeper layers if soil dries out (Reader et al. 1993). Root clumping deteriorates plant water availability, as shown by model calculations (Tardieu et al. 1992) and by measurements (Petrie and Hall 1992; Petrie et al. 1992). According to the latter study, pearl millet with a dense and clumped local root geometry was less efficient using a limited water supply than cowpea with a less dense and more evenly distributed root system, notwithstanding that the total root length of the former was twice as great as that of the latter. Root clustering increases intraspecific competition for the available water. Growth along macropores encounters reduced mechanical resistances and facilitates the access to deeper soil layers, but increases the geometric mean pathway from bulk soil to the plant (Smucker and Aiken 1992). The advancing front of grain sorghum root tips penetrates downward into the soil with a rate of 2.7 cm day^{-1} up to a maximum depth of 190 cm, which is reached shortly after anthesis (Robertson et al. 1993a,b). *Arachis hypogaea* water uptake corresponds with the root density profile as long as water is readily available; soil hydraulic conductivity controls the water extraction rate of groundnut in dry soil (Sarkar and Kar 1992). Root penetration depth alone characterizes only partly the water uptake efficiency of a root system. Besides rooting depth, Ehlers et al. (1991) enumerate as other determinant parameters root length density, root length per soil layer, and the sum of the root length, together with the demand parameters leaf area index and potential evapotranspiration. The authors give some empirical relationships relating the maximum water uptake rate with these parameters. Root growth directions depend particularly upon gravity and gradients of soil water potential (measurements of spatial and temporal variations of soil matric potentials by microtensiometry: Vetterlein et al. 1993). Hydrotropism modifies, but – as a rule – does not neutralize gravitropism (seminal roots of wheat: Oyanagi et al. 1992; nodal roots of maize: Nakamoto 1993; lateral roots of Sitka spruce: Coutts and Nicoll 1993). Low temperatures of deep soil layers in spring may reduce the time for an ample root extension growth in northern regions (Entz et al. 1992), soil composition (higher clay contents) may impede full water consumption by the plants (Meinke et al. 1993), insufficient nutrient supply leads to intensified root growth even if soil water content is favorable (Martynova and Usmanov 1991). Roots penetrate best into deep layers of a drying soil in parallel with the ontogenetic development of plant: mature cotton plants with a shallow root system developed in a wet topsoil failed to extend their roots into the depth after change of the irrigation system from superficial drip irrigation to deep-soil wetting (Carmi et al. 1993); but new roots can develop from an already established root system after relieve of drought. This gives the plants a chance for survival after water supply has become improved (Bassirirad and Caldwall 1992; Meisner and Karnok 1992; Xu and Bland 1993). Onion plants recovered in this way after being held without water for a period of 200 days, when hitherto dormant lateral root primordia throughout rejuvenated the root system (Stasovski and Peterson 1993). Successful transplantation of tree seedlings also often relies upon the sufficient development of new roots (Brissette and Chambers 1992). Compacted soil layers impede the advancement into the depth, particularly of monocot roots (Materechera et al. 1993). Hence tillage (Kovar et al. 1992) and subsoiling (Rengasamy and Reid 1993) considerably improve the exploitation of soil water reserves by crop roots.

Concentration gradients of water around roots and the water distribution in the root system are studied noninvasively by computer-assisted tomography and ^{1}H-NMR, respectively. Using the former method, Hamza and Aylmore (1992a,b) mapped water uptake patterns under the influence of NaCl salinization alongside lupin and radish roots. Under the experimental conditions, the roots behaved as nearly perfect osmometers, as it was inferred from the linear relationship between water uptake and total root-leaf water potential. Grose and Hainsworth (1992) studied by this method soil water extraction by *Phytophthora*-infested lupin plants. NMR imaging detects more easily water in roots than in soil and can be used therefore to analyze root (water) distribution patterns (Bottomley et al. 1993). The gradient echo imaging technique is superior for this purpose to the spin-echo technique (Southon and Jones 1992). For close-up analysis of the soil-root contact zone van Noordwijk and colleagues (Kooistra et al. 1992; van Noordwijk et al. 1992; Veen et al. 1992) used a thin-section technique. Root hairs were found to bridge efficiently the gaps between root and soil, and soil compaction up to a certain degree improved the contact of maize roots to the surrounding soil, and in this way water uptake to the plant. Varney and Canny (1993) quantify the average water uptake of maize roots in a moist environment to be on average $5 \mu l\ h^{-1}\ cm^{-2}$ root surface with locally eightfold higher maximum uptake rates. Macfall et al. (1991) confirmed by ^{1}H-NMR the superiority of fine roots in terms of water gain per surface area as compared with tap roots and first order lateral roots of *Pinus taeda*. The resistance against soil-root water transfer increases drastically if air gaps develop at the interface. Gaps between roots and the surrounding soil develop dynamically when roots shrink. The increased transfer resistance exerts its affect into both directions: plant water uptake is rendered more difficult, and a water loss from the plant root into a dry soil becomes reduced. This could be very beneficial for plants in arid habitats (Nobel and Cui 1992a,b). Under such drought conditions root cells undergo ultrastructural changes (reduced number of mitochondrial cristae, arrangement of ER, disintegration of Golgi apparatus, condensation of nuclei: Ciamporova and Mistrik 1993).

An increase in the water transport resistance due to such gaps must be considered also in models calculating water flow through the soil-plant-atmosphere continuum (SPAC). Flow through the unsaturated bulk soil is governed principally by Darcy's law. It will be a tolerable simplification which reduces computation efforts, to replace the nonlinear partial differential equations by nonlinear algebraic ones (Govindaraju and Kavvas 1993) and to relate soil hydraulic conductivity to soil water content characteristics, if water potential data are not available (Moldrup et al. 1992). Nonlinear parameter estimation is superior in any case to linear methods (Rasiah et al. 1992). However, as Cardon and Letey (1992) emphasize, mechanistic equations based on Darcy's Law cannot consider appropriately the solute exchange between soil and roots which occurs concomitantly to the water uptake and exerts influence upon the latter. This is particularly important with brackish soil solutions.

Coming to the plant compartment, Ardner (1991) emphasizes a nonlinear behavior of the root membranes as sinks for water. Hector et al. (1993) concentrate their calculations upon root water uptake as depending upon rooting densities, differences between soil and stem water potential and a (constant) water uptake resistance. Chen and Lieth (1992, 1993) give a model for water transport through the SPAC of potted crops, Jensen et al. (1993) simulate and verify by field measurements of soil water contents and plant transpiration rates the water use of barley. Both approaches consider the uptake processes from different soil layers in their dependence upon the particular soil water availability. Jensen et al. (1993) emphasize explicitly the influence of air gaps at the soil/root interface upon the uptake resistance. The model of Chen and Lieth (1992, 1993) calculates the root resistance as depending upon root length density, which is influenced by branching, extension, and depth, and upon soil water content. It allows for an influence of the whole plant's carbon balance upon the root growth term. Habib et al. (1991) point out that a comprehensive model of matter flow through the roots must consider mineral transport together with water flow from soil into the plant. However, such models will increase enormously in complexity, and the body of quantitative data on mineral uptake processes at the soil/root interfaces is still rather incomplete.

It is commonly assumed that water and solutes cross root exodermis and cortex following appropriate potential or concentration gradients until they reach the endodermis (analysis of electrical coupling of cells across the cortex: Cortes 1992; pressure probe and nanoliter osmometer measurements of ψ_P, and ψ_π measurements across the cortex: Zimmermann et al. 1992). Suberin layers in the endodermis canalize water and solute flow to the symplast, where a metabolic control of ion entrance into the central cylinder occurs. The barrier of the Casparian strip stops particularly solute flow; the small water molecules also permeating membranes probably will not be blocked as much (Peterson et al. 1993); but also outside the endodermis, in cortical walls near the root tips, permeability for tracers can be reduced (Enstone and Peterson 1992). Root hydraulic conductivity decreases under drought as a result of an increased soil/root interface resistance built by dry soil sheaths in the root hair zone, of drought-induced rupture of cortical cells, and of xylem embolisms (North and Nobel 1992). Salinization reduces the radial hydraulic conductivity across the cortex, calcium can, in part, compensate this effect (Michalow 1989b; Azaizeh and Steudle 1991; Azaizeh et al. 1992). At junctions between the stem or the main root and lateral roots of desert succulents, an increased radial conductivity is measured under drought; but axial water flow becomes strongly limited due to embolism of the abundantly pitted and less lignified tracheary cell walls at these sites (North et al. 1992, 1993). Because the embolism is reversible under rewetting, these junctions act as rectifiers, allowing a rapid water delivery to the aerial parts after rainfall, but reducing water loss from the succulent shoots to dry soil (Ewers et al. 1992). With greater distances from the root tip, axial conductivity per vessel cross section area deviates from that of ideal capillaries for reasons of remaining cross walls (Melchior and Steudle 1993). This, however, becomes overcompensated by the increased number of vessel elements. As a result, it is the radial resistance which controls water uptake in proximal root regions, and the axial resistance in the apical part. The latter is therefore to a certain degree hydraulically isolated from the rest of the root (Enstone and Peterson 1992; Frensch and Hsiao 1993; Melchior and Steudle 1993). Suberization accompanying both senescence and drought adaptation of roots decreases root hydraulic conductivity (Cruz et al. 1992; Huang and Nobel 1992, 1993).

The active ion import into the central cylinder at the endodermis, entailing water inflow, is influenced by temperature (Michalov 1989a, 1993). Temperatures lower than optimal for a particular species therefore also reduce root water uptake without the possibility of a substantial acclimation (Bolger et al. 1992). ABA may, however, shift the thermal optimum for root water and ion uptake and transport to lower temperatures (Bassirirad and Radin 1992). This phytohormone increases under drought. Then a changing pH gradient, particularly across the cortical plasma membranes, brings about an increased ABA import into the xylem of a stressed plant (Daeter et al. 1993). Besides the effects of ABA as a long-distance messenger for transpiration control under stress (see Sect. 3) it enhances cell expansion under low water potentials at root tips and inhibits further growth in the mesocotyl region (studies with maize: Saab et al. 1992).

As a rule, water is transported in xylem vessels acropetaly under tension. It is pushed upwards by energy-consuming root pressure which, aside from osmotic forces, can also rely upon the action of contractile proteins (Mozhaeva 1991). The energy required to build up root pressures comes from respiration processes – in wet soil only 12% of the total root respiration is used up by growth of new roots, the rest serves for steady-state metabolism (Nobel et al. 1991). Root pressure is particularly important for water delivery to the shoots of deciduous trees in the leafless state: then slightly negative temperatures trigger starch hydrolysis and sugar release to the xylem conduits of young *Juglans* trees. If winter temperatures are above zero, tensions instead of pressures of the xylem sap can be found (Ameglio and Cruiziat 1992).

Xylem sap is a diluted solution particularly of inorganic ions, soluble carbohydrates, organic and amino acids and various phytohormones. The composition varies diurnally (Ferrario et al. 1992a,b) and seasonally (Dambrine et al. 1992; Sauter and van Cleve 1992). Pruned grapevine shoots exudate over a 1-week period large amounts of bleeding sap containing inorganic ions of in total 17mM and many soluble organic compounds (13.5 mM). The release of the latter increased by recutting treatments (Glad et al. 1992). Metabolically outstanding developmental stages like bud break or flower formation are preceded by peak values of xylem sap ion contents (Havelange and Bernier 1993). Root and branch ion concentrations of an adult spruce are higher than those of the bole, and there is a dynamic cycling of a pool of mobile ions among the different parts of the tree crown (Dambrine et al. 1992). The amount of K^+ relocation from roots to aerial parts of maize plants is partly uncoupled from root K^+ uptake (Engels and Marschner 1992). NO_3^- is the preferred nitrogen compound for acropetal transport via the transpiration stream if nitrate serves as the root medium nitrogen source in *Ricinus*. If the plants are supplied with NH_4^+, glutamine is the transport form of nitrogen exported to the aboveground parts (Peuke and Jeschke 1993; see also: Plhak 1992). The relative concentration of nitrate in the xylem sap of two *Eucalyptus* species increased under water stress, as was also the case with B and Si concentrations (Stark 1992). The amounts of Ca^{2+} carried along with the xylem stream are also under metabolic control – an urgent necessity in order to avoid stomatal malfunction which otherwise would result from excess Ca^{2+} supply coming along the apoplastic pathway to the guard cells (Atkinson et al. 1992; Ruiz et al. 1993). Nearly all widespread amino acids and organic acids can be detected in xylem exsudates (Sauter and van Cleve 1992; Senden et al. 1992), and also various polypeptides and polysaccharides (Satoh et al. 1992). ABA transported via the transpiration stream from roots to leaves plays an important role in stomatal responses to soil drought stress, but ABA derived from leaves and redistributed in the plant contributes as well to the total xylem sap ABA content (Neales and McLeod 1991). Also cytokinins, acting antagonistically to ABA with respect to stomatal conductances, were found as components of the xylem sap (Fusseder et al. 1992; Bano et al. 1993). Xylem sap pH was measured in vivo using ion-sensitive field-effect transistors by Herrmann and Tesche (1992). A discussion went on about the importance of the transpiration stream for long-distance ion distribution in the plant originating from an observation that sufficient ion supply occurs also when transpiration is restricted by a vapour-saturated aerial environment (Tanner and Beevers 1991; Smith 1991).

The structural-functional relationships which govern water transport through the shoot to the leaves were reviewed by Tyree and Ewers (1991). The species-specific hydraulic architecture is chiefly characterized (1) by the vulnerability of stems to drought-induced embolism, (2) by the height of pressure gradients and sustainable water potential extremes which are in a dynamic balance with the development of the foliage of a plant, and (3) the water storage capacity of the stem. For investigation of the latter aspect, Holbrook et al. (1992) apply time-domain dielectric measurements in order to determine *Sabal* palm stem water contents. Sap flow from excised sugar maple stems correlates with their water content (Johnson and Tyree 1992).

The other two points quoted by Tyree and Ewers (1991) refer to anatomical and physiological aspects of water ascent in the shoot and were studied very intensively during the years under report. The state of knowledge on this matter at the end of the 1980s is presented comprehensively by a book of Borghetti et al. (1993). Since then, studies of xylem anatomy and vascular transport have been conducted, inter alia, with annual herbs, like the halophyte *Salicornia europaea* (Ellison et al. 1993) or rice plants (Hirasawa et al. 1992), woody vines like *Bauhinia* (Ewers and Fisher 1991) and *Vitis* (Schultz and Matthews 1993), ash seedlings (Shumway et al. 1991), conifers (Pothier et al. 1989) and broadleaved trees (Lo Gullo and Salleo 1990; Tyree and Alexander 1993). Irrespective of the particular topics of these studies they coincide, emphasizing (1) the decrease in conductivity at nodes, branch and shoot-petiole junctions, (2) the high ontogenetic plasticity of vessel numbers and diameters meeting the different requirements of supply of quite distinct leaf areas, e.g., in sun and shade, respectively, and, in general (3) the close relationships between stem hydraulic structure and the ecological situation of a plant – of course, within the genetically limited abilities of the taxa. Evolutionary progressions of such abilities become obvious from comparative studies within one genus. With such an approach, Chiu and Ewers (1992) analyzed wood structure and function of a twinner, a scrambler, and a shrub species of the genus *Lonicera*. Detailed studies on factors affecting the relationship between sap wood area and leaf area of coniferous trees were conducted by Coyea and Margolis (1992) and Shelburne et al. (1993).

Vessel diameters strongly influence the water flow through a shoot, even if in reality lower water transport rates occur than calculated by the Hagen-Poiseuille law. Hydraulic diameters of vessels having noncircular openings can be approximated for calculation purposes by the length of the short axes of an ellipse or a rectangle times a correction factor (Lewis 1992). Similarly, water flux projections for asymmetrical stems from heat pulse point measurements must account for eccentricity of the velocity profile in order to avoid substantial errors (Hatton et al. 1992; Cermak et al. 1992). Vessel length and cross wall properties are other anatomical quantities responsible to a great extent for deviations from flow rates through ideal capillaries, which determine species-specifically the transpiration stream through a shoot. Increasing thickness and angles of obliquity of scalariform perforation plates decrease xylem conductivity (Schulte and Castle 1993a,b); nonrandom vessel end distribution along a vascular bundle reduces the accuracy of calculations of average vessel length (Tyree 1993). Empirically, repeated shortening of branches in comparison with their hydraulic conductance per unit pressure gradient renders estimates of species-specific average vessel length (Chiu and Ewers 1993; Sobrado 1993).

Embolism, as a widespread phenomenon affecting stem water transport, is observed by appropriate, mostly acoustic determination approaches in more and more plant species. The frequency of occurrence of cavitations depends upon the xylem structure: out of a total of 43 northern temperate tree species, eight conifers had virtually no late-winter embolisms, cavitation-caused conductivity losses amounted to 17% in diffuse-porous hardwoods, to 26% in semi-ring-porous ones, and to 55% in ring-porous trees (Wang et al. 1992; similar findings: Ikeda and Ohtsu 1992; Sperry and Sullivan 1992). Cochard (1992) states that drought-tolerant species are less vulnerable to air embolism than drought-avoiding ones. The deciduous oaks which grow in Europe are prone to embolism in the sequence *Q. pubescens* < *Q. petraea* < *Q. robur* < *Q. rubra* (Cochard et al. 1992a). *Q. ilex* exhibits different vulnerabilities to freeze- and summer drought-induced embolism (Lo Gullo and Salleo 1993). Some evolutionary lines of cacti which grow by preference in semi-arid habitats show little protection against cavitation, combined with a limited water-storage capacity, others, less lignified, but highly succulent ones, very seldom experience cavitation of their vessels. The latter species are able to colonize extremely arid, unshaded sites (Mauseth 1993).

In situ dehydration leads to a degree of embolism similar to bench-top drying of excised shoots (Tyree et al. 1992), and positive air pressures inside a pressure chamber produce a loss of hydraulic conductivity similar to that when the shoot is under xylem tension (Cochard et al. 1992b). This observation supports the air-seeding hypothesis of the production of embolisms. At the same time, however, this finding raises some questions about possible artifacts when pV curves are produced with sap removal and lowered water potentials brought about by overpressurization of shoots inside a pressure bomb. Recovery from embolism is possible – within limits: Sobrado et al. (1992) – if gas bubbles become redissolved under increased xylem pressure. This process can be modeled, predicting recovery dynamics (Tyree and Yang 1992; Yang and Tyree 1992).

The water potentials under which first cavitation events occur are reported to range between −0.9 MPa (*Pinus tunbergii*: Ikeda and Ohtsu 1992), −1 to −1.5 MPa (*Populus, Quercus rubra*: Tyree et al. 1992) and −2.5 to −4 MPa (diverse oak species: Cochard et al. 1992a). Water potential ranges in the same order of magnitude are reported to occur in plants in countless field measurements determined with both pressure bomb and psychrometer techniques. Based on xylem pressure probe measurements (Benkert et al. 1991) Zimmermann et al. (1993) cast doubts on whether the cohesion theory can serve as the sole possible explanation of the ascent of sap. Xylem vessel pressures measured directly reached only about −0.2 MPa, and the authors argue that long-distance transport in plants might be achieved only by the combination of several forces: tension at a much smaller degree than traditionally assumed, osmotic gradients alongside a vertically compartmented xylem, capillary forces, and interfacial forces at air-water interfaces of stationary small bubbles at the vessel walls. Similarly, Kargol (1992) proposes a graviosmotic hypothesis as complementary to the cohesion theory. As Mansfield et al. (1993) emphasized, new approaches to the study of the transpiration stream could lead to reconsideration even of long-established views; future studies will show whether these unorthodox thoughts, still running alongside the main stream of the accepted opinion, will gain broader experimental basis and acceptance. The discussion whether or not the unexpected experimental findings need reconsideration of the theoretical framework began already by comments and reply by Passioura (1991) and Zimmermann et al. (1991).

Velocities and amounts of water flow through stems are measured by heat pulse and heat balance methods. Using heat pulse measurements, Smith (1992) determined *Populus deltoides* water uptake. The parabolic heat pulse velocity profile proved to be unaffected by the presence of irregularly narrow growth rings. The amount of transpiration flow was calibrated against the heat pulse velocity data by direct measurements done with a cut tree. The heat balance method, which does not need an empirical or computational calibration (Cohen et al. 1993), has meanwhile gained more widespread use. The discussion of measurement errors of the heat balance approach is focused by Shackel et al. (1992) and by Cabibel and Do (1991) on influences of changing ambient and stem temperatures. Senock and Ham (1993) apply heat balance sap flow gauges to stems <10 mm diameter. Used in conjunction with leaf transpiration measurements, D_2O transport determinations, or calculations of evapotranspiration (Dugas et al. 1993), both heat pulse and heat balance studies are useful for stand water turnover estimates. Dye and Olbrich (1993), investigating *Eucalyptus* water turnover, emphasize the need for sampling heat pulse data from a relatively large number of trees for reliable results. Cienciala et al. (1992) scaled up *Picea excelsa* sap flow rates to stand transpiration, and from this and the existing vapor pressure deficit, calculated a canopy conductance term for use in Penman-Monteith calculations. Data of evaporation, xylem sap flow, and tree transpiration of a New Zealand *Nothofagus* forest are reported by Köstner et al. (1992) and Kelliher et al. (1992). Cermak et al. (1993) measured by heat balance determinations a steep increase of the transpiration stream upon irrigation in *Fagus* trees suffering from forest dieback as contrasting to healthy controls which obviously were not as much affected by local drought.

3. Stomatal Behavior and Transpiration

Progress in knowledge of stomatal biology has advanced particularly under two aspects: (1) The electro-osmotic transport of solutes across the guard cell plasmalemma is becoming more and more understood as the crucial process of stomatal osmoregulation. A special contribution about this topic was given in the previous volume of *Progress in Botany* (**55**, 1994) by Thiel and Gradmann. (2) The plant internal coordination of stomatal conductance under the influence of soil water shortage has gained equal interest as stomatal responses to plant-external environmental factors. Root-shoot signals, most often identified as chemical (phytohormone) signals, but sometimes also of electric or hydraulic character (Fromm and Eschrich 1993; Tardieu and Davies 1993), play an outstanding role in such a coordination. This aspect also is covered by some comprehensive recent reviews (Tardieu 1993; Lösch and Schulze 1994). As a further overview on stomatal biology, that of Tallman (1992) can be cited. Therefore, this chapter of the biennial report on progress in plant-water relations can be kept shorter this time.

Guard cell carbon metabolism was analyzed with respect to key enzymes of sucrose and malate synthesis. Hite et al. (1993) found very high activities of sucrose-phosphate synthase in guard cells, but also considerable activity of the degradative enzyme sucrose synthase. Also from fructose-6-phosphate phosphorylation activities, an intensive carbon metabolism can be inferred, which meets increased demands for energy and carbon skeletons during stomatal opening (Hite et al. 1992). Phosphoenolpyruvate carboxylase (PEP-Case) activity of guard cell protoplasts is much higher than that of mesophyll or epidermis cell protoplasts (Schnabl et al. 1992; Schulz et al. 1992). A partial purification and physiological characterization of this enzyme is reported by Denecke et al. (1993). While malate from the guard cell cytoplasm is ineffective, externally applied malate induces major alterations in the voltage-dependent activity of the guard cell anion channel. Hedrich and Marten (1993) conclude from this that a feedback regulation of anion efflux through this malate-sensitive channel could provide a sensor mechanism for the leaf-internal CO_2 concentration.

Apart from the indirect effect of illumination upon stomatal apertures via photosynthetic reduction of leaf-internal CO_2, blue light exerts a direct influence affecting the polarization of the guard cell plasma membrane. This is not considered in detail in the special article of Thiel and Gradmann (1994) on the electro-physiology of stomata, and will therefore be dealt with here by citations of more recent literature. Upon blue light absorption, protons are extruded from guard cells (vanadate inhibition of the pumping ATPase: Amodeo et al. 1992; activation by synergistic fusicoccin + light effects: Assmann and Schwartz 1992). The energy needed for this proton pumping as provided from both guard cell chloroplasts and mitochondria oxygen supply is guaranteed (Gautier et al. 1992; Mawson 1993; fluorescence quenching of guard cell chloroplasts induced by blue light: Srivastava and Zeiger 1992). Deprivation of CO_2 enhances guard cell proton extrusion (Lasceve et al. 1993). Calmodulin and Ca^{2+}/calmodulin-dependent myosin light chain kinase are components of the signal transduction process in blue light-dependent proton pumping (Shimazaki et al. 1992).

Soluble sugars as possible cytoplasmic osmolytes are accumulated in guard cells by blue light-induced starch degradation and by red light-dependent photosynthetic carbon reduction. The quantities of this sugar production under 2.5

and 3 h irradiation at 25 μmol m^{-2} s^{-1} blue and 50 μmol m^{-2} s^{-1} red light, respectively, range between 300 and 500 fmol sugars per guard cell (Poffenroth et al. 1992). *Zea mays* stomata increase their conductance upon red light pulses (Assmann et al. 1992), as can also be observed with leaves of the facultative CAM plant *Portulacaria afra* in the C_3-modus. By contrast, *Portulacaria* plants performing CAM did not respond to light, either in the blue or in the red wavelength (Lee and Assmann 1992). Gas exchange measurements with inverted ivy leaves demonstrate that indirect stomatal responses to light via photosynthesis are sufficiently explained by the effect of a reduced leaf-internal CO_2 concentration (Aphalo and Jarvis 1993).

The blue-light response of stomata shows a considerable plasticity, depending upon the light environment which prevails during leaf development (Assmann 1992). More than that, plants under natural conditions respond to changing light conditions with distinct sensitivities. Subalpine plants undergo reversible decreases of conductance upon transient shade situations. Normally, high light-exposed desert plants close the stomata rapidly when shaded, but do not reopen as promptly (Knapp and Smith 1991). C_4 grasses earn more benefits in terms of an improved water use efficiency (WUE) than C_3 grasses because they undergo smaller changes of leaf conductance under a variable light regime (Knapp 1993). Differences exist also in stomatal movement dynamics between *Piper* species from sun and shade habitats of a tropical forest in Mexico under the influence of lightflecks. Longer-lasting stomatal opening enhances carbon gain of the shade plants, but a sensitive closing response to higher vapor pressure deficits (VPD) improves WUE at the cost of assimilation. This situation prevails in sun-adapted species (Tinoco-Ojanguren and Pearcy 1992, 1993).

Stomatal responses to (leaf internal) CO_2 concentration was considered in studies on the effects of increased (in the experiments: doubled) ambient CO_2 concentrations. As a rule, WUE became considerably improved with higher CO_2 concentrations if plant water supply was limited (e.g., Curtis and Teeri 1992; Nie et al. 1992; Polley et al. 1992 – the latter study compares WUE of C_3 and C_4 perennials under reduced CO_2 concentrations). Photosynthetic stimulation by high CO_2 is more obvious at the canopy than at the individual leaf level (Nijs et al. 1989). It is not only the higher gradient of CO_2 across narrowed stomatal pores which improves growth under elevated CO_2 levels. Senescence of stomatal functioning also becomes delayed with long-term exposure to a high CO_2 atmosphere (Curtis and Teeri 1992). The normally observed stomatal closure of senescent leaves does not result from higher leaf-internal CO_2 levels when the photosynthetic capacity of the mesophyll goes down. Stomata in epidermal peels of senescing *Nicotiana glauca* plants respond to light in the same manner as those from nonsenescing leaves (Thomas et al. 1991). Lauer and Boyer (1992) did not, as is usual, calculate intercellular CO_2 concentration (p_i) but measured it directly. Leaves of well-watered plants maintained a relatively constant differential between the CO_2 concentrations outside and inside the leaf. Under water shortage p_i increased despite stomatal closure, which indicates a reduced and finally zero mesophyll photosynthesis. This did not happen to leaves treated with

ABA. Differences in stomatal and mesophyll sensitivity to water stress are highly species-specific; stomatal closure under mild water stress could be important for an apparently water stress-induced inhibition of photosynthesis (Ni and Pallardy 1992). Such nonuniform behavior of the stomata of a leaf has been found by various studies (e.g., Santrucek 1991; Beyschlag et al. 1992; Epron and Dreyer 1993; Mott et al. 1993), but it does not occur in all plants (Gunasekera and Berkowitz 1992a). Wise et al. (1992) were able to induce or to suppress an inhomogeneous photosynthetic activity across cotton leaves, depending on the plant's drought acclimation.

The effects upon stomatal sizes and densities of long-lasting reduced or elevated atmospheric CO_2 levels have been controversially discussed (Woodward 1987; Körner 1988) and are investigated by different approaches. An elevated CO_2 atmosphere affecting the plant during its ontogenetic lifetime does not have significant effects (Ryle and Stanley 1992); but prevailing higher temperatures decrease stomatal sizes and densities (Beerling and Chaloner 1993a) which might be the very cause of such morphogenetic changes documented by subfossile leaf structures (*Olea europaea* leaves collected from Tutankhamun's tomb onward until present: Beerling and Chaloner 1993b; *Salix herbacea* leaves from 140 000 years onwards: Beerling and Chaloner 1993c).

As in previous years, there are again several papers dealing with stomatal responses to VPD (e.g., Guehl et al. 1991: comparison of mediterranean *Abies* species; Faensen-Thiebes 1992: *Melilotus alba*; Dai et al. 1992: castor bean; Kawamitsu et al. 1993: *Oryza sativa* (C_3) and *Panicum maximum* (C_4) stomatal responses to VPD after different humidity pretreatments; Lösch and Kazda 1992, Meinzer et al. 1993, both: VPD responses of rainforest trees; Dufrene and Saugier 1993: oil palm). The same is the case for stomatal responses to leaf or soil water potentials (e.g., Huzulak and Matejka 1992: emphasis that conductance is not directly controlled by leaf water potential; Ferreira and Katerji 1992: VPD effect overrides soil drought effect; Ekanayake and De Jong 1992: comparison of potato genotypes; Acherar et al. 1991: comparison of *Quercus* species). Some papers report investigations over the whole season and the dependencies of leaf conductances upon the joint influence of environmental effects (e.g., Maier and Teskey 1992: *Pinus strobus*; Miller et al. 1992: *Juniperus occidentalis*; Ellsworth and Reich 1992: *Acer saccharum*; Wiebel et al. 1993: *Garcinia mangostana*).

Lösch and Schulze (1994) summarize that stomatal responses to VPD and soil drought are both feedforeward responses based on mechanisms and signals which bypass the mesophyll and in this way improve the homeostasis of its water relations. More recent studies and model calculations suggest, however, that additionally to the chemical message of the ABA, as the drought-stress phytohormone, coming along the xylem sap pathway from the roots, hydraulic effects within the epidermis and from the mesophyll have to be considered (Tardieu et al. 1992, 1993; Tardieu and Davies 1993). It is the sufficiently high concentration at the guard cell sites of ABA, not as much its absolute amount supplied to the leaf, which efficiently brings about stomatal closure (Gowing et al. 1993; Trejo et al. 1993). Other chemicals, like auxins (Ricanek and Vicherkova 1992), cytokinins (Drüge and Schönbeck 1993), some amino acids (Rai and Sharma 1992) or phenolic substances (Purohit et al. 1992) can modify or even completely offset the closing effect of ABA. Slovik et al. (1992) and Slovik and Hartung (1992a,b) calculated by a model the compartmental distribution and redistribution of ABA in intact leaves.

A coordinated water flow through the SPAC must rely upon finely tuned relationships between water uptake, transport, and loss, together with proper information about the mesophyll demands sent to the sites of regulation of this throughflow, the stomata. Lee and Bowling (1992) deduce from different stomatal openings in epidermal peels floating on incubation solution with and without leachates from mesophyll or chloroplasts, that in intact leaves a chemical message moves from the mesophyll to the epidermis, probably a product of photosynthesis. The apoplastic water and solute flow through the mesophyll and the epidermis cell wall is, as a flow through leaky tubes, not more adequately described by the Hagen-Poiseuille law (Canny 1993). Dyes cannot correctly trace the source-sink relationships of this finely ramified system, because water and many solutes penetrate cell membranes more rapidly than the relatively voluminous dye molecules; ionic absorption/adsorption processes can be added, increasing the complexity. Hydathodes may serve as sites for retrieval of solutes from the transpiration stream (Wilson et al. 1991).

It is plausible meanwhile that ABA serves as an efficient messenger which brings the information on water shortage at the soil/root interface to the stomata and the mesophyll (Khalil and Grace 1993). Still less clear are the causal links bringing about the coordination between the hydraulic conductivity of the stem and the gas phase conductance of the stomata (Meinzer and Grantz 1991). This coordination must work at the level of ontogenetic development of the plant structures. By artificial manipulation of the hydraulic conductance of birch trees, Sperry et al. (1993) found two responses in order to restore homeostasis: earlier in the vegetation period, new xylem was produced so that leaf conductances and transpiration returned to the control levels. Later in the season when conductances generally declined, a premature senescence and drop of leaves reduced the transpiring leaf area per functional xylem area (similar findings: Shumway et al. 1991). Seasonal changes in the relationships between leaf water status and stomatal conductance of fruit tree leaves were found also by Yoon and Richter (1992). A regular change of leaf phenotypes could be a means by which tropical evergreen shrubs cope with variation in water loss between the rainy and dry season (Mulkey et al. 1992). Leaf rolling (e.g., Heckathorn and Delucia 1991; Premachandra et al. 1993) efficiently decreases the transpiring grass leaf area under water shortage. Winkel and Rambal (1993) state that on a diurnal scale, grapevine-internal water status is stabilized by stomatal regulation, but that on a seasonal scale both stomatal regulation and limitation of total leaf area are important. Cultivars from arid regions control their leaf areas more efficiently than cultivars whose origin is a mesic climate. Rambal (1993) integrates various root water access situations, transpiring leaf areas, the stomatal regulation, and the critical levels of tissue water potential into a complex simulation model.

Cuticular transpiration depending on VPD is influenced by the water content of the cuticle, which changes with relatively long time constants (van Gardingen and Grace 1992). Differences in cuticular transpiration of plants depend upon structure and composition of the cuticle. Particularly epicuticular waxes increase in their amounts during the course of drought hardening (Premachandra et al. 1992)

4. Water Relations of Seeds and During Early Germination Stages

Many homoiohydric plants once in their life cycle also obtain very high degrees of desiccation resistance, namely as seeds. Recalcitrant seeds tolerate desiccation only down to moisture contents of approximately 30%, regularly matured orthodox seeds do not loose viability even at low moisture contents of 3% or less. Ultradrying may even improve a long-time seed storage (Cheng et al. 1992).

a) Orthodox Seeds

The transition to the desiccation-resistant stage in orthodox seeds occurs already on the mother plant. Intermittent water stress imposed there influences the dormancy of *Sorghum halepense* caryopses. Probably water stress enhances the glume permeability to oxygen diffusion and lowers by this way the seed dormancy (Benech et al. 1992). In *Acer platanoides*, the extent of maturation drying of the immature seeds determines the degree of low moisture contents which can be endured by the ripe seeds without loss in viability. Maturation drying down to 28% moisture content permits the survival of seeds with only 3% moisture content (Hong and Ellis 1992). The water potential gradient along a developing maize cob reduces or even terminates kernel growth in the apical parts. This occurs because endosperm cell divisions are decreased there, possibly under the influence of ABA (Ober et al. 1991). With rice kernel moisture contents of above 4% , a negative logarithmic relation exists between seed longevity and moisture content, but not below this threshold (Ellis et al. 1992a). The remaining water in dry seeds is partly concentrated in a glassy state. These glassy domains can protect intracellular components against destruction in the dehydrated state (Bruni and Leopold 1992a,b). When hydration starts, first of all water dipoles become rearranged from former tight hydrogen bonds, a reorientation of CH_2O groups occurs, and glassy domains are found at sugar-water interfaces (Bruni and Leopold 1992a).- ^{1}H-NMR relaxation measurements are used to study seed water sorption and desorption, the latter proceeding faster than the former (Fleischer and Werner 1992), and to determine the self-diffusion of water and lipids (Bacic et al. 1992; Brosio 1992).

Changes of dry seed lipids were found to be associated with loss of viability (calorimetric studies by Vertucci 1992). The desiccation-tolerant stage of soybean seeds is associated with higher levels of sugars (sucrose and, particularly conspicuous, stachyose) and maturation proteins (Blackman et al. 1992). The ratio of mono- to oligosaccharides is more closely correlated with the acquisition of desiccation tolerance than the absolute amount of accumulated carbohydrates (Ooms et al. 1993). Leprince et al. (1992) found a correlation between the monosaccharide content and the activity of the alternative respiration pathway of germinating maize radicles. The relative contribution of the alternative pathway to total CO_2 release increased first upon inbibition, then decreased; but in general respiratory carbon flow predominantly goes through the cytochrome pathway. Carbon import into developing pea ovules is turgor-controlled under short-term changes of the water potential gradient (Thorpe et al. 1993). There are complex supply-demand relationships between solute efflux from the coats and the phloem assimilate supply to the developing seeds (Ellis et al. 1992a; Patrick 1993a,b).

When germination starts, seed tissue osmotic relations control protein synthesis qualitatively and quantitatively (Dell'Aquila 1992). ABA seems to be involved in the production of germination-specific polypeptides, since ABA defect mutants of *Arabidopsis* produce seeds which more or less lack such proteins and are unable both to undergo maturation drying in the immature phase and to remain alive when desiccated later on (Meurs et al. 1992). Certain mRNAs are preserved still in the dry seed, and control the production of specific proteins in the germinating seed, which probably plays a role in the ABA-mediated desiccation tolerance. Another protein, called germin, is encoded by an mRNA which is transcribed after mature embryos are soaked. This latter protein plays a role in auxin-mediated cell enlargement (Lane 1991). Low seed water potentials will inhibit the action of the far-red-sensitive form of phytochrome which induces endosperm softening as a prerequisite for successful germination (De Miguel and Sanchez 1992).

Inbibition can also occur in immature seeds. The saturation water content reached by this process is a function of the initial one. It is, therefore, higher in soaked immature seeds than in mature ones (Ibrahim et al. 1992; Egli and Tekrony 1993). Low water potentials of fruit tissue in combination with high temperatures and low oxygen supply are detrimental for muskmelon (*Cucumis melo*) seed viability (Welbaum 1993). Seed water content and storage temperature interact, increasing or decreasing germinability of seeds in a complicated manner (Welbaum and Bradford 1991; Kovach and Bradford 1992). Inbibition is canalized in the seeds of some plant families, particularly in Rosaceae (Gudin et al. 1991), Leguminosae (Marconi et al. 1993; Serrato-Valenti et al. 1993) and Malvaceae (Serrato-Valenti et al. 1992; Egley and Paul 1993), by surface structures and coat incrustations like phenoles and lipids. Acidic polysaccharides and other hydrophilic substances in the chalaza region are the sites of first water entry into the imbibing seed. Different peculiarities of testa and pericarp bring about different intensities of K^+ leakage which generally occur from inbibing seeds (Beecroft and Lott 1993). A balance between the velocity of water uptake and solute leakage is crucial to avoid inbibitional damage; it is controlled by structural properties at both the tissue (Legesse and Powell 1992) and the subcellular level (Wolswinkel et al. 1992). Inbibition injury resulting from a prolonged soaking of bean seeds could result from a depressed activity of the oxidative pentose phosphate pathway (Pretorius and Small 1992).

b) Recalcitrant Seeds

Tomato seeds do not undergo maturation drying. Here, the pattern of protein synthesis functions as a switch from the developmental to the germinative mode, and this is under control from the fruit environment (Berry and Bewley 1991, 1992). Recalcitrant seeds occur in systematically and ecologically very different groups of flowering plants (e.g., Farrant et al. 1989; Xia et al. 1992). In the South African *Landolphia kirkii*, axes from immature seeds are relatively more desiccation-resistant than those of mature ones. The advanced developmental stages differ in the degree of vacuolation and endomembrane development (Berjak et al. 1992). pV curves of *Quercus robur* embryos (Grange and Finch-Savage 1992; Poulsen and Eriksen 1992) show that, with progress of development, the specific moisture content decreases and the osmotic potential increases. Cotyledonary starch is the major site of matrix-bound water. Water potentials below −5 MPa bring about damage, as seen from reduced respiratory activities and increased leakage conductivity. A 15-fold increased proline content indicates drought hardening at such low water potentials. Damage occurs at a critical low moisture content (Finch-Savage 1992). When this declines below 47%, free radicals become built up at higher amounts (Hendry et al. 1992). Oak cotyledons counteract this rise of radicals by increased superoxide dismutase and glutathione reductase, axes by higher levels of the antioxidants ascorbic acid and α-tocopherol. ABA contents of the acorns remain high until near shedding, thereafter they decline (Finch-Savage et al. 1992). Increased oxidation damage upon cell ultrastructures were found also in *Vatica* (Dipterocarpaceae) seeds when their moisture content decreased (Liu and Xuezhi 1993). In several publications Farrant et al. (1992a,b, 1993a,b,c) described development and associated metabolic processes of the recalcitrant *Avicennia marina* seeds. Water content remains high throughout all stages of histological differentiation until abcission of the viviparous seeds (Farrant et al. 1992a). Protein synthesis increases during histodifferentiation but no particular maturation proteins can be found (Farrant et al. 1992b). Predominant reserve substances are soluble sugars. Embryonic contents of phytohormones are high until seed shedding, except for ABA. The latter increases only in the pericarp tissue, possibly as a means to prevent precocious germination (Farrant et al. 1993a,b). Full germinability is acquired with separation from the pericarp; thereafter dehydration tolerance improves somewhat (Farrant et al. 1993c). The continuously high moisture contents of recalcitrant seeds reduce or at least complicate storage possibilities. Hydrated seed axes may be cryopreserved, if by an appropriate freezing rate the formation of large ice crystals is prevented (Wesley-Smith et al. 1992). Xia et al. (1992) and Fu et al. (1993) propose the storage of lychee, longan, and jack fruit seeds aseptically in moist media so that seed moisture contents around 40% are maintained.

5. Implications of Water Stress

a) Drought Effects on Metabolism

Water stress has influences upon plant metabolism at all levels of integration. **Processes of transcription and protein synthesis** are affected in various ways. The expression of some genes is specifically altered, and synthesis of particular heat-shock and wound-response proteins has been described (Ledoigt and Coudret 1992; Guy et al. 1992). Upon changes in the osmolarity of the extracellular environment of yeast, specific genes become activated, encoding kinases which serve as intermediate enzymes in protein phosphorylation cascades. Strains lacking such genes are viable only on osmotically stabilized media (Paravicini et al. 1992; Brewster et al. 1993). Drought-stress-induced proteins were found in the 22–24 kDa range of molecular weights (Reviron et al. 1992; Reddy et al. 1993); those with the much higher molecular masses of 85 and 160 kDa are obviously stress proteins produced under hardening conditions against both cold and drought stress (Guy et al. 1992). Specific wall-extension proteins with molecular masses of 29 and 30 kDa were isolated from sufficiently turgescent growing cucumber seedlings (McQueen-Mason et al. 1992). It is possible that the stress response already occurs at the transcriptional levels, as altered RNA or even DNA amounts (Ashraf and Bray 1993; Cocucci et al. 1993). Vernon et al. (1993) give evidence that several mRNA transcripts are produced upon *Mesembryanthemum crystallinum* leaf tissue exposure to salt stress (involvement of phytohormones in the process of altered enzyme production when ice plant leaves shift under stress from C_3 to CAM: Thomas et al. 1992). The regulation can occur, on the other hand, also during or after the translation (Zhang and Pei-Hong 1991: reduced translation activity in wheat leaves under stress; Oliver 1991: translation-dependent changes in hydration and rehydration proteins of *Tortula ruralis*; Harn and Daie 1992: post-translational modification of cytosolic fructose-1,6-bisphosphate levels). By comparison of chromosome substitution lines, Bobo et al. (1992) were able to correlate the metabolically complex regulation of the minimal stomatal resistance levels of winter wheat cultivars with the presence or absence of a certain chromosome.

The pool of adenine nucleotides decreases with increasing water deficits, particularly in nonhardened leaves (Zagdanska 1989, 1991).

Due to modified enzyme availability the plant's **carbon metabolism** is influenced by water stress particularly with respect to chemical reactions and the produced substances. Water stress influences upon the biophysical processes are, as a rule, not the reason for reduced photosynthesis (Martin and Ruiz-Torres 1992); but also the processes of energy gain and dissipation become influenced by drought due to structural changes of the chloroplast. There will be a decrease in the chlorophyll/carotenoid ratio (Barry et al. 1992), and the thylakoid structures become altered (Ristic et al. 1992). This results from a changed fatty acid composition and shifted ratios between the various lipid classes (Meyer et al. 1992), as it is found generally in membranes under stress (El Sayed 1992; Navari-Izzo et al. 1993). The ratio of hydrophilic to hydrophobic proteins increases more than fivefold (Sgherri et al. 1993). Generally, proteolytic activities in different cell compartments, among them also plasma and chloroplast membranes, are enhanced under water stress (Roy-Macauley et al. 1992). As a result of all these dessiccation-dependent changes, membrane permeability increases. Production of higher amounts of free radicals has been found under water stress (Quartacci and Navari-Izzo 1992), and some (but not yet unequivocal) data suggest an increased lipid peroxidation combined with reduced amounts of radical-scavenging enzymes (Jiang et al. 1991; Irigoyen et al. 1992b; Wang et al. 1993; Xu and Qi 1993). A drought-stress-resistant maize strain responded to such oxidative stresses by the increased synthesis of glutathione reductase and other antioxidant enzymes (Pastori and Trippi 1992, 1993). A higher K^+ supply can partly offset the detrimental effects of water stress upon chloroplast membrane susceptibility (Dekov and Velichkov 1992), and polyamines act as stabilizators in thylakoid membranes of osmotically stressed oat leaves when applied externally (Kubis et al. 1991; Floryszak-Wieczorek et al. 1992a; Besford et al. 1993). Putrescine becomes produced at higher rates under osmotic stress (Kubis and Krzywanski 1991; Chen and Kao 1993; Tiburcio et al. 1993) and may therefore function also naturally as a substance protecting membranes from drought-dependent peroxidation. By stabilizing thylakoid membranes and reducing peroxidation polyamines normalize the otherwise disturbed chlorophyll fluorescence when under water stress (Floryszak-Wieczorek et al. 1992b; Merzlyak et al. 1992).

Chlorophyll fluorescence itself is slightly influenced by water stress (Selmani and Wassom 1991; Schmuck et al. 1992) and may be used as an indicator for injury to the photosynthetic apparatus from various stress conditions (heat in combination with high humidity or low soil moisture contents: Oda et al. 1993) or to distinguish between genotypes of different drought resistance (Djekoun and Planchon 1991). However, there is not a direct correlation between, e.g., variable fluorescence and crop yield: when checking the effects of insufficient water supply, a reliable selection for resistant genotypes requires measurements of yet other drought stress responses, additionally to estimates of the variable fluorescence pattern (Selmani and Wassom 1992). In soybean under water (and nitrogen) stress midday values of ratios of variable to maximum fluorescence (F_v/F_m) are strongly depressed if the leaves are prevented from a paraheliotropic orientation (Kao and Forseth 1992a,b). Reduced (F_v/F_m) values were found to occur parallel to the midday depression of gas exchange also in other field studies (Epron et al. 1992). Fluorescence measurements prove the photosystem II (PS II) to have a very high functional stability under desiccation stress (Dreyer et al. 1992 – but see the contrasting findings by van Rensburg and Kruger 1993). Drought hardening also increases the heat stability of PS II but does not desensitize it against photoinhibition (Havaux 1992). An enhanced chloroplast susceptibility to photooxidative damage may be linked to a reduced availability of oxidation-preventing enzymes such as catalase (Streb et al. 1993). The PS I-depending cyclic electron transport could, on the other hand, sustain a transthylakoid proton gradient which controls PS II activity (Katona et al. 1992).

Among the Calvin cycle enzymes, ribulose-bisphosphate-carboxylase (RubPCase) receives most interest if drought stress effects upon carbon fixation are studied. As a rule, under water stress, reduced RubPCase activities are found to be more depressed in drought-susceptible cultivars, less in drought-resistant ones (e.g., Gimenez et al. 1992; Xue et al. 1992; Holaday et al. 1992; Antolin and Sanchez-Diaz 1993; Parry et al. 1993). Often this activity and net CO_2 uptake run in the same direction, but there is no interdependence between these two parameters: other influencing factors, such as stomatal limitations of gas exchange or nutrient stress, interfere. Thus, field-grown sunflowers photosynthesize so that a balance is achieved optimizing water and nitrogen use efficiency (Fredeen et al. 1991).

More than 140 papers published during the past 2 years report on carbon gain per transpired water, the water use efficiency (WUE). Most often they compare different cultivars or agricultural treatments in their effects upon the crop production. Carbon isotope discrimination is increasingly used as a method of choice in such investigations. As examples, nearly at random selected, from a total of approximately 30 papers using this method, those of Johnson (1993), Maryland et al. (1993 – both WUE studies with grasses), Zhang et al. (1993: douglas-fir provenance test), Kirda et al. (1992: wheat productivity analysis), and Ismail and Hall (1993: comparisons of cowpea genotypes) may be cited. The method proved to be useful also when seasonal changes of WUE were followed (Ehleringer et al. 1992), transpiration and photosynthesis of hosts and hemiparasites were compared (Küppers 1992) or the improvement of barley and wheat WUE during the last 7000 years in the Mediterranean area was assessed using grain remnants from archeological sites (Araus and Buxo 1993).

Water-stressed leaves accumulate more sucrose and less starch than nonstressed ones (Zrenner and Stitt 1991; Castrillo 1992). This can bring up the beneficial secondary effect of available solutes for an efficient cold stress avoidance by a high freezing point depression (Amundson et al. 1993). In some plants, like pigeon pea, drought brings about an increase of polyols by enhanced activities of anabolic enzymes (Keller and Ludlow 1993).

Nitrogen and water use efficiency often appear to show parallels if genotypes of different efficiency are compared (Eghball and Maranville 1991). Nitrate reductase and glutamine synthetase, the enzymes of nitrogen assimilation, are both depressed in their activities by water shortage but can recover completely after rewatering (Kumar and Srivastava 1992: NH_4 assimilation, Jonas et al. 1992, Koehler et al. 1992, Widmann et al. 1993: NO_3 reduction). Sometimes a reduced nitrate supply under osmotic stress aggravates the decrease of nitrogen metabolism (Ourry et al. 1992). Proline accumulation, on the other hand, counteracts the stress-induced decrease in nitrate reduction (Bandurska 1991). Water stress depresses *Phaseolus* nodulation reversibly, during vegetative stages and irreversibly during the reproductive stage (Pena-Cabriales and Castellanos 1993). Nodule enzymes which catalyze fermentation processes become stimulated under desiccation, which suggests some reduction in O_2 availability (Irigoyen et al. 1992a). Paraheliotropic leaf orientation of soybean is particularly conspicuous in nodulated, water-stressed plants (Kao and Forseth 1993).

As in previous periods under review, the majority (more than 20) of papers concerned with water stress and nitrogen compounds report on increases in proline levels. Different species accumulate different amounts, and further ecological factors such as, e.g., fertilization, pesticide impacts, temperature and air humidity, progress of the season, and various phenological stages of the plants will modify the results. Proline becomes accumulated in above- and belowground parts, also in drought roots grown under the influence of desiccation (Vartanian et al. 1992). Proline accumulation needs energy supply. This is easily provided by photosynthesis, but the process can encounter limitations if based solely on carbohydrate oxidation (Joyce et al. 1992). The biosynthetic pathways producing the other group of compatible nitrogen-containing molecules, glycine betaine and related substances (e.g., β-alanine betaine: Hanson et al. 1991; other compounds: Koheil et al. 1992; Naidu et al. 1992; Pesci and Reggiani 1992) are analyzed by several detailed investigations. Glycine betaine is synthesized by a two-step pathway, choline > betaine aldehyde > glycine betaine. Rice lacks enzymes to metabolize choline via the aldehyde intermediate into the final quarternary ammonium compound and is therefore, unlike other Poaceae, unable to utilize glycine betaine as a protective cytoplasmatic osmolyte (Rathinasabapathi et al. 1993). Externally supplied choline does not increase, but rather decreases, the stress-dependent increase of glycine betaine – it seems that the balance between precursor and end product cannot be shifted artificially (Eckert et al. 1992). This may result from the need for a phytohormonal message probably coming from the roots, but not necessarily ABA, which brings about an increase of mRNA for betaine aldehyde dehydrogenase (McCue and Hanson 1992). This enzyme catalyzes the last step of the above-mentioned biosynthetic pathway.

ABA is the most prominent **phytohormone** involved in plant water relations. Its functions, timing of the desiccation tolerance during seed development (e.g., Iida et al. 1992) and signaling as a messenger root water shortage to the stomata, have been mentioned already in the previous chapters. It controls the translation of mRNA when stress proteins are produced (e.g., Abromeit et al. 1992), but does not seem to be the immediate or the one and only substance which transfers the stress information at the subcellular level (Butler and Cuming 1993). Morphogenetic effects are associated with local ABA accumulation under water stress (maintenance of root elongation and inhibition of mesocotyl cell elongation: Saab et al. 1992; development of shorter root hairs in *Arabidopsis*: Schnall and Quatrano 1992), tissue proline levels increase in light, but not in darkness under the influence of ABA (Pesci 1992), and a reduced K^+ leakage occurs if protoplasts under osmotic shock are protected by higher ABA levels (Markarova et al. 1991). A couple of potential precursors of ABA biosynthesis (Parry and Horgan 1992), generally belonging to the carotenoids, can be identified from root tissue (Parry and Horgan 1992). The relative carotenoid composition there differs considerably from that found in leaves. ABA redistribution in roots caused by changing pH gradients can account for up to a two- to three-fold accumulation of ABA in the xylem sap of stressed plants (model calculations by Daeter et al. 1993). While the first increase of ABA upon drought stress originates from the roots, later also leaf ABA production contributes to the overall increase of xylem ABA under stress (Neales and McLeod 1991). Concomitantly with such a stress-dependent ABA increase xylem sap cytokinin content decreases (Bano et al. 1993). Xylem sap concentration of the latter group of phytohormones undergoes daily variations which, however, cannot be correlated with the actual plant water status (Fusseder et al. 1992). Ethylene production is a turgor-sensitive process (Saftner 1992) and is considered to increase under water stress (e.g., Feng and Barker 1992). Eklund et al. (1992), however, did not find any sign of a drought-induced ethylene production in Norway spruce: the highest ethylene concentration coming from the cambial/xylem region coincided with highest predawn water potentials, sap flow rates and tree growth intensities.

b) Drought Resistance

Approximately 60 papers report on comparative screening tests with cultivars and similar applied approaches in order to select drought-resistant crops for agriculture and forestry under water-limited conditions. Fussell et al. (1991) describe the procedures of a breeding program for drought tolerance as used by international crop breeding centers. Generally, plants can avoid or tolerate water shortage. As a rule, a particular species or cultivar follows more the one or the other strategy. Only seldom, if at all, has an evolutionary optimization occurred in both directions, and also it cannot be gained by breeding efforts. Thus, for instance, among some grain crops cultivated in Japan, sorghum and Japanese millet possess a high dehydration tolerance due to a high osmotic adjustment, pearl millet has the highest water retention capacity, maize is inferior under both aspects (Inada et al. 1992). Co-occurring understory plants from a deciduous Japanese forest can be separated into those which experience low water potentials combined with slightly increased soil-to-leaf hydraulic conductances if transpirational demand increases, shrubs with decreased water potentials and distinctly increased conductances, and species which increase water gain so drastically that leaf water potentials remain high (Ishida et al. 1992). It is the strategy of avoidance responding to an imbalance of water supply and demand, if sensitive stomatal closure occurs and leaf water potentials are kept at high levels (examples: sugar maple: Ellsworth and Reich 1992; sycamore: Khalil and Grace 1992; lack of sufficient stomatal control of needle desiccation in pollution-damaged spruce trees: Maier-Maercker and Koch 1991). Alternatively, the plants tolerate a considerable water loss – resulting in low water potentials – in favour of a continuous, nearly unrestricted gas exchange (e.g., certain wheat genotypes: Gunasekera et al. 1992a). Often the two different strategies are realized within one group of closely related taxa. Even cultivars or provenances of the same species differ among each other following more the one or the other way (e.g., Gebre and Kuhns 1993; Kuhns et al. 1993). In the CAM succulent *Senecio medley-woodii* water contents of young leaves are kept high under drought at the expense of the water content of older leaves which loose their turgescence (Donatz and Eller 1993; – water migration from *Prunus* inflorescences to the bud scales as a protection mechanism against cold damage: Kader and Proebsting 1992). The biochemical reasons for osmoregulatory events – which entail locally improved water relations – can be traced back to particular gene transcriptions and translations (Newton et al. 1991). Mucilage of cacti functions, together with associated solutes, as a capacitor buffering changes of the cladode water content during initial phases of drought or directing ice formation under low temperatures to the extracellular spaces (Goldstein and Nobel 1991; Nobel et al. 1992b). A higher drought resistance can be based upon a comparatively better root water gain facilitated by mycorrhizal fungi (Boyle and Hellenbrand 1991; Davies et al. 1992). A seasonal improvement of drought resistance can result from ontogenetic changes in tissue properties (Wright et al. 1992), but most often increased levels of solutes, protective nitrogen compounds, and seasonally changed membrane structures bring about a drought stress hardening. As a rule, an improved frost tolerance is linked to such osmoregulatory and structural hardening processes (Gesch et al. 1992; Grossnickle 1992).

6. Habitat Water Relations and Plant Performance

From the many studies describing field performance of particular species under a certain water regime and their adaptation to water stress conditions, only a few will be mentioned, which highlight the interference between typical habitat conditions and plant water turnover. In some cases, the peculiar plant water relations characterize, as most prominent aspects among all interactions with environmental parameters, the whole vegetation of a macro- or mesoclimatically distinct area.

There are increasing efforts to investigate the ecophysiology of **tropical rainforest** trees as an analytical means to understand the functioning of this extremely important and worldwide extremely endangered ecosystem. Transpiration depending upon microclimate, nutrition, plant phenology, and the successional status of the taxa is analyzed by various studies. Reekie and Wayne (1992) found stomatal conductances of the tropical American pioneer species *Piper auritum* and *Cecropia obtusifolia* insensitive to reductions of water supply down to the low value of 20% soil water content, but responding to changed humidities. The latter response was absent in the co-occurring *Trichospermum mexicanum* which, on the other hand, closed the stomata gradually as soil moisture decreased. Stomatal responses sensitive to humidity were found with canopy leaves of *Anacardium excelsum* (tropical America) by Meinzer et al. (1993) and of *Saccoglottis gabonensis* (tropical Africa) by Lösch and Kazda (1992). Stomata of young leaves of the latter tree, both in the canopy and at branches of smaller trees near the floor, are much more responsive than those of fully developed leaves. On bright days in French Guyana, transpirational sap flow of young *Simarouba amara* and *Goupia glabra* trees reached up to 4 kg H_2O h^{-1} dm^{-2} sap wood crosssectional area, with reduced throughflow values around midday (Granier et al. 1992). By contrast, transpiration rates of mangrove plants are very small, possibly for reasons of increased salt concentrations in their root zone, whereby freshwater is extracted by ultrafiltration of the mangrove roots (Passioura et al. 1992; – measurements by ^{1}H-NMR and ^{23}Na-NMR of water and salt distribution in *Bruguiera* tissues: Koizumi et al. 1992). Roberts et al. (1993) scaled up leaf conductances from individual layers of an Amazonian rainforest to calculate canopy transpiration. The results agreed well with those derived from eddy correlation measurements. Distal parts of liana stems undergo considerable changes of water content between the dry and the wet season (Mooney et al. 1992). Seasonal variation of precipitation controls bursts of leaf development of some rainforest shrubs (Wright 1991). A strong annual rhythm influencing phenology and plant physiology prevails in the Amazonian inundation forests. Submerged leaves of the tree species of this area retain full functional capability even if inundated for up to 10 months (Fernandes-Correa and Furch 1992; Schlüter and Furch 1992).

Trees of **semideciduous tropical forests** (cerrados and cerradãos in Brazil) cease photosynthesizing only at low water potentials around −3 MPa (Sato and De Moraes 1992). Then the stomata are closed for most of the day (Perez and De Moraes 1991).

Broadleaved **evergreen forests, growing under a subtropical or humid-temperate climate**, which is brought about by the influence of monsoons or humidity-laden tradewinds, normally have to endure only moderate stress conditions. Mishio (1992) compared anatomical and physiological adaptations to dry conditions of five co-occurring shrubs from shallow-soil ridges of the Bonin Islands, Sun (1992) reports on photosynthetic rates under decreased water potentials of South Chinese *Ardisia* and *Rhodomyrtus* trees. The shrubs and trees which form the Canarian laurel forest are inferior to mediterranean sclerophyllous species in terms of stem hydraulic conductivity, sensitive stomatal closure upon water shortage, and desiccation tolerance (Lösch 1993). Innes and Kelly (1992) distinguished among native New Zealand shrubs, growing under such a climate, drought avoiders (maintaining high internal water potentials under water shortage), tolerators (surviving very low internal water potentials), and intermediate types.

In the more arid subtropical summer rain regions, competition for the limited and inhomogenously distributed soil moisture between trees and grasses (e.g., Gordon et al. 1989) results in **a savanna-type vegetation** and in **gallery forests**. Patterned Sahel woodland is only 40% vegetation covered; the bare soil areas serve as rainfall collectors for the vegetation which transpires approximately 70% of the precipitated water (Culf et al. 1993). *Faidherbia albida* develops deep-reaching taproots under drought, *Acacia nilotica* explores the moisture reserves of a larger, but more shallow rooting volume (Osonubi et al. 1992). In the southwestern semidesert areas and in the intermountain basin of the USA, tree vegetation is concentrated on river beds. Groundwater pumping will threaten these phreatophytic plants, which rely with their transpiration completely or mainly upon the water supply from deep aquifers (stable isotope composition ratios of xylem and soil water probes: Busch et al. 1992; correlations between stand structures and water availability: Stromberg et al. 1993). Higher in the mountains of Arizona and New Mexico, pinyon-juniper communities form a characteristic vegetation. Different stand densities compensate for changes in water availability along the altitudinal (≈ xerothermic) gradient of their occurrence (Lajtha and Getz 1993). Also the plants of this woodland partly use the groundwater reserves, and partly rely upon the actual precipitation (D/H isotope ratios: Flanagan et al. 1992). Vegetation strips following geomorphological contours which locally provide a better water availability under otherwise semi-arid conditions were described and analyzed for seedling demography and performance by Mauchamp et al. (1993). Bragg et al. (1993) measured water relations of *Quercus* stands which as gallery forests invade the Kansas prairie. Semidesert palms and shrubs of Baja California are adapted to their arid growing sites by vertical leaf orientation, reflective waxes on leaf surfaces, pubescence, low osmotic potentials, and intensive transpiration cooling (Schmitt et al. 1993). Rocky outcrops in the Sonoran desert offer moisture-storing pockets and wet soil patches which become exploited by cactus and agave root systems with several times higher densities than normal (Nobel et al. 1992a). The monocarpic *Rumex cyprius* from North African deserts responds to different water availabilities, allocating more or less biomass to vegetative growth. In this way, the life cycle of this species is safely completed even during years of less than average rainfall (Hegazy and Ismail 1992). Under water-limited conditions the reproductive phase of such annual plants starts earlier, the number of diaspores per biomass is increased, and senescence proceeds faster. Those populations of such taxa, which normally occur under a better water supply, do not change as quickly from the vegetative to the generative stage; but also these plants possess a remarkable developmental plasticity (Aronson et al. 1992, 1993).

A broad survey of grassland annuals under a **mediterranean winter rain climate** (Fernandez et al. 1993) emphasizes the predominance of plant size variation in response to water shortage also at the species level. Larger species have larger seeds, lower relative growth rates, and a lower reproductive output; 83% of the year-to-year variance of species abundance in a dry grassland in Spain is correlated with varied rainfall intensities (Figueroa and Davy 1991). An intensive partitioning of the root-accessible soil moisture permits high species diversities of such grasslands, which are mostly composed by annuals (Gordon and Rice 1992). Perennial grasses are competitive in such drought-prone mediterranean-type grasslands if they start extracting the available moisture early in the year; summed up over a year, annual grasses use only 50% of the soil water which is consumed by perennials in the borderzone between a mediterranean and a maritime-moderate climatic regime (Borman et al. 1992). Under such a climatic transition situation in Southern France, south-facing hills bear an evergreen *Quercus ilex* woodland, north-facing ones summergreen forests. There, *Tilia cordata* water potentials do not drop below −1.5 MPa, whereas close by, on the south-facing slopes, *Quercus ilex* experiences values of −3 MPa (Pigott and Pigott 1993). Acherar et al. (1991) studied with pot experiments water potentials and leaf conductances of evergreen and deciduous mediterranean oaks. They confirmed that for all species a negative exponential relation exists between predawn water potential and maximal stomatal conductance, but with distinctly different equation coefficients for the deciduous and the evergreen species. Rambal (1993) models water flow through the SPAC of a mediterranean evergreen scrubland with water loss by transpiration, soil water availability, and the hydraulic transport resistances as the determining parameters.

Ecologically different species of the genus *Quercus* were studied with respect to water relations in their North American and South-to-Central European distribution areas in so many detailed investigations that the time seems to be ripe to compile all this information comprehensively (publications from the past years only: Kwon and Pallardy 1989; Lo Gullo and Salleo 1990; Acherar et al. 1991; Cochard et al. 1992a; Dreyer et al. 1992; Kleiner et al. 1992; Kubiske and Abrams 1992; Bragg et al. 1993; Breda et al. 1993; Gordon and Rice 1993; Kuhns et al. 1993). Otherwise, among trees from **deciduous forests under a moderate climate** particularly those of flood plains and mixed forests gained higher research interests. *Acer negundo*, on shallow water table sites of Central Indiana, never experiences water stress so that gas exchange is primarily controlled by light and temperature. Nevertheless, there is a seasonal decline of leaf osmotic potentials at saturation and turgorloss point (Foster 1992). In such forests, liana competition can negatively effect the trees. *Liquidambar* trees do not suffer from competitively reduced water, but from less available light and nitrogen, if serving as hosts for various vines (Dillenburg et al. 1993). Even then, however, the situation of these plants is better than that of *Liquidambar* street trees, which suffer heavily from restricted water and nutrient availability under a much drier atmospheric environment in their urban growing places (Kjelgren and Clarkf 1993). A similar 3-year study on New York City street-tree water relations by Whitlow et al. (1992) reports maximum temperatures $>41\,^{\circ}C$ and peak vapor pressure deficits of 7.5 kPa. Such conditions, of course, led to substantial tree water deficits, but with a lower frequency than commonly expected, and the investigated species, *Tilia cordata* and *Fraxinus pennsylvanica*, remained standing in a relatively good condition. Also the European ash shows good adaptive plasticity under less than optimal water supply: on a mesoxerophilic site, predawn water potentials dropped down to −4.8 MPa seasonally, and even under good water supply conditions, maximal leaf conductances were lower than those of ash leaves in alluvial forests. *Fraxinus excelsior* survives under stress by good desiccation tolerance, not by efficient avoidance of water shortage (Carlier et al. 1992). Sugar maple, by contrast, follows a “conservative” water use pattern and has low drought tolerance. This became obvious to Ellsworth and Reich (1992) when investigating *Acer saccharum* seedlings which grew in a forest understory, in small gaps, or on large clearings, respectively. Year-to-year precipitation variability and site differences between the forest edge and the interior influenced also the water potential relations of a species-rich old-growth deciduous forest in Indiana (Jacquart and Armentano 1992). Kriebitzsch (1992) completed his comprehensive analysis of the ecophysiology of understory plants growing in a

Central European submontane beech forest on limestone by a quantitative treatment of the CO_2 balance and net primary productivity of the six most important herbs.

The relationships between meadow soil moisture dynamics and plant communities in northwestern Croatia were analyzed by Ilijanic (1992). The establishment and performance of *Rumex* species (*R. acetosa*, *R. crispus*, *R. palustris*) on Dutch floodplains was investigated by Voesenek and Blom (1992) and Voesenek et al. (1992). Male *Rumex acetosella* plants have generally low water consumption rates, females show considerable population differences: those from only recently established young populations acclimate to water shortage by decreased water potentials and sustained high transpiration rates. Plants from a 15-year-old population maintained high water potentials throughout a stress period at the cost of a reduced gas exchange (Houssard et al. 1992).

The water balance of a newly established montane spruce forest in Czechia was determined in the first years by weed evapotranspiration; later, transpiration of the spruce treelets predominated (Kantor 1992). Canadian **boreal forests** offer good conditions for *Larix laricina* and *Picea mariana*, if the soils are well aerated; at flooded localities *Betula glandulosa* is superior. The photosynthesis of these trees is normally not gas exchange-limited (Dang et al. 1991). An elevated ambient CO_2 supply ameliorates detrimental flooding effects, but to different degrees for different species (Miao et al. 1992).

The arctic-alpine dioecious shrub *Salix arctica* shows at its growing sites in the **arctic-alpine tundra** a significant spatial segregation of sexes, with male individuals colonizing the more xeric and nutrient-poor locations. Both genders differ in the sensitivity of stomatal control of transpiration and in their water relations parameters. Additionally, there are intraspecific differences between arctic and alpine populations of this wide-ranging species, and intra-individual gradients are found for leaf conductances, photosynthetic and respiration rates, carbohydrate and nitrogen reserves, and drought-stress sensitivity of the ramets (Dawson and Bliss 1989a,b, 1993).

7. Water Relations of Poikilohydric Plants

Phycophyta. The Characean cells are very suitable objects for studies of electro-osmotic ion and water transport; 29 water molecules coupled to one potassium ion can be transported through a *Chara corallina* K^+ channel with a pore radius of 1.5 nm (Homble and Very 1992). Ca^{2+} ions control polarity and permeability of the *Chara* plasma membrane and influence by this way the cell turgor under NaCl influence (Reid et al. 1993). Zn^{2+} ions were also shown to decrease the membrane permeability. They behave as competitors to Ca^{2+}, but the mechanisms by which both ions influence membrane conductivity are not the same (Rygol et al. 1992). Another alga used in many laboratories for studies of osmotic adaptation under salt stress is *Dunaliella*. The preferred osmoregulatory solute of this genus is glycerol, which upon hypotonic stress can be released to the medium via membrane channels (Fujii and Hellebust 1992). Also yeast cells and other heterotrophic microorganisms use glycerol and other polyols as osmoregualtory substances (Van Zyl et al. 1991; Van Zyl and Kilián 1993; Yagi 1992; Kelavkar and Chhatpar 1993). Higher thallophytes accumulate by preference sugars or simple sugar alcohols such as, e.g., cryoprotective substances (comparative studies of antarctic cryptogams: Roser et al. 1992)

Trentepohlia air algae use the high humidity and moderate light levels at midmorning for photosynthesis; in dry state, later during the day, they are well protected from photoinhibition (Ong et al. 1992).

Supralitoral algae experience periods of desiccation of various lengths and occasional inundation by sea water. *Prasiola crispa* becomes irreversibly damaged if water loss exceeds 90%. Thick cell walls are important prerequisites of this alga in order to endure prolonged desiccation periods; vacuoles are formed only under hyperosmotic stress (Jacob et al. 1992a,b).*Cladophora* shows only incomplete turgor adjustment if subjected to extremely different salinities (Wiencke et al. 1992). Levitt and Bolton (1991) found very different photosynthetic rates of intertidal seaweed species, depending on thallus structure. Thin, sheet-like *Porphyra* thalli and fleshy forms had high CO_2 uptake rates, terete branched algae reached only low photosynthetic rates. In submersed thalli a seasonal pattern with highest CO_2 uptake in spring and lowest in winter could be found. Recovery of *Fucus* photosynthesis upon resubmergence depends on desiccation during emergence; the critical threshold is reached with less than 30% water content (Beer and Kautsky 1992). The cell walls of brown algae are involved in cell water and ion regulation on account of many negatively charged matrix polysaccharides. There is a clear correlation between the occurrence of the individual species within the intertidal vertical gradient and the wall contents of sulfated fucans (Kloareg 1991). With miniature agarose beads, Brawley and Johnson (1993) determined wetness duration and evaporation conditions within and between intertidal microhabitats where sites, more or less prone to water stress, often determine the survival and colonization possibilities of the first developmental stages of seaweeds.

Mycophyta. Substrate moisture conditions control establishment and virulence of phytopathogenic fungi. More than a dozen papers report, as in previous years, on particular wetness duration and water potential conditions for maximal production and germination of fungal spore types, conidia and other propagules. Mycelial growth of wood-colonizing fungi is best at 60–80% substrate water content (Mata 1992). Sclerotia formation by *Morchella esculenta* is enhanced by hexoses which increase the mycelial turgor (Amir et al. 1992). *Achlya* hyphae continue growing under reduced turgor by extension of the apical cell walls, which become softened under osmotic stress (Money and Harold 1992). Similarly, turgor is necessary, but not sufficient to explain *Saprolegnia* apical extension rates, cytoskeletal regulation is involved in the growth process as well (Kaminskyi et al. 1992). Osmosensitive *Saccharomyces cerevisiae* cells undergo rapid and reversible changes in actin filament organization (Chowdhury et al. 1992). An actin-binding protein restores the filaments which were disassembled under osmotic stress. Enzymatic proteins, kinases, became formed after particular genes were activated upon an increased external osmolarity (Paravicini et al. 1992; Brewster et al. 1993). Simultaneously, polyol accumulation occurs in osmoregulating yeast cells (Yagi and Nishi 1993). Ortega et al. (1992) describe a pressure clamp method which allows transpiration measurements of growing *Phycomyces* cells. A nutrient, but not a water translocation exists from *Glomus* hyphae to the host root (George et al. 1992). Possibly, nutrient transport becomes facilitated via subcellular compartments, as was demonstrated with exchange of substances inside *Pisolithus* terminal cells and between them and the penultimate cells (Shepherd et al. 1993). Otherwise, mass flow of water and nutrients along hyphae and hyphal complexes, like rhizomorphs, of higher fungi follows concentration gradients, partly with separation of different transported substances between apoplastic and symplastic parts of the hyphal strands (Cairney 1992). Infection by wood-destroying *Armillaria* species brings about decreases in stomatal gas exchange of susceptible trees (Loreto et al. 1993).

Lichens. A water canalization is found alongside the rhizine strands of *Cladonia sulphurina* thalli (Ott et al. 1993). Maximum thallus water contents of the Umbilicariaceae depend upon thallus porosity, which was determined comparatively among 13 species by Valladares et al. (1993). In general, a radial plectenchymatic medulla confers larger porosity to the thallus than an arachnoidal one. The prevailing humidity conditions can considerably influence the anatomy and morphology of a lichen thallus (*Hypogymnia physodes*: Hyvarinen 1992). An increased or decreased enzyme production of *Xanthoria*, depending upon thallus water status, has been reported (Rodriguez and Vicente 1991). Lichen photosynthesis as influenced by thallus imbibition was measured in the field under quite different climatic conditions, most often under boreal or subpolar climates (CO_2 gas exchange data of frozen *Usnea sphacelata* in the Antarctic: Kappen 1989; *Buellia frigida* in situ fluorescence pattern analysis by a pulse amplitude modulated system: Schröter et al. 1992; thallus moisture and temperature interactions on CO_2 gas exchange of Greenlandic *Peltigera aphthosa* thalli: Schröter et al. 1991, of *Cetraria nivalis* and *C. delisei* from Svalbard: Schipperges 1992; lichen photosynthesis during summer and fall in the Brooks Range, Alaska, tundra: Hahn et al. 1993; carbon balance calculations of *Cladonia portentosa* under the climatic conditions of a dune habitat on a North Sea island: Bruns-Strenge and Lange 1992; photosynthesis of endolithic South African lichens depending in part upon water condensed due to temperature gradients between air and rocks: Wessels and Kappen 1993). Lange et al. (1993) offer further evidence that activation of net photosynthesis by dry cyanobacterial lichens requires liquid water. This fact is not restricted to the symbiontic situation, but is also the case if the thallus structure is destroyed or the phycobiont is isolated. In cyanobacteria, desiccation induces a functional detachment of the phycobilisomes from PS II. Energy transfer from the former to the latter is restored only if rehydration occurs with liquid water. In green alga symbionts, desiccation induces a functional interruption of energy transfer between the light harvesting chlorophyll a/b pigment complex and PS II. Rehydration by vapor is sufficient to restore this transfer (Bilger et al. 1989). Very fascinating is the finding that a CO_2 concentration mechanism exists in lichens (Badger et al. 1993; Palmquist 1993): it is based on active dissolved inorganic carbon uptake, a process which is energized by photosynthetic electron transport. Either CO_2 or HCO_3^- are captured from the external environment. Carbonic anhydrase, contained in carboxysomes, mediates this CO_2-concentrating mechanism in free-living and symbiontic cyanobacteria. In unicellular green algae it seems to be located in the pyrenoid – species lacking a pyrenoid obviously also lack the CO_2-concentrating mechanism and attain lower net photosynthesis rates. CO_2 limitation of the latter is controlled therefore not only by the hydration-dependent physical diffusion resistance, but also by intrinsic biochemical properties.

Bryophyta. Mosses from xeric and hydric habitats can differ in their desiccation resistance (Seel et al. 1992a). The more resistant *Tortula ruraliformis* has a higher superoxide dismutase, but similar peroxidase activities as the drought-sensitive *Dicranella palustris*. Antioxidants with high chemical O_2 affinity seem to protect the mosses better from oxidative damage under desiccation and high irradiance than oxygen-processing enzymes (Seel et al. 1992b). Recovery of protein synthesis after desiccation is controlled at the translational, not the transcriptional level (Oliver 1991); closely related species of *Tortula* differ in the degree of this damage limitation and cellular repair facilities (Oliver et al. 1993). ABA influences this protein resynthesis upon rehydration (Werner et al. 1991; – on the occurrence of ABA in bryophytes and other lower plant groups see Hartung and Gimmler (1994) in *Progress in Botany*, Vol. 55. Photosynthetic recovery of boreal forest mosses is higher and gets ahead better than that of peat moss species (Silvola 1991). *Sphagnum* species from hummocks and depressions differ in their $\delta^{18}O$ values as a result of different relief-dependent water supply and demand balances (Aravena and Warner 1992). Under natural conditions, the gas exchange of mosses is most often far below the optimum due to insufficient irradiance, adverse temperatures, and – most important – insufficient thallus hydration (*Hylocomium splendens*: Sonesson et al. 1992).

The osmotic potentials of fully hydrated mosses range between −0.5 and −1 MPa (Rütten and Santarius 1993): the extremely low osmotic values usually determined by plasmolysis do not reflect the real osmolalities and may be caused by structural peculiarities of the moss leaf cells. Sugars (mostly sucrose: Smirnoff 1992) constitute roughly 20–30% of the total osmolality of the cell solutions; they increase and decrease seasonally (Rütten and Santarius 1992). Freeze-thaw cycles (performed with Antarctic bryophytes) increased rates of sucrose leakage from moss cushions (Melick and Seppelt 1992), tropical canopy bryophytes released sugars and polyols to stemflow and canopy leachate when rewetted during the naturally occurring desiccation/rehydration cycles (Coxson et al. 1992).

Poikilohydric cormophytes. The combined influence of desiccation and high light intensity increases the potential danger of light-induced damage also in desiccation-resistant cormophytes. A normal recovery of the photochemical efficiencies of *Polypodium polypodioides* fronds occurs only when the photoinhibition affected fully hydrated leaves. If it occurred to desiccated ones, repair was impossible (Muslin and Homann 1992). Under natural conditions this danger is avoided by leaf rolling, so that light reflection from the abaxial leaf surface is possible. Stem curling of *Selaginella lepidophylla* serves the same purpose (Lebkuecher and Eickmeier 1992, 1993). Turgescent plants are protected against photoinhibitory damage by a zeaxanthin-associated mechanism which also protects PS II when thylakoid membranes undergo water loss (Eickmeier et al. 1993). The occurrence of this spikemoss, by preference on north-facing slopes, may confer additionally to favorable habitat light conditions, even if correlative analyses suggest competitive influences by other plants as the main reason for the particular distribution pattern (Van Auken and Bush 1992). *Polypodium virginianum* cannot absorb water through its fronds, but needs rehydration via the soil-root pathway (Matthes-Sears et al. 1993). Specific proteins, produced during rehydration, were found in this fern by Reynolds and Bewley (1993).

By contrast, the monocotyledonous *Xerophyta scabrida* is able to rehydrate also by leaf water uptake, a process which even precedes the development of new adventitious roots (Tuba et al. 1993). Greening of the leaves is based upon de-novo chlorophyll and carotenoid synthesis. Proteins synthesized during *Craterostigma plantagineum* desiccation could serve as protective substances (Schneider et al. 1993); an increase of fatty acids and a disappearance of glucose and fructose during the course of desiccation are obvious (Bianchi et al. 1992). Similar trends were found also in desiccating *Myrothamnus* leaves (Bianchi et al. 1993). Stevanovic et al. (1992) determined in desiccated *Ramonda* leaves extremely reduced lipid levels which recover after rewetting; Stefankov et al. (1992) emphasize that the changes are most intensive when the leaves are in an intermediate hydration state.

References

Abromeit M, Askman P, Sarnighausen E, Doerffling K (1992) J Plant Physiol 140: 617–622. **Acherar M**, Rambal S, Lepart J (1991) Ann Sci For (Paris) 48:561–574. **Ameglio T**, Cruiziat P (1992) C R Acad Sci Ser III Sci Vie 315:429–434. **Amir R**, Levanon D, Hadar Y, Chet I (1992) Mycol Res 96:943–948. **Amodeo G**, Srivastava A, Zeiger E (1992) Plant Physiol 100:1567–1570. **Amundson RG**, Kohut RT, Laurence JA, Fellows S, Colavito LJ (1993) Environ Exp Bot 33:383–390. **Andreev IM** (1992) Biol Membr 9:19–25. **Antolin MC**, Sanchez-Diaz M (1993) J Exp Bot 44:1341–1349. **Aphalo PJ**, Jarvis PG (1993) J Exp Bot 44:791–800. **Araus JL**, Buxo R (1993) Aust J Plant Physiol 20:117–128. **Aravena R**, Warner BG (1992) Bryologist 95:445–448. **Ardner WR** (1991) Irrig Sci 12:109–114. **Aronson J**, Kigel J, Shmida A, Klein J (1992) Oecologia 89:17–26. **Aronson J**, Kigel J, Shmida A (1993) Oecologia 93:336–342. **Ashraf M**, Bray CM (1993) Seed Sci Res 3:15–23. **Assmann SM** (1992) Oecologia 92:188–195. **Assmann**

SM, Schwartz A (1992) Plant Physiol 98:1349–1355. **Assmann SM**, Lee DM, Malkus P (1992) Photochem Photobiol 56:685–689. **Atkinson CJ**, Ruiz LP, Mansfield TA (1992) J Exp Bot 43:1315–1324. **Azaizeh H**, Steudle E (1991) Plant Physiol 97:1136–1145. **Azaizeh H**, Gunse B, Steudle E (1992) Plant Physiol 99:886–894.

Bacic G, Srejic R, Lahajnar G, Zupancic I, Ratkovic S (1992) Seed Sci Technol 20: 233–240. **Badger MR**, Pfanz H, Buedel B, Heber U, Lange OL (1993) Planta 191: 57–70. **Bandurska H** (1991) Acta Physiol Plant 13:3–11. **Bano A**, Doerffling K, Bettin D, Hahn H (1993) Aust J Plant Physiol 20:109–115. **Barker DJ**, Sullivan CY, Moser LE (1993) Agron J 85:270–275. **Barry P**, Evershed RP, Young A, Prescott MC, Britton G (1992) Phytochemistry 31:3163–3168. **Bassirirad H**, Caldwell MM (1992) Physiol Plant 86:525–531. **Bassirirad H**, Radin JW (1992) Plant Physiol 99:34–37. **Beecroft P**, Lott JNA (1993) Seed Sci Res 3:105–110. **Beer S**, Kautsky L (1992) Bot Mar 35:487–491. **Beerling DJ**, Chaloner WG (1993a) Ann Bot (Lond) 71:231–235. **Beerling DJ**, Chaloner WG (1993b) Ann Bot (Lond) 71:431–435. **Beerling DJ**, Chaloner WG (1993c) Biol J Linn Soc 48:343–353. **Benech ARL**, Fenner M, Edwards PJ (1992) Funct Ecol 6:596–605. **Benkert R**, Balling A, Zimmermann U (1991) Bot Acta 104:423–432. **Berjak P**, Pammenter NW, Vertucci C (1992) Planta 186:248–261. **Berry T**, Bewley JD (1991) Planta 186:27–34. **Berry T**, Bewley JD (1992) Plant Physiol 100:951–957. **Besford RT**, Richardson CM, Campos JL, Tiburcio AF (1993) Planta 189:201–206. **Beyschlag W**, Pfanz H, Ryel RJ (1992) Planta 187:546–553. **Bianchi G**, Gamba A, Murelli C, Salamini F, Bartels D (1992) Phytochemistry 31:1917–1922. **Bianchi G**, Gamba A, Limiroli R, Pozzi N, Elster R, Salamini F, Bartels D (1993) Physiol Plant 87:223–226. **Bilger W**, Rimke S, Schreiber U, Lange OL (1989) J Plant Physiol 134:261–268. **Blackman SA**, Obendorf RL, Leopold AC (1992) Plant Physiol 100:225–230. **Bleiss W**, Ehwald R (1993) Physiol Plant 88:541–548. **Bobo MS**, Planchon C, Morris R (1992) Euphytica 62:59–62. **Bolger TP**, Upchurch DR, McMichael BL (1992) Environ Exp Bot 32:49–54. **Borghetti M**, Grace J, Raschi A (1993) Water transport in plants under climatic stress. Cambridge Univ Press, Cambridge. **Borman MM**, Johnson DE, Krueger WC (1992) Agron J 84:897–904. **Bottomley PA**, Rogers HH, Prior SA (1993) Plant Cell Environ 16:335–338. **Boyer JS** (1992) Curr Top Plant Biochem Physiol 11:1–17. **Boyle CD**, Hellenbrand KE (1991) Can J Bot 69:1764–1771. **Bragg WK**, Knapp AK, Briggs JM (1993) For Ecol Manage 56:29–41. **Brawley SH**, Johnson LE (1993) J Phycol 29:528–535. **Breda N**, Cochard H, Dreyer E, Granier A (1993) Can J For Res 23:1136–1143. **Brewster JL**, De Valoir T, Dwyer ND, Winter E, Gustin MC (1993) Science (Wash DC) 259:1760–1763. **Brissette JC**, Chambers JL (1992) Tree Physiol 11:289–303. **Brosio E**, Di Nola A, Fracassi M, Carnovale E, Marconi E (1992) Cell Mol Biol 38:693–697. **Bruni F**, Leopold AC (1992a) Biophys J 63:663–672. **Bruni F**, Leopold AC (1992b) Seed Sci Res 2:251–253. **Bruns-Strenge S**, Lange OL (1992) Flora (Jena) 186:127–140. **Busch DE**, Ingraham NL, Smith SD (1992) Ecol Appl 2:450–459. **Butler WM**, Cuming AC (1993) Planta 189:47–54.

Cabibel B, Do F (1991) Agronomie (Paris) 11:669–678. **Cairney JWG** (1992) Mycol Res 96:135–141. **Canny MJ** (1993) Philos Trans R Soc Lond B Biol Sci 341:87–100. **Cardon GE**, Letey J (1992) Soil Sci Soc Am J 56:1876–1880. **Carlier G**, Peltier JP, Gielly L (1992) Ann Sci For (Paris) 49:207–223. **Carmi A**, Plaut Z, Sinai M (1993) Irrig Sci 13:177–182. **Castrillo M** (1992) J Exp Bot 43:1557–1561. **Cermak J**, Cienciala E, Kucera J, Hallgren J-E (1992) Tree Physiol 10:367–380. **Cermak J**, Matyssek R, Kucera J (1993) Tree Physiol 12:281–290. **Chen CT**, Kao CH (1993) Plant Growth Regul 13:197–202. **Chen D-X**, Lieth JH (1992) Agric For Meteorol 59:129–148. **Chen D-X**, Lieth JH (1993) J Am Soc Hortic Sci 118:181–187. **Chen Z**, Jiao X-Z, Hong L (1991) Acta Phytophysiol Sin 17:333–341. **Cheng H-Y**, Zheng G-H, Tao KLJ (1992) Chin J Bot 4:20–25. **Chiou J-S**, Krishna PR, Kamaya H, Ueda I (1992) Biochim Biophys Acta 1110:225–233. **Chiu S-T**, Ewers FW (1992) Trees (Berl) 6:216–224. **Chiu S-T**, Ewers

FW (1993) J Exp Bot 44:175–181. **Chiu S-W**, Novotny JA, Jakobsson E (1993) Biophys J 64:98–109. **Chowdhury S**, Smith KW, Gustin MC (1992) J Cell Biol 118:561–571. **Ciamporova M**, Mistrik I (1993) Environ Exp Bot 33:11–26. **Cienciala E**, Lindroth A, Cermak J, Hallgren J-E, Kucera J (1992) Trees (Berl) 6:121–127. **Cochard H** (1992) Tree Physiol 11:73–83. **Cochard H**, Breda N, Granier A, Aussenac G (1992a) Ann Sci For (Paris) 49:225–233. **Cochard H**, Cruiziat P, Tyree MT (1992b) Plant Physiol 100:205–209. **Cocucci SM**, Morgutti S, Abruzzese A, Alisi C, Yahye AH, Espen L (1993) Physiol Plant 87:609–615. **Cohen Y**, Takeuchi S, Nozaka J, Yano T (1993) Agron J 85:1080–1086. **Colombo SJ**, Blumwald E (1992) Plant Cell Environ 15:837–842. **Colombo SJ**, Teng Y (1992) Oecologia 92:410–415. **Cortes PM** (1992) Plant Cell Environ 15:351–363. **Coutts MP**, Nicoll BC (1993) New Phytol 124:277–281. **Coxson DS**, McIntyre DD, Vogel HJ (1992) Biotropica 24:121–133. **Coyea M**, Margolis HA (1992) Can J For Res 22:1684–1693. **Cruz RT**, Wayne RJ, Drew MC (1992) Plant Physiol 99:203–212. **Culf AD**, Allen SJ, Gash JHC, Lloyd CR, Wallace JS (1993) Agric For Meteorol 66:65–80. **Curtis PS**, Teeri JA (1992) Can J For Res 22:1320–1325.

Daeter W, Slovik S, Hartung W (1993) Philos Trans R Soc Lond B Biol Sci 341:49–56. **Dai Z**, Edwards GE, Ku MSB (1992) Plant Physiol 99:1426–1434. **Dambrine E**, Carisey N, Pollier B, Girard S, Granier A, Lu P, Biron P (1992) Ann Sci For (Paris) 49:489–510. **Dang QL**, Liefers VJ, Rothwell RL, MacDonald SE (1991) Oecologia 88:317–324. **Davies FT** Jr, Potter JR, Lindermann RG (1992) J Plant Physiol 139:289–294. **Dawson TE**, Bliss LC (1989a) Oecologia 79:322–331. **Dawson TE**, Bliss LC (1989b) Oecologia 79:332–343. **Dawson TE**, Bliss LC (1993) Funct Ecol 7:293–304. **Dekov I**, Velichkov D (1992) Fiziol Rast (Sofia) 18:3–9. **Dell Aquila A** (1992) Ann Bot (Lond) 69:167–171. **De Miguel L**, Sanchez RA (1992) J Exp Bot 43:969–974. **Denecke M**, Schulz M, Fischer C, Schnabl H (1993) Physiol Plant 87:96–102. **Dillenburg LR**, Whigham DF, Teramura AH, Forseth IN (1993) Am J Bot 80:244–252. **Dixon MA** (1987) Hort Science 22:285–287. **Dixon MA**, Tyree MT (1984) Plant Cell Environ 7:693–697. **Djekoun A**, Planchon C (1991) Euphytica 55:247–254. **Donatz M**, Eller BM (1993) J Plant Physiol 141:750–756. **Dreyer E**, Epron D, Matig OEY (1992) Ann Sci For (Paris) 49:615–625. **Drory A**, Borochov A, Mayak S (1992) J Plant Physiol 140:116–120. **Drüge U**, Schönbeck F (1993) J Plant Physiol 141:40–48. **Dufrene E**, Saugier B (1993) Funct Ecol 7:97–104. **Dugas WA**, Wallace JS, Allen SJ, Roberts JM (1993) Agric For Meteorol 64:47–62. **Dye PJ**, Olbrich BW (1993) Plant Cell Environ 16:45–53.

Eckert H, Bergmann H, Eckert G, Müller H (1992) Angew Bot 66:124–129. **Eghball B**, Maranville JW (1991) Commun Soil Sci Plant Anal 22:1367–1382. **Egley GH**, Paul RN Jr (1993) Seed Sci Res 3:119–127. **Egli DB**, Tekrony DM (1993) Seed Sci Technol 21:139–148. **Ehleringer JR**, Phillips SL, Comstock JP (1992) Funct Ecol 6:396–404. **Ehlers W**, Hamblin AP, Tennant D, Van der Ploeg RR (1991) Irrig Sci 12:115–124. **Eickmeier WG**, Casper C, Osmond CB (1993) Planta 189:30–38. **Ekanayake IJ**, De Jong JP (1992) Ann Bot (Lond) 70:53–60. **Eklund L**, Cienciala E, Hallgren J-E (1992) Physiol Plant 86:297–300. **Ellis EC**, Turgeon R, Spanswick RM (1992a) J Exp Bot 43:1235–1241. **Ellis RH**, Hong TD, Roberts EH (1992b) Ann Bot (Lond) 69:53–58. **Ellison AM**, Niklas KJ, Shumway S (1993) Aquat Bot 45:325–339. **Ellsworth DS**, Reich PB (1992) Tree Physiol 10:1–20. **El Sayed H** (1992) J Agron Crop Sci 169:161–168. **Engels C**, Marschner H (1992) Physiol Plant 86:263–268. **Enstone DE**, Peterson CA (1992) Can J Bot 70:1502–1512. **Entz MH**, Gross KG, Fowler DB (1992) Can J Plant Sci 72:1109–1120. **Ephrath JE**, Hesketh JD (1991) Photosynthetica (Prague) 25:607–619. **Epron D**, Dreyer E (1993) Tree Physiol 13:107–117. **Epron D**, Dreyer E, Breda N (1992) Plant Cell Environ 15:809–820. **Eriksson P-O**, Lindblom G (1993) Biophys J 64:129–136. **Evans RD**, Black RA, Loescher WH, Fellows RJ (1992) Plant Cell Environ 15:49–60. **Ewers FW**, Fisher JB (1991) Oecologia 88:233–237. **Ewers FW**, North GB, Nobel PS (1992) New Phytol 121:377–385.

Faensen-Thiebes A (1992) Flora (Jena) 186:341–357. **Fan T W-M**, Colmer TD, Lane AN, Higashi RM (1993) Anal Biochem 214:260–271. **Farrant JM**, Pammenter NW, Berjak P (1989) Planta 178:189–198. **Farrant JM**, Pammenter NW, Berjak P (1992a) Ann Bot (Lond) 70:75–86. **Farrant JM**, Berjak P, Pammenter NW (1992b) Plant Growth Regul 11:257–265. **Farrant JM**, Pammenter NW, Berjak P (1993a) Seed Sci Res 3:1–13. **Farrant JM**, Berjak P, Pammenter NW (1993b) Ann Bot (Lond) 71:405–410. **Farrant JM**, Berjak P, Cutting JGM, Pammenter NW (1993c) Seed Sci Res 3:55–63. **Feng J**, Barker AV (1992) J Plant Nutr 15:2471–2490. **Fernandes-Correa AF**, Furch B (1992) Amazoniana 12:71–84. **Fernandez AR**, Laffarga JM, Ortega F (1993) J Veg Sci 4:313–322. **Ferrario S**, Agius I, Morisot A (1992a) J Plant Nutr 15:69–84. **Ferrario S**, Agius I, Morisot A (1992b) J Plant Nutr 15:85–98. **Ferreira MI**, Katerji N (1992) Oecologia 92:104–107. **Figueroa ME**, Davy AJ (1991) J Ecol 79:925–941. **Finch-Savage WE** (1992) J Exp Bot 43:663–669. **Finch-Savage WE**, Clay HA, Blake PS, Browning G (1992) J Exp Bot 43:671–679. **Flanagan LB**, Ehleringer JR, Marshall JD (1992) Plant Cell Environ 15:831–836. **Fleischer G**, Werner A (1992) Biochim Biophys Acta 1116:305–308. **Floryszak-Wieczorek J**, Grabiikowski E, Kubis J, Krzywanski Z (1992a) Acta Physiol Plant 14:3–10. **Floryszak-Wieczorek J**, Murkowski A, Kubis J, Krzywanski Z (1992b) Acta Physiol Plant 14:119–126. **Foster JR** (1992) Tree Physiol 11:133–149. **Fredeen AL**, Gamon JA, Field CB (1991) Plant Cell Environ 14:963–970. **Frensch J**, Hsiao TC (1993) Planta 190:263–270. **Fromm J**, Eschrich W (1993) J Plant Physiol 141:673–680. **Fu JR**, Xia QH, Tank LF (1993) Seed Sci Technol 21:85–95. **Fujii S**, Hellebust JA (1992) Can J Bot 70:1313–1318. **Fusseder A**, Wartinger A, Hartung W, Schulze E-D, Heilmeier H (1992) New Phytol 122:45–52. **Fussell LK**, Bidinger FR, Bieler P (1991) Field Crops Res 27:183–200.

Gautier H, Vavasseur A, Lasceve G, Boudet AM (1992) Plant Physiol 98:34–38. **Gawrisch K**, Ruston D, Zimmerberg J, Parsegian VA, Rand RP, Fuller N (1992) Biophys J 61:1213–1223. **Gebre GM**, Kuhns MR (1993) Can J For Res 23:1291–1297. **George AP**, Nissen RJ (1992) Hortic Sci 67:445–455. **George E**, Haeussler K-U, Vetterlein D, Gorgus E, Marschner H (1992) Can J Bot 70:2130–2137. **Gesch GW**, Kenefick DG, Koepke JA (1992) Crop Sci 32:180–186. **Gimenez C**, Mitchell VJ, Lawlor DW (1992) Plant Physiol 98:516–524. **Glad C**, Regnard JL, Querou Y, Brun O, Morot-Gaudry JF (1992) Am J Enol Vitic 43:275–282. **Goldstein G**, Nobel PS (1991) Plant Physiol 97:954–961. **Gordon DR**, Rice KJ (1992) Am J Bot 79:967–976. **Gordon DR**, Rice KJ (1993) Ecology 74:68–82. **Gordon DR**, Welker JM, Menke JW, Rice KJ (1989) Oecologia 79:533–541. **Govindaraju RS**, Kavvas ML (1993) J Hydrol (Amst) 146:321–339. **Gowing DJG**, Jones HG, Davies WJ (1993) Plant Cell Environ 16:453–459. **Grange RI**, Finch-Savage WE (1992) J Exp Bot 43:657–662. **Granier A**, Huc R, Colin F (1992) Ann Sci For (Paris) 49:17–24. **Grose MJ**, Hainsworth JM (1992) J Exp Bot 43:121–128. **Gross K**, Koch W (1991a) Physiol Plant 83:290–295. **Gross K**, Koch W (1991b) Physiol Plant 83:296–303. **Grossnickle SC** (1992) Tree Physiol 11:229–240. **Grossnickle SC** (1993) Trees (Berl) 7:148–155. **Gudin S**, Arene L, Chavagnat A, Bullard C (1991) Biol Plant (Prague) 33:262–267. **Guehl JM**, Aussenac G, Bouachrine J, Zimmermann R, Pennes JM, Ferhi A, Grieu P (1991) Can J For Res 21:1507–1515. **Gunasekera D**, Berkowitz GA (1992a) Plant Physiol 98:660–665. **Gunasekera D**, Berkowitz GA (1992b) Plant Sci (Limerick) 86:1–12. **Guy C**, Haskell D, Neven L, Klein P, Smelser C (1992) Planta 188:265–270.

Ha K-S, Thompson GA Jr (1992) Biochemistry 31:596–603. **Habib R**, Pages L, Jordon MO, Simonneau T, Sebillotte M (1991) Agronomie (Paris) 11:623–644. **Hahn SC**, Tenhunen JD, Popp PW, Meyer A, Lange OL (1993) Flora (Jena) 188:125–143. **Hamza MA**, Aylmore LAG (1992a) Plant Soil 145:187–196. **Hamza MA**, Aylmore LAG (1992b) Plant Soil 145:197–205. **Hanson AD**, Rathinasabapathi B, Chamberlin B, Gage DA (1991) Plant Physiol 97:1199–1205. **Harn C**, Daie J (1992) Plant Cell Physiol 33:763–770. **Hartung W**, Gimmler H (1994) Prog Bot 55:157–173. **Hatton TJ**, Greenslade D, Dawes WR (1992)

Tree Physiol 11:185–196. **Havaux M** (1992) Plant Physiol 100:424–432. **Havelange A**, Bernier G (1993) Physiol Plant 87:353–358. **Heckathorn SA**, Delucia EH (1991) Bot Gaz 152:263–268. **Hector DJ**, Gregson K, McGowan M (1993) Field Crops Res 32:287–304. **Hedrich R**, Marten I (1993) Embo 12:897–901. **Hegazy AK**, Ismail SM (1992) Acta Oecol 13:193–202. **Hendry GAF**, Finch-Savage WE, Thorpe PC, Atherton NM, Buckland SM, Nilsson KA, Seel WE (1992) New Phytol 122:273–279. **Herrmann V**, Tesche M (1992) Trees (Berl) 6:13–18. **Hirasawa T**, Gotou T, Ishihara K (1992) Jpn J Crop Sci 61:153–158. **Hite DRC**, Bodson MJ, Outlaw WH Jr (1992) Histochem J 24:368–374. **Hite DRC**, Outlaw WH Jr, Tarczynski MC (1993) Plant Physiol 101:1217–1221. **Ho C**, Stubbs CD (1992) Biophys J 63:897–902. **Hoekstra FA**, Crowe JH, Crowe LM, Van Roekel T, Vermeer E (1992) Plant Cell Environ 15:601–606. **Hohl M**, Schopfer P (1992) Planta 187:209–217. **Holaday A**, Scott A, Ritchie SW, Nguyen HT (1992) Environ Exp Bot 32:403–410. **Holbrook NM**, Burns MJ, Sinclair TR (1992) J Exp Bot 43:111–120. **Homble F**, Very AV (1992) Biophys J 63:996–999. **Hong TD**, Ellis RH (1992) Seed Sci Res 2:169–172. **Houssard C**, Escarre J, Vartanian N (1992) New Phytol 120:551–559. **Hu Z-L**, Li L, Jin J-H, Jiao X-Z (1993) Acta Phytophysiol Sin 19:124–130. **Huang B**, Nobel PS (1992) J Exp Bot 43:1441–1449. **Huang B**, Nobel PS (1993) New Phytol 123:499–507. **Huzulak J**, Matejka F (1992) Biol Plant (Prague) 34:291–296. **Hyvarinen M** (1992) Lichenologist (Lond) 24:267–279.

Ibrahim AE, Tekrony DM, Egli DB, Van Sanford DA (1992) Seed Sci Technol 20:39–46. **Iida Y**, Watabe K-I, Kamada H, Harada H (1992) J Plant Physiol 140:356–360. **Ikeda T**, Ohtsu M (1992) Ecol Res 7:391–395. **Ilijanic L** (1992) Acta Bot Croat. 51:41–60. **Inada K**, Matsuura A, Yamane M (1992) Jpn J Crop Sci 61:87–95. **Innes KPC**, Kelly D (1992) N Z J Bot 30:81–94. **Irigoyen JJ**, Emerich DW, Sanchez-Diaz M (1992a) Physiol Plant 84:61–66. **Irigoyen JJ**, Emerich DW, Sanchez-Diaz M (1992b) Physiol Plant 84:67–72. **Ishida A**, Yamamura Y, Hori Y (1992) Ecol Res 7:213–223. **Ismail AM**, Hall AE (1993) Crop Sci 33:498–503. **Ito S**, Gyokusen K (1993) Ecol Res 8:159–166. **Ito T**, Yamazaki M, Ohnishi S-I (1989) Biochemistry 28:5626–5630.

Jacob A, Wiencke C, Lehmann H, Kirst GO (1992a) Bot Mar 35:297–303. **Jacob A**, Lehmann H, Kirst GO, Wiencke C (1992b) Bot Acta 105:41–46. **Jacquart EM**, Armentano TV (1992) Am Midl Nat 127:158–171. **Jensen CR**, Andersen MN, Lösch R (1992) Plant Soil 140:225–239. **Jensen CR**, Svendsen H, Andersen MN, Lösch R (1993) Plant Soil 149:1–26. **Jiang M-Y**, Jing J-H, Wang S-T (1991) Acta Phytophysiol Sin 17:80–84. **Johnson RC** (1993) Crop Sci 33:169–174. **Johnson RW**, Tyree MT (1992) Plant Physiol 100:853–858. **Jonas OA**, Pereyra M, Cabeza C, Golberg AD, Ledent JF (1992) Cereal Res Commun 20:13–18. **Joyce PA**, Aspinall D, Paleg LG (1992) Aust J Plant Physiol 19:249–261.

Kader SA, Proebsting EL (1992) J Am Soc Hortic Sci 117:955–960. **Kaminskyj SGW**, Garrill A, Heath IB (1992) Exp Mycol 16:64–75. **Kantor P** (1992) Lesnictvi (Prague) 38:823–838. **Kao W-Y**, Forseth IN (1992a) Environ Exp Bot 32:343–349. **Kao W-Y**, Forseth IN (1992b) Plant Cell Environ 15:703–710. **Kao WY**, Forseth IN (1993) Am J Bot 80:886–891. **Kappen L** (1989) Antarct Sci 1:31–34. **Kargol M** (1992) Gen Physiol Biophys 11:469–487. **Katona E**, Neimanis S, Schoenknecht G, Heber U (1992) Photosynth Res 34:449–464. **Kawamitsu Y**, Yoda S, Agata W (1993) Plant Cell Physiol 34:113–119. **Kelavkar UP**, Chhatpar HS (1993) World J Microbiol Biotechnol 9:579–582. **Keller F**, Ludlow MM (1993) J Exp Bot 44:1351–1359. **Kelliher FM**, Koestner BMM, Hollinger DY, Byers JN, Hunt JE, McSeveny TM, Meserth R, Weir PL, Schulze E-D (1992) Agric For Meteorol 62:53–73. **Khalil AAM**, Grace J (1992) J Exp Bot 43:1591–1602. **Khalil AAM**, Grace J (1993) J Exp Bot 44:1127–1134. **Kikuta SB**, Richter H (1992) J Exp Bot 43:1039–1044. **Kirda C**, Mohamed ARAG, Kumarasinghe KS, Montenegro A, Zapata F (1992) Plant Soil 147:217–223. **Kjelgren RK**, Clarkf JR (1993) Trees (Berl) 7:195–

201. **Kleiner KW**, Abrams MD, Schultz JC (1992) Tree Physiol 11:271–287. **Kloareg B** (1991) Bull Soc Bot Fr Actual Bot 138:305–318. **Knapp AK** (1993) Ecology 74:113–123. **Knapp AK**, Smith WK (1991) Bot Gaz 152:269–274. **Koheil MAH**, Hilal SH, El-Alfy TS, Leistner E (1992) Phytochemistry 31:2003–2008. **Köhler K-H**, Schmerder B, Sheikhany H (1992) J Plant Physiol 139:528–532. **Koizumi M**, Ishida N, Takagishi H, Shirata K, Kano H (1992) Bot Mag Tokyo 105:1–11. **Kooistra MJ**, Schoonderbeek D, Boone FR, Veen BW, Van Noordwijk M (1992) Plant Soil 139:119–130. **Koppenaal RS**, Tschaplinski TJ, Colombo SJ (1991) Can J Bot 69:2522–2528. **Körner C** (1988) Flora 181:253–257. **Koster KL**, Lynch DV (1992) Plant Physiol 98:108–113. **Köstner BMM**, Schulze E-D, Kelliher FM, Hollinger DY, Byers JN, Hunt JE, McSeveny TM, Meserth R, Weir PL (1992) Oecologia 91:350–359. **Kovach DA**, Bradford KJ (1992) J Exp Bot 43:747–757. **Kovar JL**, Barber SA, Kladivko EJ, Griffith DR (1992) Soil Tillage Res 24:11–27. **Kriebitzsch W-U** (1992) Flora (Jena) 187:135–158. **Kubis J**, Krzywanski Z (1991) Acta Physiol Plant 13:13–20. **Kubis J**, Skoczek H, Krzywanski Z (1991) Acta Physiol Plant 13:139–146. **Kubiske ME**, Abrams MD (1991) Physiol Plant 83:107–116. **Kubiske ME**, Abrams MD (1992) Can J For Res 22:1402–1407. **Kueppers M** (1992) Trees (Berl) 7:8–11. **Kuhns MR**, Stroup WW, Gebre GM (1993) Can J For Res 23:387–393. **Kumar R**, Srivastava GC (1992) Indian J Plant Physiol 35:140–144. **Kutschera U**, Froehlich M (1992) J Plant Physiol 139:519–522. **Kwon KW**, Pallardy SG (1989) Can J For Res 19:622–626.

Lajtha K, Getz J (1993) Oecologia 94:95–101. **Lane BG** (1991) Faseb (Fed Am Soc Exp Biol) J 5:2893–2901. **Lange OL**, Büdel B, Meyer A, Kilian E (1993) Lichenologist (Lond) 25:175–189. **Lasceve G**, Gautier H, Jappe J, Vavasseur A (1993) Physiol Plant 88:453–459. **Lauer MJ**, Boyer JS (1992) Plant Physiol 98:1310–1316. **Lebkuecher JG**, Eickmeier WG (1992) Can J Bot 70:205–211. **Lebkuecher JG**, Eickmeier WG (1993) Ecology (Tempe) 74:1073–1080. **Lecoeur J**, Wery J, Turc O (1992) Plant Soil 144:177–189. **Ledoigt G**, Coudret A (1992) Bull Soc Bot Fr Lett Bot 139:175–190. **Lee DM**, Assmann SM (1992) Physiol Plant 85:35–42. **Lee J**, Bowling DJF (1992) J Exp Bot 43:951–957. **Legesse N**, Powell AA (1992) Seed Sci Technol 20:173–180. **Leprince O**, Van der Werf A, Deltour R, Lambers H (1992) Physiol Plant 85:581–588. **Levitt GJ**, Bolton JJ (1991) Bot Mar 34:403–410. **Lewis AM** (1992) Am J Bot 79:1158–1161. **Li D-Q**, Zou Q, Cheng B-S (1992) Acta Phytophysiol Sin 18:37–44. **Li X**, Feng Y, Boersma L (1993a) Ann Bot (Lond) 71:303–310. **Li Z-C**, Durachko DM, Cosgrove DJ (1993b) Planta 191:349–356. **Linnenbrink M**, Lösch R, Kappen L (1992) Flora (Jena) 187:121–133. **Liu W**, Xuezhi S (1993) For Res 6:162–166. **Liu Q**, Katou K, Okamoto H (1992) Plant Cell Physiol 33:915–919. **Livingston NJ**, Von Aderkas P, Fuchs EE, Reaney MJT (1992) Plant Physiol 100:1304–1309. **Lo Gullo MA**, Salleo S (1990) G Bot Ital 124:601–614. **Lo Gullo MA**, Salleo S (1993) Plant Cell Environ 16:511–519. **Loreto F**, Burdsall HH Jr, Tirro A (1993) Hortscience 28:222–224. **Lösch R** (1993) Water relations of Canarian Laurel forest trees. In: **Borghetti M**, Grace J, Raschi A (eds) Water transport in plants under climatic stress. Cambridge Univ Press, Cambridge, pp 243–246. **Lösch R**, Kazda M (1992) Water turnover of leaves in the rainforest canopy, the understorey and in a regrowth site of the Reserve de Campo/S. Cameroon. In: **Hallé F**, Pascal O (eds) Biologie d une canopée de forêt équatoriale II Fondation ELF, Paris, pp 206–216. **Lösch R**, Schulze E-D (1994) Ecol Stud 100:185–204.

Macfall JS, Johnson GA, Kramer PJ (1991) New Phytol 119:551–560. **Maier CA**, Teskey RO (1992) Can J For Res 22:1387–1394. **Maier-Maercker U**, Koch W (1991) Trees (Berl) 5:164–170. **Mansfield TA**, Davies WJ, Leigh RA (1993) Philos Trans R Soc Lond B 341:1–4. **Marconi E**, Carnovale E, Di Nola A, Brosio E (1993) Int J Food Sci Technol 28:25–33. **Markarova EN**, Pustovoitova TN, Aleshina MP (1991) Fiziol Rast (Mosc) 38:560–566. **Martin B**, Ruiz-Torres NA (1992) Plant Physiol 100:733–739. **Martynova AV**, Usmanov IY (1991) Ukr Bot Zh 48:75–79. **Maryland HF**, Johnson DA, Asay KH,

Read JJ (1993) Aust J Plant Physiol 20:361–369. **Mata G** (1992) Cryptogam Bot 2:387–390. **Materechera SA**, Alston AM, Kirby JM (1993) Plant Soil 149:149–158. **Matthes-Sears U**, Kelly PE, Larson DW (1993) Oecologia 95:9–13. **Matyssek R**, Maruyama S, Boyer JS (1991a) Plant Cell Environ 14:917–924. **Matyssek R**, Tang A-C, Boyer JS (1991b) Plant Cell Environ 14:925–930. **Mauchamp A**, Montana C, Lepart J, Rambal S (1993) Oikos 68:107–116. **Mauseth JD** (1993) Ann Bot (Lond) 72:81–89. **Mawson BT** (1993) Planta 191:293–301. **McCue KF**, Hanson AD (1992) Aust J Plant Physiol 19:555–564. **McIntosh TJ**, Simon SA (1993) Biochemistry 32:8374–8484. **McQueen-Mason S**, Durachko DM, Cosgrove DJ (1992) Plant Cell 4:1425–1433. **Meier CE**, Newton RJ, Puryear JD, Sen S (1992) J Plant Physiol 140:754–760. **Meinke H**, Hammer GL, Want P (1993) Field Crops Res 32:59–81. **Meinzer FC**, Grantz DA (1991) Physiol Plant 83:324–329. **Meinzer FC**, Goldstein G, Holbrook NM, Jackson P, Cavelier J (1993) Plant Cell Environ 16:429–436. **Meisner CA**, Karnok KJ (1992) Agron J 84:159–165. **Melchior W**, Steudle E (1993) Plant Physiol 101:1305–1315. **Melick DR**, Seppelt RD (1992) Antarct Sci 4:399–404. **Mellier A** (1989) Chem Phys Lipids 51:23–30. **Merzlyak MN**, Pavlov VK, Tatyana V (1992) J Plant Physiol 139:629–631. **Meurs C**, Basra AS, Karssen CM, Van Loon LC (1992) Plant Physiol 98:1484–1493. **Meyer WS**, Barrs HD (1991) Irrig Sci 12:125–134. **Meyer S**, Nhu-Hung SP, Tremolieres A, De Kouchkovsky Y (1992) Photosynth Res 32:95–107. **Miao SL**, Wayne PM, Bazzaz FA (1992) Oecologia 90:300–304. **Michalov J** (1989a) Biol Plant (Prague) 31:292–301. **Michalov J** (1989b) Biol Plant (Prague) 31:302–311. **Michalov J** (1993) Biol Plant (Prague) 35:251–259. **Miller PM**, Eddleman LE, Miller JM (1992) For Ecol Manage 48:185–215. **Mishio M** (1992) Aust J Plant Physiol 19:539–553. **Moldrup P**, Rolston DE, Hansen JA, Yamaguchi T (1992) Soil Sci 153:87–93. **Money NP**, Harold FM (1992) Proc Natl Acad Sci USA 89:4245–4249. **Mooney HA**, Chu C, Bullock SH, Robichaux R (1992) Biotropica 24:134–139. **Mott KA**, Cardon ZG, Berry JA (1993) Plant Cell Environ 16:25–34. **Mozhaeva LV** (1991) Izv Timiryazev S-KH Akad 0:69–81. **Mulkey SS**, Smith AP, Wright SJ, Machado JL, Dudley R (1992) Proc Natl Acad Sci USA 89:9084–9088. **Munoz FJ**, Labrador E, Dopico B (1993a) Physiol Plant 87:544–551. **Munoz FJ**, Dopico B, Labrador E (1993b) Physiol Plant 87:552–560. **Muslin EH**, Homann PH (1992) Plant Cell Environ 15:81–90.

Naidu BP, Paleg LG, Jones GP (1992) Phytochemistry 31:1195–1197. **Nakamoto T** (1993) Plant Soil 152:261–267. **Navari-Izzo F**, Quartacci MF, Melfi D, Izzo R (1993) Physiol Plant 87:508–514. **Neales TF**, McLeod AL (1991) Plant Cell Environ 14:979–986. **Newton RJ**, Funkhouser EA, Fong F, Tauer CG (1991) For Ecol Manage 43:225–250. **Ni B-R**, Pallardy SG (1992) Plant Physiol 99:1502–1508. **Nie D**, He H, Mo G, Kirkham B, Kanemasu ET (1992) Agric For Meteorol 61:205–217. **Nijs I**, Impens I, Behaeghe T (1989) J Exp Bot 40:353–360. **Nishi T**, Yagi T (1992) FEMS (Fed Eur Microbiol Soc) Microbiol Lett 99:95–99. **Nobel PS**, Cui M (1992a) Ann Bot (Lond) 70:485–491. **Nobel PS**, Cui M (1992b) Plant Soil 145:157–166. **Nobel PS**, Lopez FB, Alm DM (1991) J Exp Bot 42:1215–1224. **Nobel PS**, Miller PM, Graham EA (1992a) Oecologia 92:90–96. **Nobel PS**, Cavelier J, Andrade JL (1992b) J Exp Bot 43:641–648. **Nonami H**, Boyer JS (1993) Plant Physiol 102:13–19. **Norberg P**, Engstrom L, Nilsson R (1992) Biochim Biophys Acta 1112:52–56. **North GB**, Nobel PS (1992) New Phytol 120:9–19. **North GB**, Ewers FW, Nobel PS (1992) Am J Bot 79:1039–1050. **North GB**, Huang B, Nobel PS (1993) Bot Acta 106:126–135.

Ober ES, Setter TL, Madison JT, Thompson JF, Shapiro PS (1991) Plant Physiol 97:154–164. **Oda M**, Li Z, Tsuji K, Ichimura K, Sasaki H (1993) J Jpn Soc Hortic Sci 62:399–405. **Oliver MJ** (1991) Plant Physiol 97:1501–1511. **Oliver MJ**, Mishler BD, Quisenberry JE (1993) Am J Bot 80:127–136. **Ong B-L**, Lim M, Wee Y-C (1992) J Phycol 28:768–772. **Ooms JJJ**, Leon-Kloosterziel KM, Bartels D, Koornneef M, Karssen CM (1993) Plant Physiol 102:1185–1191. **Ortega JKE**, Bell SA, Erazo AJ (1992) Plant Physiol 100:1036–1041. **Osonubi O**, Bakare ON, Mulongoy K (1992) Biol Fertil Soils 14:159–

165. **Ott S**, Mechmann A, Jahns HM (1993) Symbiosis 15:151–164. **Ourry A**, Mesle S, Boucaud J (1992) New Phytol 120:275–280. **Oyanagi A**, Sato A, Wada M (1992) Jpn J Crop Sci 61:119–123.

Palmqvist K (1993) Planta 191:48–56. **Paravicini G**, Cooper M, Friedli L, Smith DJ, Carpentier J-L, Klig LS, Payton MA (1992) Mol Cell Biol 12:4896–4905. **Parry AD**, Horgan R (1992) Planta 187:185–191. **Parry MAJ**, Delgado E, Vadell J, Keys AJ, Lawlor DW, Medrano H (1993) Plant Physiol Biochem (Paris) 31:113–120. **Passioura JB** (1991) Botan Acta 104:405–411. **Passioura JB**, Fry SC (1992) Aust J Plant Physiol 19:565–576. **Passioura JB**, Ball MC, Knight JH (1992) Funct Ecol 6:476–481. **Pastori GM**, Trippi VS (1992) Plant Cell Physiol 33:957–961. **Pastori GM**, Trippi VS (1993) Physiol Plant 87:227–231. **Patrick JW** (1993a) Physiol Plant 87:345–352. **Patrick JW** (1993b) Physiol Plant 89:87–96. **Pena A**, Uribe S, Clemente M, Sanchez N (1992) Arch Microbiol 158:75–80. **Pena-Cabriales JJ**, Castellanos JZ (1993) Plant Soil 152:151–155. **Perez SCJG De A**, De Moraes JAPV (1991) Rev Bras Biol 51:805–811. **Pesci P** (1992) Physiol Plant 86:209–214. **Pesci P**, Reggiani R (1992) Physiol Plant 84:134–139. **Peterson CA**, Murrmann M, Steudle E (1993) Planta 190:127–136. **Peterson RL** (1992) Can J Bot 70:661–675. **Petrie CL**, Hall AE (1992) Aust J Plant Physiol 19:591–600. **Petrie CL**, Kabala ZJ, Hall AE, Simunek J (1992) Soil Sci Soc Am J 56:1686–1694. **Peuke AD**, Jeschke WD (1993) J Exp Bot 44:1167–1176. **Pezeshki SR**, Choi HS, Delaune RD (1993) Biol Plant (Prague) 35:43–51. **Pigott CD**, Pigott S (1993) J Ecol 81:557–566. **Plhak F** (1992) Biol Plant (Prague) 34:109–113. **Poffenroth M**, Green DB, Tallman G (1992) Plant Physiol 98:1460–1471. **Polley HW**, Johnson HB, Mayeux HS (1992) Funct Ecol 6:693–703. **Pothier D**, Margolis HA, Waring RH (1989) Can J For Res 19:432–439. **Poulsen KM**, Eriksen EN (1992) Seed Sci Res 2:215–221. **Poxleitner M**, Seitz-Beywl J, Heinzinger K (1993) Z Naturforsch Sect C Biosci 48:654–665. **Premachandra GS**, Joly RJ (1992) J Plant Physiol 139:355–360. **Premachandra GS**, Saneoka H, Fujita K, Ogata S (1992) J Exp Bot 43:1569–1576. **Premachandra GS**, Saneoka H, Fujita K, Ogata S (1993) J Agron Crop Sci 170:195–201. **Pretorius JC**, Small JGC (1992) Seed Sci Res 2:33–39. **Pritchard J**, Hetherington PR, Fry SC, Tomos AD (1993) J Exp Bot 44:1281–1289. **Purohit S**, Laloraya MM, Bharit S, Nozzolillo C (1992) J Exp Bot 43:103–110.

Quartacci MF, Navari-Izzo F (1992) J Plant Physiol 139:621–625. **Quick WP**, Chaves MM, Wendler R, David M, Rodrigues ML, Passaharinho A, Pereira JS, Adcock MD, Leegood RC, Stitt M (1992) Plant Cell Environ 15:25–36.

Rai VK, Sharma UD (1992) Beitr Biol Pflanz 66:393–405. **Rambal S** (1993) Plant Cell Environ 16:35–43. **Rasiah V**, Carlson GC, Kohl RA (1992) Soil Sci Soc Am J 56:1267–1271. **Rathinasabapathi B**, Gage DA, Mackill DJ, Hanson AD (1993) Crop Sci 33:534–538. **Reader RJ**, Jalili A, Grime JP, Spencer RE, Matthews N (1993) J Ecol 81:543–550. **Reddy KRK**, Rao AH, Sree BK, Reddy AR (1993) J Plant Physiol 141:373–375. **Reekie EG**, Wayne P (1992) Can J Bot 70:2334–2338. **Reid RJ**, Tester M, Smith FA (1993) Plant Cell Environ 16:547–554. **Rengasamy JI**, Reid JB (1993) Field Crops Res 33:197–215. **Reviron M-P**, Vartanian N, Sallantin M, Huet J-C, Pernollet J-C, De Vienne D (1992) Plant Physiol 100:1486–1493. **Reynolds TL**, Bewley JD (1993) J Exp Bot 44:921–928. **Ricanek M**, Vicherkova M (1992) Biol Plant (Prague) 34:259–265. **Ristic Z**, Gifford DJ, Cass DD (1992) J Plant Physiol 139:467–473. **Roberts J**, Cabral OMR, Fisch G, Molion LCB, Moore CJ, Shuttleworth WJ (1993) Agric For Meteorol 65:175–196. **Robertson MJ**, Fukai S, Ludlow MM, Hammer GL (1993a) Field Crops Res 33:81–97. **Robertson MJ**, Fukai S, Ludlow MM, Hammer GL (1993b) Field Crops Res 33:99–112. **Rodrigues ML**, Chaves MM, Wendler R, David MM, Quick WP, Leegood RC, Stitt M, Pereira JS (1993) Aust J Plant Physiol 20:309–321. **Rodriguez M**, Vicente C (1991) Symbiosis 11:255–262. **Roser DJ**, Melick DR, Ling HU, Seppelt RD (1992) Antarct Sci 4:413–420. **Roy-Macauley H**, Zuily-Fodil Y, Kidric M, Phamthi AT, Da Silva V (1992)

Physiol Plant 85:90–96. **Ruiz LP**, Atkinson CJ, Mansfield TA (1993) Philos Trans R Soc Lond B Biol Sci 341:67–74. **Ruiz-Sanchez MC**, Sanchez-Blanco MJ, Planes J, Alarcon JJ, Torrecillas A (1993) J Agric Sci 120:347–351. **Rütten D**, Santarius KA (1992) Oecologia 91:260–265. **Rütten D**, Santarius KA (1993) J Plant Physiol 14:739–744. **Rygol J**, Arnold WM, Zimmermann U (1992) Plant Cell Environ 15:11–24. **Ryle GJA**, Stanley J (1992) Ann Bot (Lond) 69:563–565.

Saab IN, Sharp RE, Pritchard J (1992) Plant Physiol 99:26–33. **Saftner RA** (1992) Physiol Plant 85:157–166. **Saliendra NZ**, Meinzer FC (1991) J Exp Bot 42:1251–1260. **Santrucek J** (1991) Photosynthetica (Prague) 25:527–536. **Sarkar S**, Kar S (1992) Agric Water Manage 21:137–144. **Sato A**, De Moraes JAPV (1992) Arq Biol Tecnol (Curitiba) 35:763–775. **Satoh S**, Iizuka C, Kikuchi A, Nakamura N, Fujii T (1992) Plant Cell Physiol 33:841–847. **Sauter JJ**, Van Cleve B (1992) Trees (Berl) 7:26–32. **Schipperges B** (1992) Polar Res 11:47–68. **Schlüter U-B**, Furch B (1992) Amazoniana 12:51–69. **Schmitt AK**, Martin CE, Löschen VS, Schmitt A (1993) J Arid Environ 24:155–164. **Schmuck G**, Moya I, Pedrini A, Van der Linde D, Lichtenthaler HK, Stober F, Schindler C, Goulas Y (1992) Radiat Environ Biophys 31:141–151. **Schnabl H**, Denecke M, Schulz M (1992) Bot Acta 105:367–369. **Schnall JA**, Quatrano RS (1992) Plant Physiol 100:216–218. **Schneider K**, Wells B, Schmelzer E, Salamini F, Bartels D (1993) Planta 189:120–131. **Schroeter B**, Jacobsen P, Kappen L (1991) Symbiosis 11:131–146. **Schroeter B**, Green GA, Seppelt RD, Kappen L (1992) Oecologia 92:457–462. **Schulte PJ**, Castle AL (1993a) J Exp Bot 44:1143–1148. **Schulte PJ**, Castle AL (1993b) J Exp Bot 44:1135–1142. **Schulte PJ**, Henry LT (1992) Tree Physiol 10:381–389. **Schultz HR**, Matthews MA (1993) Planta 190:393–406. **Schulz M**, Hunte C, Schnabl H (1992) Physiol Plant 86:315–321. **Seel WE**, Hendry GAF, Lee JA (1992a) J Exp Bot 43:1023–1030. **Seel WE**, Hendry GAF, Lee JA (1992b) J Exp Bot 43:1031–1037. **Selmani A**, Wassom CE (1991) Trans Kans Acad Sci 94:85–94. **Selmani A**, Wassom CE (1992) Field Crops Res 31:173–184. **Senden MHMN**, Van der Meer AJGM, Limborgh J, Wolterbeek HT (1992) Plant Soil 142:81–89. **Senock RS**, Ham JM (1993) Plant Cell Environ 16:593–601. **Serpe MD**, Matthews MA (1992) Plant Physiol 100:1852–1857. **Serrato-Valenti G**, Cornara L, Lotito S, Quagliotti L (1992) Ann Bot (Lond) 69:313–321. **Serrato-Valenti G**, Cornara L, Ferrando M, Modenesi P (1993) Can J Bot 71:834–840. **Sgherri CLM**, Pinzino C, Navari-Izzo F (1993) Physiol Plant 87:211–216. **Shackel KA**, Greve C, Labavitch JM, Ahmadi H (1991) Plant Physiol 97:814–816. **Shackel KA**, Johnson RS, Medawar CK, Phene CJ (1992) J Am Soc Hortic Sci 117:351–356. **Shelburne VB**, Hedden RL, Allen RM (1993) For Ecol Manage 58:193–209. **Shepherd VA**, Orlovich DA, Ashford AE (1993) J Cell Sci 105:1173–1178. **Shimazaki K-I**, Kinoshita T, Nishimura M (1992) Plant Physiol 99:1416–1421. **Shumway DL**, Steiner KC, Abrams MD (1991) Can J Bot 69:2158–2164. **Silvola J** (1991) Lindbergia 17:5–10. **Slovik S**, Hartung W (1992a) Planta 187:26–36. **Slovik S**, Hartung W (1992b) Planta 187:37–47. **Slovik S**, Baier M, Hartung W (1992) Planta 187:14–25. **Smirnoff N** (1992) J Bryol 17:185–191. **Smith JAC** (1991) Bot Acta 104:416–421. **Smith RE** (1992) S Afr J Bot 58:100–104. **Smucker AJM**, Aiken RM (1992) Soil Sci 154:281–289. **Snaar JEM**, Van As H (1992) Biophys J 63:1654–1658. **Sobrado MA** (1993) Oecologia 96:19–23. **Sobrado MA**, Grace J, Jarvis PG (1992) J Exp Bot 43:831–836. **Sonesson M**, Gehrke C, Tjus M (1992) Oecologia 92:23–29. **Southon TE**, Jones RA (1992) Physiol Plant 86:322–328. **Sperry JS**, Sullivan JEM (1992) Plant Physiol 100:605–613. **Sperry JS**, Alder NN, Eastlack SE (1993) J Exp Bot 44:1075–1082. **Srivastava A**, Zeiger E (1992) Plant Physiol 100:1562–1566. **Stark N** (1992) Trees (Berl) 6:7–12. **Stasovski E**, Peterson CA (1993) Can J Bot 71:700–707. **Stefanov K**, Markovska Y, Georgi K, Popov S (1992) Phytochemistry 31:2309–2314. **Stevanovic B**, Thu PTA, De Paula FM, Da Silva JV (1992) Can J Bot 70:107–113. **Streb P**, Michael-Knauf A, Feierabend J (1993) Physiol Plant 88:590–598. **Stromberg JC**, Wilkins SD, Tress JA (1993) Ecol Appl 3:307–314. **Sun G-C** (1992) Acta Bot Yunnanica 14:307–313.

Talbott LD, Ray PM (1992) Plant Physiol 98:369–379. **Tallman G** (1992) Crit Rev Plant Sci 11:35–57. **Tan W**, Blake TJ, Boyle TJB (1992) Physiol Plant 85:645–651. **Tanner W**, Beevers H (1991) Bot Acta 104–422. **Tardieu F** (1993) Hil Trans R Soc Lond B 341:57–66. **Tardieu F**, Davies WJ (1992) Plant Physiol 98:540–545. **Tardieu F**, Davies WJ (1993) Plant Cell Environ 16:341–349. **Tardieu F**, Bruckler L, Lafolie F (1992) Plant Soil 140:291–301. **Tardieu F**, Zhang J, Gowing DJG (1993) Plant Cell Environ 16:413–420. **Thiel G**, Gradmann D (1994) Prog Bot 55:59–78. **Thomas H**, Evans C (1991) Grass Forage Sci 46:391–398. **Thomas C**, Davis SD, Tallman G (1991) Plant Cell Environ 14:971–978. **Thomas JC**, McElwain EF, Bohnert HJ (1992) Plant Physiol 100:416–423. **Thorpe MR**, Minchin PEH, Williams JHH, Farrar JF, Tomos AD (1993) J Exp Bot 44:937–945. **Tiburcio AF**, Kaur-Sawhney R, Galston AW (1993) Plant Growth Regul 13:103–109. **Tinoco-Ojanguren C**, Pearcy RW (1992) Oecologia 92:222–228. **Tinoco-Ojanguren C**, Pearcy RW (1993) Oecologia 94:388–394. **Trejo CL**, Davies WJ, Ruiz LDMP (1993) Plant Physiol 102:497–502. **Tuba Z**, Lichtenthaler HK, Csintalan Z, Pocs T (1993) J Plant Physiol 142:103–108. **Tyree MT** (1993) Can J Bot 71:297–302. **Tyree MT**, Alexander JD (1993) Trees (Berl) 7:156–159. **Tyree MT**, Ewers FW (1991) New Phytol 119:345–360. **Tyree MT**, Yang S (1992) Plant Physiol 100:669–676. **Tyree MT**, Alexander J, Machado J-L (1992) Tree Physiol 10:411–415.

Valladares F, Wierzchos J, Ascaso C (1993) Am J Bot 80:263–272. **Van Auken OW**, Bush JK (1992) Southwest Nat 37:274–279. **Van Gardingen PR**, Grace J (1992) J Exp Bot 43:1293–1299. **Van Noordwijk M**, Kooistra MJ, Boone FR, Veen BW, Schoonderbeek D (1992) Plant Soil 139:109–118. **Van Rensburg L**, Kruger GH (1993) Plant Physiol 141:357–365. **Van Zyl PJ**, Kilian SG (1993) Appl Microbiol Biotechnol 39:235–241. **Van Zyl PJ**, Prior BA, Kilian SG (1991) Appl Microbiol Biotechnol 36:369–374. **Varney GT**, Canny MJ (1993) New Phytol 123:775–786. **Vartanian N**, Hervochon P, Marcotte L, Larher F (1992) J Plant Physiol 140:623–628. **Veen BW**, Van Noordjijk M, De Willigen P, Boone FR, Kooistra MJ (1992) Plant Soil 139:131–138. **Vernon DM**, Ostrem JA, Bohnert HJ (1993) Plant Cell Environ 16:437–444. **Vertucci CW** (1992) Plant Physiol 99:310–316. **Vetterlein D**, Marschner H, Horn R (1993) Plant Soil 149:263–273. **Voesenek LACJ**, Blom CWPM (1992) Acta Bot Neerl 41:319–329. **Voesenek LACJ**, De Graaf MCC, Blom CWPM (1992) Acta Bot Neerl 41:331–343.

Walker S, Hsiao TC (1993) Environ Exp Bot 33:447–456. **Wan C**, Sosebee RE, McMichael BL (1993) Environ Exp Bot 33:323–330. **Wang B-X**, Sun L, Huang J-C (1993) Sci China Ser B Chem Life Sci Earth Sci 36:45–51. **Wang J**, Ives NE, Lechowicz MJ (1992) Funct Ecol 6:469–475. **Wang Z**, Stutte GW (1992) J Am Soc Hortic Sci 117:816–823. **Webb MS**, Hui SW, Steponkus PL (1993) Biochim Biophys Acta 1145:93–104. **Welbaum GE** (1993) J Exp Bot 44:1245–1252. **Welbaum GE**, Bradford KJ (1991) J Exp Bot 42:1137–1146. **Werner O**, Espin RMR, Bopp M, Atzorn R (1991) Planta 186:99–103. **Wesley-Smith J**, Vertucci CW, Berjak P, Pammenter NW, Crane J (1992) Plant Physiol 140:596–604. **Wessels D**, Kappen L (1993) Koedoe 36:27–48. **Whitlow TH**, Bassuk NL, Reichert L (1992) J Appl Ecol 29:436–450. **Widmann K**, Gebauer G, Rehder H, Ziegler H (1993) Oecologia 94:146–152. **Wiebel J**, Eamus D, Chacko EK, Downton WJS, Ludders P (1993) Tree Physiol 13:55–69. **Wiencke C**, Gorham J, Tomos D, Davenport J (1992) Estuarine Coastal Shelf Sci 34:413–427. **Wilder GJ**, Johansen JR (1992) Can J Bot 70:2384–2404. **Wilson TP**, Canny MJ, McCully ME (1991) Physiol Plant 83:225–232. **Winkel T**, Rambal S (1993) Aust J Plant Physiol 20:143–157. **Wise RR**, Ortiz-Lopez A, Ort DR (1992) Plant Physiol 100:26–32. **Wolswinkel P**, Ammerlaan A, Koerselman-Kooij J (1992) J Exp Bot 43:681–693. **Woodward FI** (1987) Nature 327:617–618. **Wright SJ** (1991) Ecology 72:1643–1657. **Wright SJ**, Machado JL, Mulkey SS, Smith AP (1992) Oecologia 89:457–463. **Wu JV** (1992) Biophys J 61:1316–1331.

Xia QH, Chen RS, Fu JR (1992) Seed Sci Technol 20:269–279. **Xu C-C**, Qi Z (1993) Acta Agron Sin 19:359–364. **Xu X**, Bland WL (1993) Agron J 85:697–702. **Xue S**, Wang P-H, Xu D-Q, Li L-R (1992) Acta Phytophysiol Sin 18:1–7.

Yagi T (1992) Microbios 70:93–102. **Yagi T**, Nishi T (1993) Microbios 74:155–166. **Yang S**, Tyree MT (1992) Plant Cell Environ 15:633–643. **Yoon TM**, Richter H (1992) J Korean Soc Hortic Sci 33:318–323.

Zagdanska B (1989) J Plant Physiol 134:320–326. **Zagdanska B** (1991) Plant Physiol Biochem (Paris) 29:409–414. **Zhang B**, Archbold DD (1993) J Am Soc Hortic Sci 118:280–285. **Zhang H**, Pei-Hong W (1991) Acta Phytophysiol Sin 17:259–266. **Zhang J**, Marshall JD, Jaquish BC (1993) Oecologia 93:80–87. **Zhu GL**, Boyer JS (1992) Plant Physiol 100:2071–2080. **Zhu J-J**, Beck E (1991) Plant Physiol 97:1146–1153. **Zimmermann U**, Balling A, Rygol J, Link A, Haase A (1991) Bot Acta 104:412–415. **Zimmermann U**, Haase A, Langbein D, Meinzer F (1993) Philos Trans R Soc Lond B Biol Sci 341:19–31. **Zimmermann U**, Rygol J, Balling A, Kloeck G, Methler A, Haase A (1992) Plant Physiol 99:186–196. **Zrenner R**, Stitt M (1991) Plant Cell Environ 14:939–946.

Prof. Dr. Rainer Lösch
Abt. Geobotanik der
Heinrich-Heine-Universität
Universitätsstr. 1/26.13
40225 Düsseldorf, Germany

II. Humic Substances and Plant Nutrition

By Zeno Varanini and Roberto Pinton

1. Introduction

Since the studies by Liebig (1856), it is well known that plants, as long as they are adequately supplied with light and mineral nutrients, can live in the absence of the organic and inorganic structural components of the soil. Nowadays the use of hydroponics is popular not only among plant physiologists but also in certain commercial activities.

Nevertheless, in a natural environment, the root apparatus of plants has evolved, lives, and grows in contact with the soil matrix, which cannot be simply considered as a nutrient solution. The soil is composed of an organic and mineral solid phase and an aqueous phase. In these phases the nutrients are distributed in a way that is, besides other factors, determined also by the action of the root apparatus.

In temperate climates, 1–3% of the soil's weight consists of organic matter. Soil fertility is strictly related to its organic matter content, which can vary from less than 1% in young soils or in those which are exhausted by over-intensive cropping, to 95% in some deep peats (Nannipieri and Sequi 1982). The mass of organic carbon in the soil (30.1×10^{14} kg) exceeds the amount of carbon present in other pools of the terrestrial part of the planet (Bohn 1976).

Due to the heterogeneous composition of the soil's organic matter it is necessary to define the different constituents to understand the role they play in plant nutrition.

According to Stevenson (1982), the term organic matter describes the organic substances present in soil, excluding unaltered animal and plant tissues, the products of their partial decomposition and the soil biomass. The term organic matter therefore refers to:

- organic polymers of known structure;
- simple compounds such as sugars and amino-acids, and other small molecules;
- humic substances.

In this chapter, only the action of the humic substances on plant nutrition will be considered. As humic substances are the object of study of soil scientists and of a relatively small number of plant nutritionists, the essential facts on the

Progress in Botany, Vol. 56

nature and chemical-structural characteristics of humic substances present in the soil will be summarized here.

2. Definitions, Chemical and Structural Characteristics of Humic Substances

Even if it appears rather vague, a very appropriate definition of humic substances has been proposed by Aiken et al. (1985), who considered them as a generical "category of naturally occurring, biogenic, heterogeneous organic substances that can generally be characterized as being yellow to black in color, of high molecular weight, and refractory". About 60% of soil organic matter is represented by these substances; they are the result of chemical and biological decay of animal, microbial, and plant residues, due to the activity of soil microorganisms. These products tend to link into complex chemical structures that are more stable than the original compounds (Schnitzer 1978).

Jenkinson and Rayner (1977) have, in fact, estimated that, unlike the other organic fractions present in the soil, humic substances have a half-life of about 2000 years.

In the soil, humic substances are linked to the mineral components through various chemical-physical interactions (Theng 1979) and, as already stated, mixed with other organic compounds of different nature. It is necessary to define the chemical and physical nature of humic substances in order to understand their effects on plant nutrition. The first problem to be dealt with must then be that of their extraction from soil and separation from the mineral components and from nonhumic substances. Many methods have been developed; one of the most popular consists of extraction using strong alkaline solutions (e.g., NaOH 0.1–0.5 N). Although this method has frequently been used in the past decades for the high yield of humic substances extracted (up to 80%), its use has been questioned, as the humic molecules are probably altered during the extraction, giving origin to possible artifacts (Tinslay and Salam 1961). More recently, the use of less drastic and more selective extractants, such as $Na_4P_2O_7$, organic chelating agents, aprotic dipolar solvents, has been recommended (Stevenson 1982). Notwithstanding the lower yield, which can range from 5 to 50%, these extractants are preferred, as they modify to a lesser extent the humic polymers during the isolation.

Different extractants can also determine selective isolation of humic fractions with respect to the molecular weight. Cameron et al. (1972) showed that the size of the extracted humic molecules, as determined by gel-filtration and ultrafiltration, varies depending on the type of extractant used: the molecular weight of extracted fractions is lower with $Na_4P_2O_7$ than with NaOH.

The definition of the chemical changes which occur when using different methods of extraction is still today the object of exhaustive research by soil scientists. By means of sophisticated analytical techniques it is necessary to clarify this first point, to be able to determine in future the true structure of humic substances in the soil.

Humic substances can be further fractionated into components having different chemical and physical properties. Many fractionation methods have been suggested, but none of them has lead to the isolation of fractions that can be considered chemically "pure" and representative of the native soil humic substances. Hayes et al. (1989) recently presented a thorough discussion of the subject. The most common and accepted fractionation technique is based on the different solubility in water at various pH values. Aiken et al. (1985) therefore distinguished:

Humic acid: the fraction that is not soluble in water under acidic conditions (pH$<$2) but soluble at higher pH values.
Fulvic acid: the fraction that is soluble in water under all pH conditions.
Humins: the humic fraction that is not soluble in water at any pH value.

The humic and fulvic fractions can be considered "pure" (although not in a chemical sense) once the salts and, in the case of fulvic acids, the known organic compounds (simple sugars, amino acids, etc.) that are not structurally bound, have been removed. Extraction and isolation techniques for humic and fulvic acids are described in detail by Stevenson (1982).

A great effort has been made to identify structure and composition of humic and fulvic acids. Elemental analyses of these compounds has shown that, in general, 98–100% of the elements (not considering ashes) are C, H, O, N, S, P. The distribution of these elements is given in Table 1. Generally, P accounts for 0.1–1.0% of humic and fulvic acids.

Details on the reactivity of these compounds have been obtained from the determination of oxygen-containing functional groups in humic substances (Table 2).

In addition, the presence of nitrogen-containing groups has been extensively reported. The problem of defining the structure of humic substances has been deeply discussed and has attracted the attention of many soil chemists. Different structural models for humic molecules (Flaig 1960; Schnitzer and Khan 1972; Stevenson 1982) have been proposed, but none of them, as noted by Hayes (1991), can be considered completely satisfactory (Fig. 1).

Table 1. Elemental contents of humic and fulvic acids (Stevenson 1982)

	Dry and de-ashed (%)	
Element	Fulvic acids	Humic acids
C	40–50	50–60
O	44–50	30–35
H	4–6	4–6
N	<1–3	2–6
S	0–2	0–2

Table 2. Range of distribution of oxygen-containing functional groups in humic and fulvic acids isolated from soils of widely different climatic zones (in mEq/100 g) (After Schnitzer 1977)

	Humic acids	Fulvic acids
Total acidity	560–890	640–1420
COOH	150–570	520–1120
Acidic OH	210–570	30–570
Weakly acidic + alcoholic OH	20–490	260–950
Quinone and ketonic C=O	10–560	120–420
OCH_3	30–80	30–120

In addition, a structure similar to polymaleic acid has been proposed for fulvic acids (Anderson and Russell 1976) and the aromatic nature of these acids has been questioned (Farmer and Pisaniello 1985).

Recently, sophisticated analytical methods such as ^{13}C nuclear magnetic resonance (NMR), electron spin resonance (ESR), pyrolysis-mass spectroscopy (PMS), and supercritical gas extraction, have been used to gain a more precise and satisfactory structural definition of humic substances. None of these methods, taken individually, seems capable of giving definitive results, even though the use of ^{13}C NMR appears to be quite promising, as it does not require the extraction of the humic material from the soil, so that the former is analyzed in its native form (Schnitzer 1990).

Some information on the structural features of humic substances is available. It has, in fact, been established that these molecules contain considerable amounts of single-ring aromatic structures that can interact with each other and with aliphatic chains, therefore forming the skeleton of these macromolecules. The aggregation of these "building blocks" causes the formation of complexes with molecular dimensions that can vary from a few hundreds or a few thousands of Da (fulvic acids) (Linehan 1977) up to hundreds of thousands (humic acids) (Swift and Posner 1971). Furthermore, some distinctive features of humic and fulvic acids have been recently defined. In particular, it has been shown that the latter have a smaller percentage of aromatic C (25% as compared to 35–40%), and higher negative charge and polarity. It has also been proved that at normal pH values of cultivated soils, humic substances are found in the form of polyanions with a more or less slackened spherical or elliptic structure (Ghosh and Schnitzer 1980). Although it is clear from this description that it is not yet possible to define the general structure of humic substances, and although this scope represents an important step for further developments in this field, as observed by Hayes (1991): "it is not necessary to know the structure accurately in order to obtain an acceptable understanding of important reactions involving humic substances".

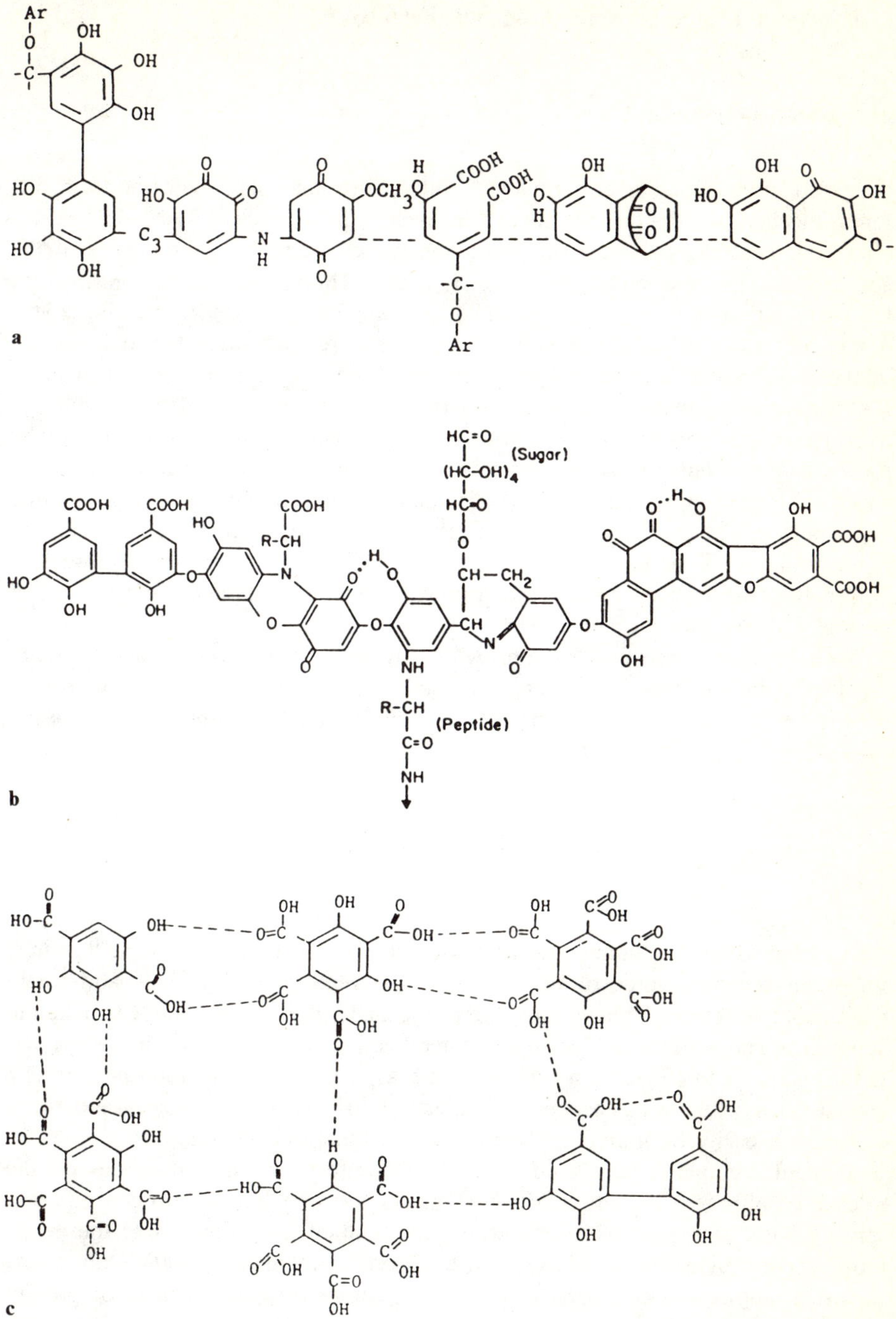

Fig. 1a–c. Hypothetical structure of humic acid. **a** According to Flaig (1960). **b** According to Stevenson (1982). **c** Structure of fulvic acid as proposed by Schnitzer and Khan (1972)

3. Effects of Humic Substances on Soil Fertility

a) Physical Properties of Soil

Humified organic matter can influence soil fertility and plant nutrition with different mechanisms that cannot always be clearly distinguished. The more general effects concern the influence of humic substances on physical properties of the soil such as structure, color, and water retention. Humic substances contribute to the stabilization of the soil structure (Piccolo and Mbagwu 1990), creating a sufficient percentage (40–50% in volume) in micropores that favor the maintenance of the soil's water regime and gas exchange (O_2), and enhance root growth. It has been proved that humic acids contribute to cement clay particles through hydrogen bonds and coordination with polyvalent cations (Stevenson 1982). The presence of the latter in soils seems to be essential for a positive action on the structure (Giovannini and Sequi 1976), and, in the absence of cations, humic substances could have an opposite effect: there would, in fact, be an increased repulsion between the negative charges of the clays and those of humic substances whose functional groups, such as carboxyl acids, are completely dissociated at normal soil pH values (Oades 1984).

Well-structured soils possess considerable water-holding capacity, and this fact, together with the typical dark color of soils which are rich in organic matter, contributes to the increase of thermal capacity and absorption of solar energy, especially in the case of bare ground.

b) Source of Nutrients

As mentioned earlier, humic substances contain N, S, P. Even though the most important organic source of plant nutrients is represented by fresh and easily degradable residues of plants and microorganisms, there is no doubt that humic substances can also be subjected to mineralization, thus releasing inorganic nutrients which can be used by plants. This aspect is of great relevance if one considers that humic substances can release nutrients in such places and times that cannot easily be mimicked by the use of mineral fertilizers.

Particular emphasis has been given to the role of humic substances on the release of nitrogen. In particular, the topic concerning humic nitrogen and soil fertility has been extensively reviewed quite a number of times over the years, most recently by Schnitzer (1985) and by Stevenson and He (1990). Only a few important points will be emphasized in this chapter. Nitrogen seems to be present in humic substances in different forms (Haworth 1971). Peptides and proteins can be bound to the central humic core of humic substances through hydrogen bonds. A fraction of nitrogen can be covalently bound to phenolic or quinonic rings by aminic groups. Another considerable fraction (up to 60% of total nitrogen) is thought to be present as an integral component of humic substances in the form of heterocyclic rings or as bridges between quinones. A quantitative characterization of these forms was obtained mainly by acid hydrolysis of the humic substances (Stevenson and He 1990). The soluble fraction probably contains labile forms of nitrogen such as peptides, amino sugars and amino acids that are peripherical to the aromatic ring; it has been considered as the most easily available nitrogen pool for plant nutrition. Nowadays, this approach no longer fully satisfies soil chemists, as the distribution of the different forms of nitrogen does not seem to vary with soil management practices such as long-term cropping or the addition of organic amendments (Stevenson 1982). This observation suggests that humic substances are being constantly modified under the action of soil microorganisms. In a steady state, humus mineralization is counterbalanced by the synthesis of new humic matter (humification). During humification, the nitrogen present in amino acids and in other amino compounds is incorporated into the more stable structure of humic acids, and thereby becomes less available for mineralization (Stevenson and He 1990). It is quite likely that a more detailed knowledge of the humic substances' structure and of the processes of interconversion (recycling: mineralization/humification cycle) will lead to a better definition of humic N availability. This fact is particularly important if one considers that mineralization is not uniform throughout the soil profile, but is enhanced in the rhizosphere, where microbial activity due to root exudates strongly influences this process (Kuikman and van Veen 1989).

Only little information is available on forms of S present in humic substances, and their eventual availability for plant nutrition. Nevertheless, it has been ascertained that humic substances contain S in the form of proteins and sulfate esters, and possibly in relatively stable thiazine rings. These compounds seem to be decomposed very slowly in soil, as proved also by their resistance to enzymatic hydrolysis (Houghton and Rose 1976).

Due to the low solubility of inorganic P in soil and its low mobility – Pi interacts with many mineral components, such as clay, iron and aluminum oxides, and carbonates – contribution of humic P could be relevant, especially at the rhizosphere, where there is a higher phosphatase activity as compared to bulk soil (Helal and Sauerbeck 1984; Tarafdar and Jungk 1987). This enzyme activity has been found to be related to organic P depletion and to inorganic P accumulation around plant roots, thus suggesting that mineralization is greater than plant absorption (Tarafdar and Jungk 1987). Organic P is especially concentrated in humic acids. The knowledge of chemical structure of P in humic substances is increased after the use of ^{31}P Nuclear Magnetic Resonance (Wilson 1990). Bedrock et al. (1994) have shown, using ^{31}P NMR, that different forms of P can be associated to humic fractions with different molecular dimensions. In fact, the P present in high-molecular fractions of humic acids was mainly in the form of esaphosphate inositole, while in the fractions with lower molecular weight, P was mainly detected as inorganic orthophosphate. The latter appears to be the compound more easily used by plants.

c) Influence on Cation Exchange Capacity

An essential role played by humic substances in plant nutrition is that concerning soil cation exchange capacity. It has been proved that humified organic matter contributes for about 50% to soil cation exchange capacity (Allison 1973), and favors retention of cationic nutrients in an exchangeable form that can become available to plant nutrition. Moreover, as shown by Tan (1978), humic and fulvic acids can contribute to the release of potassium fixed in the interlayers of 2:1 clay minerals, thereby making this macronutrient more readily available.

Due to the ability to exchange cations, humic substances can act as buffer against changes of pH in the soil. This feature is of great relevance: as stated by Allison (1973) "the buffering capacity of soils is of great practical significance. If this did not exist it would be difficult to conceive agriculture as we know it today. We would have a situation similar to that encountered in hydroponics. Plants could be grown, but at great expense and attention to details".

d) Complexing Properties

The importance of humic substances lies also in their ability to complex metals in soil, especially cationic micronutrients, thus modifying their solubility. This is especially true for those metals whose solubility is regulated by the formation of insoluble hydroxides produced through a progressive polynucleation of hydrated ions as the pH value increases (Baes and Mesmer 1976); in soils with pH above neutrality, micronutrients in soil solution would, in fact, be present at concentrations estimated to be insufficient for the plant's needs (Lindsay 1991).

The ability of humic substances to form complexes with cationic micronutrients depends on their contents of electron-donor functional groups (Linehan 1985). Either weak bonds such as water bridges, electrostatic attraction due to cation exchange capacity, or strong bonds involving coordination with single groups (e.g., –COOH) or formation of ring structures (chelates) with carboxyl, alcohol, and aminic groups, can thereby be formed (Boyd et al. 1979; 1981). The formation of the different types of bonds depends on the latters' degree of saturation; the weaker bonds play a more important role when the stronger ones become saturated. The stability order of the complexes formed between metals and humic acids has been determined through potentiometric titration and follows the Irwing-Williams series ($Pb^{2+} > Cu^{2+} > Ni^{2+} > Co^{2+} > Zn^{2+} > Cd^{2+} > Fe^{2+} > Mn^{2+} > Mg^{2+}$). On the other hand, as shown by van Dijk (1971), at a pH value of 5 there was no great difference in the strength of bonds between humic acids and metals such as Ca^{2+}, Mg^{2+}, Mn^{2+}, Co^{2+}, Ni^{2+} and Zn^{2+}, whilst Pb^{2+}, Cu^{2+} and Fe^{3+} were more strongly bound to humic acids; this behavior indicates that at different pH values, metal-humic substances complexes of different stability are formed in soil.

The solubility of the complexes formed by humic acids or fulvic acids with micronutrients depends on the pH, on the presence of salts, and on the saturation degree of binding sites. In relation to the last point, it is important to note that the complexes formed with fulvic acids are generally more soluble than those formed with humic acids, and could be considered as natural chelates able to reach the plant roots (Mirave and Orioli 1987). Fulvic acids are generally present in the solution of forest soils (Dawson et al. 1981), although information on their concentration in cultivated soils is up to now still scarce (Linehan 1977). Moreover, it has been proved that the quantity of fulvic acids present in soil solution does not depend on the total concentration of soil organic matter (Linehan 1985).

On the other hand, insoluble humic acids can withhold considerable amounts of microelements (Linehan 1985), thereby representing a potential reservoir of such elements in soil. In the presence of high quantities of metals, single molecules of fulvic acids can form interchain bonds producing insoluble macromolecules (Sequi et al. 1975).

As humic substances have the ability to form insoluble complexes with metals, they therefore play an important role in detoxifying soils from those elements (such as Cd and Al) that can be present at exceedingly high concentrations in soil solution. The complexation operated by humic and fulvic acids reduces the concentration of noxious ions (not complexed free metals), thus preventing their rapid absorption by plant roots. It is a well-known fact that humic substances can reduce aluminum toxicity, as the latter depends on the concentration of the undissociated species (Gjessing et al. 1989; Asp and Berggren 1990). Similar results have been reported also for cadmium (Cabrera et al. 1988) and zinc (van der Werff and Out 1981).

The determination of the stability constants for humic acid-metal and fulvic acid-metal complexes have been reported (Schnitzer and Hansen 1970; van Dijk 1971; Stevenson and Ardakani 1972). Due to the heterogeneity of humic substances concerning molecular weight, content of functional groups, variety of bonding sites, and changes in conformation of these macromolecules with pH and salt concentration, it is difficult to reach an unequivocal state of knowledge. Nevertheless, there is sufficient evidence to generally conclude that humic-metal complexes have a lower stability constant as compared to complexes between metallic ions and synthetic chelating agents (e.g., EDTA) (Schnitzer and Hansen 1970) and to many organic compounds of biological origin naturally present in the bulk soil and in the rhizosphere (Martell 1975; Mench et al. 1988). The ability of humic substances to complex cationic nutrients can have a great importance in phosphorous nutrition (Stevenson 1991). These substances can, in fact, increase the availability of insoluble phosphates, as they complex Fe^{3+} and Al^{3+} in acidic soils, and Ca^{2+} in calcareous soils. Phospho-humic complexes can also be formed by bridging with Fe and/or Al ions (Stevenson and Ardakani 1972), although it is not yet clear whether they play a role in increasing phosphorous availability. It is also possible that higher availability of the macronutrient might be due to competition between polyanionic humates and phosphate ions for the adsorption surfaces.

It is important to note that most of the information available on the equilibria between metals and humic substances has been obtained in simplified systems, using humic fractions isolated from the soil. Therefore the results of these studies can refer to situations occurring in the bulk soil. On the other hand, recent studies (Marschner et al. 1986) clearly show that, in the rhizosphere, pH, redox potential, microbiological and chemical composition, and concentration of nutrients in the solution, can be quite different than in bulk soil. Therefore, the role of humic substances as micronutrient reservoir and/or possible source of natural chelates might be reconsidered at the rhizosphere. With regard to this subject, it is interesting to note that root exudates, whose stability constant for metals is higher than that of humic substances (Martell 1975), are much more labile than humic molecules, because they are easily degraded by soil microbes (Römheld 1991). Furthermore, it has been proved (Dell'Agnola and Nardi 1987; Albuzio and Ferrari 1989) that the presence of organic acids, like those excreted from roots, can lead to the dissociation of the humic macrostructure and the release of low molecular weight fractions.

4. Effect on Growth and Nutrient Contents of Plants

Together with the actions described in the previous paragraphs, many studies show that humic substances can stimulate the growth of plant tissues and the total quantity of absorbed nutrients (for a recent review on this subject see Vaughan and Malcolm 1985; Chen and Aviad 1990).

Rauthan and Schnitzer (1981) found that cucumber plants grown for 6 weeks in a nutrient solution more than doubled their dry weight when $300 mg^{-1}$ of fulvic acids were present in the solution. Concomitantly, an increase of the total contents of N, P, K, Ca, Mg, Cu, Fe, and Zn was observed in the shoots. Likewise, Tan and Nopamornbodi (1979) showed that shoots of maize plants grown for 16 days in a nutrient solution in which $640\,mg\,l^{-1}$ of humic acids had been added, tripled their dry weight and increased their total contents of N, P, K, Ca, Mg, Cu, Fe, Zn, and Mn. Other authors report effects of lesser importance (Mylonas and McCants 1980a; Vaughan and Malcolm 1985) with differences in relation to the plant organ considered, as shown by Dormaar (1975) in *Festuca scabrella* and by Ernst et al. (1987) in *Scrophularia nodosa*; in these plants the effects were particularly visible in the shoots rather than in the roots. The variability in the results can be due to differences of the humic substances' origin (Lee and Bartlett 1976) and of the humic fractions used (Ernst et al. 1987). On the other hand, various plant species treated with the same humic fraction have shown different responses (Dormaar 1975). This observation would suggest the existence of an adaptability to humus-rich soil conditions in plants that only slightly react to the treatment with the humic substances (Ernst et al. 1987). This has already been suggested for soil microorganisms (Visser 1985). The effects caused by humic substances on growth and nutrition were observed not only in hydroponics, but also when plants were grown on sand (Levesque 1970) or soil, especially those with low organic matter contents (Lee and Barlett 1976; Fagbenro and Agboola 1993), and in callis in vitro cultures (Irintoto et al. 1993).

Independently of all experimental conditions, the response to humic substances is always related to the concentration applied. Even though it is not possible to define an unequivocal optimal concentration, almost all the reports considered showed the existence of a level at which the positive effect of humic substances was highest; above and below this value, the effect was lower and, in some cases, even negative.

The effects on the growth of plant tissues have been explained by assuming a hormonal action of humic substances, as shown by some experiments in which some humic fractions showed a hormone-like activity (O'Donnell 1973; Linehan 1977; Cacco and Dell'Agnola 1984). This hypothesis has led many authors to study the action of humic substances on enzymatic activities which are relevant for plant growth. For example, Mato et al. (1971, 1972a,b) proved that humic substances are able to inhibit the activity of IAA-oxidase; this would contribute to keeping a higher contents of IAA in the tissue, therefore stimulating growth. Inhibition and/or stimulation have been observed when other enzymatic activities (peroxidase, invertase, choline esterase, phosphatase) were examined (Vaughan et al. 1978; Malcolm and Vaughan 1979a,b; De Almeida et al. 1980; Muscolo et al. 1993). These results are not easily interpreted, and may also be due to the use of different fractions at sometimes high concentrations of humic substances and use of crude plant tissue homogenates. Thus, it is not possible to draw a clear conclusive picture on specific interactions occurring between humic substances and the enzymatic activities related to growth stimulation.

Another possible mechanism through which humic substances can stimulate plant growth seems to invove Fe-chelation (Schnitzer and Poapst 1967). Vaughan (1974) suggested that the promotion of root lengthening could be caused by Fe^{2+} chelation, therefore inhibiting the formation of cell-wall bound hydroxyproline.

From a nutritional point of view, it is interesting to note that the response of plant roots to the exposure to humic substances involves also morphological changes such as proliferation of lateral and adventitious roots (Linehan 1976; Mylonas and McCants 1980b), as well as root hairs (Rauthan and Schnitzer 1981). It is noteworthy that the changes in the morphology of the root apparatus and the increase of the surface area can influence the plant's ability to absorb soil nutrients (Marschner et al. 1986).

5. Effects on Mechanisms of Nutrient Uptake

a) Uptake Rates

Various studies carried out on excised roots or on whole plants have proved that, as a rule, uptake of cationic and anionic macronutrients is increased when the roots are placed in contact with suitable concentrations of humic substances.

Guminski et al. (1983) and Maggioni et al. (1987) observed in tomato and in oat roots respectively, that following a treatment with humic substances, the uptake rate for K^+ increased by 20–30%. Similar increases were reported for Ca^{2+} uptake in tomato roots, while Mg^{2+} uptake increased to a lesser extent (Guminski et al. 1983). The effects observed on sulfate and nitrate uptake rates were even more relevant. The uptake rate of sulfate in fact showed almost a 100% increase in oat roots after contact for 8 h with different fractions of humic substances at 50 mg l^{-1} org C (Maggioni et al. 1987), and a 40% increase in barley roots treated for 6 h with 50 mg l^{-1} of humic substances (Dell'Agnola and Ferrari 1971) at the same concentration. Similar values were measured for net nitrate influx in both barley (Albuzio et al. 1986) and oat (Nardi et al. 1991) roots. Testing the effect of humic fractions obtained by various soil extractants and from different sources, Piccolo et al. (1992) showed that the fraction having the highest degree of acidic functionality and the smallest molecular size was the more active in promoting the enhancement of net nitrate influx.

Unlike the effects on sulfate and nitrate uptake, strong inhibition to chloride uptake was observed (Vaughan and McDonald 1971; Guminski et al. 1983).

An essential requirement for a stimulatory action of humic substances on the uptake rate of cations and anions is the contact between humic substances and roots usually for more than 6 h at physiological temperatures (Dell'Agnola and Ferrari 1971; Albuzio et al. 1986; Maggioni et al. 1987). When the roots were kept in contact with humic substances for short periods (up to 2 h), lower stimulation was observed (Dell'Agnola and Ferrari 1971; Vaughan et al. 1978; Maggioni et al. 1987). At concentrations higher than 50 mg l^{-1}, sulfate uptake was even inhibited (Dell'Agnola and Ferrari 1971; Maggioni et al. 1987). This latter effect was considered to be due to a competitive inhibition, somehow related to the amount of carboxyl groups in the humic fractions (Dell'Agnola and Ferrari 1971). An incubation for longer periods, but at 0–2 °C, did not lead to any stimulatory action (Dell'Agnola and Ferrari 1971). These phenomena can be explained by taking into consideration the possibility of humic substances of having both a direct action of negative modulation of sulfate carriers, and stimulatory action on metabolic processes. This last point was studied through the use of transcription and translation inhibitors (actinomycin D and cycloheximide, respectively). Both these compounds neutralized the stimulatory action of humic substances without influencing the inhibition effects, therefore suggesting that the former action is somehow related to the protein synthesis mechanisms (Dell'Agnola and Ferrari 1971). Furthermore, studies with ^{14}C-leucine and ^{3}H-uridine proved that humic substances were able to stimulate the incorporation of ^{14}C-leucine in proteins, but not the incorporation of ^{3}H-uridine in RNA, showing that humic substances could influence carrier protein synthesis at the translation level (Dell'Agnola and Ferrari 1975). Changes in protein synthesis after contact between humic substances and beet discs were also reported by Vaughan and McDonald (1976), and related to selective changes in ion uptake.

b) Energy Metabolism

Changes in the capacity of accumulating nutrients can also be related to variations in the energy metabolism of the cell (Poole 1978). Quite a number of studies show the influence of humic and fulvic acids on respiration and photosynthesis. Sladky (1959) proved that tomato roots grown in nutrient solutions in the presence of humic acids or fulvic acids consumed 24 or 38% more oxygen, respectively. A similar result was reported by Vaughan (1967) for beet root storage tissue under axenic conditions. The mechanisms that lead to the increase of oxygen consumption have still not been sufficiently cleared. Flaig and Riemer (1971), using a synthetic humic acid, suggested that humic substances partially uncouple oxidative phosphorylation, thereby decreasing the cellular ATP content. Visser (1987) showed that short-time contact between rat liver mitochondria and humic acids caused uncoupling of respiration from oxidative phosphorylation. A 30–40% decrease of the cellular ATP content, without a parallel increase in oxygen consumption, has recently been observed by Nardi et al. (1991) in roots of oat seedlings which had been treated with low and high molecular weight humic fractions.

On the other hand, Visser (1987) demonstrated that 60–120-min incubation of rat liver mitochondria with 40 mg l^{-1} humic substances maintained a high efficiency of oxidative phosphorylation. Khristeva et al. (1980), based on experiments with natural humic substances suggested that the respiratory increase is accompanied by a higher production of ATP.

The influence of humic substances on photosynthesis is even less clear. Various works report a positive effect on chlorophyll contents when humic substances are added to the nutrient solution or are sprayed on the leaves (Sladky 1959; Sladky and Tichy' 1959). The increase in the chlorophyll content has been considered the cause of the increase in the leaf dry weight (Sladky and Tichy' 1959). On the other hand, humic substances can have a positive influence on the uptake and translocation of cations, such as Fe and Mg, involved in chlorophyll synthesis (Fernandez 1968; Lobartini and Orioli 1988).

c) Uptake of Humic Substances

Part of the effects described above can be rationally explained if one takes into account the possibility that humic fractions may enter into the root free space, be absorbed by cells, and then translocated. Proof of this has been found by various authors through the use of ^{14}C-labeled humic substances (Prat and Pospìsil 1959; Fuehr and Sauerbeck 1967a,b; Vaughan and Linehan 1976). These studies clearly show that, although most of the radioactivity associated with humic substances was irreversibly adsorbed to the cell walls, plant tissues were able to absorb into the symplast a measurable amount of radiolabeled humic substances which, to a small extent, could be translocated from the roots to the shoots. In a more recent report, Vaughan and Ord (1981) showed that pea roots were capable of absorbing humic acids and, to a greater extent, fulvic acids, by a mechanism which partly depends on the metabolism. After 18 h of incubation, 60% of ^{14}C-fulvic acids was found in the soluble cell fraction, as compared to 45% of ^{14}C-humic acids. This difference seems to depend on the low molecular weight of fulvic acids (< 2500 Da). In fact, when the plants were treated for 18 h with fractions of fulvic acids or humic acids with a molecular weight lower than 2500 Da, 70% of the radioactivity was found in the soluble cell fraction, therefore suggesting that humic-substance components of low molecular weight were responsible for the direct effects on metabolism and mineral nutrition. Carpita et al. (1979) showed that cell walls of plants generally have pores with a width ranging from 3.5 to 5.2 nm, which permit the passage of macromolecules. It has, in fact, been estimated that globular proteins or polysaccharides with a molecular weight up to 17000 and 6500 Da respectively are able to pass through these pores. Schnitzer and Kodama (1975), in an electronmicroscopic study, showed that in fulvic acid fractions at pH 2.5 spherical particles with a diameter 1.5–2.0 nm are present. Chen and Schnitzer (1976), by measurements of viscosity, pointed out that particles of fulvic acids with a molecular weight of about 2000 Da have a rod shape with a diameter of 0.6–0.7 nm at pH values ranging from 1 to 10. Humic acid fractions of molecular weight around 3000 Da have a similar structure when pH values ranged from 7.0 to 10.5. Various studies have pointed out that low molecular weight humic fractions have the greatest biological effects, as estimated by increase in enzymatic activities (Vaughan et al. 1974) and ionic uptake (Vaughan and McDonald 1976; Dell'Agnola and Nardi 1987), or hormone-like action (Nardi et al. 1988). Interestingly enough, as reported above, it has been shown that high molecular weight humic fractions treated with acidic solutions and/or organic acids released low molecular weight components (Albuzio and Ferrari 1989). Root exudates could have a similar de-polycondensating effect on humic substances, therefore releasing low molecular weight fractions capable of exercising biological effects on plants (Albuzio et al. 1993).

d) Membrane Properties

The fact that humic fractions can pass through the cell wall and can be absorbed into the symplast implicates a direct interaction with the plasma membrane, the major barrier between root cell cytoplasm and soil environment. Therefore this membrane could be one primary target of humic substance activity.

Various authors have suggested that humic substances can alter membrane permeability to solutes through a surfactant effect of these molecules. Humic substances can indeed be considered surface-active compounds, as they have diphilic properties and are therefore able to decrease surface tension of aqueous solutions (Visser 1964; Chen and Schnitzer 1978; Yonebayashi and Hattori 1987). Visser (1985) demonstrated that growth stimulation of certain microorganisms in the presence of humic substances was similar to the stimulation observed when commercial surfactant agents, such as Tween 20–80 (polyoxyethylene sorbitan monolaurate, stearate, oleate) and Brij 30 (polyoxyethylene laurylether) were used. This similarity in growth stimulation suggests that humic-acid action is at least partly due to their surface activity. Vakhmistrov et al. (1989) were able to prove that by adding humic substances or Tween 20 to the nutrient solution, also the growth of cucumber plants under sterile conditions was stimulated. The effect of the two surfactants was similar, although not identical, as the optimal value of the nutrient solution's surface tension was lower for Tween than for humic substances, and growth decrease at suboptimal concentrations was more gradual for the commercial detergent. Samson and Visser (1989) suggested that, additionally, humic acids might alter membrane fluidity. These authors observed an increase in K^+ efflux from potato cells in the presence of humic acids in the culture medium at a concentration corresponding to their critical micelle concentration; the effect was considerably reduced when cells were incubated in the presence of cholesterol. This compound can affect membrane permeability and stability (Yeagle 1987).

Electrophysiological studies have shown that in the presence of 1 mM KCl humic acids induce a hyperpolarization of transmembrane potential in wheat roots (Slesak et al. 1983; Slesak and Jurek 1988). Humic acids were also able to stimulate proton extrusion in wheat coleoptiles (Slesak 1992). In the presence of N,N′-dicyclohexylcarbodi-imide (DCCD) or diethylstilbestrol (DES), inhibitors of plasma membrane H^+-ATPase, an almost complete disappearance of the stimulatory effects caused by humic acids was observed. The absence of stimulation of proton extrusion under conditions of H^+-ATPase inhibition suggests a possible involvement of the plasma membrane proton pump in the action of humic substances. The possibility of an interaction between humic substances and plasma membrane H^+-ATPase has been recently supported using isolated plasma membrane vesicles. Varanini et al. (1993) showed that low molecular weight ($<$ 5000 Da) humic substances at concentrations (0.375 mg l^{-1}) much lower than those commonly used for studying biological effects stimulate the K^+-dependent phosphohydrolytic activity of plasma membrane ATPase up to four to five times. A stimulation of the initial rate of ATP-dependent intravesicular proton accumulation, although to a lower degree, was also recorded. An increase in passive permeability of membrane vesicles to protons was observed, therefore suggesting both a direct stimulation by humic substances of plasma H^+-ATPase, and an alteration of membrane permeability. In a similar way, low molecular weight humic substances were shown to modify tonoplast H^+-ATPase activity (Pinton et al. 1992), although in this case the stimulatory effect did not depend on a direct activation of the enzyme, but on a change of permeability and/or transmembrane electric potential. The interaction between humic substances and membrane transport mechanisms which are relevant for cell growth and metabolism could lead to at least a partial knowledge of the role played by humic substances on growth and plant nutrition.

6. Role of Humic Substances as Natural Chelates

Humic substances have been extensively considered as natural chelates for cationic micronutrients. A direct proof of this possibility and the mechanisms through which these complexes can be used by the plants, however, are still scarce. In addition, the variability of experimental approaches to this subject often does not permit a generalization of the results. Lobartini and Orioli (1988) showed that Fe-deficient sunflower plants could use Fe-humate added to the nutrient solution as a source of iron leading to the disappearance of chlorosis symptoms. The response of plants to the treatment was observable even if the quantity of iron absorbed was only one tenth of that obtained by using Fe-EDTA as a source of iron. The same authors showed that barley was able to accumulate ^{59}Fe from solutions with Fe-humate. The ^{59}Fe accumulation did not decrease at low temperatures and with the use of metabolic inhibitors. These results have been explained by assuming a considerable adsorption of Fe-humate in the free space of the roots.

Interesting results for the distribution of ^{59}Fe absorbed by the plants in presence of humic substances have been described by Guminski et al. (1983), Linehan and Shepard (1979) and by Aso and Takenaga (1975). Their experiments clearly show that the accumulation of iron in wheat, tomato and soybean roots decreased in the presence of humic substances in the nutrient solution, while in the shoots the concentration of ^{59}Fe remained unchanged or slightly increased. Although the mechanism of this phenomenon has not yet been understood, Aso and Takenaga (1975) suggested, on the basis of their uptake experiments using ^{59}Fe-humate and Fe-^{3}H-humate, that the translocation of iron from the root to the shoot did not directly involve the translocation of the Fe-humate complex, as the ratio between the two labeled species found in the shoots was decisively different, therefore suggesting that Fe and humic acids moved separately.

Humic substances can accept electrons from a wide spectrum of donors and transfer these to a wide spectrum of acceptors (Schindler et al. 1976; Zimmerman 1981). This aspect should be taken into account when iron uptake, which involves an obligatory reductive step in some plant species (Bienfait 1985), is considered. Furthermore, it should be kept in mind that plasma membrane of root cells is endowed with a number of oxidoreductases involved in growth and nutrition

(Crane et al. 1991), whose activity might be influenced by soil humic molecules. Surprisingly, to our knowledge, no report on this subject is available in the literature.

Mirave and Orioli (1989) observed that sunflower plants were able to absorb also ^{65}Zn from Zn-humate complexes. Higher levels of absorption were found with Zn complexed to humic acids with a molecular weight between 10 000 and 100 000 Da, as compared to those with a molecular weight of over 100 000 Da. This result supports further that micronutrient-humic substance complexes can play a role in the processes of micronutrient uptake.

7. Conclusion

Soil substances can influence plant mineral nutrition at different levels.

By using simplified experimental approaches effects on soil fertility, nutrient availability, and plant metabolism have been observed.

In order to clarify the different levels of interaction of humic molecules with soil components and plants, one of the problems which needs to be solved is to obtain definitions of structural characteristics of humic substances. In particular, as far as plant nutrition is concerned, in order to shed light on direct effects of humic substances on nutrient bioavailability at the rhizosphere and on metabolic processes of plant nutrition in the natural environment, it seems crucial to improve our knowledge on the characteristics and actual concentration of humic fractions present in soil solution and around the root systems.

In addition, it is necessary to identify the components able to reach and accumulate in the apoplast and those capable of interacting with the plasma membrane and possibly entering the cell. This should allow us to understand the nature and relevance of the effects observed on cell metabolism in conditions closer to the natural ones.

Some recent studies on the influence of humic fractions on cell membrane functions seem to support the idea that primary targets of the action of humic molecules exist at the level of root plasma membranes, which might also be related to hormone-like effects and reactions leading to changes in cell metabolism.

However, the contribution of these latter effects to the overall action of humic substances on plant mineral nutrition must be precisely defined in relation to the role of this class of substances in improving root habitability. The necessary specialization of the different research fields may lead scientists to overestimate the relevance for the general problem of some aspect of their own research, without giving due consideration to others. Thus, once again it is apparent that aspects involved in plant nutrition, especially when the soil-root system is concerned, must be approached cooperatively by scientists belonging to different areas. This is particularly lacking in the studies of the effects of soil humic substances on plant mineral nutrition.

Acknowledgement. The authors thank Professor Paolo Nannipieri and Dr. Elke Fischer-Schliebs for critical reading of the manuscript. Research supported by National Research Council of Italy, Special Project RAISA, Sub-project N. 2, Paper N. 1978.

References

Aiken GR, McKnight DM, Wershaw RL, MacCarthy P (1985) In: Aiken GR et al. (eds) Humic substances in soil sediment and water. Wiley New York, pp 1–9. **Albuzio A**, Ferrari G (1989) Plant Soil 113:237–241. **Albuzio A**, Ferrari G, Nardi S (1986) Can J Soil Sci 66:731–736. **Albuzio A**, Dell'Agnola G, Dibona D, Concheri G, Nardi S (1993) In: Paoletti MG, Foissner W, Coleman D (eds) Soil biota, nutrient cycling, and farming systems. Lewis, Boca Raton, pp 15–25. **Allison FE** (1973) Soil organic matter and its role in crop production. Elsevier, Amsterdam. **Anderson HA**, Russell JD (1976) Nature 260:597. **Aso S**, Takenaga H (1975) Trans Int Symp Humus et Planta VI, Prague, pp 1–17. **Asp H**, Berggren D (1990) Physiol Plant 80:307–314.

Baes CF, Mesmer RE (1976) The hydrolysis of cations. Wiley, New York. **Bedrock CN**, Cheshire MV, Chudek JA, Goodman BA, Shand CA (1994) Sci Total Environ (in press). **Bienfait HF** (1985) J Bioenerg Biomembr 17:73–83. **Bohn HL** (1976) J Soil Sci Soc Am Proc 40:468–470. **Boyd SA**, Sommers LE, Nelson DW (1979) Soil Sci Soc Am J 43:893–899. **Boyd SA**, Sommers LE, Nelson DW (1981) Soil Sci Soc Am J 45:1241–1242.

Cabrera D, Young SD, Rowell DL (1988) Plant Soil 105:195–204. **Cacco G**, Dell'Agnola G (1984) Can J Soil Sci 64:225–228. **Cameron RS**, Thornton BK, Swift RS, Posner AM (1972) J Soil Sci 23:394–408. **Carpita N**, Sabularse D, Montezinos D, Delmer DP (1979) Science 205:1144–1147. **Chen Y**, Aviad T (1990) In: MacCarthy P et al. (eds) Humic substances in soil and crop sciences: selected readings. American Society of Agronomy and Soil Science Society of America, Madison, pp 161–186. **Chen Y**, Schnitzer M (1976) Soil Sci Soc Am J 40:866–872. **Chen Y**, Schnitzer M (1978) Soil Sci 125:7–15. **Crane FL**, Morre' DJ, Loew HE, Boettger M (1991) In: Crane FL et al. (eds) Oxidoreduction at the plasma membrane: relation to growth and transport, vol II. CRC Press, Boca Raton, pp 21–33.

Dawson HJ, Hrutfiord BF, Zasoski RJ, Ugolini FC (1981) Soil Sci 132:191–198. **De Almeida RM**, Pospisil F, Vackova K, Kutacek M (1980) Biol Plant 22:167–175. **Dell'Agnola G**, Ferrari G (1971) Proc VIII Int Symp Agrochem, Venice, pp 298–304. **Dell'Agnola G**, Ferrari G (1975) Trans Int Symp Humus et Planta VI, Prague, pp 389–393. **Dell'Agnola G**, Nardi S (1987) Biol Fertil Soils 4:115–118. **Dormaar JF** (1975) Can J Soil Sci 55:111–118.

Ernst WHO, Kraak MHS, Stoots L (1987) J Plant Phys 127:171–175.

Fagbenro JA, Agboola AA (1993) J Plant Nutr 16:1465–1483. **Farmer VC**, Pisaniello DL (1985) Nature 313:474–475. **Fernandez VH** (1968) Pontif Acad Sci Scr Varia 32:805–850. **Flaig W** (1960) Sci Proc Royal Dublin Soc 4:49–62. **Flaig W**, Riemer H (1971) Trans Int Symp Humus et Planta V, Prague, pp 519–523. **Fuehr F**, Sauerbeck D (1967a) In: Report FAO/IAEA Meeting, Vienna, Pergamon Press, Oxford pp 73–82. **Fuehr F**, Sauerbeck D (1967b) In: Report FAO/IAEA Meeting, Vienna, Pergamon Press, Oxford pp 317–327.

Ghosh K, Schnitzer M (1980) Soil Sci 129:266–276. **Giovannini G**, Sequi P (1976) J Soil Sci 27:148–153. **Gjessing ET**, Riise G, Petersen RC, Andruchow E (1989) Sci Total Environ 81/82:683–690. **Guminski S**, Sulej J, Glabiszewski J (1983) Acta Soc Bot Pol 52:149–164.

Haworth RD (1971) Soil Sci 111:71–79. **Hayes MHB** (1991) In: Wilson WS (ed) Advances in soil organic matter research: the impact on agriculture and the environment. Royal Society of Chemistry, Cambridge, pp 3–22. **Hayes MHB**, MacCarthy P, Malcolm R, Swift RS (eds) (1989) Humic substances II. In search of structure. Wiley Interscience, Chichester. **Helal HM**, Sauerbeck DR (1984) Plant Soil 76:175–182. **Houghton C**, Rose FA (1976) Appl Environ Microbiol 31:969–976.

Irintoto B, Tan KH, Sommer HE (1993) J Plant Nutr 16:1109–1118.

Jenkinson DS, Rayner JH (1977) Soil Sci 123:298–305.

Khristeva LA, Galushko AM, Gorovaya AI, Kolbassin AA, Shortshai LP, Tkatshanko LK, Fot LV, Luk'yianenco NV (1980) Proc VI Int Peat Congr, Minnesota, pp 403–404. **Kuikman PJ**, van Veen AJ (1989) Biol Fertil Soil 8:13–18.

Lee YS, Bartlett RJ (1976) Soil Sci Soc Am J 40:876–879. **Levesque M** (1970) Can J Soil Sci 50:385–395. **Liebig J** (1856) J R Agric Soc 17:284–326. **Linehan DJ** (1976) Soil Biol Biochem 8:511–517. **Linehan DJ** (1977) J Soil Sci 28:369–378. **Linehan DJ** (1985) In: Vaughan D, Malcolm RE (eds) Soil organic matter and biological activity. Martinus Nijhoff/Dr W Junk, The Hague, Netherlands. pp 403–421. **Linehan DJ**, Shepherd H (1979) Plant Soil 52:281–289. **Linsday WL** (1991) In: Mortvedt JJ et al. (eds) Micronutrients in agriculture. Soil Science Society of America, Madison, pp 89–112. **Lobartini JC**, Orioli GA (1988) Plant Soil 106:153–157.

Maggioni A, Varanini Z, Nardi S, Pinton R (1987) Sci Total Environ 62:355–363. **Malcolm RE**, Vaughan D (1979a) Plant Soil 51:117–126. **Malcolm RE**, Vaughan D (1979b) Soil Biol Biochem 11:65–72. **Marschner H**, Roemheld V, Horst WJ, Martin P (1986) Z Pflanzenernaeh Bodenkd 149:441–456. **Martell AE** (1975) Pure Appl Chem 44:81–113. **Mato MC**, Fabregas R, Mendez J (1971) Soil Biol Biochem 3:285–288 **Mato MC**, Olmedo MG, Mendez J (1972a) Soil Biol Biochem 4:469–473. **Mato MC**, Gonzalez-Alonso LM, Mendez J (1972b) Soil Biol Biochem 4:475–478. **Mench M**, Morel JL, Guckert A, Guillet B (1988) J Soil Sci 39:521–527. **Mirave JP**, Orioli GA (1987) Plant Soil 104:214–219. **Mirave JP**, Orioli GA (1989) Sci Total Environ 81/82:679–682. **Muscolo A**, Felici M, Concheri G, Nardi S (1993) Biol Fertil Soils 15:127–131. **Mylonas VA**, McCants CV (1980a) J Plant Nutr 2:377–393. **Mylonas VA**, McCants CV (1980b) Plant Soil 54:485–490.

Nannipieri P, Sequi P (1982) In: Evolution du niveau de fertilité des sols dans différents systèmes de culture. Critères pour mesurer cette fertilité. Bari Italy Istituto sperimentale Agronomico, pp 141–151. **Nardi S**, Arnoldi G, Dell'Agnola G (1988) Can J Soil Sci 68:563–567. **Nardi S**, Concheri G, Dell'Agnola G, Scrimin P (1991) Soil Biol Biochem 23:833–836.

O'Donnell RV (1973) Soil Sci 116:106–112. **Oades JM** (1984) In: Tinslay J, Darbyshire JF (eds) Biological processes and soil fertility. Martinus Nijhoff/Dr W Junk, The Hague, pp 319–338.

Piccolo A, Mbagwu JSC (1990) Plant Soil 123:27–37. **Piccolo A**, Nardi S, Concheri G (1992) Soil Biol Biochem 24:373–380. **Pinton R**, Varanini Z, Vizzotto G, Maggioni

A (1992) Plant Soil 142:203–210. **Poole RJ** (1978) Plant Physiol 29:437–460. **Prat S**, Pospisil F (1959) Biol Plant 1:71–80.

Rauthan BS, Schnitzer M (1981) Plant Soil 63:491–495. **Roemheld V** (1991) In: Chen Y, Hadar Y (eds) Iron nutrition and interactions in plants. Kluwer, Dordrecht pp 159–166.

Samson G, Visser SA (1989) Soil Biol Biochem 21:343–347. **Schindler JE**, Williams DJ, Zimmerman AP (1976) Environ Biochem 1:109–115. **Schnitzer M** (1977) In: Proc Symp Soil Org Matter Studies, Braunschweig, Int Atomic Energy Agency, Vienna, pp 117–131. **Schnitzer M** (1978) In: Schnitzer M, Khan SV (eds) Soil organic matter. Elsevier, Amsterdam, pp 1–64. **Schnitzer M** (1985) In: Aiken GR et al. (eds) Humic substances in soil sediment and water. Wiley, New York, pp 303–325. **Schnitzer M** (1990) In: MacCarthy P et al. (eds) Humic substances in soil and crop sciences: selected readings. American Society of Agronomy and Soil Science Society of America, Madison, pp 65–90. **Schnitzer M**, Hansen EH (1970) Soil Sci 109:333–340. **Schnitzer M**, Khan SU (1972) Humic substances in the environment. Marcel Dekker, New York. **Schnitzer M**, Kodama H (1975) Geoderma 13:279–287. **Schnitzer M**, Poapst PA (1967) Nature 213:598–599. **Sequi P**, Guidi G, Petruzzelli G (1975) Geoderma 13:153–161. **Sladky Z** (1959) Biol Plant 1:142–150. **Sladky Z**, Tichy' V (1959) Biol Plant 1:9–15. **Slesak E** (1992) Electrophysiology of plant mineral nutrition, Wroclaw University, Breslau. **Slesak E**, Jurek J (1988) Acta Univ Wratislaviensis 888:13–19. **Slesak E**, Tretyn A, Jurek J (1983) Trans Int Symp Humus et Planta VIII Prague, pp 260–262. **Stevenson FJ** (1982) Humus chemistry: genesis, composition, reactions. Wiley, New York. **Stevenson FJ** (1991) In: Mortvedt JJ et al. (eds) Micronutrients in agriculture. Soil Science Society of America, Madison, pp 145–186. **Stevenson FJ**, Ardakani MS (1972) In: Mortvedt JJ et al. (eds) Micronutrients in agriculture. Soil Science Society of America, Madison, pp 79–114. **Stevenson FJ**, He X–T (1990) In: MacCarthy et al. (eds) Humic substances in soil and crop sciences: selected readings. American Society of Agronomy and Soil Science Society of America, Madison, pp 91–109. **Swift RS**, Posner AM (1971) J Soil Sci 22:237–249.

Tan KH (1978) Goederma 21:67–74. **Tan KH**, Nopamornbodi V (1979) Plant Soil 51:283–287. **Tarafdar JC**, Jungk A (1987) Biol Fertil Soil 3:199–204. **Theng BKG** (1979) Formation and properties of clay-polymer complexes. Hilger, London. **Tinslay J**, Salam A (1961) Soils Fert 21:81–90.

Vakhmistrov DB, Mishustina NE, Zverkova OA, Debets EYu (1989) Fiziol Rast 36:980–989. **Van der Werf MM**, Out T (1981) Biochem Physiol Pflanz 176:274–282. **Van Dijk H** (1971) Geoderma 5:53–67. **Varanini Z**, Pinton R, De Biasi MG, Astolfi S, Maggioni A (1993) Plant Soil 153:61–69. **Vaughan D** (1967) Soil Biol Biochem 1:15–28. **Vaughan D** (1974) Soil Biol Biochem 6:241–247. **Vaughan D**, McDonald IR (1971) J Exp Bot 22,400–410. **Vaughan D**, McDonald IR (1976) Soil Biol Biochem 8:415–421. **Vaughan D**, Linehan DJ (1976) Plant Soil 44:445–449. **Vaughan D**, Malcolm RE (1979) Soil Biol Biochem 11:57–63. **Vaughan D**, Malcolm RE (1985) In: Vaughan D, Malcolm RE (eds) Soil organic matter and biological activity. Martinus Nijhoff/Dr W Junk, The Hague, Netherlands, pp 37–75. **Vaughan D**, Ord BG (1981) J Exp Bot 32:679–687. **Vaughan D**, Cheshire MV, Mundie CM (1974) Biochem Soc Trans 2:126–129. **Vaughan D**, Ord BG, Malcolm RE (1978) J Exp Bot 29:1337–1344. **Visser SA** (1964) Nature 204:581. **Visser SA** (1985) Soil Biol Biochem 17:457–462. **Visser SA** (1987) Sci Total Environ 62:347–354.

Wilson MA (1990) In: MacCarthy P et al. (eds) Humic substances in soil and crop sciences: selected readings. American Society of Agronomy and Soil Science Society of America, Madison, pp 221–260.

Yeagle P (1987) The membranes of cells. Academic Press, Orlando. **Yonebayashi K**, Hattori T (1987) Sci Total Environ 62:55–64.

Zimmerman AP (1981) Hydrobiologia 78:259–265.

Prof. Zeno Varanini
Dr. Roberto Pinton
Dipartimento di Produzione Vegetale
e Tecnologie Agrarie
Via Delle Scienze 208
I-33100 Udine, Italy

III. Selected Topics of Polyketide Biosynthesis

By Horst-Robert Schütte

1. Introduction

It is well known that many species of bacteria, fungi, and higher plants produce a variety of secondary metabolites from acetyl-CoA and malonyl-CoA; these compounds are considered to be formed via a polyoxomethylene chain, a hypothetical intermediate (Mosbach 1969; Schütte 1971, 1982; Turner 1972; Bu'Lock 1979; Herbert 1981; Leistner 1981; Turner and Aldridge 1983; Simpson 1984, 1985, 1987a, b, 1991; Gill and Steglich 1987; Luckner 1990; O'Hagan 1992). In many cases, they exhibit antibiotic, antitumor, or mycotoxic activities.

As first suggested by Birch and Donovan (1953) and Birch (1967), the individual polyketide biosynthetic pathways, in spite of a remarkable variety of end products, apparently follow a common basic reaction scheme which resembles in many of its constituent steps the fatty acid biosynthetic reaction sequence. In all cases, an enzyme-bound primer (e.g., acetate) is elongated, by several successive rounds of condensation with malonate, to an enzyme bound polyketide acid which is subsequently modified, depending on the specific synthase, by one or several of the following reactions: ketoreduction, dehydration, enoyl reduction, cyclization, or aromatization. The various different polyketide synthases therefore appear to use basically the same set of component enzymes, but they differ in the frequency and sequential order in which the individual component enzymes act upon the reaction intermediates. Another structural difference refers to the fact that polyketide synthases occur either as large, multifunctional enzymes (type I PKS) or as perhaps aggregated systems of individual component enzymes (type II PKS).

The specific molecular structure of each synthase, i.e., the topological order of its constituent catalytic domains, together with certain synthase-specific functions, is likely to control the type of end product formed in each case. The elucidation of these structure/function interrelationships in multienzyme systems with different end product specificities but similar modular composition, represents an extraordinary challenge to the enzymologist. It may provide insight into the structural requirements of an enzyme complex when channeling a sequence of catalytic reactions into a specific direction.

Progress in Botany, Vol. 56

In secondary metabolism, our understanding of polyketide biosynthesis is most advanced in the actinomycete bacteria, where the structural genes of many polyketide-producing strains have been cloned, and information is forthcoming on the structure and organization of the polyketide synthases (e.g., Bibb et al. 1989; Sherman et al. 1989; Bartel et al. 1990; Otten et al. 1990).

2. 6-Methylsalicylic Acid and Orsellinic Acid

6-Methylsalicylic acid synthase (Dimroth et al. 1970) from *Pencillium patulum* is the best-defined polyketide synthase. It is a multifunctional Type I PKS which has been purified to homogeneity and has a subunit molecular mass of 188 kDa. The complete nucleotid sequence of the gene and the amino acid sequence of the enzyme have been determined (Beck et al. 1990). The gene consists of an open reading frame 5322 base pairs long, and encodes for 1774 amino acids. Most polyketide synthases share at least four common activities, i.e., acyl and malonyl transferases, an acyl carrier protein, and the condensing enzyme 2β-ketoacyl synthase. Based on sequence similarities with other polyketide synthases, these domains were identified within the 6-methylsalicylic acid gene. It also possesses a β-ketoreductase, and this domain has similarly been located. Interestingly, the level of sequence homology through these domains was greatest with genes encoding for mammalian Type I fatty acid synthase, and homologies are limited when comparisons are made to fungal Type II fatty acid synthases, even from the same organism, and also to plant chalcone and resveratrol synthases. (Schröder et al. 1988; Wiesner et al. 1988).

Studies probing the absolute stereospecificity of hydrogen atom removal during the biosynthesis of 6-methylsalicylic acid (IV) have been reported (Abel and Staunton 1981, 1984; Abel et al. 1982; Floss et al. 1984; Jordan and Spencer 1990a, b). These studies have exploited the use of chiral malonate (Huang et al. 1986; Jordan et al. 1986). Figure 1 illustrates the stages in 6-methylsalicylic acid (IV) biosynthesis. The first condensation between acetyl-CoA and malonyl-CoA proceeds with inversion of configuration and generates an acetoacetyl-thioester (I). The prochiral hydrogen atoms are labeled from malonate. A similar malonyl-CoA condensation extends the polyoxomethylene chain by another acyl group to give (II). After this condensation, an NADPH-dependent β-ketoreductase activity operates to generate β-hydroxy alcohol (III), and then elimination of water affords the required *cis* double bond. The final malonyl-CoA condensation generates (V) and then an intramolecular condensation and enolization provides the aromatic nucleus of 6-methylsalicylic acid (IV). The studies were aimed at probing the absolute stereochemistry of hydrogen removal from (III) during the elimination to the *cis* double bond, and of hydrogen removal C-3 of (V) during the enolization to generate the aromatic ring.

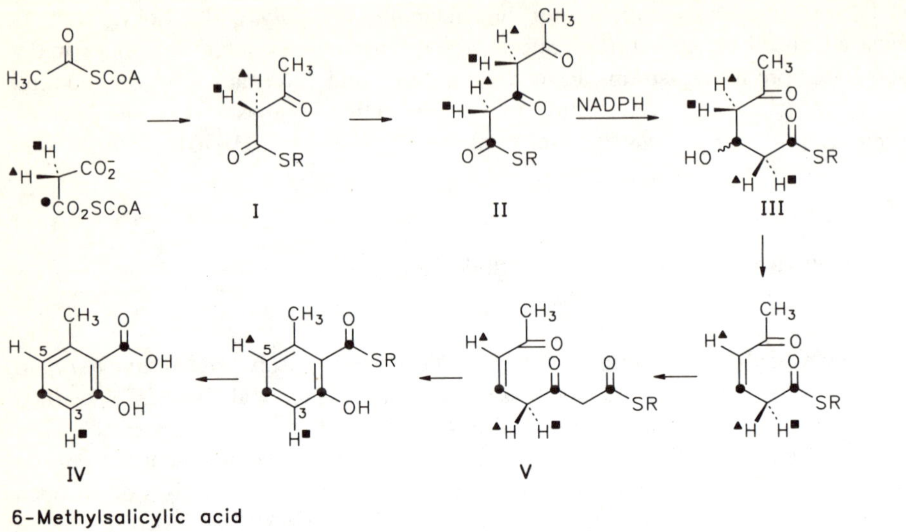

Fig. 1. Biosynthesis of 6-methylsalicylic acid

It was demonstrated that opposite prochiral hydrogens were eliminated, possibly consistent, with a single base sitting above the folded/cyclized polyketide chain as shown in Fig. 1. The chiral malonates were incubated in the enzyme system, this time with acetoacetyl-CoA (I), that only one more malonate contributing hydrogen atoms to the aromatic nucleus is required. Results revealed that it is the H_{Re} proton that is retained at C-3 of 6-methysalicylic acid (IV) and the H_{Si} proton that is lost. It is therefore deduced that the hydrogen at C-5 must derive from the opposite prochiral hydrogen, H_{Si}.

A similar analysis was carried out for the biosynthesis of orsellinic acid (VI) using a synthase from *Penicillium cyclopium* (Woo et al. 1989). Orsellinic acid synthase does not possess a reductase enzyme, and requires only the condensation of three malonyl-CoA on to an acetyl-CoA starter unit, followed by intramolecular condensation and two enolizations, to furnish the aromatic nucleus (Fig. 2). In direct contrast to the study on 6-methylsalicylic acid, the results suggest that hydrogen removal during the enolizations is non stereospecific in the case of orsellinic acid (VI).

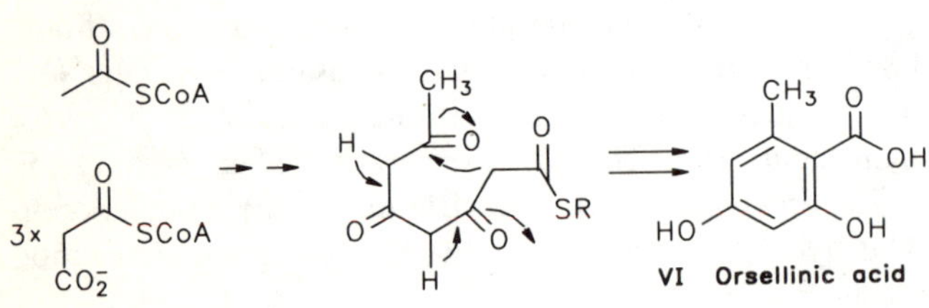

Fig. 2. Biosynthesis of orsellinic acid

3. Melleine, Aspyrone, and Related Compounds

Carrot cells accumulate a phytoalexin, 6-methoxymellein (VIII), when they are challenged by pathogenic microorganisms as one of the active defense responses (Condon and Kuc 1960; Kurosaki and Nishi 1983, 1984; Kurosaki et al. 1985a; Marinelli et al. 1990). Induction of synthetic activity of this compound was also observed on treatment of the cells with several kinds of chemical substances (elicitors) including metal ions (Kurosaki et al. 1987b), oligogalacturonides (Kurosaki et al. 1984, 1985b, 1987a) and 2-chloroethylphosphonic acid (2-CEPA), and ethylene-generating reagent (Coxon et al. 1973; Jaworski and Kuc 1974). These elicitors showed specific inducing activity of phytoalexin production and did not affect fatty acid synthesis in carrot tissues, which may share some common properties with 6-hydroxymellein biosynthesis.

From feeding experiments, 6-methoxymellein has been considered to be synthesized via a polyoxomethylene intermediate (Fig. 3) (Condon et al. 1963; Stoessl and Stothers 1978). However, enzymes involved in the biosynthesis of polyketide compounds are generally very unstable, and there have been few reports which demonstrate synthetic activity of this class of compounds in cell-free extracts. An enzyme activity for synthesis of 6-hydroxymellein (VII), a direct precursor of 6-methoxymellein (VIII), could be detected in all cell-free extracts prepared from elicitor-treated carrot root tissues (Kurosaki et al. 1989a). An O-methyltransferase activity could be found which catalyzes the final step of 6-methoxymellein (VIII) biosynthesis (Kurosaki and Nishi 1988). The activity of the O-methyltransferase was not found in fresh carrot roots, but increased when they were tested with either type of elicitor. Irrespective of the elicitor such as a metal ion, ethylene, or oligogalacturonides, 6-methoxymellein (VIII) production took place at almost the same rate as the O-methyl group transfer reaction to a hydroxyl group of 6-hydroxymellein (VII). Synthetic activity of 6-hydroxymellein was ca. two to fourfold that of the O-methyltransferase. These results suggest

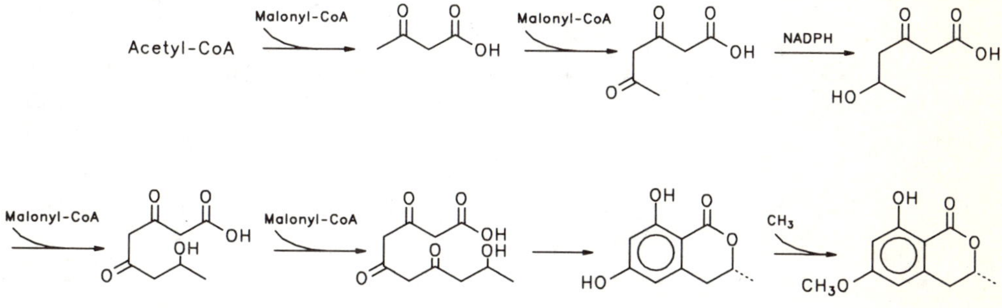

VII 6-Methylmellein VIII 6-Methoxymellein

Fig. 3. Hypothetical scheme of 6-methoxymellein biosynthesis

that O-methylation of the didydroiso-coumarin precursor is a rate-limiting step of 6-methoxymellein (VIII) biosynthesis (Kurosaki and Nishi 1991).

Furthermore, it has been shown that 6-hydroxymellein (VII) production was an NADPH-dependent process and, in the absence of the reagent, triacetic acid lactone was produced as a derailment product of the reaction process (Kurosaki et al. 1989b). This finding suggested that the reduction of the double bond at the 3,4-position of the phytoalexin takes place during the elongation of the poly-oxomethylene chain. This NADPH-dependent reduction seems to occur at the triacetate stage before the condensation of the third malonyl-CoA as the conversion of carbonyl to hydroxyl group.

Mellein and related compounds have been isolated also from a fungus source. *Aspergillus melleus* is a rich source of polyketide metabolites (Fig. 4), including mellein (IX) (Blair and Newbold 1955; Arakawa 1968; Aldridge et al. 1971; Devys et al. 1980, Parisi et al. 1993), hydroxymellein (X) (Cole et al. 1971), aspyrone (XI) (Holker and Simpson 1981; Rosenbrook and Carney 1970), and penicillic acid (XII) which is known to be produced by cleavage of an aromatic system (Birch et al. 1958; Mosbach 1960; Bentley and Keil 1962). Although substantial amounts of aspyrone (ca. 350 mg/l) are produced during the first 10–12 days of incubation, a new compound, asperlactone (XIII), appears to be produced at the expense of aspyrone (XI) from day 10 onwards (Garson et al. 1984). Aspyrone (XI) is a weak broad spectrum antibiotic, and was first isolated from fermentations of *Aspergillus melleus* (Mills and Turner 1967). Later, it was isolated from other *Aspergillus* species (Natori et al. 1970; Moore et al. 1974).

With its unusual branched carbon skeleton aspyrone (XI), the major product of short-term fermentation in *Aspergillus melleus* is not obviously biosynthetically related to other simpler pyrone metabolites such as asperline (XV) (Argoudelis and Zieserl 1966) (Fig. 4). The co-metabolites, mellein (IX) and hydroxymellein (X) can, however, originate from standard cyclization of a linear polyketone

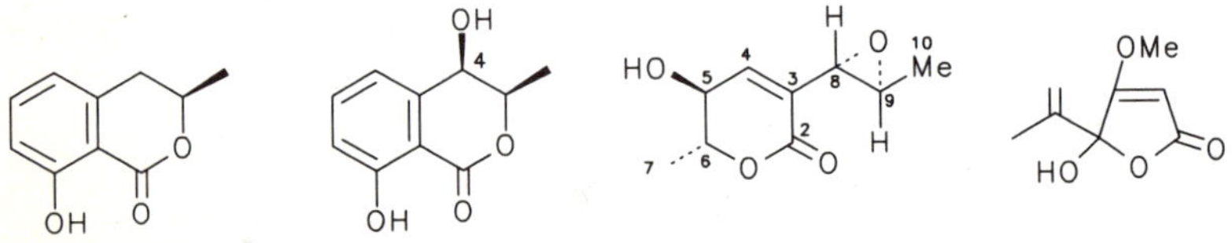

IX Mellein X 4-Hydroxymellein XI Aspyrone XII Penicillic acid

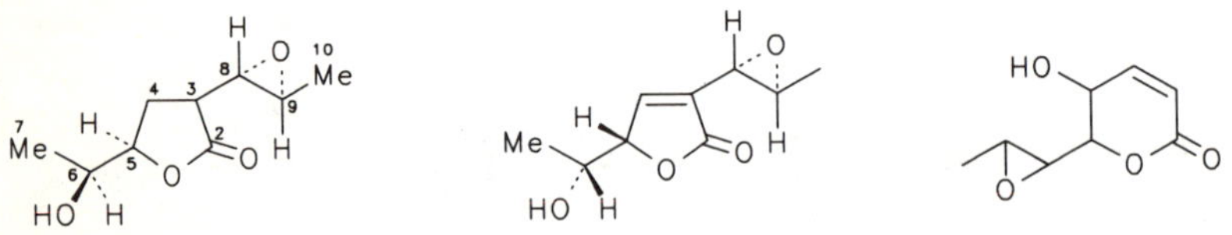

XIII Asperlactone XIV Isoasperlactone XV Asperline

Fig. 4. Metabolites of *Aspergillus melleus*

(Abell et al. 1983). Biosynthetic studies with labeled acetate have established that aspyrone (XI) is an extensively modified polyketide (Simpson 1975; Simpson and Holker 1975; Tanabe et al. 1976; Holker and Simpson 1981), and C-10 of aspyrone has been identified as chain starter unit (Copeland et al. 1984). Five carbons are derived from the methyl group of acetate, and the remaining four from the carboxy group.

The pattern of incorporation of tritium from [2-^{3}H]acetate is inconsistent with a biosynthesis from aromatic precursors of the mellein (IX) type. The evidence suggests that the carbon skeleton is produced like that of the co-metabolite asperlactone (XIII), by decarboxylation and rearrangement of a linear pentaketide intermediate. The pattern of incorporation of [1, 2-$^{13}C_2$]acetate, and especially, for C-2 and C-8 indicated that these two carbons derive from the same acetate unit. It was therefore suggested that the biosynthesis of aspyrone (XI) proceeds via a Favorski-type rearrangement of a linear polyoxomethylene chain such as XVI to the aspyrone skeleton (XVII) (Fig. 5). In the first phase of the pathway, it is proposed that a linear C_{20} precursor is build up by repeated head-to-tail condensation of acetate units on a polyketide synthase system of enzymes. In the second phase, this linear skeleton is rearranged, with loss of one carboxy-derived carbon, to give the aspyrone (XI) skeleton.

It was presented evidence that the C_9-diepoxide (XXIII) is a late intermediate in the second phase of the pathway (Staunton and Sutkowski 1991a; Garson et al. 1984). Based on the widely accepted hypothesis that the polyketide synthases will resemble a fatty acid synthase, it is reasonable to suppose that the structure of the intermediate it assembles will consist of a linear C_{10} fatty acid chain with an array of functional groups appropriate for subsequent modification of the structure to (XXIII), using standard biosynthetic reactions. All preceding intermediates would be bound to the polyketide synthases by covalent links. A proposal for this key first enzyme-free intermediate is the hydroxy acid (XVIII) (Fig. 6). A feature of Fig. 6 is the removal of C-1 of the first free intermediate by decarboxylation using a distant carbonyl as an electron sink. In Fig. 6 it is an aldehyde group (XX), formed by rearrangement of an epoxide group (XIX) in the later stages of the biosynthetic pathway. An important feature is that it allow the retention of both hydrogens at C-2 of the C_{10} intermediate in the course of the removal of C-1. Incorporation studies with deuterium-labeled compounds have shown that a C_{10} acid, (4E,6E,8E)-3-hydroxydeca-4,6,8-trienoic acid (XVIII), is incorporated intact into aspyrone

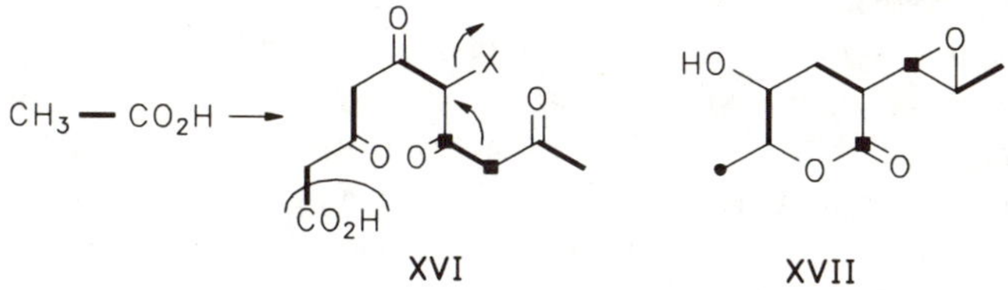

Fig. 5. Biosynthesis of the aspyrone skeleton

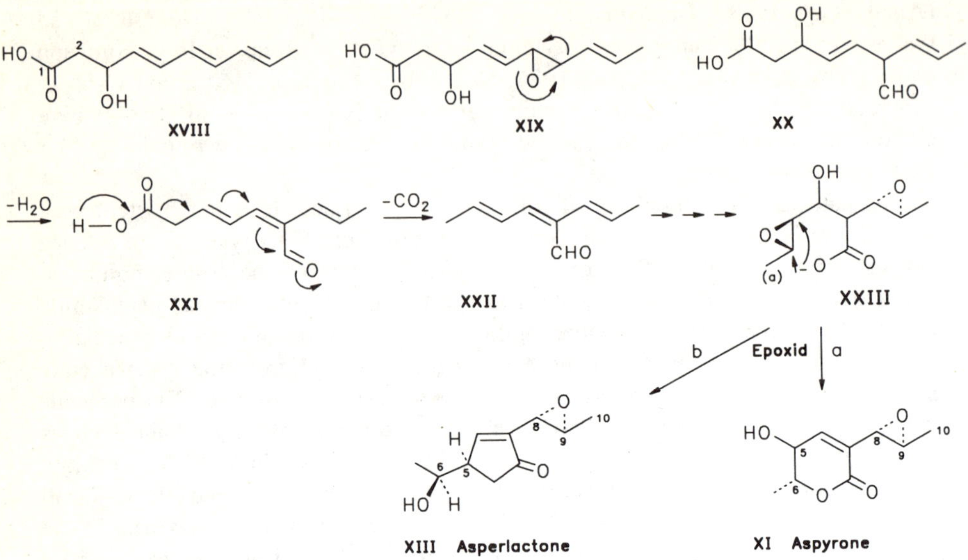

Fig. 6. Biosynthesis of asperlactone and aspyrone

(XI) when administered to growing cells of *Aspergillus melleus* (Staunton and Sutkowski 1991b).

Furthermore, the mode of operation of the aspyrone polyketide synthase is interesting. Despite great efforts, very little is known about the detailed mechanism of any such enzyme system. Current speculations are guided by the widely held view that polyketide synthase systems probably resemble fatty acid synthases. In fatty acid biosynthesis, the chain extension is carried out by means of repeated operation of the standard set of reactions. Initially, the starter acid (usually acetyl coenzyme A) is bound as a thioester to a thiol at the active site of a ketone synthase which catalyzes step 1. Both free acids and N-acylcysteamine thioesters were investigated. The utility of these thioester analogues for the study of polyketide synthase systems was demonstrated in the pioneering work on the synthase responsible for 6-methylsalicylic acid (IV) biosynthesis (Dimroth et al. 1970). An N-acylcysteamine residue is capable of mimicking the thiol terminus of the pantotheine unit which is present in CoA and the ACP components of the polyketide synthase systems. In suitable cases, an acyl group attached to the thiol group of the N-acylcysteamine can therefore be infiltrated onto the ACP thiol group of the polyketide synthase under investigation, and is then ready to be carried through subsequent chain extension steps. Recently, there have been a number of reports that these thioster derivatives are capable of passing intact across cell walls in a number of microorganisms, thus opening the way to incorporation experiments with intact organisms (Cane and Yang 1987; Yue et al. 1987; Spavold and Robinson 1988; Yoshizawa et al. 1990). The intact incorporation of labeled N-acylcysteamine thioesters of crotonic, (2E,4E,)-hexadienoic, and (2E,4E,6E)-octatrienoic acids into aspyrone (XI) support the hypothesis that the aspyrone polyketide synthase resembles a fatty acid synthase in its mode of operation, and that it follows a processive mode of operation (Staunton and Sutkowski 1991c).

The C_3 residue of crotonate, C-2, C-3 and C-4, is incorporated without rearrangement into aspyrone (XI) to provide carbons, C-8, C-9, and C-10. Investigating the biosynthetic steps leading from acetate, the chain starter acid, to crotonate, incorporation studies, using deuterium-labeled precursors has shown that the enzyme-bound intermediates generated in the first two steps of the chain extension cycle catalyzed by the aspyrone polyketide synthase are thioesters of acetoacetic acid and (R)-3-hydroxybutyric acid (Jacobs et al. 1991), i.e., that the aspyrone polyketide synthase used the same isomer (3-R) of 3-hydroxybutyrate as the fatty acid synthase systems investigated to date. It may be that the aspyrone polyketide synthase is an incomplete fatty acid synthase which lacks the active site, enoyl reductase, employed in the final step of the normal fatty acid synthase chain extension cycle.

The common occurrence of asperlactone (XIII) and isoasperlactone (XIV) with aspyrone (XI) in *A. melleus* suggested that aspyrone and the γ-lactone metabolites were biosynthetically related. The biosynthesis of asperlactone (XIII) has been investigated using [1,2-$^{13}C_2$]-,[2-^{13}C,2-$^2H_3$]- and [$^2H_3$]-acetate (Brereton et al. 1980,1984; Garson and Staunton 1981). The pattern of incorporation of intact acetate units shows that also in this case C-2 and C-8, originally joined in a C_2 unit derived from acetate, become separated by a rearrangement, and so finish in a 1,3 relationship. The carbon skeleton is therefore built up in identical fashion to that of the co-metabolite aspyrone (XI). The retention of two acetate-derived hydrogens from [2-^{13}C,2-$^2H_3$]acetate at C-7 rules out the intermediacy of structures in which this carbon forms part of an aromatic ring. This experiment also confirms that C-10 can retain three deuterium atoms and so is a chain starter methyl group. To account for these results, a biosynthetic model involving partially reduced polyoxomethylene intermediates is suggested (Fig. 6). Mechanistically feasible decarboxylation and rearrangement steps are followed by stereospecific opening of an epoxide ring to generate aspyrone (XI) (way a) or asperlactone (XIII) (way b). The possible derivation of the aromatic co-metabolites, mellein (IX) and hydroxymellein (X), from the same polyoxomethylene intermediate, generated on the same polyketide synthase, is possible.

In addition, some results with oxygen-17 are reported. In comparison with ^{18}O, ^{17}O has been relatively little used as a biosynthetic tracer despite its advantage of being observable directly by ^{17}O NMR spectroscopy, rather than being limited to indirect observation through an isotopic effect on a neighboring NMR active nucleus (^{13}C). The principal reason for this neglect is not expense but the difficulty of acquiring good quality ^{17}O NMR spectra. On the positive side, the low natural abundance of ^{17}O (0.04%), compared with ^{18}O (0.20%) and ^{13}C (1.1%), can be an important advantage in biosynthetic studies.

Evidence that hydroxymellein (X) arise by direct hydroxylation of the benzylic methylene group of mellein (IX) has been obtained by incorporation of ^{17}O-

labeled acetate and oxygen, and from direct conversion of ^{2}H-labeled mellein to hydroxymellein (Abell et al. 1987).

For the three isomeric metabolites aspyrone (XI), asperlactone (XIII), and isoasperlactone (XIV), it has been suggested that they derive from a common biosynthetic precursor, the diepoxide (XXIII) (Fig. 6). The absolute configurations of the chiral centers in aspyrone and asperlactone are satisfactorily explained by this scheme. Studies of the origin of the oxygen atoms have established that in the major metabolite, aspyrone (XI), all four oxygens are labeled by molecular oxygen, but not uniformly: the level of enrichment in the two oxygens of the lactone group is half that in the remaining two sites. Therefore, in individual aspyrone molecules, three of the four oxygens derive from molecular oxygen: the hydroxy and epoxide oxygens, together with one or the other of the oxygen atoms of the lactone group. No oxygen isotope was incorporated from the carboxy group of acetate, so it can be assumed that the fourth oxygen comes from water. The labeling pattern is consistent with the proposed derivation of (XI) from (XXIII), assuming that the two epoxide residues of (XXIII) are derived from molecular oxygen in the standard way, that one of the two carboxylate oxygens is introduced at some stage by an oxidation using molecular oxygen and that all three isomeric metabolites, aspyrone (XI) asperlactone (XIII) and isoasperlactone (XIV), arise from a common diepoxide intermediate (Ahmed et al. 1985; Staunton and Sutkowski 1991a).

4. Nectriapyrone and Related Pyrones

Nectriapyrone (XXIV) has been isolated from *Gliocladium vermoesenii* (Avent et al. 1992b). This α-pyrone is one of an increasing number of C-6 substituted pyrones to be isolated from fungi. The structure of nectriapyrone may be dissected into two isoprene units, and when it was first isolated (Nair and Carey 1975), it was assumed to be a monoterpenoid. However, a similar metabolite, fusalanipyrone (XXVI) obtained from *Fusarium solani* (Abraham and Arfmann 1988), was shown to incorporate two trideuteriomethyl units from [Me-$^2H_3$]methionine, consistent with its biosynthesis from a tetraketide with two methylations (Abraham et al. 1990).

A minor metabolite, vermopyrone (XXV), co-occurs with nectriapyrone (XXIV) in *Gliocladium vermoesenii*. This compound possesses an acetyl side-chain in place of the butenyl side-chain of nectriapyrone. The results of feeding experiments with acetate and methionine lead to labeling patterns for nectriapyrone and vermopyrone which are indicative of a tetraketide biosynthesis with two C-alkylations and an O-methylation (Avent et al. 1992c) (Fig. 7). Vermopyrone (XXV) may arise through the cleavage of nectriapyrone (XXIV). It is interesting to note that the acetyl side-chain arises from the methylation of an acetate unit

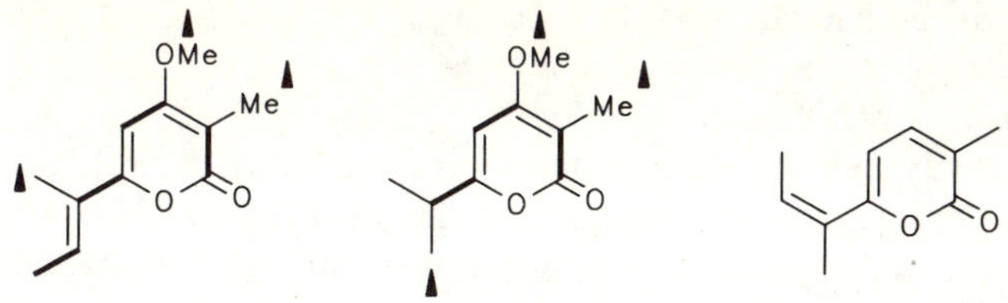

XXIV Nectriapyrone XXV Vermopyrone XXVI Fusalanipyrone

Fig. 7. Nectriapyrone and related pyrones

and a subsequent oxidation, i.e., not all CH_3CO units in polyketides are, a priori, derived biosynthetically from an acetate unit.

5. Harzianolide

Harzianolide (XXVII) is an antifungal metabolite of the biological control agent, *Trichoderma harzianum*, (Almassi et al. 1991; Claydon et al. 1991). It possesses a butenolide ring system and it joins a small group of other microbial γ-butyrolactones such as seiridin (XXVIII) (Evidente et al. 1986), Butalactin (XXIX) (Franco et al. 1991) and asperlactone (XIII) (Fig. 8). Biosynthetically, the compound is interesting in that, although apparently a polyketide, it possesses an odd number of carbon atoms with both chains terminating in methyl groups whilst in the butenolide ring, it possesses two contiguous branches.

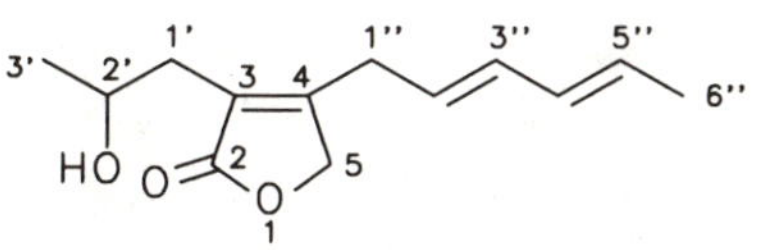

XXVII Harzianolide

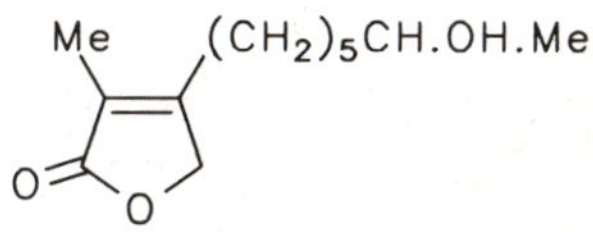

XXVIII Seiridin

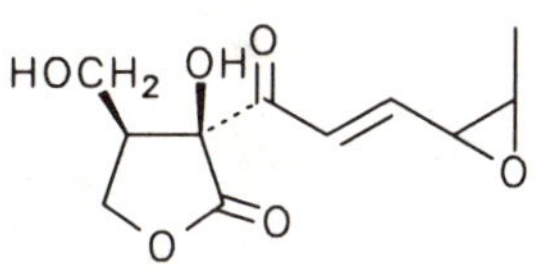

XXIX Butalactin

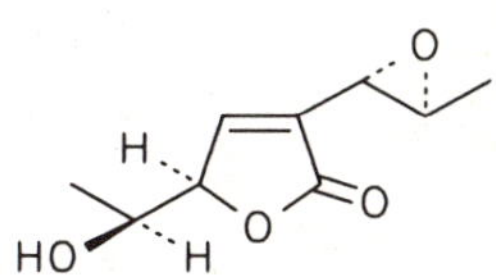

XXX Asperlactone

Fig. 8. Harzianolide and related compounds

The labeling pattern of the harzianolide after feeding [1-^{13}C]- and [2-^{13}C] acetate showed that it is derived from seven acetate units with the loss of one carboxyl carbon (Avent et al. 1992a). However, the [1,2-$^{13}C_2$]acetate incorporation experiment revealed that only four acetate units have remained intact. As only one methyl group has completely lost its carboxy partner, the other two acetate units must be involved in rearrangements. The interpretation of this labeling pattern follows the precedent of the epoxide rearrangement in aspyrone (XI) biosynthesis. In the case of harzianolide (XXVII), a diepoxide of a C-14-unsaturated acid (XXXI) undergoes rearrangement with the extrusion of two aldehyde groups (XXXII) (Fig. 9). The double bonds move into conjugation with an aldehyde to generate a vinylogous β-keto-acid (XXXIII) which undergoes decarboxylation to (XXXIV). An internal Cannizaro reaction may generate the lactone (XXXV) and a further double bond migration then affords harzianolide (XXVII).

A further antifungal metabolite of this fungus is harzianopyridone (XXXVI), which is formed in small amounts (ca. 1–5 mg/l culture broth) when the fungus is grown on a 2% malt extract medium in surface culture for 30 days. The polyketide origin of the side-chain was established by [1-^{13}C]- and [1,2-$^{13}C_2$]acetate feeding experiments (Dickinson et al. 1989). These afforded harzianopyridone (XXXVI) with the enrichment and coupling pattern shown in Fig. 10. Hence the molecule contains a tetraketide unit. The origin of the extra carbon atom in the side-chain and of the methoxy groups was established by [Me-^{13}C]methionine feeding experiments.

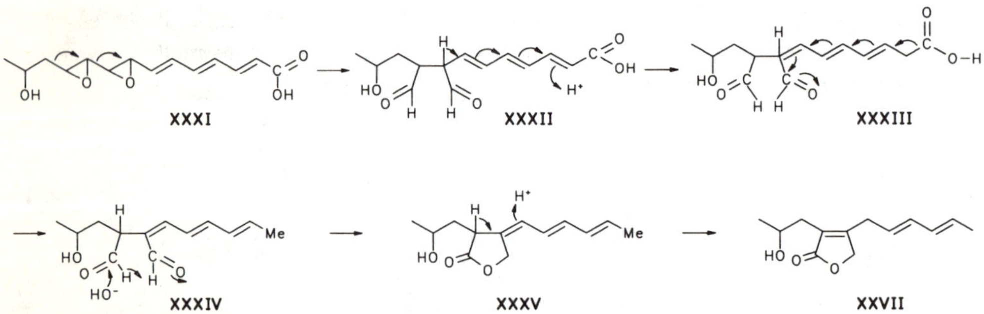

Fig. 9. Hypothetical pathway to harzianolide

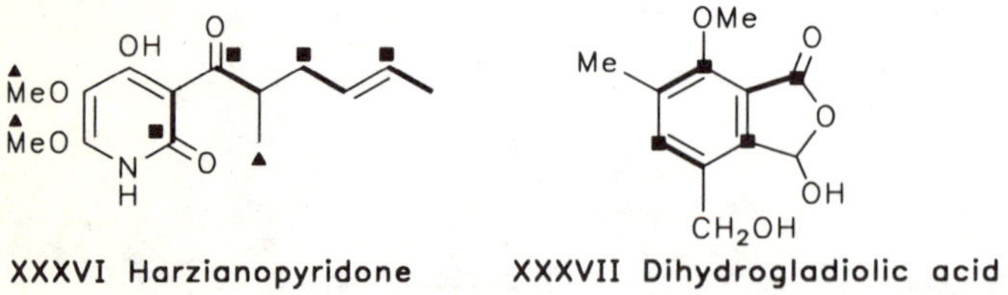

Fig. 10. Biosynthesis of harzianopyridone and dihydro-gladiolic acid

6. Dihydrogladiolic Acid

The fungus, *Penicillium gladioli*, produces antifungal metabolites, gladiolic acid and dihydrogladiolic acid (XXXVII) (Duncanson et al. 1953). The results of feeding experiments with ^{13}C-acetate and methionine show that dihydrogladiolic acid is a tetraketide with an acetate labeling pattern in accord with the accepted biosynthetic route (Fig. 10) (Avent et al. 1990). The additional carbon atoms arise from methionine.

7. Spicifernin

A strain of *Cochliobolus spicifer* Nelson (D-5), a pathogen of leaf spot disease in wheat, produces several phytotoxins, i.e., spiciferones A (XXXIX), B (XL), and C (XLI), and spiciferinone (XLII), and a plant growth promoter, i.e., spicifernin (XXXVIII) (Fig. 11) (Nakajima et al. 1989, 1990, 1991, 1992a). In spite of their different carbon skeletons, they have the following unique structural features in common: a quaternary carbon bearing an ethyl, a methyl and a ketonic carbonyl, and vicinal methyls. This suggests strongly that these unique metabolites have

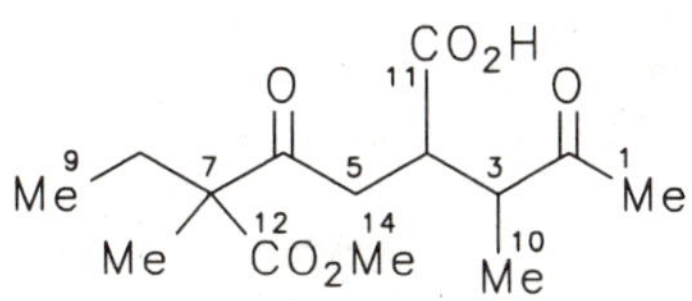

XXXVIII Spicifernin

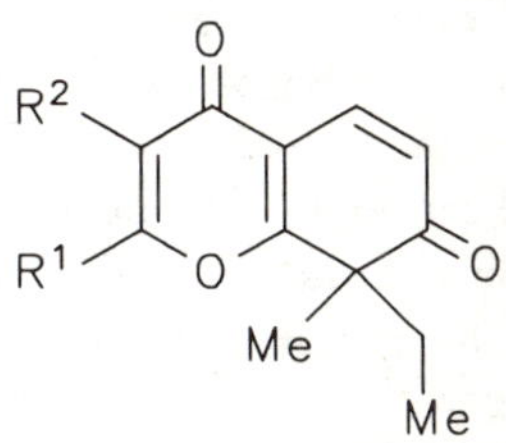

XXXIX R^1=Me, R^2=Me Spiciferone A
XL R^1=CH_2OH, R^2=CH_2OH Spiciferone B
XLI R^1=Me, R^2=CH_2OH Spiciferone C

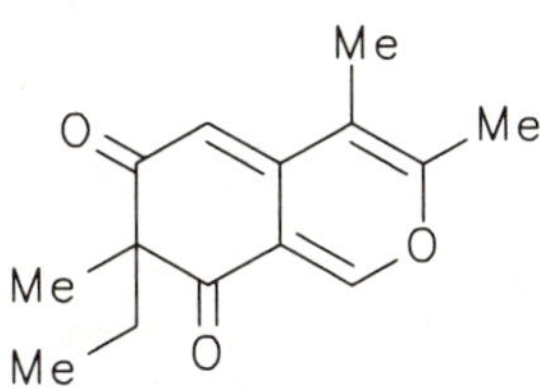

XLII Spiciferinone

Fig. 11. Metabolites of *Cochliobolus spicifer*

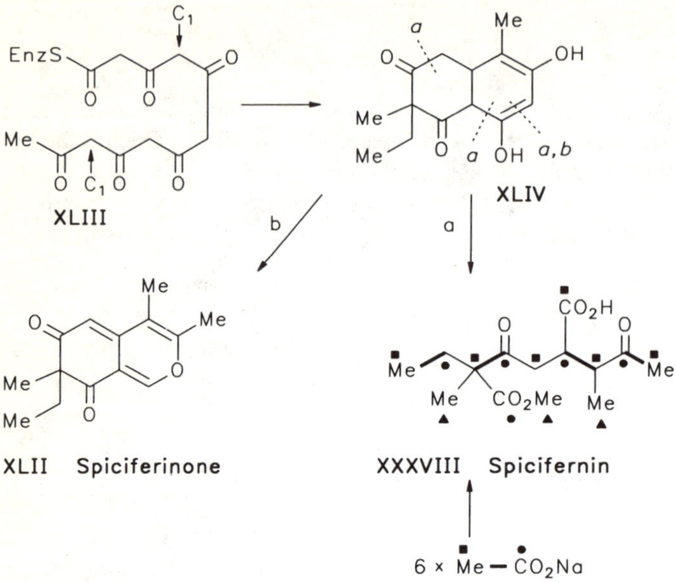

Fig. 12. Hypothetical pathway for the biosynthesis of spicifernin and spiciferinone

the same origin. In order to explore this possibility, the biogenetic origins of spicifernin, a major metabolite of this fungus, is investigated. The results of the feeding experiments are consistent with a proposal that a single hexaketide chain (such as XLIII) bearing two C-methyl groups from C_2 units is folded to give a bicyclic aromatic intermediate (XLIV), which is then subject to double oxidative C–C bond cleavage, decarboxylation (path a), and further introduction of a C_1 unit into the carboxyl function to produce spicifernin (XXXVIII) (Fig. 12) (Nakajima et al. 1992b). If the bicyclic intermediate is modified by an oxidative C–C bond cleavage and subsequent cyclization (path b), then spiciferinone (XLII) will form.

References

Abell C, Staunton J (1981) Chem Commun 856–858. **Abell C**, Staunton J (1984) Chem Commun 1005–1007. **Abell C**, Garson MJ, Leeper FJ, Staunton J (1982) Chem Commun 1011–1013. **Abell C**, Doddrell DM, Garson MJ, Laue ED, Staunton J (1983) Chem Commun 694–696. **Abell C**, Sutkowski AC, Staunton J (1987) Chem Commun 586–587. **Abraham WR**, Arfmann HA (1988) Phytochemistry 27:3310–3311. **Abraham WR**,

Knoch I, Witte L (1990) Phytochemistry 29:2877–2878. **Ahmed SA**, Simpson TJ, Staunton J, Sutkowski AC, Trimble LA, Vederas JC (1985) Chem Commun 1685–1687. **Aldridge DC**, Galt S, Giles D, Turner WB (1971) J Chem Soc C 1623–1627. **Almassi F**, Ghisalberti EL, Narbey MJ, Sivasithamparam K (1991) J Nat Prod 54:396–402. **Arakawa H** (1968) Bull Chem Soc Jpn 41:2451. **Argoudelis AD**, Zieserl JF (1966) Tetrahedron Lett 1969–1973. **Avent AG**, Hanson JR, Truneh A (1990) Phytochemistry 29:2133–2134. **Avent AG**, Hanson JR, Truneh A (1992a) Phytochemistry 31:791–793. **Avent AG**, Hanson JR, Truneh A (1992b) Phytochemistry 31:1065–1066. **Avent AG**, Hanson JR, Truneh A (1992c) Phytochemistry 31:3447–3449.

Bartel PL, Zhu C, Lampel JS, Dosch DC, Conners NC, Strohl WR, Beale JM, Floss HG (1990) J Bacteriol 172:4816–4826. **Beck J**, Ripka S, Siegner A, Schiltz E, Schweizer E (1990) Eur J Biochem 192:487–498. **Bentley R**, Keil JG (1962) J Biol Chem 237:867–873. **Bibb MJ**, Biro S, Motamedi H, Collins JF, Hutchinson CR (1989) EMBO 8:2727–2736. **Birch AJ**, Donovan FW (1953) Aust J Chem 6:360–368. **Birch AJ** (1967) Science 156:202–206. **Birch AJ**, Blance GE, Smith H (1958) J Chem Soc 4582–4583. **Blair J**, Newbold GT (1955) J Chem Soc 2871–2875. **Brereton RG**, Garson MJ, Staunton J (1980) Chem Commun 1165–1167. **Brereton RG**, Garson MJ, Staunton J (1984) J Chem Soc Perkin Trans I 1027–1033. **Bu'Lock** (1979) In: Barton DHR, Ouis WD (eds) Comprehensive organic chemistry. Vol 5. Pergamon Press, Oxford, pp 927–987.

Cane DE, Yang CC (1987) J Am Chem Soc 109:1255–1257. **Claydon N**, Hanson JR, Truneh A, Avent AG (1991) Phytochemistry 30:3802–3803. **Cole RJ**, Moore JH, Davis ND, Kirksey JW, Diener U (1971) J Agric Food Chem 19:909–911. **Condon P**, Kuc J, (1960) Phytopathology 50:267–270. **Condon P**, Kuc J, Draudt HN (1963) Phytopathology 53:1244–1250. **Copeland RJ**, Hill RA, Hinchliffe DJ, Staunton J (1984) J Chem Soc Perkin Trans I 1013–1019. **Coxon DT**, Curtis FR, Price KR, Levett G (1973) Phytochemistry 12:1881–1885.

Devys M, Bousquet JF, Kollmann A, Barbier M (1980) Phytochemistry 19:2221–2222. **Dickinson JM**, Hanson JR, Hitchcock PB, Claydon N (1989) J Chem Soc Perkin Trans I 1885–1887. **Dimroth P**, Walter H, Lynen F (1970) Eur J Biochem 13:98–110. **Duncanson LA**, Grove JF, Zealley J (1953) J Chem Soc 3637–3645.

Evidente A, Randazza G, Ballio A (1986) J Nat Prod 49:593–601.

Floss HG, Tsai MD, Woodward RW (1984) Top Stereochem 15:253–321. **Franco CMM**, Borde UP, Vijayakumar EKS, Chatterjee S, Blumbach J, Ganguli BN (1991) J Antibiot 44:225–231.

Garson MJ, Staunton J (1981) Chem Commun 708–710. **Garson MJ**, Staunton J, Jones PG (1984) J Chem Soc Perkin Trans I 1021–1026. **Gill M**, Steglich W (1987) Progr Chem Org Nat Prod 51:1–286.

Herbert RB (1981) The biosynthesis of secondary metabolites. Chapman and Hall, London, pp 28–49. **Holker JSE**, Simpson TJ (1981) J Chem Soc Perkin Trans I 1397–1400. **Huang S**, Beale JM, Keller PJ, Floss HG (1986) J Am Chem Soc 108:1100–1101.

Jacobs A, Staunton J, Sutkowski AC (1991) Chem Commun 113–114. **Jaworski JG**, Kuc J (1974) Plant Physiol 53:331–336. **Jordan PM**, Spencer JB (1990a) Chem Commun 238–242. **Jordan PM**, Spencer JB (1990b) Chem Commun 1704–1706. **Jordan PM**, Spencer JB, Corina DL (1986) Chem Commun 911–913.

Kurosaki F, Nishi A (1983) Phytochemistry 22:669–672. **Kurosaki F**, Nishi A (1984) Physiol Plant Pathol 24:169–176. **Kurosaki F**, Nishi A (1988) FEBS Lett 227:183–186. **Kurosaki F**, Nishi A (1991) Phytochemistry 30:1823–1825. **Kurosaki F**, Matsui K, Nishi A (1984) Physiol Plant Pathol 25:313–322. **Kurosaki F**, Tsurusawa Y, Nishi A (1985a) Phytochemistry 24:1479–1480. **Kurosaki F**, Futamura K, Nishi A (1985b) Plant Cell Physiol 26:693–700. **Kurosaki F**, Tsurusawa Y, Nishi A (1987a) Plant Physiol 85:601–604. **Kurosaki F**, Tsurusawa Y, Nishi A (1987b) Phytochemistry 26:1919–1923. **Kurosaki F**, Kizawa Y, Nishi A (1989a) Phytochemistry 28:1843–1845. **Kurosaki F**, Kizawa Y, Nishi A (1989b) Eur J Biochem 185:85–89.

Leistner E (1981) In: Stumpf PK, Conn EE (eds) The biochemistry of plants, vol 7. Academic Press, New York, pp 403–423. **Luckner M** (1990) In: Secondary metabolism in plants, microorganisms and animals. Gustav Fischer, Jena.

Marinelli F, Ronchi VN, Pini D, Salvadori P (1990) Phytochemistry 29:849–851. **Mills SD**, Turner WB (1967) J Chem Soc C 2242–2244. **Moore JH**, Murray TP, Marks ME (1974) J Agric Food Chem 22:697–698. **Mosbach K** (1960) Acta Chem Scand 14:457–464. **Mosbach K** (1969) Angew Chem 81:233–244.

Nair MSR, Carey ST (1975) Tetrahedron Lett 1655–1658. **Nakajima H**, Hamasaki T, Kimura Y (1989) Agric Biol Chem 53:2297–2299. **Nakajima H**, Hamasaki T, Maeta S, Kimura Y, Takeuchi Y (1990) Photochemistry 29:1739–1743. **Nakajima H**, Hamasaki T, Kohno M, Kimura Y (1991) Phytochemistry 30:2563–2565. **Nakajima H**, Kimura Y, Hamasaki T (1992a) Phytochemistry 31:105–107. **Nakajima H**, Matsumoto R, Kimura Y, Hamasaki T (1992b) Chem Commun 1654–1656. **Natori S**, Sakaki S, Kurata H, Udagawa S, Inchinoe M, Saito M, Umeda M (1970) Chem Pharm Bull 18:2259–2268.

O'Hagan D (1992) Nat Prod Rep 5:447–479. **Otten SL**, Stutzman-Engwall KJ, Hutchinson CR (1990) J Bacteriol 172:3427–3437.

Parisi A, Piatelli M, Tringali E, Lio GMDS (1993) Phytochemistry 32:865–867.

Rosenbrook W, Carney RE (1970) Tetrahedron Lett 1867–1870.

Schröder G, Brown JWS, Schröder J (1988) Eur J Biochem 172:161–169. **Schütte HR** (1971) Fortschr Bot 33:105–119. **Schütte HR** (1982) Biosynthese niedermolekularer Naturstoffe. Gustav Fischer, Jena, pp 50–83. **Sherman DH**, Malpartida F, Bibb MJ, Kieser HM, Bibb MJ, Hopwood DA (1989) EMBO 8:2717–2725. **Simpson TJ** (1975) Tetrahedron Lett 175–178. **Simpson TJ** (1984) Nat Prod Rep 1:281–297. **Simpson TJ** (1985) Nat Prod Rep 2:321–347. **Simpson TJ** (1987a) Nat Prod Rep 4:339–376. **Simpson TJ** (1987b) Chem Soc Rev 16:123–160. **Simpson TJ** (1991) Nat Prod Rep 8:573–602. **Simpson TJ**, Holker JSE (1975) Tetrahedron Lett 4693–4696. **Spavold ZM**, Robinson JA (1988) Chem Commun 4–6. **Staunton J**, Sutkowski AC (1991a) Chem Commun 1106–1108. **Staunton J**, Sutkowski AC (1991b) Chem Commun 1108–1110. **Staunton J**, Sutkowski AC (1991c) Chem Commun 1110–1112. **Stoessl A**, Stothers JB (1978) Can J Bot 56:2589–2593.

Tanabe M, Uramoto M, Hamasaki T, Cary L (1976) Heterocycles 5:355–365. **Turner WB**, (1972) Fungal metabolites, Academic Press, London. **Turner WB**, Aldridge CD (1983) Fungal metabolites, Academic Press, London.

Wiesner P, Beck J, Beck KF, Ripka S, Müller G, Lücke S, Schweizer E (1988) Eur J Biochem 177:69–79. **Woo ER**, Fujii I, Ebizuka Y, Sankawa U, Kawaguchi A, Huang S, Beale JM, Shibuya M, Mocek U, Floss HG (1989) J Am Chem Soc 111:5498–5500.

Yoshizawa Y, Li Z, Reese PB, Vederas JC (1990) J Am Chem Soc 112:3212–3213. **Yue S**, Duncan JS, Yamamoto Y, Hutchinson CR (1987) J Am Chem Soc 109:1253–1255.

Prof. Dr. Horst-R. Schütte
Inst. f. Pflanzenbiochemie
PSF 250
06018 Halle/Saale, Germany

IV. Photosynthesis

Carbon Metabolism: The Chloroplast's Sesquicentenary, and Some Thoughts on the Limits to Plant Productivity

By Grahame J. Kelly and Erwin Latzko

"At ten p.m. or thereabout,
the little bean began to sprout.
By morning it had grown so tall
you couldn't see the top at all".
(Anonymous 1734)

1. Introduction

The anonymous author of the above quotation, which is taken from her (or his) legendary children's story "Jack and the Beanstalk", commented that it is the kind of thing that "proceeds from the maggots in a madman's brain". It does indeed require little worldly experience to recognise that such a prodigious rate of plant growth is unheard of, and the addition of a primary-school level of education would (hopefully!) impart a sufficient knowledge of photosynthesis as to help one realise that, assuming the bean seed was of normal size, such growth would be impossible during the darkness between 22.00 h and morning. Nevertheless, if an innocent child (just exposed to the Jack-and-the-beanstalk tale) was to reflect upon reality and ask why plants do not grow faster than they do, then even a good photosynthesis researcher would find it difficult to frame a concise answer.

Indeed, *we* are not ashamed to confess that we are unsure of the precise answer. What we do know is that photosynthesis research, since its inception in 1648 by Van Helmont (Walker 1992a; Huzisige and Ke 1993), has revealed hundreds of components of the green cell's photosynthetic machinery, and that the characteristics of these are now sufficiently well understood that a certain few can be identified as being intimately involved in determining "photosynthetic capacity". Moreover, current research on photosynthesis (including photosynthetic carbon metabolism) is directed towards obtaining more refined explanations, apparently for two reasons: (1) to discover whether some form of biotechnology, e.g. genetic engineering, can be used to increase photosynthetic capacity and, consequently, agricultural productivity, and (2) to ascertain the responses of plants to predicted changes in the global environment – especially to the increased level

Progress in Botany, Vol. 56

of atmospheric CO_2. Ironically, it may be argued that these investigations are at the expense of those aimed at discovering more components of the photosynthetic machinery: the literature (for the past 2 years) on chloroplasts and the Calvin cycle is meagre. Alternatively, perhaps the remaining components to be discovered are few.

This review covers research published since our last chapter in Volume 54 of Progress in Botany, and concludes by attempting to relate these studies to a consideration of the biophysical and biochemical factors that limit plant productivity. Meantime, within the review, we pay particular attention to events in and on the chloroplast, not only because this assists our interpretation of the limits to productivity, but also because this year (according to Huzisige and Ke 1993) we celebrate the 150th anniversary of the discovery of chloroplasts by von Mohl (1844).

2. New Locations and Pathways

Without doubt, essentially all photosynthetic CO_2 fixation is accomplished by the Calvin cycle in the green cells of plants, algae and cyanobacteria. The leaf is the major location of plants' green cells, but other localities are not uncommon, e.g. potentially photosynthetic chloroplasts have been found under the bark and wood of poplar twigs, in the dimly illuminated pith cells (van Cleve et al. 1993), and they have been induced to develop in cultured *Acmella oppositifolia* roots which, consequently, became green and autotrophic (Flores et al. 1993). More exotic are reports of genetically engineered *Escherichia coli* producing ribulose-1,5-bisphosphate (RuBP) and sucrose-P (Hudson et al. 1992; Sonnewald et al. 1993), metabolites that in nature are the proprietary of photosynthetic cells.

Obscure anaerobic photosynthetic bacteria are the source of three alternatives to the Calvin cycle: the reductive citric acid cycle, the reductive acetyl CoA pathway, and the reductive 3-hydroxypropionate cycle; the latter was recently proposed to accommodate photosynthesis by *Chloroflexus aurantiacus* (Strauss and Fuchs 1993). For higher plants, Janave et al. (1993) have further evidence that potato tuber chloroplasts can directly reduce CO_2 to formate and thence glyoxylate, but the H-donor for the reduction is still unclear. Interestingly, a major soluble protein in mitochondria from *non*photosynthetic plant cells appears to have the opposite function, i.e. formate oxidation (Colas des Francs-Small et al. 1993).

3. Uptake of Inorganic Carbon

The uptake of CO_2 and/or HCO_3^- by photosynthetic cells has always focussed on aquatic species, but within this group the tide has turned: algae, rather than cyanobacteria, have received most attention in the past 2 years. For cyanobacteria, an NADH dehydrogenase seems to be part of the mechanism supplying ATP to pump inorganic carbon (Ci) into cells (Ogawa 1992). Either CO_2 or HCO_3^- may be pumped in, although the activity of the HCO_3^- pump seems to be highest (Yu et al. 1994). Once inside the cell, the HCO_3^- becomes the source of CO_2 for RuBP carboxylase. Recent work has shown that the efficient generation of a high concentration of CO_2 from this HCO_3^- occurs in intracellular structures, called carboxysomes, that contain RuBP carboxylase and a distinctive carbonic anhydrase (Badger and Price 1992; Marcus et al. 1992; Price et al. 1992).

More complex Ci-concentrating mechanisms are possessed by some eukaryotic microalgal cells. External and internal carbonic anhydrases (Shiraiwa et al. 1993; Williams and Colman 1993; Palmqvist et al. 1994a), three unidentified chloroplast-envelope proteins (Thielmann et al. 1992; Ramazanov et al. 1993), and a plasma-membrane ATPase that appears to be part of an active CO_2-uptake mechanism (Rotatore et al. 1992; Karlsson et al. 1994), all seem to be important when Ci accumulation is carried out by these cells. A new twist is that links to photorespiration and stress responses have appeared: Ramazanov and Cárdenas (1992a) found that a functional photorespiratory glycolate pathway was required before the induction of a Ci-transport mechanism occurred in *Chlamydomonas reinhardtii*, while Sültemeyer et al. (1993) concluded that the ATP to power active Ci uptake is generated by a "pseudocyclic" electron transport in which O_2 is the electron acceptor and, consequently, superoxide radical ions are produced. This reactive form of oxygen can be removed by enzymes of the ascorbate/glutathione system (Sect. 4g), and indeed the activities of these enzymes increased 4- to 20-fold when *C. reinhardtii* cells were induced to begin active Ci uptake.

Experiments with marine algae indicate that a variety of approaches to Ci uptake exist. The green microalga *Dunaliella* has almost no external carbonic anhydrase (Goyal et al. 1992), and probably transports CO_2 across the plasma membrane (Ramazanov and Cárdenas 1992b). However, the green macroalga *Ulva* appears to use an anion transporter to permit entry of HCO_3^- (which, at 2 mM, is abundant in seawater) (Drechsler et al. 1993). Studies with seven species of red macroalgae indicate that only some species possess active Ci-uptake mechanisms, and that only some can use HCO_3^-, although all can use CO_2 (Haglund et al. 1992; Israel and Beer 1992; Johnston et al. 1992). Finally, chromophyte microalgae, which make up a good portion of the ocean's phytoplankton, may be limited by Ci supply despite the abundance of total Ci in seawater: both diatoms and coccolithophorids depended on simple diffusive entry of CO_2 for photosynthesis, and since the concentration of CO_2 in seawater is the same as in air, viz. $10 \mu M$, but its rate of diffusion in seawater is less than 0.1% of that in air, slow growth ensues (Nimer et al. 1992; Riebesell et al. 1993). These conclusions are supported by the generally low levels of carbonic anhydrase in these microalgae – the enzyme was restricted to the chloroplast in one examined coccolithophorid (Quiroga and González 1993) – and contrast with the reported active uptake of CO_2 by the freshwater diatom *Navicula pelliculosa* (Rotatore and Colman 1992). However, this contrast may not reflect any particular inability of the marine microalgae to synthesise carbonic anhydrase. Rather, it may be a consequence of the fact that carbonic anhydrase is a zinc-containing protein, but there is not enough zinc in ocean waters to support the synthesis of adequate amounts of the enzyme. In a very recent report, Morel et al. (1994) provide strong evidence that this is the case. Their report is of special significance because it positions carbonic anhydrase as a central component in the major issues of (1) the global carbon cycle, and (2) the determinants of the standing biomass of oceanic phytoplankton.

On land, plants obtain Ci via CO_2 diffusion through air. There is no recent evidence for any active Ci-uptake mechanism on the plasma membrane of photosynthetic cells of higher plants (Loreto et al. 1992; Palmqvist et al. 1994b). Nevertheless, the terrestrial plant community is not totally devoid of the Ci-uptake phenomenon, in that the cyanobacterial photobionts of lichens have been shown to operate Ci-concentrating mechanisms (Badger et al. 1993; Palmqvist 1993). In addition, carbonic anhydrase is in the leaves of all land plants, and this represents the presence of one component of the algal systems that assist the delivery of CO_2 to RuBP carboxylase. We are still impressed by the observation of Porter and Grodzinski (1984) that bean leaves grown in a high-CO_2 atmosphere contained only one-tenth of the carbonic anhydrase activity found in leaves from plants grown in a normal atmosphere; now we are intrigued to learn that wheat leaves *normally* contain a low activity of carbonic anhydrase, equivalent to about one-tenth of that in the leaves of rice, spinach, pea and bean (Makino et al. 1992).

4. Events In and On the Chloroplast

As mentioned in the Introduction, this year marks the 150th anniversary of the publication of one of the reports from the discoverer of chloroplasts, von Mohl (1844). Given this anniversary, it is fitting that we emphasise the obvious central role of chloroplasts to photosynthesis. A good beginning is the reminder from Winter et al. (1993) that these photosynthetic organelles are not only qualitatively, but also quantitatively important: in barley leaf mesophyll cells, the combined volume of the chloroplasts is easily twice the volume of the cytosol.

a) The Calvin Cycle

Essentially all global CO_2 assimilation occurs via the Calvin cycle. Further evidence has appeared that the enzymes catalysing the 13 reactions of this cycle are not randomly distributed through the chloroplast stroma, but rather organised in some fashion. This concept is attractive, given the high protein concentration and consequent limited water availability in the stroma (Sainis and Srinivasan 1993). A multienzyme complex of the five enzymes that use ATP and NADPH to convert ribose-5-P and CO_2 to glyceraldehyde-3-P (GAP) was detected by Gontero et al. (1993), while Süss et al. (1993) found a similar complex closely associated with the surface of thylakoid membranes, especially in the vicinity of thylakoid membrane proteins involved in the generation of the NADPH and ATP used to drive the Calvin cycle. On the other hand, Scagliarini et al. (1993) have reason to doubt that GAP dehydrogenase could be part of any membrane-bound multienzyme complex.

b) RuBP Carboxylase

The most abundant protein of the chloroplast, the CO_2-fixing enzyme RuBP carboxylase, has continued to attract researchers' attention, principally with respect to details of previously identified regulatory mechanisms (see Fig. 1 in our 1990 review). These mechanisms have been reviewed more recently by Portis (1992). One mechanism, viz. activation by interaction with a regulatory protein called rubisco activase, clearly can be important in vivo: tobacco plants genetically engineered to contain a low content of activase were found to photosynthesise poorly, presumably because the activase-mediated promotion of RuBP carboxylase carbamylation (i.e. covalent attachment of an activator CO_2) was incomplete (Mate et al. 1993). However, the activase does more than promote carbamylation. It also promotes dissociation of the inhibitors RuBP and carboxyarabinitol-1-P (CA1P) from the carboxylase (Wang and Portis 1992; Mate et al. 1993) and somehow is itself activated in a light-dependent process (Woodrow and Mott 1992) that may be related to its ability to hydrolyse ATP (Campbell and Ogren 1992). Furthermore, self-association of activase enhances these functions, and may well occur in the chloroplast stroma (Salvucci 1992).

A second carboxylase regulator of contemporary interest is the sugar-P CA1P that inhibits enzyme activity. As noted above, activase causes it to dissociate from the enzyme, and it is then destroyed by a phosphatase (Kingston-Smith et al. 1992) that releases carboxyarabinitol. This carboxy-sugar now has been identified in leaves, and appears to be a precursor of CA1P resynthesis in darkened leaves (Moore and Seemann 1992) although the responsible enzymes(s) has not yet been detected. Overall, the contribution of CA1P to carboxylase regulation is species-variable. Its role is minor in C_4 plants such as *Chenopodium album* (Sage et al. 1993). Even in C_3 plants where it has a major role, e.g. in *Phaseolus vulgaris*, this role is predominately one of maintaining the carboxylase in an inactive state in darkened leaves; adjustment of the carboxylase activity to variable light intensities during the day still involves modulation of the extent of carbamylation (Sage 1993; Sage et al. 1993).

Other studies of RuBP carboxylase regulation refer to activation by Pi that may be more significant in sink-limited plants (Sawada et al. 1992); activation when 6-P-gluconate binds to regulatory sites which otherwise can bind RuBP (Yokota et al. 1992, 1994); and a novel phosphorylation of *Cicer arietinum* RuBP carboxylase that stabilised the enzyme's quaternary structure – comparable phosphorylation of the enzyme from tobacco was neither detected, nor required for stability (Aggarwal et al. 1993).

In normal air, RuBP carboxylase sometimes utilises O_2 instead of CO_2. This so-called oxygenase activity initiates photorespiration (see Sect. 4g) and reduces plant productivity. An enticing objective of research is, therefore, to improve the enzyme's specificity for CO_2. However, initial efforts in this direction have met with only limited success. Two approaches have been tried. In the first, Read and Tabita (1992) constructed "hybrid" enzymes composed of large subunits from a cyanobacterial enzyme and small subunits from either of two marine microalgae, *Cylindrotheca* sp. or *Olisthodiscus luteus* (normal RuBP carboxylase is composed of eight large subunits and eight small subunits). Indeed, the cyanobacteria/*Cylindrotheca* hybrid enzyme showed a 60% better specificity than the cyanobacterial enzyme. However, its specificity was still lower than that of the native *Cylindrotheca* enzyme, and the overall catalytic activity of the hybrid enzyme was greatly reduced. In the second approach, site-directed mutagenesis was used. The first report of an improvement in specificity resulting from this technique is that from Harpel and Hartman (1992) who replaced serine-368 with alanine in the *Rhodospirillum rubrum* RuBP carboxylase. However, only 2% of the enzyme's activity remained. Consequently, the outlook for increasing the specificity of crop-plant RuBP carboxylases, while maintaining their catalytic capacity, is not especially promising (Spreitzer 1993).

c) Other Calvin Cycle Enzymes

Minor attention has been paid to the remaining enzymes of the Calvin cycle. Chloroplast glycerate-3-P kinase (and its cytosolic counterpart) have now been purified from a eukaryotic alga, viz. *Selenastrum minutum* (Lin and Turpin 1993). The in vivo state of the next Calvin cycle enzyme, GAP dehydrogenase, has been considered by Scagliarini et al. (1993), who concluded that dark/light-mediated association/dissociation of four enzyme molecules possibly occurs, the changes being assisted by regulator molecules that include the product glycerate-1,3-P_2 (Trost et al. 1993).

Incidentally, some interesting concepts concerning GAP dehydrogenases have arisen from studies with nonchloroplast forms of the enzyme. The cytosolic NADP-linked *non*reversible GAP dehydrogenase, present in all photosynthetic organisms except cyanobacteria (Mateos and Serrano 1992), seems to be more important to photosynthetic carbon metabolism than previously envisaged. Specifically, it is thought to have a central role in the cytosolic biosynthesis of sugar alcohols, such as mannitol (Fig. 1) and sorbitol (Everard et al. 1993), and since these sugar alcohols constitute some 30% of the global translocated carbon, its role is quite significant.

The nonreversible GAP dehydrogenase does not seem to be closely related to the more conventional, reversible GAP dehydrogenases that use NADP (photosynthetic cells) or NAD (cells of all organisms) (Michels et al. 1994). It is thus fascinating that Corbier et al. (1992), using site-directed mutagenesis, were able to transform the conventional

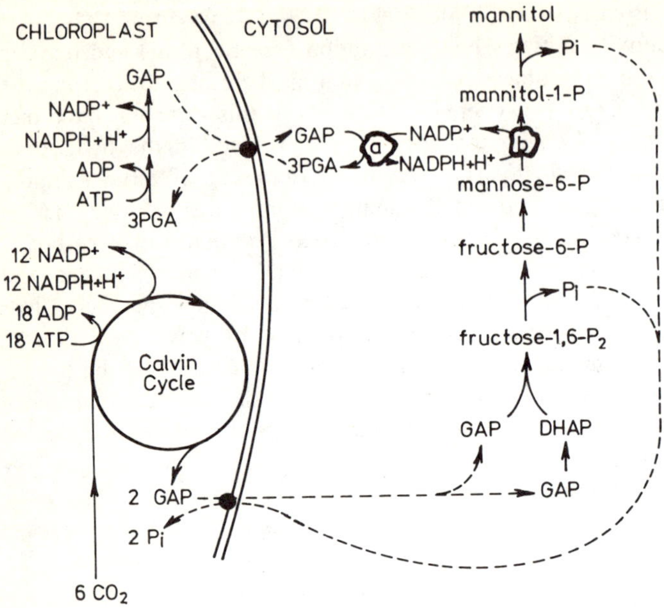

Fig. 1. Central role of nonreversible GAP dehydrogenase *a* in the synthesis of sugar alcohols for translocation in the phloem sap. The enzyme participates in a shuttle system for the indirect transport of photosynthetically generated NADPH from the chloroplast to the cytosol. The cytosolic NADPH is, in turn, used by mannose-6-P reductase. *b* ● Phosphate translocator of the inner membrane of the chloroplast envelope (see Sect. 4e). *DHAP* Dihydroxyacetone-P; *3PGA* glycerate-3-P. (After Everard et al. 1993)

NAD-linked enzyme of *Escherichia coli* into an NADP-linked nonreversible form simply by altering cysteine-149 to alanine. In other studies with the NAD-linked enzyme from animal cells, additional possible functions for the enzyme have been revealed: it may act as a "binding protein" for the export of tRNA from the nucleus (Singh and Green 1993), and it may be involved in cellular signalling processes after an unusual modification in which the greater portion of the substrate NAD (specifically, ADP-ribose; i.e. NAD after removal of the nicotinamide base) is covalently attached to it (Kots et al. 1993). It remains to be seen whether or not plant GAP dehydrogenases possess these functions.

Pea chloroplasts have been found to contain not one, but two, aldolase isoenzymes, and although they are very similar (Razdan et al. 1992), it is intriguing to reflect that the Calvin cycle includes two different aldolase-catalysed reactions, for the synthesis of (respectively) fructose-1,6-P_2 and sedoheptulose-1,7-P_2. These sugar-P_2s are then hydrolysed by specific bisphosphatases. The chloroplast fructose-1,6-bisphosphatase undergoes light-mediated activation in a process that involves the small protein thioredoxin; the properties of this protein seem to vary between plant species (Prado et al. 1992). Separation of chloroplast fructose-1, 6-bisphosphatase from its cytosolic counterpart has been achieved for *Chlorella kessleri* (Grotjohann et al. 1993) and for spinach and pea (Fonollá et al. 1994);

in the latter case, the two forms of the enzyme were shown to be immunologically distinct. Finally, a notable observation is that the deactivation of stromal enzymes such as the bisphosphatases during the brief but often frequent exposures of leaves to low light (as is typical in a field plant canopy) results in a definite limitation on photosynthetic carbon uptake (Sassenrath-Cole and Pearcy 1992).

d) Starch Metabolism

When chloroplasts in plant leaves are well illuminated, photosynthetic CO_2 fixation can exceed the immediate capacity of sucrose synthesis in the cytosol and/or the capacity of sink tissues to utilise sucrose. If this occurs, the excess photosynthate may be diverted into the synthesis of a transitory reserve of starch that can be remobilised, after the sun goes down, to extend net sucrose synthesis into the night. An endogenous circadian shift ensures that the "sunset shift" away from starch synthesis occurs effectively (Li et al. 1992a). The metabolic pathway for the daytime starch synthesis is GAP →→ hexose-P →→ ADP-glucose → starch. Regulation of this pathway may involve several enzymes, including starch synthase which can undergo a Ca^{2+}/calmodulin-dependent phosphorylation and consequent alteration in activity (Dreier et al. 1992). However, the best-known regulatory enzyme of starch synthesis is ADP-glucose pyrophosphorylase (Stark et al. 1992). The pivotal role of this enzyme has been used by Okita (1992) to argue against any alternative starch synthesis pathway not involving it, including the proposed pathway in which ADP-glucose is generated in the cytosol by sucrose synthase, and then transported into the chloroplast by a chloroplast-envelope located adenylate translocator (see Fig. 3 in our preceding review). In any event, this pathway no longer can be seriously considered because, although all types of plastids appear to be capable of importing ADP-glucose (Ardila et al. 1993), sucrose synthase in fact cannot directly generate ADP-glucose (Perata et al. 1992). In addition, the *affinity* of the envelope's adenylate translocator for ADP-glucose (as determined with pea-root plastids) appears to be too low to permit a reasonable rate of ADP-glucose import (Schunemann et al. 1993).

Initial steps in the degradation of chloroplast starch probably use a debranching enzyme and at least four amylases (Li et al. 1992b; Ghiena et al. 1993), as well as phosphorylase (see our last review).

Starch is also synthesised and degraded in plastids other than chloroplasts, and these nongreen plastids have received more attention than chloroplasts during the past 2 years. Etioplasts, the immediate precursors of chloroplasts, synthesise starch from triose-P imported by a typical chloroplast-type phosphate translocator of the inner envelope membrane (Batz et al. 1992). However, other nongreen plastids, including maize-seed, potato-tuber and cauliflower-bud amyloplasts lack fructose-1,6-bisphosphatase and therefore could not generate starch from triose-P; instead, glucose-P is imported as the direct substrate for starch synthesis (Kossmann et al. 1992; Tobias et al. 1992; Neuhaus et al. 1993) via a recently identified translocator with a high affinity for hexose-P (Batz et al. 1993). Furthermore, a similar mechanism of hexose-P import operates on other plastids that form little or no starch, such as leucoplasts and root plastids. These oxidise the hexose-P to CO_2: root plastids use the oxidative pentose-P pathway, thereby generating NADPH that, in turn, reduces NO_2^- to NH_4^+ (Borchert et al. 1993; Coates and ap Rees 1994).

e) Envelope Transporters

Certain proteins of the inner membrane of plastids act as transporters, controlling the entry and exit of metabolites to and from the plastids. The best known is the chloroplast phosphate translocator that, during photosynthesis, allows exit of triose-P in exchange for imported Pi. As mentioned above in Sect. 4g, the etioplast phosphate translocator operates similarly, but the equivalent amyloplast translocator also accepts hexose-P's. Interestingly, Overlach et al. (1993) have found that chloroplasts in guard cells are like amyloplasts in that their phosphate translocators also accept hexose-P; in addition, these chloroplasts (like amyloplasts) lack fructose-1,6-bisphosphatase. It would appear that the specificity of this translocator is easily adjusted – indeed, Fischer et al. (1994) have concluded that only minor exchanges of amino acids are required for the translocator to accept P-enolpyruvate, as is required in C_4 plants.

The phosphate translocator can also mediate triose-P/glycerate-3-P shuttles for the indirect export of ATP and NAD(P)H from chloroplasts to the cytosol. This shuttle ensures that, upon illumination, the cytosolic ATP/ADP ratio rises rapidly in a matter of seconds, reflecting the similar rise that occurs in the chloroplast (Gardeström 1993). The phosphate translocator, which in evolutionary terms is as old as the chloroplast (Schlichting and Bothe 1993), is not the only envelope translocator. Others include the adenylate translocator (see Sect. 4d above), a recently identified pyrophosphate (PPi) translocator (Lunn and Douce 1993) that may import cytosolic PPi generated during sucrose synthesis (Sect. 6), an ATP-driven H^+ pump that moves H^+ from the chloroplast stroma to the cytosol, thus facilitating a suitable stroma:cytosol ΔpH (Berkowitz and Peters 1993), and a transporter that exchanges outgoing glycolate for incoming glycerate during photorespiration (Yu and Woo 1992; Young and McCarty 1993).

f) Photoinhibition

It sometimes happens that the rate of CO_2 assimilation into carbohydrates during photosynthesis is markedly less than the capacity for photosynthetic electron transport from water to NADP. Examples include sun plants exposed to low-water or low-temperature stress, and shade plants exposed to full sunlight. A consequence is that the photosynthetic apparatus of the chloroplast thylakoids becomes over-reduced, which causes it to become altered in such a way that, upon return of the plant to normal conditions, CO_2 fixation is depressed, i.e. photoinhibited. Recent examples include protoplasts of the brown alga *Laminaria saccharina* exposed to high light (Benet et al. 1994) and *Sphagnum* mosses exposed to moderately high light intensities, similar to those that they often experience in the field (Murray et al. 1993).

The alteration that the thylakoid undergoes has several components. One is that some photosystem II reaction centres and/or light-harvesting antennae become inactive and aggregate somewhat in the membrane; another is that one of the proteins of the photosystem II reaction centre, termed the D1 protein, fails to undergo the relatively rapid rate of turnover (i.e. degradation and resynthesis) that is normal for it (Aro et al. 1993, 1994; Schnettger et al. 1994). Interference with turnover of the D1 protein leads to severe photoinhibition, but inactivation of photosystem II centres can be viewed as a normal adjustment of electron flow to match a lower demand for NADPH (Walker 1992b; Huner et al. 1993). The down regulation of photosystem II centres is in turn promoted by increased acidity in the thylakoid lumen (due to accumulation of H^+ which has difficulty exiting via the coupling factor because insufficient ADP + P_i is returned from a slowly turning Calvin cycle), and by the carotenoid pigment zeaxanthin which is formed from violaxanthin and ascorbate in this low pH environment (Bailey and Walker 1992; Gilmore and Yamamoto 1992; Choudhury et al. 1993; Demmig-Adams and Adams 1993; Havaux and Gruszecki 1993; Johnson et al. 1993; Jahns and Krause 1994; Neubauer and Yamamoto 1994). All of these changes favour the dissipation of excess harvested light energy as heat, and therefore help to avoid damage to the photochemical apparatus.

Incidentally, in a report just published, Quiñones and Zeiger (1994) provide evidence that zeaxanthin may be the long-sought blue light photoreceptor of plants.

g) The Mehler Reaction, and Photorespiration

It may be questioned why the chloroplast needs to resort to the above largely biophysical mechanisms for energy dissipation under stress situations when two alternative biochemical mechanisms exist, viz. O_2 reduction (Mehler reaction) and photorespiration. The short answer relates to capacities: neither of these processes can proceed as fast as CO_2 fixation (Heber and Walker 1992), and therefore neither can prevent photoinhibition if the light intensity is high but CO_2 is unavailable, as is possible if water stress leads to stomatal closure on a sunny day. Ultimately, both processes must be viewed primarily as responses to the "interference" by O_2 in metabolism, and the fact that those responses involve the (possibly beneficial) dissipation of harvested light energy is, presumably, a secondary consequence.

The product of the Mehler reaction is the superoxide anion. This anion is processed in the chloroplast by four enzymes of the so-called ascorbate/glutathione system (Fig. 2, enzymes e, b, c and d; see also Fig. 7 of our 1982 review). The flow of electrons from H_2O to O_2 ("pseudocyclic" electron transport) and subsequent processing of superoxide anions by the ascorbate/glutathione system not only directly dissipate harvested light energy, but also influence zeaxanthin-dependent energy dissipation (see Sect. 4f above) in two somewhat opposing ways: on the one hand, it supports electron transport that, at the plastoquinone step, moves protons from the stroma into the thylakoid lumen, thereby creating the low pH that favours zeaxanthin-dependent energy dissipation (Neubauer and

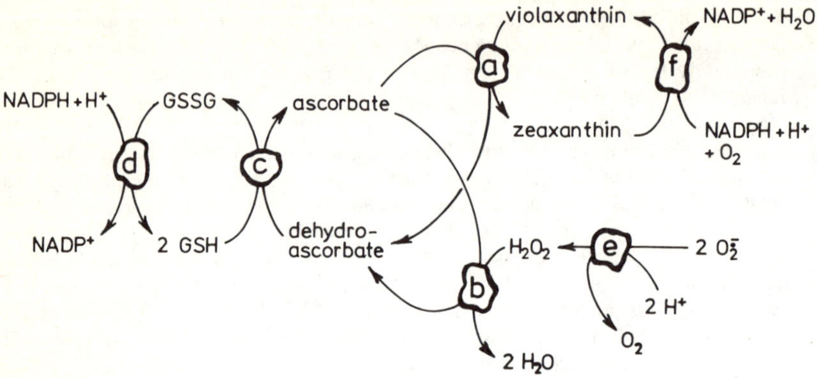

Fig. 2. Competition between *a* violaxanthin deepoxidase and *b* ascorbate peroxidase for chloroplast ascorbate. Other enzymes: *c* dehydroascorbate reductase; *d* glutathione reductase; *e* superoxide dismutase and *f* zeaxanthin epoxidase. *GSH* and *GSSG* reduced and oxidized glutathione respectively. (After Neubauer and Yamamoto 1994)

Yamamoto 1992); on the other hand, if "locks up" a good portion of the chloroplast's ascorbate, thereby restricting the deepoxidase that also needs ascorbate to generate the zeaxanthin in the first place (Neubauer and Yamamoto 1994; see Fig. 2).

One point that has become dramatically clear in the past 2 years is that plants respond to stress situations by increasing the activities of the ascorbate/glutathione-system enzymes, and/or increasing the pool sizes of ascorbate and glutathione. Recent examples include drought stress of pea and other species (Smirnoff 1993; Mittler and Zilinskas 1994), high light/low temperature stress of tomato, wheat and pea (Mishra et al. 1993; Walker and McKersie 1993; Edwards et al. 1994), low nutrient stress of bean and spruce (Cakmak and Marschner 1992; Polle et al. 1992), pollutant or herbicide stress of pumpkin and maize (Ranieri et al. 1993; Pastori and Trippi 1993), and O_2 stress of rice and wheat (Ushimaru et al. 1992; Albrecht and Wiedenroth 1994). Moreover, in two instances where the activity of an enzyme was boosted by genetic engineering, the resultant transgenic tobacco plants displayed increased tolerance to stress (Aono et al. 1993; Gupta et al. 1993). Clearly, there is scope for biotechnologists to increase the productivity of plants growing in stressful environments.

A final fascinating report is that from Sültemeyer et al. (1993). These authors report one of the largest jumps in the activities of ascorbate/glutathione system enzymes (two- to ten-fold) when *Chlamydomonas reinhardtii* cells were transferred from 5% CO_2 to air-level CO_2. It was proposed that the extra activities can deal with the extra superoxide anions formed when the rate of "pseudocyclic" electron transport is increased to supply (via the associated pseudocyclic photophosphorylation) ATP to drive this alga's CO_2 uptake (see Sect. 3).

The second process that reflects the "interference" of O_2 in the metabolism of photosynthetic cells is photorespiration. An intriguing proposal is that inhibition of photorespiration may be the biochemical explanation for the increased growth recorded following a brief application of methanol to the leaves of plants in an arid environment (Nonomura and Benson 1992).

The initial reaction of photorespiration is that in which O_2 substitutes for CO_2 on RuBP carboxylase (see Sect. 4b), leading to the formation of one glycerate-3-P and one glycolate-P. The latter is hydrolysed in the chloroplast to glycolate and Pi. The remainder of the process occurs outside the chloroplast, and achieves, overall, the conversion of two molecules of glycolate to one each of CO_2 and glycerate; the latter is moved into the chloroplast (Sect. 4e) where it is phosphorylated prior to reentering the Calvin cycle. However, this may not be the only major metabolic role of glycerate; considerable amounts are produced and simply stored in *Vicia faba* (El Bouhmadi and Larher 1993).

Minor attention has been given to the extra-chloroplastic reactions of photorespiration. Most occur in the peroxisome, mediated by enzymes that appear to be organised into multienzyme complexes (Heupel and Heldt 1994). However, the CO_2 evolving reaction, viz. 2 glycine + NAD $\rightarrow$ serine + CO_2 + NH_4^+ + NADH, takes place in the mitochondria. Reoxidation of the NADH is coupled to ATP formation (Wigge et al. 1993). This ATP, and also the NH_4^+, in turn contribute to inactivation of the Krebs cycle's pyruvate dehydrogenase complex (Gemel and Randall 1992), implying that active photorespiration retards the classical "dark" respiration in illuminated photosynthetic cells.

There has been some confusion concerning the nature and location of the first extra-chloroplast reaction of photorespiration (i.e. glycolate oxidation) in green algae, but now Stabenau et al. (1993) claim to have settled the matter by reporting that only the charophycean line contains, like higher plants, glycolate oxidase. Members of the other two lines (Chlorophyceae and Ulvophyceae), including *Dunaliella* spp. previously reported to contain this oxidase, in fact contain a mitochondrial glycolate dehydrogenase only.

5. Mitochondrial Respiration

The perceived importance of mitochondrial respiration to photosynthesis continues to fluctuate. As mentioned above, Gemel and Randall (1992) have evidence that entry of pyruvate into the Krebs cycle might be inhibited in the light, and indeed Kent et al. (1992) have calculated that illumination somehow reduces mitochondrial respiration by about 20%. Nevertheless, features of mitochondrial respiration that seem to be essential to photosynthesis continue to be reported. The most popular is that it might act as an "energy release valve", thereby lessening the likelihood of photoinhibition (Saradadevi and Raghavendra 1992; Krömer et al. 1993; Shyam et al. 1993). This role is easier to accept if the alternative, cyanide-resistant system for NADH oxidation is utilised (Weger and Dasgupta 1993). Another role is that vital metabolic links exist between chloroplasts, mitochondria and the cytosol (Hanning and Heldt 1993; Krömer et al. 1993), especially in stomatal guard cells (Mawson 1993). The most comprehensive picture is that from Hanning and Heldt (1993) who envisage an abbreviated Krebs cycle

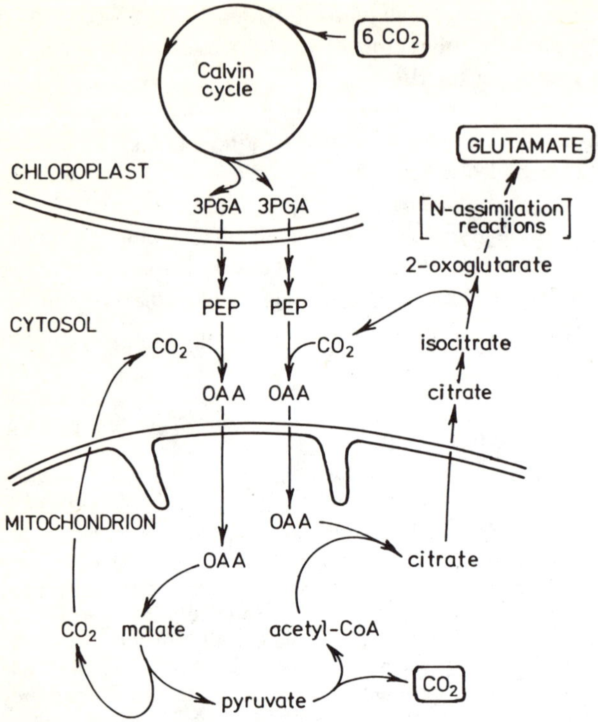

Fig. 3. Carbon balance for the biosynthesis of glutamate during the cooperation of chloroplasts mitochondria and cytosol in photosynthetic cells. The sum of the reactions is: 6 $CO_2 \rightarrow$ glutamate + CO_2. *OAA* Oxaloacetate; *PEP* phosphoenolpyruvate; *3PGA* glycerate-3-P. (Adapted from Hanning and Heldt 1993)

that assists in the manufacture of amino-acid carbon skeletons from the products of the chloroplast's CO_2 fixation (Fig. 3).

6. Sucrose Biosynthesis

Sucrose is the major form of carbohydrate exported out of photosynthetic tissues and translocated to sink tissues. It is manufactured in the cytosol from GAP exported from the chloroplast via the phosphate translocator. The biosynthesis pathway from GAP to fructose-6-P is as shown in Fig. 1 for mannitol synthesis; subsequent reactions are (in order): fructose-6-P $\rightarrow$ glucose-6-P $\rightarrow$ glucose-1-P, glucose-1-P + UTP $\rightarrow$ UDP-glucose + PPi, UDP-glucose + (a second) fructose-6-P $\rightarrow$ sucrose-P + UDP (this reaction catalysed by sucrose-P synthase [SPS]), sucrose-P $\rightarrow$ sucrose + Pi.

The final reaction is catalysed by a phosphatase that Echeverria and Salerno (1993) confirmed to be, as expected, in the cytosol. This report, and another indicating that the amount and activity of the cytosolic fructose-1,6-bisphosphatase required to generate fructose-6-P might be light modulated (Khayat et al. 1993), are two of the few studies that do not deal with SPS. However, before reviewing the research on SPS, a few words relating to the preceding reaction that generates PPi are in order. It had been thought that efficient conversion of this PPi to 2 Pi might be one role for the PPi-linked P-fructokinase in the cytosol of all higher plant photosynthetic cells so far investigated [including C_4-plant bundle sheath cells (Clayton et al. 1993), previously reported to lack the enzyme]. However, this role is now more dubious for three reasons. Firstly, there is a translocator for importing PPi into chloroplasts (Lunn and Douce 1993) where pyrophosphatase is located – indeed, the PPi contents of cytosol and chloroplast are similar (Eberl et al. 1992). Secondly, when *Escherichia coli* pyrophosphatase was genetically engineered into the cytosol of tobacco-leaf and potato-leaf cells, carbon flow increased in the direction of sucrose, at the expense of chloroplast starch (Jelitto et al. 1992). Thirdly, plants such as celery export much of the newly fixed carbon from their leaves in the form of sugar alcohols which are synthesised via a pathway that does not involve PPi formation (Everard et al. 1993; see Fig. 1). However, we are not aware of any reports concerning the PPi-linked P-fructokinase status of celery leaves. This matter clearly deserves attention, if only to clarify the role of the enzyme. The only recent reports on the enzyme relate to nonphotosynthetic tissues, viz. the potato tuber enzyme's net catalysis is in the glycolytic direction (Hajirezaei et al. 1994), and the tomato fruit enzyme needs a reducing environment for fructose-2,6-P_2-activation to be effective (Kiss et al. 1991).

The further investigations of SPS have confirmed that the enzyme is activated upon illumination by a dephosphorylation catalysed by a phosphatase (Weiner et al. 1992, 1993), and is inactivated upon darkness by a phosphorylation catalysed by a kinase; the critical phosphorylation/dephosphorylation site is serine-158 (Huber and Huber 1992; McMichael et al. 1993). The modifying enzymes are in turn influenced by light and metabolites, especially Pi (Weiner et al. 1992, 1993); Pi-supply also markedly influences the amount of SPS in leaves (Usuda and Shimogawara 1993).

Two other enzymes regulated through a phosphorylation/dephosphorylation cycle are nitrate reductase and phosphoenolpyruvate (PEP) carboxylase. The former is similar to SPS, undergoing activation via a phosphatase-catalysed dephosphorylation (Huber et al. 1992), but PEP carboxylase shows the opposite response, being activated via a kinase-catalysed phosphorylation (see Sect. 7d). Links have been established by Quy and Champigny (1992) who found that NO_3^- enhances the activities of the kinases that regulate both SPS and PEP carboxylase, but since the effects are opposite (inactivation and activation, respectively) carbon flow in the cytosol might be expected to occur in the direction of PEP and amino-acid C-skeletons (thus the NO_3^- could be put to good use) rather than in the direction of sucrose. The NO_3^- effect on PEP carboxylase seems to be mediated by glutamine activation of the kinase that catalyses its activation (Manh et al. 1993).

The fate of sucrose manufactured in the cytosol is export into vacuoles via a transporter protein (Thom et al. 1992) or export into the phloem via an apoplastic and/or symplastic route, depending on species (van Bel et al. 1994). In the former, a H^+/sucrose cotransporter found only on phloem cells loads sucrose into the phloem (Riesmeier et al. 1993a). When the amount of this cotransporter in potato leaves was reduced using cDNA-based antisense technology, the effect was dramatic: leaves displayed a 20-fold increase in soluble carbohydrate content, a 5-fold increase in chloroplast starch, but reduced photosynthesis. Clearly, translocation was mostly blocked; indeed, root and tuber growth was poor (Riesmeier et al. 1994).

Similar results were obtained when Heineke et al. (1992) expressed a foreign enzyme, viz. yeast invertase, in the cell walls of potato leaves. Once again, photosynthesis dropped after sucrose exported from cells was hydrolysed to glucose and fructose that, not being recognised by the phloem's H^+/sucrose contransporter, simply accumulated, raising the osmolality of the cell sap. The authors attributed the drop in photosynthesis to this osmolality increase, consistent with another study in which invertase secreted by a pathogenic fungus on barley leaves also caused an accumulation of sugars and decreased photosynthesis (Scholes et al. 1994).

Interference with the normal flow of carbon inside the photosynthetic cell also causes major (and somewhat predictable) changes. For example, when the amount of the potato chloroplast's phosphate translocator was reduced using antisense technology, photosynthesis was halved and, as expected, fixed carbon was trapped in the chloroplast, boosting the chloroplast starch level three-fold (Riesmeier et al. 1993b). Similarly a *Flaveria linearis* mutant that lacked most of the normal cytosolic fructose-1,6-bisphosphatase activity was also found to be incapable of moving all fixed carbon into sucrose, with a comparable result: more carbon moved to chloroplast starch (Sharkey et al. 1992).

These drastic changes to photosynthetic physiology usually resulted in reduced photosynthesis, which is not unexpected, but identifying the principal regulatory step(s) that mediate such a reduction is not easy. In fact, even with normal plants it is known that a period of rapid photosynthesis can lead to an accumulation of sucrose (and/or other major photosynthetic products) that somehow inhibits subsequent photosynthesis. Goldschmidt and Huber (1992) do not believe that sucrose, starch or hexoses are directly responsible (consistent with the observation that the leaves of some plants, e.g. cassava, are avid accumulators of sucrose; Angelov et al. 1993). They suggest that some intermediate metabolic step is critical, a point that might implicate SPS. However, earlier suggestions that this enzyme might be regulated by sucrose appear to be unfounded (Mitra and Srivastava 1992), and when the activity was boosted in tomato leaves by genetic engineering, photosynthesis in a normal environment was little affected (Galtier et al. 1993). Incidentally, the activity is also increased by low temperature (Guy et al. 1992) and by application of gibberellins (Cheikh et al. 1992). In the latter case, protein synthesis was involved. Perhaps also the sucrose-induced drop in photosynthesis is a consequence of altered protein synthesis/gene expression (Krapp et al. 1993). But, unless sucrose directly interacts with genes, unknown intermediate regulatory steps must still exist.

7. C_4 Metabolism

a) C_4 Plants

The first thing that happens to CO_2 upon entering the leaf of a C_4 plant is its conversion to HCO_3^- in the mesophyll cells. This is catalysed by carbonic anhydrase, a good portion of which is in the plasma membrane of these cells (Utsunomiya and Muto 1993). The consequent low concentration of CO_2 results in the cytostic pH being higher than usual (Yin et al. 1993). In the cytosol, HCO_3^- is combined with PEP (Sect. 7d) to yield oxaloacetate that is rapidly converted to a 4C-compound (malate or aspartate). This is moved into inner bundle sheath cells where it is decarboxylated (by one of three possible enzymes), releasing and concentrating the initial CO_2 at the site of the Calvin cycle (which is restricted to these cells). The remaining 3C-piece is moved back into mesophyll cells to regenerate PEP.

C_4 plants are subtyped according to the decarboxylation enzyme. Hattersley (1992) has shown interesting correlations between subtype and rainfall for Australian C_4 grasses: NADP-malic enzyme and PEP carboxykinase subtypes were slightly more plentiful in higher rainfall areas, but the proportional representation of the NAD-malic enzyme subtype was much greater in low rainfall areas.

Regeneration of PEP takes place in mesophyll cell chloroplasts. The returning 3C-piece must first be converted to pyruvate that, in turn, must be transported into the chloroplast. This is effected by a chloroplast-envelope pyruvate transporter that is stimulated by Na^+ (or, less usually, H^+; Aoki et al. 1992); direct Na^+ stimulation of the protein that transports pyruvate may occur, or alternatively, the protein is a Na^+/pyruvate symporter (Murata et al. 1992). Once inside the chloroplast, pyruvate is converted to PEP by the enzyme pyruvate Pi dikinase. Like SPS and other enzymes mentioned above (Sect. 6), this enzyme also undergoes light-mediated regulation via a phosphorylation/dephosphorylation system; the kinase/phosphatase that acts on the enzyme constitutes only a tiny fraction of the total chloroplast protein (Smith et al. 1994), not a sizeable fraction as reported earlier.

The classical picture of enzyme distribution between mesophyll and bundle-sheath cells in C_4 plant leaves has been challenged by Ueno (1992), who describes a C_4 grass, *Aristida latifolia*, that has two bundle sheaths. Cells of the outer sheath contain *both* RuBP carboxylase and pyruvate Pi dikinase, a rare and possibly unique event since these enzymes usually occur separately (in bundle-sheath cells and mesophyll cells, respectively).

The benefits of C_4 photosynthesis have been considered for today's environment, and for those of other geological periods. In the current climate, a major advantage is that the act of concentrating CO_2 at the site of RuBP carboxylase reduces the propensity for RuBP oxygenation and, hence, "wasteful" photorespiration. Further studies have confirmed that photorespiration is low or absent in C_4-plant leaves (Brown and Byrd 1993; Dai et al. 1993; Oberhuber and Edwards 1993), even though photorespiratory enzymes are present (e.g. P-glycolate phosphatase; Baldy 1993) and the potential for photorespiration exists, e.g. when the CO_2 supply is very low (Dai et al. 1993; Raghavendra et al. 1993).

It might be envisaged that the "CO_2 pump" of the C_4 plants could keep photorespiration at an insignificant level even if atmospheric CO_2 was considerably reduced below the current level of 357 ppm; of course C_3 plants, which already photorespire under 357 ppm CO_2, would photorespire even more and thus be at a competitive disadvantage. This line of thought has been used to explain (1) the unexpected abundance of C_4 plants in a temperate and moist Himalayan location (where CO_2 is low due to altitude; Rama Das and Vats 1993), (2) the rapid expansion of C_4 plants some 7 million years ago when the atmospheric CO_2 level became relatively low (Cerling et al. 1993; Morgan et al. 1994), and (3) the partial decline of C_4 plants 7000–9000 years ago when the retreat of the last Ice Age was accompanied by a rapid increase in atmospheric CO_2 from about 200 to 280 ppm (Cole and Monger 1994); quite probably, the decline in the number of C_4 plants was matched by an equivalent increase in the number of C_3 plants as they benefited from the increased CO_2 (Polley et al. 1993). Finally, readers may be intrigued to learn that, because of C_4 plants, the terrestrial biosphere will continue for at least another 0.9 billion years, even though C_3 plants will cease to grow when atmospheric CO_2 falls below 150 ppm about 0.1 billion years from now (Caldeira and Kasting 1992).

b) C_3–C_4 Intermediate Plants

Minor attention has been given to those plants with photosynthetic characteristics intermediate between true C_3 plants and true C_4 plants. Rawsthorne (1992), in a review of what is known, was unhappy with the term "C_3–C_4" but could not find an all-embracing alternative. The difficulty is apparent from the report by Brown et al. (1993) that *Flaveria brownii*, a C_3–C_4 species that is so close to being C_4 that it is termed "C_4-like", has been crossed with *Flaveria linearis*, a more conventional C_3–C_4.

A salient feature of C_3–C_4 plants is that photorespiration is reduced. The reduction is probably assisted by efficient recapture of photorespired CO_2 by PEP carboxylase (a process that increases the photon requirement of these plants; Osborne 1994), by overall reductions in the activities of photorespiratory enzymes (Devi and Raghavendra 1993), and by restriction of the CO_2-evolving enzyme, glycine decarboxylase, to bundle-sheath cells (Morgan et al. 1993).

c) Crassulacean Acid Metabolism (CAM)

The classical CAM plant captures CO_2 at night, then photosynthetically reduces it to carbohydrate the next day, behind closed stomates. Transpirational water loss is thereby minimised. Nocturnal CO_2 capture is catalysed by PEP carboxylase and malate dehydrogenase (Sect. 7d), and as the resultant malate accumulates it is stored in the large vacuole typical of these plants. Daytime decarboxylation of malate releases the CO_2 for the Calvin cycle, while the remaining 3C-piece is metabolised to starch which becomes the source of the next night's PEP.

This diurnal metabolic cycle persists even if the plant is disturbed by continuous light or darkness. Continuation of the disturbance results, however, in a damping of the rhythm due to excessive accumulation of malate in darkness (Wilkins and Thomas 1993), or to a drift toward chaotic metabolism in light (Lüttge and Beck 1992).

Normal CAM requires high activities of starch degrading enzymes to generate the required amount of PEP at night (Paul et al. 1993). It also seems to require respiratory activity: by night to permit malate accumulation (Watanabe et al. 1992), and by day respiration is increased by engagement of the alternative, cyanide-resistant pathway, perhaps to allow some mitochondrial metabolism of malate without upsetting the cell's ATP/ADP ratio (Robinson et al. 1992).

Tonoplast proteins that contribute to the movement of malate into and out of the vacuole have received more attention. Active malate uptake is driven by an H^+/ATPase, as evidenced from experiments with the orchid *Epidendrum ibaguense* (White and Smith 1992) and the facultative CAM plant *Mesembryanthemum crystallinum*; in this latter case, a previously identified H^+/pyrophosphatase appeared to have minor importance for malate uptake (Bremberger and Lüttge 1992). The H^+/ATPase has been solubilized and then reconstituted, in an active state, into proteoliposomes (Behre et al. 1992). It even has been visualised and quantified on freeze-fracture replicas of *M. crystallinum* tonoplasts; its density in the membrane increased after CAM was induced, further establishing its central role in CAM (Klink and Lüttge 1992).

While H^+/ATPase presumably establishes an H^+ gradient across the tonoplast and thereby drives malate movement, less is known of the proteins that actually translocate the malate. Two channel proteins for malate^{2-}, possibly one for uptake and the other for efflux, have been detected in *Graptopetalum paraguayense* by Iwasaki et al. (1992). One clear aspect is that control of malate movement and storage is dependent on a relatively low level of tonoplast fluidity, which in turn can be induced by environmental factors that promote CAM (Kliemchen et al. 1993).

Although CAM is designed to equip plants for survival under harsh conditions, especially those where water deficit is common, its capacity to protect the plant from stress is still limited, thus it is not unusual to find CAM plants with metabolism altered as a response to severe stress. For example, the Madagascan

CAM plant *Xerosicyos dunguyi* switches to CAM-idling when water stress becomes particularly severe: stomata shut permanently, but a low diurnal fluctuation in organic acids continues internally (Bastide et al. 1993). A similar response is seen when water stress is experienced by Central American species of *Clusia* that can quickly (within 24 h) switch from C_3 to CAM photosynthesis (Zotz and Winter 1993). A notable feature is a marked accumulation of citrate (Borland et al. 1993) that could play a pivotal role in preventing photoinhibition during prolonged drought (Franco et al. 1992) by participating in a CAM-idling metabolic cycle that consumes energy. One possibility is suggested in Fig. 4, in which 10 ATP are consumed for each molecule of citrate cycled. However, the system could additionally consume the equivalent of 4 NADH per citrate if the NADH and $FADH_2$ generated in the Krebs cycle during the citrate-to-pyruvate conversion was reoxidised by the alternative electron transport pathway that releases heat instead of ATP. As mentioned above, this pathway was found to be engaged by day in *Kalanchoë daigremontiana* (Robinson et al. 1992).

Experiments with the facultative CAM plant *M. crystallinum* have shown that the switch from C_3-photosynthesis to CAM photosynthesis that is induced by water stress cannot occur until young plants have passed from the seedling to the juvenile growth phase (Herppich et al. 1992). The switch is accompanied

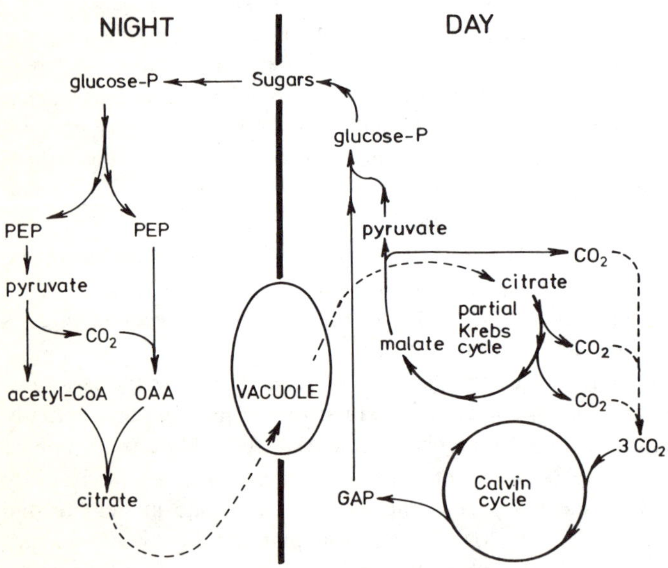

Fig. 4. Carbon flow in stressed CAM plants. By night, net conversion of reserve polysaccharide to citrate occurs. By day, each molecule of citrate is converted to three of CO_2 and one of pyruvate; the 3 CO_2 are then photosynthetically reduced to one of GAP which, with the pyruvate, replenishes the pool of reserve polysaccharide. There is no *net* fixation of CO_2 in these stressed plants, and therefore no growth. Overall, the entire diurnal cycle dissipates the energy of 10 ATP per molecule of citrate cycled. Scheme developed from observations of Franco et al. (1992)

by marked increases in the activities of starch-degrading enzymes (Paul et al. 1993), as is required to initiate the generation of PEP, and in the activity of PEP carboxylase. A new observation is that events between the environmental signal (e.g. water stress) and the synthesis of new protein (e.g. PEP carboxylase) involve well-known plant hormones such as abscisic acid and cytokinin (Schmitt and Piepenbrock 1992; Dai et al. 1994). Hormonal effects are possibly also involved in "reversing" stomatal behaviour, as observed when CAM was induced in the facultative CAM plant *Portulaca afra* (Lee and Assman 1992).

d) PEP Carboxylase and Malate Dehydrogenase

A central feature of C_4 metabolism is the capture of CO_2 by PEP carboxylase. The product is the 4C acid oxaloacetate which, in the majority of cases, is rapidly reduced to malate by malate dehydrogenase (although a significant alternative is transamination to asparatate). Both the carboxylase and dehydrogenase have recently received considerable attention.

The light-mediated activation of the cytosolic enzyme PEP carboxylase in C_4 plant leaves has been a primary point of interest. The activation is achieved by attachment of a phosphate group to a serine residue near the N-terminus of the protein; the phosphate is transferred from ATP in a reaction catalysed by a "PEP-carboxylase kinase" (PEPC-kinase). It is the negative charge on the phosphate that is critical for activation (Wang et al. 1992). Light-induced alkalisation of the cytosol, ATP formation, and Ca^{2+} all appear to be important for the activation to proceed (Devi and Raghavendra 1992; Pierre et al. 1992). The exact mechanism by which light is connected to PEPC-kinase activity is still unclear, although it is likely that a light-induced signal originates in the chloroplast, and cytosolic protein synthesis is involved (Bakrim et al. 1992; Li and Chollet 1993). The role of Ca^{2+} is also unclear: a Ca^{2+}-dependent PEPC-kinase has been identified (Ogawa et al. 1992), but Ca^{2+}-independent forms of the enzyme have also been obtained and are believed to be more important (Bakrim et al. 1992; Wang and Chollet 1993a).

A similar phosphorylative activation of PEP carboxylase occurs in CAM plants, but in darkness (see our last review); the activation has now been demonstrated for the facultative CAM plant *M. crystallinum* (Baur et al. 1992). Recent evidence indicates that the activation is widespread, in that phosphorylation of PEP carboxylase from tobacco leaves (Wang and Chollet 1993b), wheat leaves (Manh et al. 1993), bean stomatal cells (Schnabl et al. 1992), and even soybean root nodules (Schuller and Werner 1993) has been reported. However, the extent of the resultant activation may be less for the C_3-plant enzyme compared to the C_4-plant enzyme (Rajagopalan et al. 1993).

Other regulatory aspects of PEP carboxylase have been briefly considered. Reversible association/dissociation of enzyme subunits may be involved in inactivation of the enzyme upon binding to chloroplast membranes (Wu and Wedding 1992) and in kinetic hysteresis in vitro (Avilan 1993), but is unlikely to be important for light/dark regulation (Weigend and Hincha 1992). The CAM-plant enzyme easily associates with malate dehydrogenase, resulting in a change in activity (Queiroz-Claret and Queiroz 1992). Several similar, but not identical, forms of PEP carboxylase have been reported for *M. crystallinum* leaves (Slocombe et al. 1993) and for bean guard cells, but only one form was in bean-leaf mesophyll cells (Schulz et al. 1992). Interestingly, two C_4 species of *Panicum* contained PEP carboxylases that totally differed with respect to cold inactivation in vitro (Krall and Edwards 1993).

The chloroplast's NADP-linked malate dehydrogenase has been the object of site-directed mutagenesis aimed at cysteine residues implicated in the light-dependent reductive activation process. Results show that the reduction of two disulfides per enzyme monomer are involved for the sorghum enzyme (Issakidis et al. 1992) and probably also for the pea enzyme (Reng et al. 1993); other approaches led to a comparable conclusion for the maize enzyme (Hatch and Agostino 1992). Prevention of disulfide formation by replacing cysteines with serines results in a "permanently active" enzyme, unable to undergo light-mediated regulation (Issakidis et al. 1993). Incidentally, light-mediated activation of the native enzyme increases its hydrophobicity, an event that has practical and, perhaps, physiological consequences (Scheibe et al. 1993).

8. The Limits to Plant Productivity

We return to our question: why can't plants grow faster than they do? We have addressed this question previously in our 1988 review, where two levels were distinguished: (1) external factors that limit growth in the field – principally temperature and the availabilities of light, water, CO_2 and nutrients, and (2) plant physiological/biochemical factors that determine the photosynthetic capacity, i.e. the growth rate when temperature is optimum and the supplies of the other four external factors are adequate. In trying to identify the factor(s) that ultimately determine maximum plant productivity, we confess to finding the literature somewhat confusing. However, before explaining our predicament, it is worthwhile to briefly review some of the recent relevant research, since this provides clues as to what the ultimate determinant(s) of plant productivity might be.

a) CO_2

Much attention has been given to CO_2 supply because the level of CO_2 in the atmosphere is well below that required to saturate photosynthesis, but *less* so each year because fossil fuel combustion adds a further 10 billion tonnes to the atmosphere annually (Walker 1992a). The number of reports confirming that this increased CO_2 supply is stimulating photosynthesis and plant growth is itself increasing exponentially; recent studies revealed positive responses for crops and grasses (Nie et al. 1992; Baker and Allen 1993; Holbrook et al. 1993; Betsche 1994), the CAM plant *Opuntia ficus-indica* (Cui et al. 1993), and trees (Gunderson et al. 1993; Idso and Kimball 1992, 1993; Tissue et al. 1993). It is not inconceivable that the stimulation of tree growth is occurring on a global scale (Gifford 1994). Two comprehensive reviews of CO_2-stimulation have been published (Bowes 1993; Poorter 1993); in the latter, the growth stimulation of 156 plant species by doubled atmospheric CO_2 was, on average, 37%.

However, responses to increased CO_2 are not straightforward, as we emphasised earlier (see our 1990 review). Stimulation is less if there are inadequate sinks for photosynthate to travel to (Poorter 1993); in many of these cases, excess photosynthetic product simply accumulates as chloroplast starch (Körner and

Arnone 1992; Morin et al. 1992; Holbrook et al. 1993) or some other nonstructural carbohydrate (Smart et al. 1994). Another point is that growth stimulation and altered assimilate partitioning may not be entirely due to increased photosynthesis: increased CO_2 also reduces dark respiration (Idso and Kimball 1993; Bunce 1994) which, of course, translates into extra growth, and it reduces photorespiration which, in turn, appears to influence whether carbon flows to starch or sucrose (Morin et al. 1992). A third point concerns diffusion of CO_2 from the atmosphere to the chloroplast stroma. In the past, the rate of this diffusion was thought to be so high that it could not significantly limit photosynthesis. While this may be true for some plants, it may not apply to others, especially woody species for which rather low conductances have been measured (Lloyd et al. 1992; Parkhurst 1994). In particular, *Citrus aurantium* had a very low conductance, which may explain why it has a low growth rate (Baker and Allan 1993), and why its growth is so effectively stimulated by increased CO_2 (Idso and Kimball 1992). A final point concerns the CO_2-response of plants in an ecosystem context: in two recent studies, CO_2-stimulation of plant growth was followed by an increase in soil microflora activity that led, in one instance, to a doubling of CO_2 evolution from the soil (Körner and Arnone 1992), and in the other, to a sequestration of soil nutrients that retarded further plant growth (Diaz et al. 1993).

b) Water

Although it is well known that severe and prolonged water stress results in stomatal closure, lack of CO_2, inhibited photosynthesis and onset of photoinhibition, the consequences of milder water stress are less clear-cut. It seems that the availability of CO_2 for photosynthesis is little affected, but photosynthesis is nevertheless reduced via altered biochemistry; however, the alteration does not directly involve RuBP carboxylase (Lauer and Boyer 1992; Gunasekera and Berkowitz 1993).

c) Light

Concepts concerning the effect of light intensity on photosynthesis are not as clear as might be thought. One common misconception is that, at low intensities, the photosynthetic light-response curve (i.e. curve of photosynthesis versus light intensity) for "shade" plants is steeper than that for "sun" plants. In fact, Walker (1992a,b) has pointed out that, given the almost invariable quantum yield of 9 (i.e. 9 photons utilised per molecule of CO_2 fixed) for C_3 plants when not stressed and under good conditions, the slopes of these light-response curves at low light intensities are identical and simply reflect the fact that both types of plant are using all available light as best they can, and close to the maximum efficiency dictated by the Z-scheme of photochemistry, i.e. 8 photons required to generate the 2 NADPH needed to reduce 1 CO_2 to carbohydrate in the Calvin cycle.

At high light intensities, explanations are more difficult. A common observation is that the slope of the light-response curve becomes zero when light saturation is reached. This occurs at about one-half of full sunshine for sun plants, but at a much lower light intensity for shade plants (Walker 1992b). The lower value for shade plants reflects the lower capacity to fix CO_2, due at least in part to a reduced content of RuBP carboxylase; this reduction in the carboxylase-protein content spares available nitrogen, which is invested instead into additional pigment-protein complexes of light-harvesting antennae (Woodward 1990). An explanation, in biophysical terms, of why several to many hundreds of accessory light-harvesting pigment molecules and their associated proteins are required per reaction centre chlorophyll, is given by Glazer (1989).

Sun plants, on the other hand, are free to invest a greater proportion of their nitrogen into the provision of RuBP carboxylase (Makino et al. 1992), thus the question arises as to what limits their photosynthetic ability. In normal air, their light-response curves do show light saturation, but when CO_2 is increased, either artificially (Bugbee and Salisbury 1988; Fig. 5) or naturally (C_4 plants), light saturation is difficult to detect. Results from the impressive experiments of Bugbee and Salisbury (1988; see also Bugbee and Monje 1992) led these authors to conclude that, in essence, the dry matter production of wheat plants growing under ideal conditions was limited by an *external* factor, i.e. light supply, *even when this supply was well in excess of full sunshine*. Presumably, all of the plants

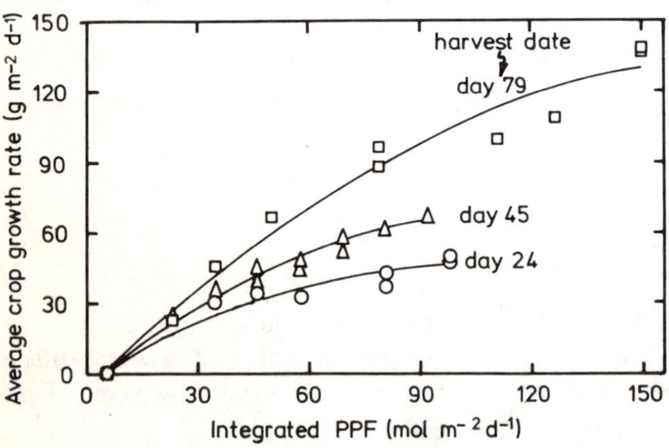

Fig. 5. Effect of daily PPF (photosynthetic photon flux 400–700 nm) on average wheat growth rate at three different harvest dates. An integrated PPF of 150mol m^{-2} d^{-1} equates to a light intensity of 2080 μmol photons m^{-2} s^{-1} for 20 h. This is more than twice the PPF from summer sunlight in the field. Redrawn from Bugbee and Salisbury (1988)

internal physiological/biochemical components could accomodate the prodigious growth rates achieved (see below).

d) Determinants of Photosynthetic Rate

How, then, should we view experiments in which the capacity of one or other *biochemical* component is considered in the context of the photosynthetic rate? Three components are often considered: (1) the rate of photosynthetic electron transport (as detemined by its limiting step, the rate of plastoquinol diffusion within the thylakoid membrane; Blackwell et al. 1994), (2) the activity of RuBP carboxylase, and (3) the utilisation of exported GAP in the cytosol and the return of the Pi, liberated from the GAP during this process, back to the chloroplast; proteins that are contenders for the rate-limiting step in this process are the cytosolic fructose-1,6-bisphosphatase (Daie 1993), SPS (Galtier et al. 1993), and the chloroplast envelope's phosphate translocator (Riesmeier et al. 1993b).

There are numerable instances where one or other of these components has been recognised as imposing a primary limitation on photosynthesis (Bowes 1993; Ögren and Evans 1993; Scholes et al. 1994). It is beyond the scope of this review to give further consideration to electron transport, but we shall briefly consider the other two components. Firstly, the evidence that RuBP carboxylase activity is relevant. Certainly, the carboxylation of RuBP in plants grown in *normal* air can contribute to the determination of photosynthetic rate (Jiang et al. 1993; Lauerer et al. 1993; Masle et al. 1993; Stitt and Schulze 1994), but this is as much a reflection of the low CO_2 supply as it is of the RuBP carboxylase level (Bowes 1993; Woodrow 1994). There is now ample evidence that, when plants are grown in a high-CO_2 atmosphere, photosynthesis increases (Poorter 1993; see Sect. 8a). However, while some species appear to make full use of the extra CO_2 by allowing their normal RuBP carboxylase (i.e. that present in air-grown plants) to operate correspondingly faster, at least an equal number reduce the activity by regulatory means (Sect. 4b), or by synthesising less of it (Bowes 1993; Socias et al. 1993; Tissue et al. 1993; Sicher et al. 1994). One is left with the impression that the RuBP carboxylase activity, while not excessive, is still adequate.

The case for the other option (cytosolic utilisation of GAP and return of Pi to the chloroplast) as being the determinant of photosynthetic rate under high light and saturating CO_2 is stronger (Bowes 1993), but less attention has been devoted to it. Scholes et al. (1994) concluded that it applied to barley leaves. The activity of SPS appeared to be important in bean leaves (Socias et al. 1993). Notably, the photosynthetic capacity of tomato leaves was increased by 20% after the SPS activity was elevated by expressing a maize SPS gene in the leaves; however, the activity of the maize enzyme could not be regulated by the tomato leaf cells (Galtier et al. 1993), and it is questionable whether a similar increase would be measured if the extra SPS was from spinach (Sonnewald et al. 1994).

Finally, we conclude by addressing the fundamental question of whether or not the capacities of the above three biochemical components exceed those required to fully utilise the light energy available in full sunshine. If they do [and the above-mentioned experiments of Bugbee and Salisbury (1988) suggest that they do in wheat], then the biochemical/biophysical component that determines photosynthetic *capacity* will not be found by growing plants under full sunshine, because the light level will still be below saturation.

In addition, if these plants convert light energy into biomass with an efficiency close to the theoretical maximum, then we may conclude that the component determining photosynthetic capacity is not yet limiting photosynthesis; rather, light supply is. Indeed, measurements consistent with appropriately high efficiencies have been made. Bugbee and Salisbury (1988) report that the photosynthetic machinery of wheat was able to convert light energy into biomass at close to the maximum expected efficiency, even when the daily supply of light energy was more than twice that in normal summer sunshine. They calculated an efficiency of 7 to 10% for wheat grown up to 79 d. This compares with the proposed theoretical maximum of 15.6% (Bugbee and Monje 1992). The associated growth rate under the equivalent of full sunshine was about 60 g dry matter m^{-2} d^{-1} (Fig. 5). An important point is that these are not isolated values; the best reported growth rates for plants in the field (and presumably at ambient CO_2) were at least one-half of the CO_2-saturated rate for wheat (Bugbee and Salisbury 1988), and more recent reports include growth rates for cacti in Mexico (Nobel et al. 1992) and for a C_4 plant on an Amazon floodplain (Piedade et al. 1994) that we calculate to be equivalent to 13 and 30 g dry matter m^{-2} d^{-1}. Furthermore, leaf photosynthesis rates of over 60 μmol CO_2 fixed m^{-2} s^{-1} have been observed for castor bean (Dai et al. 1992) and calculated feasible for other species (Wullschleger 1993); such rates are comparable to a growth rate of 60 g dry matter m^{-2} d^{-1}. Finally, it is worth recalling that rapidly growing microalgal cultures under high light and saturating CO_2 (the latter due to their CO_2-concentrating mechanism; Sect. 3) have been reported to have high photon-to-biomass conversion efficiencies, e.g. 13% for *Phaeodactylum tricornutum* (Thomas et al. 1984).

Collectively, these reports indicate that, for some sun plants at least, growth is ultimately limited by the availability of solar energy. The arrangement of the photochemical apparatus and the nature of light (Walker 1992b) dictate that the maximum growth rate possible (and demonstrated to be practicable) is 60 g dry matter m^{-2} d^{-1}.

Therefore, if we knew the area upon which Jack's proverbial beanstalk grew, we could now predict the maximum rate at which it could have grown upwards, and we could explain that it could not have grown faster because the sun is not bright enough to power faster growth!

References

Aggarwal KK, Saluja D, Sachar RC (1993) Phytochemistry 34:329–335. **Albrecht G**, Wiedenroth E-M (1994) J Exp Bot 45:449–455. **Angelov MN**, Sun J, Byrd GT, Brown RH, Black CC (1993) Photosyn Res 38:61–72. **Aoki N**, Ohnishi J, Kanai R (1992) Plant Cell Physiol 33:805–809. **Aono M**, Kubo A, Saji H, Tanaka K, Kondo N (1993) Plant Cell Physiol 34:129–135. **Ardila F**, Pozueta-Romero J, Akazawa T (1993) Plant Cell Physiol 34:237–242. **Aro E-M**, McCaffery S, Anderson JM (1993) Plant Physiol 103:835–843. **Aro E-M**, McCaffery S, Anderson JM (1994) Plant Physiol 104:1033–1041. **Avilan L** (1993) Photosynthetica 28:317–320.

Badger MR, Price GD (1992) Physiol Plant 84:606–615. **Badger MR**, Pfanz H, Büdel B, Heber U, Lange O (1993) Planta 191:57–70. **Bailey KJ**, Walker DA (1992) Plant Physiol 99:124–129. **Baker JT**, Allen LH (1993) Vegetatio 104/105:239–260. **Bakrim N**, Echevarria C, Cretin C, Arrio-Dupont M, Pierre JN, Vidal J, Chollet R, Gadal P (1992) Eur J Biochem 204:821–830. **Baldy P** (1993) J Plant Physiol 141:633–635. **Bastide B**, Sipes D, Hann J, Ting IP (1993) Plant Physiol 103:1089–1096. **Batz O**, Scheibe R, Neuhaus HE (1992) Plant Physiol 100:184–190. **Batz O**, Scheibe R, Neuhaus HE (1993) Biochem J 294:15–17. **Baur B**, Dietz K-J, Winter K (1992) Eur J Biochem 209:95–101. **Behre B**, Ratajczak R, Lüttge U (1992) Bot Acta 105:260–265. **Benet H**, Bruss U, Duval J-C, Kloareg B (1994) J Exp Bot 45:211–220. **Berkowitz GA**, Peters JS (1993) Plant Physiol 102:261–267. **Betsche T** (1994) Environ Exp Bot 34:75–86. **Blackwell M**, Gibas C, Gygax S, Roman D, Wagner B (1994) Biochim Biophys Acta 1183:533–543. **Borchert S**, Harborth J, Schünemann D, Hoferichter P, Heldt HW (1993) Plant Physiol 101:303–312. **Borland AM**, Griffiths H, Broadmeadow MSJ, Fordham MC, Maxwell C (1993) Oecologia 95:444–453. **Bowes G** (1993) Annu Rev Plant Physiol Plant Mol Biol 44:309–332. **Bremberger C**, Lüttge U (1992) Planta 188:575–580. **Brown RH**, Byrd GT (1993) Plant Physiol 103:1183–1188. **Brown RH**, Byrd GT, Bouton JH, Bassett CL (1993) Plant Physiol 101:825–831. **Bugbee B**, Monje O (1992) Bioscience 42:494–502. **Bugbee BG**, Salisbury FB (1988) Plant Physiol 88:869–878. **Bunce JA** (1994) Physiol Plant 90:427–430.

Cakmak I, Marschner H (1992) Plant Physiol 98:1222–1227. **Caldeira K**, Kasting JF (1992) Nature 360:721–723. **Campbell WJ**, Ogren WL (1992) Plant Cell Physiol 33:751–756. **Cerling TE**, Wang Y, Quade J (1993) Nature 361:344–345. **Cheikh N**, Brenner ML, Huber JL, Huber SC (1992) Plant Physiol 100:1238–1242. **Choudhury NK**, Choe HT, Huffaker RC (1993) J Plant Physiol 141:551–556. **Clayton H**, Ranson J, ap Rees T (1993) Arch Biochem Biophys 301:151–157. **Coates SA**, ap Rees T (1994) Phytochemistry 35:881–883. **Colas des Francs-Small C**, Ambard-Bretteville F, Small ID, Remy R (1993) Plant Physiol 102:1171–1177. **Cole DR**, Monger HC (1994) Nature 368:533–536. **Corbier C**, Della Seta F, Branlant G (1992) Biochemistry 31:12532–12535. **Cui M**, Miller PM, Nobel PS (1993) Plant Physiol 103:519–524.

Dai Z, Edwards GE, Ku MSB (1992) Plant Physiol 99:1426–1434. **Dai Z**, Ku MSB, Edwards GE (1993) Plant Physiol 103:83–90. **Dai Z**, Ku MSB, Zhang D, Edwards GE (1994) Planta 192:287–294. **Daie J** (1993) Photosyn Res 38:5–14. **Demmig-Adams B**, Adams WW (1993) Plant Physiol 103:1413–1420. **Devi MT**, Raghavendra AS (1992) J Plant Physiol 139:431–435. **Devi MT**, Raghavendra AS (1993) J Exp Bot 44:779–784. **Diaz S**, Grime JP, Harris J, McPherson E (1993) Nature 364:616–617. **Drechsler Z**, Sharkia R, Cabantchik ZI, Beer S (1993) Planta 191:34–40. **Dreier W**, Preusser E, Gründel M (1992) Biochem Physiol Pflanzen 188:81–96.

Eberl D, Preissler M, Steingraber M, Hampp R (1992) Physiol Plant 84:13–20. **Echeverria E**, Salerno G (1993) Physiol Plant 88:434–438. **Edwards EA**, Enard C, Creissen GP, Mullineaux PM (1994) Planta 192:137–143. **El Bouhmadi K**, Larher F (1993) Plant Sci 94:99–107. **Everard JD**, Franceschi VR, Loescher WH (1993) Plant Physiol 102:345–356.

Fischer K, Arbinger B, Kammerer B, Busch C, Brink S, Wallmeier H, Sauer N, Eckerskorn C, Flügge U-I (1994) Plant J 5:215–226. **Flores HE**, Dai Y, Cuello JL, Maldonado-Mendoza IE, Loyola-Vargas VM (1993) Plant Physiol 101:363–371. **Fonollá J**, Hermoso R, Carrasco JL, Chueca A, Lázaro JJ, Prado FE, López-Gorgé J (1994) Plant Physiol 104:381–386. **Franco AC**, Ball E, Lüttge U (1992) Plant Cell Environ 15:821–829.

Galtier N, Foyer CH, Huber J, Voelker TA, Huber SC (1993) Plant Physiol 101:535–543. **Gardeström P** (1993) Biochim Biophys Acta 1183:327–332. **Gemel J**, Randall DD (1992) Plant Physiol 100:908–914. **Ghiena C**, Schulz M, Schnabl H (1993) Plant Physiol 101:73–79. **Gifford RM** (1994) Aust J Plant Physiol 21:1–15. **Gilmore AM**, Yamamoto HY (1992) Proc Natl Acad Sci USA 89:1899–1903. **Glazer AN** (1989) J Biol Chem 264:1–4. **Goldschmidt EE**, Huber SC (1992) Plant Physiol 99:1443–1448. **Gontero B**, Mulliert G, Rault M, Giudici-Orticoni M-T, Ricard J (1993) Eur J Biochem 217:1075–1082. **Goyal A**, Shiraiwa Y, Husic HD, Tolbert NE (1992) Mar Biol 113:349–355. **Grotjohann N**, Schneider G, Kowallik W (1993) Z Naturforsch 48c:22–27. **Gunasekera D**, Berkowitz GA (1993) Plant Physiol 103:629–635. **Gunderson CA**, Norby RJ, Wullschleger SD (1993) Plant Cell Environ 16:797–807. **Gupta AS**, Heinen JL, Holaday AS, Burke JJ, Allen RD (1993) Proc Natl Acad Sci USA 90:1629–1633. **Guy CL**, Huber JLA, Huber SC (1992) Plant Physiol 100:502–508.

Haglund K, Björk M, Ramazanov Z, García-Reina G, Pedersén M (1992) Planta 187:275–281. **Hajirezaei M**, Sonnewald U, Viola R, Carlisle S, Dennis D, Stitt M (1994) Planta 192:16–30. **Hanning I**, Heldt HW (1993) Plant Physiol 103:1147–1154. **Harpel MR**, Hartman FC (1992) J Biol Chem 267:6475–6478. **Hatch MD**, Agostino A (1992) Plant Physiol 100:360–366. **Hattersley PW** (1992) In: Chapman GP (ed) Desertified grasslands: their biology and management. Linnean Soc Symp Ser No 13. Academic Press, London, pp 181–212. **Havaux M**, Gruszecki WI (1993) Photochem Photobiol 58:607–614. **Heber U**, Walker DA (1992) Plant Physiol 100:1621–1626. **Heineke D**, Sonnewald U, Büssis D, Günter G, Leidreiter K, Wilke I, Raschke K, Willmitzer L, Heldt HW (1992) Plant Physiol 100:301–308. **Herppich W**, Herppich M, von Willert DJ (1992) Bot Acta 105:34–40. **Heupel R**, Heldt HW (1994) Eur J Biochem 220:165–172. **Holbrook GP**, Hansen J, Wallick K, Zinnen TM (1993) Environ Exp Bot 33:313–321. **Huber JLA**, Huber SC (1992) Biochem J 283:877–882. **Huber SC**, Huber JL, Campbell WH, Redinbaugh MG (1992) Plant Cell Physiol 33:639–646. **Hudson GS**, Morell MK, Arvidsson YBC, Andrews TJ (1992) Aust J Plant Physiol 19:213–221. **Huner NPA**, Öquist G, Hurry VM, Krol M, Falk S, Griffith M (1993) Photosyn Res 37:19–39. **Huzisige H**, Ke B (1993) Photosyn Res 38:185–209.

Idso SB, Kimball BA (1992) Plant Physiol 99:341–343. **Idso SB**, Kimball BA (1993) J Plant Physiol 141:166–171. **Israel A**, Beer S (1992) Mar Biol 112:697–700. **Issakidis E**, Miginiac-Maslow M, Decottignies P, Jacquot J-P, Cretin C, Gadal P (1992) J Biol Chem 267:21577–21583. **Issakidis E**, Decottignies P, Miginiac-Maslow M (1993) FEBS Lett 321:55–58. **Iwasaki I**, Arata H, Kijima H, Nishimura M (1992) Plant Physiol 98:1494–1497.

Jahns P, Krause GH (1994) Planta 192:176–182. **Janave MT**, Ramaswamy NK, Nair PM (1993) Eur J Biochem 214:889–896. **Jelitto T**, Sonnewald U, Willmitzer L, Hajirezeai M, Stitt M (1992) Planta 188:238–244. **Jiang C-Z**, Rodermel SR, Shibles RM (1993) Plant Physiol 101:105–112. **Johnson GN**, Young AJ, Scholes JD, Horton P (1993) Plant Cell

Environ 16:673–679. **Johnston AM**, Maberly SC, Raven JA (1992) Oecologia 92:317–326.

Karlsson J, Ramazanov Z, Hiltonen T, Gardeström P, Samuelsson G (1994) Planta 192:46–51. **Kent SS**, André M, Cournac L, Farineau J (1992) Plant Physiol Biochem 30:625–637. **Khayat E**, Harn C, Daie J (1993) Plant Physiol 101:57–64. **Kingston-Smith AH**, Major I, Parry MAJ, Keys AJ (1992) Biochem J 287:821–825. **Kiss F**, Wu M-X, Wong JH, Balogh A, Buchanan BB (1991) Arch Biochem Biophys 287:337–340. **Kliemchen A**, Schomburg M, Galla H-J, Lüttge U, Kluge M (1993) Planta 189:403–409. **Klink R**, Lüttge U (1992) Bot Acta 105:414–420. **Körner C**, Arnone JA (1992) Science 257:1672–1675. **Kossmann J**, Müller-Röber B, Dyer TA, Raines CA, Sonnewald U, Willmitzer L (1992) Planta 188:7–12. **Kots AY**, Sergienko EA, Bulargina TV, Severin ES (1993) FEBS Lett 324:33–36. **Krall JP**, Edwards GE (1993) Plant Cell Physiol 34:1–11. **Krapp A**, Hofmann B, Schäfer C, Stitt M (1993) Plant J 3:817–828. **Krömer S**, Malmberg G, Gardeström P (1993) Plant Physiol 102:947–955.

Lauer MJ, Boyer JS (1992) Plant Physiol 98:1310–1316. **Lauerer M**, Saftic D, Quick WP, Labate C, Fichtner K, Schulze E-D, Rodermel SR, Bogorad L, Stitt M (1993) Planta 190:332–345. **Lee DM**, Assmann SM (1992) Physiol Plant 85:35–42. **Li B**, Chollet R (1993) Arch Biochem Biophys 307:416–419. **Li B**, Geiger DR, Shieh W-J (1992a) Plant Physiol 99:1393–1399. **Li B**, Servaites JC, Geiger DR (1992b) Plant Physiol 98:1277–1284. **Lin M**, Turpin DH (1993) J Phycol 29:777–786. **Lloyd J**, Syvertsen JP, Kriedemann PE, Farquhar GD (1992) Plant Cell Environ 15:873–899. **Loreto F**, Harley PC, Di Marco G, Sharkey TD (1992) Plant Physiol 98:1437–1443. **Lunn JE**, Douce R (1993) Biochem J 290:375–379. **Lüttge U**, Beck F (1992) Planta 188:28–38.

Makino A, Sakashita H, Hidema J, Mae T, Ojima K, Osmond B (1992) Plant Physiol 100:1737–1743. **Manh CT**, Bismuth E, Boutin J-P, Provot M, Champigny M-L (1993) Physiol Plant 89:460–466. **Marcus Y**, Berry JA, Pierce J (1992) Planta 187:511–516. **Masle J**, Hudson GS, Badger MR (1993) Plant Physiol 103:1075–1088. **Mate CJ**, Hudson GS, von Caemmerer S, Evans JR, Andrews TJ (1993) Plant Physiol 102:1119–1128. **Mateos MI**, Serrano A (1992) Plant Sci 84:163–170. **Mawson BT** (1993) Plant Cell Environ 16:207–214. **McMichael RW**, Klein RR, Salvucci ME, Huber SC (1993) Arch Biochem Biophys 307:248–252. **Michels S**, Scagliarini S, Della Seta F, Carles C, Riva M, Trost P, Branlant G (1994) FEBS Lett 339:97–100. **Mishra NP**, Mishra RK, Singhal GS (1993) Plant Physiol 102:903–910. **Mitra S**, Srivastava GC (1992) Photosynthetica 26:87–89. **Mittler R**, Zilinskas BA (1994) Plant J 5:397–405. **Moore BD**, Seemann JR (1992) Plant Physiol 99:1551–1555. **Morel FMM**, Reinfelder JR, Roberts SB, Chamberlain CP, Lee JG, Yee D (1994) Nature 369:740–742. **Morgan CL**, Turner SR, Rawsthorne S (1993) Planta 190:468–473. **Morgan ME**, Kingston JD, Marino BD (1994) Nature 367:162–165. **Morin F**, André M, Betsche T (1992) Plant Physiol 99:89–95. **Murata S**, Kobayashi M, Matoh T, Sekiya J (1992) Plant Cell Physiol 33:1247–1250. **Murray KJ**, Tenhunen JD, Nowak RS (1993) Oecologia 96:200–207.

Neubauer C, Yamamoto HY (1992) Plant Physiol 99:1354–1361. **Neubauer C**, Yamamoto HY (1994) Photosyn Res 39:137–147. **Neuhaus HE**, Henrichs G, Scheibe R (1993) Plant Physiol 101:573–578. **Nie D**, He H, Kirkham MB, Kanemasu ET (1992) Photosynthetica 26:189–198. **Nimer NA**, Dixon GK, Merrett MJ (1992) New Phytol 120:153–158. **Nobel PS**, García-Moya E, Quero E (1992) Plant Cell Environ 15:329–335. **Nonomura AM**, Benson AA (1992) Proc Natl Acad Sci USA 89:9794–9798.

Oberhuber W, Edwards GE (1993) Plant Physiol 101:507–512. **Ogawa T** (1992) Plant Physiol 99:1604–1608. **Ogawa N**, Okumura S, Izui K (1992) FEBS Lett 302:86–88. **Ögren E**, Evans JR (1993) Planta 189:182–190. **Okita TW** (1992) Plant Physiol 100:560–564.

Osborne BA (1994) Plant Cell Environ 17:143–152. **Overlach S**, Diekmann W, Raschke K (1993) Plant Physiol 101:1201–1207.

Palmqvist K (1993) Planta 191:48–56. **Palmqvist K**, Yu J-W, Badger MR (1994a) Physiol Plant 90:537–547. **Palmqvist K**, Ögren E, Lernmark U (1994b) Plant Cell Environ 17:65–72. **Parkhurst DF** (1994) New Phytol 126:449–479. **Pastori GM**, Trippi VS (1993) Physiol Plant 87:227–231. **Paul MJ**, Loos K, Stitt M, Ziegler P (1993) Plant Cell Environ 16:531–538. **Perata P**, Pozueta-Romero J, Yamaguchi J, Akazawa T (1992) FEBS Lett 309:283–287. **Piedade MTF**, Long SP, Junk WJ (1994) Oecologia 97:193–201. **Pierre JN**, Pacquit V, Vidal J, Gadal P (1992) Eur J Biochem 210:531–537. **Polle A**, Chakrabarti K, Chakrabarti S, Seifert F, Schramel P, Rennenberg H (1992) Plant Physiol 99:1084–1089. **Polley HW**, Johnson HB, Marino BD, Mayeux HS (1993) Nature 361:61–64. **Poorter H** (1993) Vegetatio 104/5:77–97. **Porter MA**, Grodzinski B (1984) Plant Physiol 74:413–416. **Portis AR** (1992) Annu Rev Plant Physiol Plant Mol Biol 43:415–437. **Prado FE**, Lázaro JJ, Hermoso R, Chueca A, López Gorgé J (1992) Planta 188:345–353. **Price GD**, Coleman JR, Badger MR (1992) Plant Physiol 100:784–793.

Queiroz-Claret C, Queiroz O (1992) J Plant Physiol 139:385–389. **Quiñones MA**, Zeiger E (1994) Science 264:558–561. **Quiroga O**, González EL (1993) J Phycol 29:321–324. **Quy LV**, Champigny M-L (1992) Plant Physiol 99:344–347.

Raghavendra AS, Yin Z-H, Heber U (1993) Planta 189:278–287. **Rajagopalan AV**, Devi MT, Raghavendra AS (1993) Photosyn Res 38:51–60. **Rama Das VS**, Vats SK (1993) Photosynthetica 28:91–97. **Ramazanov Z**, Cárdenas J (1992a) Physiol Plant 84:502–508. **Ramazanov Z**, Cárdenas J (1992b) Physiol Plant 85:121–128. **Ramazanov Z**, Mason CB, Geraghty AM, Spalding MH, Moroney JV (1993) Plant Physiol 101:1195–1199. **Ranieri A**, Lencioni L, Schenone G, Soldatini GF (1993) J Plant Physiol 142:286–290. **Rawsthorne S** (1992) Plant J 2:267–274. **Razdan K**, Heinrikson RL, Zurcher-Neely H, Morris PW, Anderson LE (1992) Arch Biochem Biophys 298:192–197. **Read BA**, Tabita FR (1992) Biochemistry 31:5553–5560. **Reng W**, Riessland R, Scheibe R, Jaenicke R (1993) Eur J Biochem 217:189–197. **Riebesell U**, Wolf-Gladrow DA, Smetacek V (1993) Nature 361:249–251. **Riesmeier JW**, Hirner B, Frommer WB (1993a) Plant Cell 5:1591–1598. **Riesmeier JW**, Flügge U-I, Schulz B, Heineke D, Heldt H-W, Willmitzer L, Frommer WB (1993b) Proc Natl Acad Sci USA 90:6160–6164. **Riesmeier JW**, Willmitzer L, Frommer WB (1994) EMBO J 13:1–7. **Robinson SA**, Yakir D, Ribas-Carbo M, Giles L, Osmond CB, Siedow JN, Berry JA (1992) Plant Physiol 100:1087–1091. **Rotatore C**, Colman B (1992) J Exp Bot 43:571–576. **Rotatore C**, Lew RR, Colman B (1992) Planta 188:539–545.

Sage RF (1993) Photosyn Res 35:219–226. **Sage RF**, Seemann JR (1993) Plant Physiol 102:21–28. **Sage RF**, Reid CD, Moore BD, Seemann JR (1993) Planta 191:222–230. **Sainis JK**, Srinivasan VT (1993) J Plant Physiol 142:564–568. **Salvucci ME** (1992) Arch Biochem Biophys 298:688–696. **Saradadevi K**, Raghavendra AS (1992) Plant Physiol 99:1232–1237. **Sassenrath-Cole GF**, Pearcy RW (1992) Plant Physiol 99:227–234. **Sawada S**, Usuda H, Tsukui T (1992) Plant Cell Physiol 33:943–949. **Scagliarini S**, Trost P, Pupillo P, Valenti V (1993) Planta 190:313–319. **Scheibe R**, Geissler A, Rother T (1993) Arch Biochem Biophys 300:635–640. **Schlichting R**, Bothe H (1993) Bot Acta 106:428–434. **Schmitt JM**, Piepenbrock M (1992) Plant Physiol 99:1664–1669. **Schnabl H**, Denecke M, Schulz M (1992) Bot Acta 105:367–369. **Schnettger B**, Critchley C, Santore UJ, Graf M, Krause GH (1994) Plant Cell Environ 17:55–64. **Scholes JD**, Lee PJ, Horton P, Lewis DH (1994) New Phytol 126:213–222. **Schuller KA**, Werner D (1993) Plant Physiol 101:1267–1273. **Schulz M**, Hunte C, Schnabl H (1992) Physiol Plant 86:315–321. **Schünemann D**, Borchert S, Flügge U-I, Heldt HW (1993) Plant Physiol 103:131–137. **Sharkey TD**, Savitch LV, Vanderveer PJ, Micallef BJ (1992) Plant

Physiol 100:210–215. **Shiraiwa Y**, Goyal A, Tolbert NE (1993) Plant Cell Physiol 34:649–657. **Shyam R**, Raghavendra AS, Sane PV (1993) Physiol Plant 88:446–452. **Sicher RC**, Kremer DF, Rodermel SR (1994) Plant Physiol 104:409–415. **Singh R**, Green MR (1993) Science 259:365–368. **Slocombe SP**, Whitelam GC, Cockburn W (1993) Plant Cell Environ 16:403–411. **Smart DR**, Chatterton NJ, Bugbee B (1994) Plant Cell Environ 17:435–442. **Smirnoff N** (1993) New Phytol 125:27–58. **Smith CM**, Duff SMG, Chollet R (1994) Arch Biochem Biophys 308:200–206. **Socias FX**, Medrano H, Sharkey TD (1993) Plant Cell Environ 16:81–86. **Sonnewald U**, Quick WP, MacRae E, Krause K-P, Stitt M (1993) Planta 189:174–181. **Sonnewald U**, Lerchl J, Zrenner R, Frommer W (1994) Plant Cell Environ 17:649–658. **Spreitzer RJ** (1993) Annu Rev Plant Physiol Plant Mol Biol 44:411–434. **Stabenau H**, Winkler U, Säftel W (1993) Planta 191:362–364. **Stark DM**, Timmerman KP, Barry GF, Preiss J, Kishore GM (1992) Science 258:287–292. **Stitt M**, Schulze D (1994) Plant Cell Environ 17:465–487. **Strauss G**, Fuchs G (1993) Eur J Biochem 215:633–643. **Sültemeyer D**, Biehler K, Fock HP (1993) Planta 189:235–242. **Süss K-H**, Arkona C, Manteuffel R, Adler K (1993) Proc Natl Acad Sci USA 90:5514–5518.

Thielmann J, Goyal A, Tolbert NE (1992) Plant Physiol 100:2113–2115. **Thom M**, Getz HP, Maretzki A (1992) Physiol Plant 86:104–114. **Thomas WH**, Seibert DLR, Alden M, Neori A, Eldridge P (1984) Biomass 5:181–209. **Tissue DT**, Thomas RB, Strain BR (1993) Plant Cell Environ 16:859–865. **Tobias RB**, Boyer CD, Shannon JC (1992) Plant Physiol 99:146–152. **Trost P**, Scagliarini S, Valenti V, Pupillo P (1993) Planta 190:320–326.

Ueno O (1992) Physiol Plant 85:189–196. **Ushimaru T**, Shibasaka M, Tsuji H (1992) Plant Cell Physiol 33:1065–1071. **Usuda H**, Shimogawara K (1993) Plant Cell Physiol 34:767–770. **Utsunomiya E**, Muto S (1993) Physiol Plant 88:413–419.

van Bel AJE, Ammerlaan A, van Dijk AA (1994) Planta 192:31–39. **van Cleve B**, Forreiter C, Sauter JJ, Apel K (1993) Planta 189:70–73. **Von Mohl H** (1844) Botan Z II:273–277.

Walker D (1992a) Energy, plants and man. Oxygraphics Ltd, Brighton, 277 pp. **Walker D** (1992b) New Phytol 121:325–345. **Walker MA**, McKersie BD (1993) J Plant Physiol 141:234–239. **Wang Y-H**, Chollet R (1993a) Arch Biochem Biophys 304:496–502. **Wang Y-H**, Chollet R (1993b) FEBS Lett 328:215–218. **Wang ZY**, Portis AR (1992) Plant Physiol 99:1348–1353. **Wang Y-H**, Duff SMG, Lepiniec L, Crétin C, Sarath G, Condon SA, Vidal J, Gadal P, Chollet R (1992) J Biol Chem 267:16759–16762. **Watanabe S**, Tamai N, Hoshina S, Sanada Y, Nishida K, Wada K (1992) J Plant Physiol 140:129–133. **Weger HG**, Dasgupta R (1993) J Phycol 29:300–308. **Weigend M**, Hincha DK (1992) J Plant Physiol 140:653–660. **Weiner H**, McMichael RW, Huber SC (1992) Plant Physiol 99:1435–1442. **Weiner H**, Weiner H, Stitt M (1993) FEBS Lett 333:159–164. **White PJ**, Smith JAC (1992) J Plant Physiol 139:533–538. **Wigge B**, Krömer S, Gardeström P (1993) Physiol Plant 88:10–18. **Wilkins MB**, Thomas SL (1993) J Exp Bot 44:901–906. **Williams TG**, Colman B (1993) Plant Physiol 103:943–948. **Winter H**, Robinson DG, Heldt HW (1993) Planta 191:180–190. **Woodrow IE** (1994) Plant Cell Environ 17:277–286. **Woodrow IE**, Mott KA (1992) Plant Physiol 99:298–303. **Woodward FI** (1990) Trends Ecol Evol 5:111–115. **Wu M-X**, Wedding RT (1992) Plant Physiol 100:382–387. **Wullschleger SD** (1993) J Exp Bot 44:907–920.

Yin Z-H, Heber U, Raghavendra AS (1993) Planta 189:267–277. **Yokota A**, Higashioka M, Wadano A (1992) Eur J Biochem 208:721–727. **Yokota A**, Higashioka M, Taira T, Usuda H, Wadano A, Murayama H (1994) Plant Cell Physiol 35:317–321. **Young XK**, McCarty RE (1993) Plant Physiol 101:793–799. **Yu J-W**, Woo KC (1992) Aust J Plant

Physiol 19:671–678. **Yu J-W**, Price GD, Badger MR (1994) Aust J Plant Physiol 21:185–195.

Zotz G, Winter K (1993) Plant Physiol 102:835–841.

Dr. Grahame J Kelly
Centre for Molecular Biotechnology
School of life Science
Queensland University of Technology
Brisbane, Queensland 4000, Australia

Prof. Dr. Dr. Erwin Latzko
Institut für Allgemeine Chemie und Biochemie
Technische Universität München
Freising Weihenstephan, Germany

V. Plant Stress. The Adaptive Potential of Dynamic Systems

By E. Pahlich

1. Introduction

Growth and development of plants under natural conditions are the results of the interference of two major regimes of influences: the gene potential of the individual and the external environment.

In wild plants, both regimes have been harmonized during the process of evolutionary genome adaptation. The adapted genome thus provides an optimized genetic form of an organism which guaranties optimal performance in accord with the respective environment. The potential for optimal performance is expressed in particular metabolic patterns which also produce peculiarities of growth and development. These peculiarities are found in the phylogenetically adapted "functional groups" (Körner 1993) like hygrophytes, xerophytes, halophytes, termophiles, etc. Their respective metabolic strategies and structural specifications enable them to cope optimally with their environments.

The natural sites of "functional groups", however, are characterized by daily and seasonal variations of the external environments, which must be buffered by plants through flexible and ontogenetically adaptive functional responses. The question therefore arises, which boundary conditions mark the limits of those adaptive responses, and how differences of species-specific adaptive potentials can arise from a given genome. Answers to these questions should help to solve problems concerned with the elucidation of the species-specific mechanism of plant stress responses to which increasing attention is being given in basic and applied botancial research (Boyer 1982; Jones et al. 1989; Katterman 1990; Somero et al. 1992; Close and Bray 1993).

The adaptive boundaries of growth and development based on biochemical and physiological potentials, will be treated in this chapter with the responses of populations of wild plants transferred to ecologically quite different or even adverse stands. A "classical" text book example for this type of experiment was performed with *Achillea millefolium* (Clausen et al. 1945) and will be demonstrated here.

Progress in Botany, Vol. 56

2. Morphological Plasticity of *Achillea* Genotypes and Environment

Figure 1 shows the result of a "classical" response experiment with populations and clones of *Achillea millefolium* which were transferred from their natural stand to different environments. The morphological patterns of a mountain population of the plant from its natural site in Mather (California) are shown in the center part of the figure. The individuals express differences in their habit under these environmental conditions which arise from small deviations in their respective genome potentials. The differences relate to the height of the plants, the length of the internodes, the number and shape of the leaves, the size and number of flowers and seeds. Each genome generates its individual adapted morph. In this

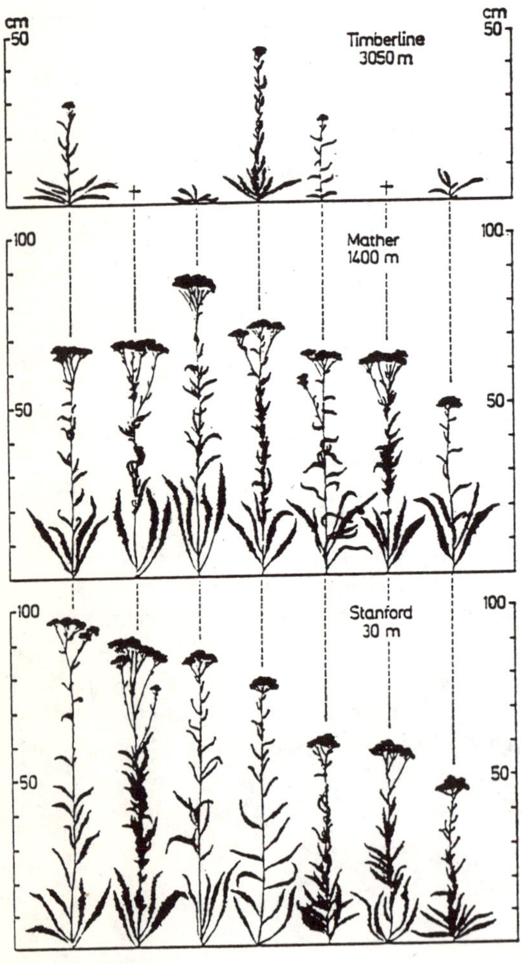

Fig. 1. The adaptive patterns of a population and of clones of this population of *Achillea lanulosa* (tatraploid). The individuals of the natural site Mather (Sierra Nevada, California, *center part of the figure*) were cloned by vegetative propagation and then transferred to Stanford (*lower part*) and Timberline (*upper part*). (Clausen et al. 1945). Differences in the morphological patterns result from respective metabolic adaptive patterns

respect, ontogenetic adaptation is the "specification of a particular genome in a particular environment (which) uniquely determines the corresponding phenotype" (Rosen 1970).

If clones of the individuals are transferred to environments different from their natural stand (lower and upper part of Fig. 1), details of environmental interactions of the clonal genome become apparent (vertical rows). New sets of population patterns arise which touch not only the variation of the general habit of the plants (lower part of Fig. 1) but also the developmental trait (upper part): in the latter extreme case, the whole life cycle is fulfilled (i.e., plants that develop flowers and seeds); only vegetative parts are developed or survival can be endangered. If clones of the plants are confronted with different environments, then obviously an either-or decision is provoked for formation of differences of internode length, leaf shape, number of flowers, or even parts of the entire ontogenetic program. The question must be solved as to how this either-or decision can be explained in terms of a quantitative concept of plant environmental interactions.

The flexible response of the developmental trait of plants towards a specific environment, which is also recognized in physiological and morphological changes, is named plasticity (Bradshaw 1965; Jennings and Trewavas 1986). Plasticity is a measure by which the expression of an individual genotype is changed by different environments. Therefore, plasticity is a quality which is shown more or less by all genotypes of evolutionary adapted wild plants.

The question might be asked, which of the *Achillea* individuals endures stress under different climatic regimes (sea coast – Fig. 1, lower part) or under the adverse exterior of the high mountain climate (Fig. 1, upper part)? Under both conditions the population patterns differ from those of the natural site. Some clones at the natural site Mather develop to a lower extent than at Stanford. According to the stress definition of Grime, (1979) these plants should endure stress because they produce matter on a level lower than their "genetic potential" allows. Other clones which fail to develop might intuitively be presumed to be under stress. The question of stress will not be discussed further here, but will be raised again later. It should, however, be noted: gene-environmental interaction arise from purely physicochemical interactions, and the plant plays no active part in this puzzle in terms of active production of compatible solutes, protease inhibitors, stress proteins, etc. Neither does it actively resist or tolerate, nor become actively hardened. It is the genetic potential which, under the influence of the respective environmental constraints, gives rise to unavoidable traits of self-organization, which as a consequence results in the deposition of metabolites in the vacuole, the extrusion of ions to the exterior via pumps, the accumulation of compatible solutes to high concentrations, or in the shift of rates of protein synthesis or degradation. Appropriately classifications describe teleological categories, but are not helpful in formulating quantitative concepts of ontogenetic functioning (Jones et al. 1989).

Of particular importance, for instance for plant breeding, is knowledge of plastic responses to the environmental conditions of crop plants which have been selected mainly for their productivity under moderate environments (Boyer 1982).

It is well documented that lines of crop plants respond differentially to their environments on the physiological level (Paleg and Aspinall 1981). It is a surprising feature of these plants that "the best-performing genotypes in stress environments, be they extremely dry, cold, hot or saline, may never have experienced stress conditions during their selection" (Richards 1993). The physiology of stress responses is still not understood, especially in terms of the underlying environmental signal-response interactions. This lack of insight is further challenged by the present view that environmentally induced physiological changes are complex and multi-faceted so that many events combine on different physiological levels (Katterman 1990) and it is not simply that "stress genes" are switched on or off. The multi-faceted response of plants leads to the development of diverse metabolic response patterns which are in accord with and typical for their environment. It has been shown, for instance, that identical lines of crop plants adapt to their "ecological or conventional" farming environments so that they produce highly structured and site-specific protein patterns (Thye and Pahlich 1994). In our opinion, this effect is a typical plastic response showing adaptivity on the metabolic level.

In this respect, the genetic lines of cultivars provide well-suited experimental material for the analysis of stress effects which are not based on entirely different biochemical properties. Despite the "quasi-clonal" character of lines of plant cultivars, these plants differentially tolerate, resist, or respond sensitively to drought, high or low temperatures, salty soils, and other adverse environments. In contrast to the clearly evident differences of phylogenetically adapted gene potentials of plants with "typical strategies" (particular biochemical pathways and transport properties), these crop plants show only gradual genome differences and yet have entirely different characters in terms of metabolic-environmental interactions. Differences in response towards environmental impacts must result from a metabolic flexibility, which most likely results from quantitative rather than qualitative genome differences. This chapter will pay particular attention to those metabolic facets which may help to understand adaptive qualities on the basis of small genetic differences.

3. Epigenetic Processes and Metabolic Plasticity

It is generally agreed that the phenotype of plants has its equivalent in corresponding biochemical and physiological patterns: "... fundamentally all plasticity is physiological" (Bradshaw 1965). This insight concentrates the problem of adaptation on a given environment, preferentially on those biochemical and metabolic actions which determine the epigenetic processes.

Epigenesis must be considered as distinct from purely molecular genetics in terms of its paradigms, methods, and assertions. Molecular genetics is based on a linear process of subsequent transformations which delivers the primary gene product – the protein. This is a straightforward process which allows one to determine which differences in gene composition exist between species, between genetic lines of crops, or even between clones of plants. Using these methods can also answer the question as to which genes are active under a given set of environmental conditions, and how these gene expressions change with changing environment or during development. Molecular genetics describes "which genes are involved in adaptive and ontogenetic processes". It is successful "at describing gene products, but fails to demonstrate specific functions for them in the larger process" (Veit et al. 1991).

The "larger process" is the dynamic interactions in the complex metabolic reaction network which, of course, is based on the primary gene products. To unravel these interactions is the first task of the epigenetic approach. The thinking and the methods are based on differential equations which describe the dynamic interactions within the metabolic network and their interactions with the physical environment. From this interaction "originates a dynamical process of self-evaluation" (Eigen 1976) which processes the environmental information in the system. The processing occurs through flexible rearrangements of the parameter structures of a cell (which is the "hardware" for metabolic functioning) in such a way that optimal metabolic performance occurs. In this respect, adaptation describes "the fitting of parameters in a system to a new situation. Gene expression and/or protein degradation may be involved" (Reich and Sel'kov 1981). This, of course, is exactly sought in adaptation, and results in metabolic states of multiple steadiness and of regular and irregular oscillations. The environment-dependent parameter creates cell-internal timing and spacing via instabilities of the dynamic metabolic structures (Boiteux et al. 1980). Epigenetic research elucidates which functional responses can occur in the face of a given set of active genes in a given environment: "how a system functions and interacts".

The genetic and epigenetic approaches can be fruitfully combined, as can be seen from the definition given below, which considers also the dynamic aspects of recent developmental and evolutionary concepts (Eigen 1988a, b).

Adaptation evolves from metabolic network interactions in complex systems. The information for the evolutionary ontogenetic process is encoded in the linear gene structures and is disentangled into chemical reaction processes ("dissipative structures", Nicolis and Prigogine 1989) which – in accordance with the respective environment – coherently and interactively self-organise. Self-organization delivers temporarily stable and optimally adapted spatio-temporal patterns.

Adaptation to an environment results not only from the additive combination of fixed gene product functions, but is a systemic property. The consequence of the parameter shaping by the external environment is a disturbance of the hitherto established kinetic interactions. Therefore the tool for analysis of adaptive responses is the kinetic approach. To master the kinetic approaches is therefore a prerequisite for analyzing and elucidating epigenetic and hence adaptive responses on the ontogentic level.

Kacser (1960) elaborated the analogy of the plastic and adaptive responses of kinetic models (of chemical reaction systems) with living organisms in respect to temperature, pleiotropy, developmental abnormalities, size effects, irreversible traits, and buffering capabilities. From these lists, it can be recognized that knowledge of the kinetic theory and its equivalent in biological processes – metabolic control concepts – are of outstanding importance for the elaboration and understanding of plasticity and adaptibility in plants.

4. Gene Potential and Metabolic Adaptibility

Figure 2 summarizes those components of the genetic potential of a plant which, after gene expression, can give rise to adaptive differences, especially between clones of populations of plants. The concept considers that differences in the gene

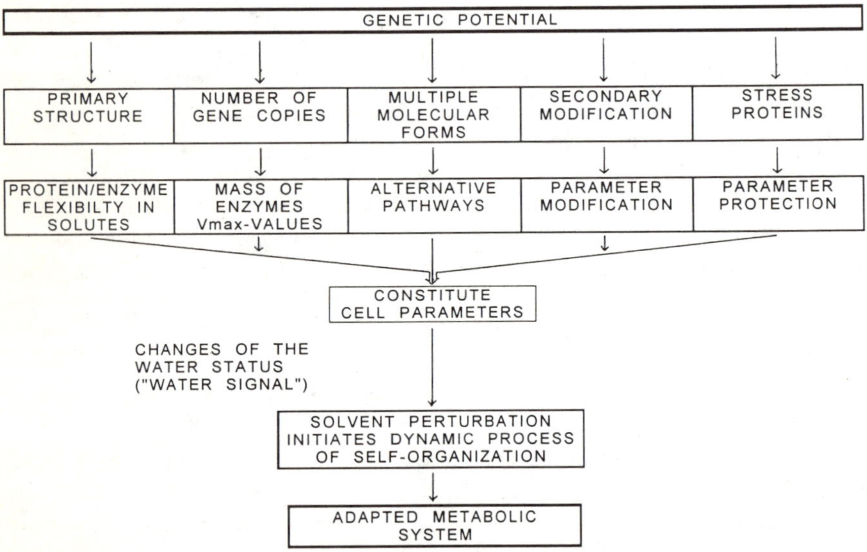

Fig. 2. Gene potential and metabolic adaptibility of individuals of plant populations. Particular metabolic adaptivity patterns are bound to respective differences in the genetic potential which are listed in the *second line of the figure.* These genetic potential differences are found in the primary structure of proteins, in the number of gene copies in the existence of multiple molecular forms of enzymes, in the potential for secondary modification of amino acid side chains of proteins and in the ability to produce protective "stress proteins". The functional consequences of these potential differences are depicted in the *third line of the figure.* All the genetic potential differences contribute to the establishment of the metabolic parameter structure of a cell. Environmental signals which induce a solvent perturbation of the cell water fractions influence this parameter structure and trigger a process of metabolic self-organization. Even far-reaching consequences in terms of dynamic state patterns can arise from this perturbation: qualitatively different regimes of homoiostasis, sensitivity, and resistance can arise

potential result in rather small shifts of the metabolic parameter structure, which in turn gives rise to major macroscopic effects of the network dynamics (the epigenetic system response). The components of the gene potential are: primary structures of proteins, number of gene copies, multiple molecular forms of enzymes, potential for secondary modification (kinases, phosphatases), and "stress proteins". Some keywords are included here which characterize the roles of these components for adaptation to different environments.

Primary structure: the primary gene products are polypeptides. According to the "thermodynamic hypothesis" of Anfinsen (1973), biologically functional tertiary structures are obtained under "normal conditions of the cell milieu". This process has been classified as a "second translation of the genetic message" (Goldberg 1985). In plant cells the internal milieu can change drastically through water loss, extreme temperatures, or by salt burdens. The cell solute therefore must be viewed as a medium which can stabilize or destabilize protein structure (Zaccai and Eisenberg 1990). [More advanced models of structure formation (Ellis and van der Vlies 1991; Shinde and Inouye 1993) are ignored here because they do not violate the general aspect of the milieu-structure relationship]. Minor changes in the primary structure of single genes of cultivars might produce protein structures with slightly different tertiary structures and flexibilities which are equipped with different enzyme characters (V_{max}, K_m, K_s, K_i) and therefore also with different kinetic properties. A difference in gene potential is apparent.

Number of gene copies: genetic lines of cereals are known to differ in ploidity level. If there is correlation between the number of gene copies in the genome and the amount of the expressed protein, then de facto the V_{max} of an enzyme is increased. As will be shown, this change can produce major consequences concerning the flux and stability properties in reaction networks. Another "minor difference" with major consequences for metabolic functioning arises from this.

Multiple molecular forms of enzymes: multiple molecular forms of enzymes open the possibility of switching to an alternative pathway in accordance with the external situation. A well-documented example is the switch from fatty acid biosynthesis to the formation of unsaturated fatty acids in *E. coli* (de Mendoza and Cronan Jr 1983), which occurs if a low temperature threshold is passed. Differences in cultivars in this respect can produce different adaptibilities in the genetic lines.

Potential for secondary modification: protein modification is an important source for switching in cells and in developmental functions (Fischer 1993; Krebs 1993). Very well studied is the role of phosphokinases and phosporylases, which initiate externally (hormonally) induced reaction cascades in cells. It is not well established whether or not environmental factors can also replace the hormone signal. Some evidence in favor of this might be found in *Arabidopsis* research (Braam and Davis 1990), where it was shown that physical factors of the external environment result in a calmodulin-mediated differential gene expression.

"Stress proteins": the role of the "stress proteins" for structure conservation is widely agreed upon and it is not impossible that also "resistant crop lines" use this strategy (Ho and Sachs 1989).

Conclusion: the genetic potential provides protein structures of varying flexibilities (which is concomitant with parameter flexibility), polyploidy of genes (which increases the V_{max} values), different ways of secondary protein modification (which is usually accompanied with on-off switches of reaction steps), and options for alternative pathways and parameter protective capacities [compatible solutes (Paleg et al. 1984), "stress proteins"]. Any one of these potentials can function in a differential way in different genotypes in a changing environment. If this happens, the parameter structure of a cell is changed and also its potential for adaptive metabolic functioning. Therefore "minor differences" on the level of the primary gene products in genetic lines of plants contribute to a particular system's character with an individual adaptive potential. (It should not be overseen that the existence of alternative pathways and stress proteins is already a trend towards the development of specific biochemical machineries. Their functioning in adaptation is self-evident and will not be discussed further here.)

5. Metabolic Network Organization and Flux Control

(In this section, some key .words concerning established kinetic theory and metabolic control concepts are repeated and commented. Experienced readers therefore may choose to pass on to Sect. 6.)

a) Towards a Quantitative Concepts

Adaptive changes of a metabolism are usually observed from in vitro measurements of changes in pool concentrations of metabolites, enzyme activities, and enzyme parameters (K_m, K_s, K_i, V_{max}). The question arises, which information can be gained from these data in terms of metabolic functioning and adaptive qualities. Some comments on the kinetic approach should help to clarify the problem.

α) The Michaelis-Menten Formalism

Metabolic processes (reaction sequences, reaction cycles, transmembrane transport) are often viewed and interpreted with the help of the so-called Michaelis–Menten formalism; but this concept was designed and can be used only in order to perform enzyme kinetics (Michaelis and Menten 1913), which is the determination of the mentioned enzyme parameters. If one substrate interacts with the active site of an enzyme and is transformed to a product under strictly defined reaction solute conditions, and if product interference with the enzyme is excluded, then the differential Eq. (1a) is the appropriate rate equation for the reaction process. With the help of this equation, the values of the enzyme parameters K_m and V_{max} of this irreversible "one enzyme – one substrate reaction" can be gained. (Instructions for the determination of the additional enzyme parameters can be found in the kinetic literature, for instance, Laidler 1958; Segel 1975; Wong 1975.)

$$dS/dt = v_0 = \frac{V_{max} \cdot S}{K_m + S} \tag{1a}$$

$$dS/dt = v_0 = \frac{V_{max}}{K_m \cdot S} \tag{1b}$$

where v_0 = initial reaction velocity, S = substrate concentration, V_{max} = maximal reaction velocity at substrate saturation, and K_m = substrate concentration at half-saturation.

At very low substrate concentrations, the catalytic rate depends nearly linearily on the substrate concentration with V_{max}/K_m being the rate constant. This is expressed in Eq. (1b). This linear dependence between enzyme concentration and flux rate is often thought to exist in metabolic processes. Under substrate saturation, which is often the assumed function with "bottle neck enzymes", is the catalytic rate independent of substrate concentration changes and is represented by V_{max} only.

The adaptive value of such an enzyme can be elucidated from the responsiveness towards medium changes of both the V_{max} and K_m values. If low substrate concentrations exist under physiological conditions (which is the most likely situation), then the metabolically representative "flexibility parameter" is the term V_{max}/K_m and not the K_m. Under substrate saturation conditions, V_{max} would be the adequate parameter for making suggestions.

Yet enzyme kinetics and enzyme functioning in metabolism are entirely different matters. Kinetic experiments are valid only under a broad set of very restricted assumptions (Webb 1963) which are hardly fulfilled under cell physiological conditions. It is not possible to try to guess the metabolic functioning of an enzyme merely from kinetic results.

Under cell physiological conditions usually the product of the reaction is also present. In the quantitative approach, this fact leads to a changed numerator in the Michaelis–Menten equation, which for low substrate concentrations must then be written as in Eq. (2). [For the sake of simplification, Eq. (1b) has been used for this discussion].

$$dS/dt = v_0 = V_{max}/K_m \times \Delta S \tag{2}$$

$$\Delta S = S\left(1 - \frac{P/S}{K_{eq}}\right),$$

where ΔS = thermodynamic driving force of the unidirectional reaction S $\Rightarrow$ P, v_0 = the initial velocity, P and S = actual concentrations of product and substrate, and K_{eq} = chemical equilibrium constant of the reaction S $\Leftrightarrow$ P.

From ΔS it is seen that the influence of mass action can dramatically alter the flux through this reaction and even reverse the reaction process. This mass action effect on its own is a fact which makes extrapolations from merely kinetic data on metabolic functioning of an enzyme impossible.

Under metabolic conditions, most of the enzymes have two or even three substrates and products. The rate equation for an isolated enzyme, considering this situation, is already very complex (for a more extended discussion see Segel 1975). This can be seen from the complete rate equation of an ordered enzyme reaction with two substrates and products as catalyzed by ethanol or malate dehydrogenase [Eq. (3)]

$$v_{nett} = \frac{Vmax_f \cdot \left(AB - \frac{PQ}{K_{eq}}\right)}{\left\{K_{ia}Km_b + Km_bA + Km_aB + AB + \frac{Km_aBQ}{K_{iq}} + \frac{ABP}{K_{ip}} + \frac{Vm_f}{Vm_r \cdot K_{eq}}\left[\frac{Km_q}{K_{ia}} \cdot AP + \frac{BPQ}{K_{ib}} + Km_qP + Km_pQ + PQ\right]\right\}}. \qquad (3)$$

The symbols in Eq. (3) follow the usual nomenclature. A, B, P and Q are the substrates and products, Km_a, Km_b, Km_p and Km_q are their K_m values; K_{ia}, K_{ib}, K_{iq} and K_{ip} are their dissociation constants and $Vmax_f$ and V_{max_r} are the V_{max} values for the forward and reverse direction of the reaction.

Such an equation allows the closest approximation to metabolic functioning of a single enzyme because it considers all possible interactions with substrates and products. (Additionally, also regulatory effects will occur which are not included here.) The net rate of such a reaction is highly nonlinear and exerts a biphasic velocity (v) – substrate saturation (s) curve or v/s profile (Fig. 3). Figure 3 examines the flux rate of the ethanol dehydrogenase reaction as the function of the coenzyme concentration (which is the first substrate to complex with the enzyme) with the help of Eq. (3). "Physiological concentrations" of ethanol (A and B) and acetaldehyde are considered at varying redox charges of the cell. The latter is expressed as the NAD mole fraction (which is the percentage of NAD in comparison to the total fraction of coenzyme [NAD+NADH]). Depending on this coenzyme ratio, which is a source for metabolic signaling (Atkinson 1986), an insensitive branch of the catalytic flow arises at low (<0.8 in A) and a sensitive branch at high mole fractions (> 0.8) of NAD. The effect on the profile of a ten fold decrease of ethanol is shown in B. (Sensitivity and insensitivity refers to coenzyme caused flux changes.) The insensitive plateau phase provides a buffer capacity because major changes of the coenzyme ratio are without significant influence on the flux rate. In the range of the sensitive branch, however, even very small changes of the coenzyme ratio result in major flux changes, including flux reversion. This flux reversion (ethanol oxidation) is possible only if the NAD/NADH ratio is far on the NAD side, i.e., if it is about 99%. Ethanol dehydrogenase therefore catalyzes its reverse reaction – the reduction of acetaldehyde – under "normal physiological conditions" in plants. It should be noted from

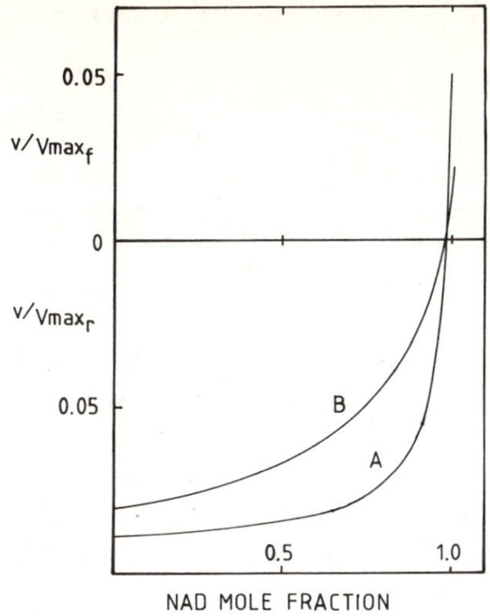

Fig. 3. The relative rates of the foreward (v/Vmax$_f$) and the reverse (v/Vmax$_r$) catalytic activities of the ethanol dehydrogenase as functions of the NAD mole fraction (NAD/NAD+ NADH) are shown at 55 mM (*A*) and at 5.5 mM (*B*) ethanol in the presence of 24.5 μM acetaldehyde. Simulations were performed with Eq. (3). (After Purich and Fromm 1972)

Fig. 3 that over the whole range of the NAD mole fractions, the absolute flux rates (ordinate) are extremely low and far from the maximal value of unity.

With Eq. (3) and Fig. 3 in mind, the reader is invited to make suggestions on the effects on flux rate changes which result from parameter or from metabolite and coenzyme concentration changes, which might have been measured as a consequence of plant adaptation to a changed environment. This situation was ment with the remark that it is impossible to make suggestions from single enzyme data on flux changes. Only rigorous calculation of the response curve, which includes all parameters and variables, can give some clues concerning the function of an enzyme in its metabolic environment (Purich and Fromm 1972).

β) The Elasticity Coefficient

From the discussion in the previous section, it becomes evident that enzyme reaction rates can be modified either by the changed ratio of substrates and products ΔS or by parameter shifts. Under adaptive constraints, both changes might occur in plants; but the respective system's character will depend primarily on the parameter flexibility in its immediate reaction milieu and therefore is of particular importance in adaptibility research. A suitable measure of enzyme flexibility (besides the "classical" parameters) is the "elasticity coefficient", ε (Kacser and Burns 1973, 1979; Kacser 1983). The multifaceted character of this coefficient can be studied in the literature cited. Only some general aspects are depicted here which can be deduced from ordinary v/s profiles.

The interdependence of a rate change of an enzyme and the concomitant substrate concentration change is formulated in Eq. (4a), which, after rearrangement, gives the elasticity in dependence to this particular substrate concentration [Eq. (4b)].

$$\frac{\delta v_1}{v_1} = \varepsilon^1_{S_1} * \frac{\delta S_1}{S_1}, \qquad (4a)$$

$$\varepsilon^1_{S_1} = \frac{\dfrac{\delta v_1}{v_1}}{\dfrac{\delta S_1}{S_1}}, \qquad (4b)$$

where $\delta S_1/S_1$ and $\delta v_1/v_1$ are small fractional changes of the substrate S which cause a small fractional change of the catalytic rate v. ε_1 is the elasticity coefficient in respect to this change of S.

Equation (4b) shows that ε is the slope of the tangent to the v/s profile at S_1. Transferred to the hyperbolic v/s profile of the Michaelis–Menten function, this means that at low substrate concentration, ε has a high value which decreases with increasing substrate concentrations and finally approaches zero at substrate saturation.

However, there exist as many elasticity coefficients as there are direct interferences of substrate-, product- or effector molecules. Therefore the elasticity coefficient is a measure of the flexibility of an enzyme in any given solute environment. The net change of the catalytic rate is due to the sum of all partial changes [Eq. (4c)]. These partial changes include alterations such as caused by enzyme parameter shifts, substrate (and/or product) changes, etc.

$$\frac{\delta v_1}{v_1} = \varepsilon^1_M * \frac{\delta E_1}{E_1} + \varepsilon^1_{S_1} \frac{\delta S_1}{S_1} + \varepsilon^1_{S_2} * \frac{\delta S_2}{S_2} \cdots . \qquad (4c)$$

This "summation effect" can be elaborated from the v/s profiles of Fig. 3. Under identical redox charges of the milieu, the elasticities of the ethanol dehydrogenase are quite different under high (A) and low (B) ethanol concentrations. In addition to the enzyme parameters V_{max}, K_m, K_s and K_i the elasticity coefficient is another enzyme parameter which especially reflects the influence of the reaction milieu on the flexible enzyme response. Unfortunately, no such elasticity measurements have been performed as yet with enzymes under conditions which mirror the cell-internal milieu of plants of different adaptive potentials.

[Regulatory enzymes with sigmoid v/s profiles are not considered here. Their rate equations (Levitzki 1978) are much more complex than the Michaelis–Menten formula and include some additional parameter terms. Their kinetic characterization and their proposed dynamics of functioning in metabolism need sophisticated mathematical treatment (for example Hess and Markus 1987) and therefore are excluded from the present discussion.]

γ) Rate Equation of a Metabolic Reaction Sequence

Up to now, only isolated single enzyme reactions have been considered; but in metabolism, enzymes function in metabolic units which include several or even many reaction steps. A rate equation for the entire reaction process therefore must include all rate equations of the single reaction steps. In Eq. (5), such a rate equation has been formulated for the easy to handle (because very short) biosynthetic sequence of proline (Scheme 1) under the assumption that a steady state is established. The equation was developed according to the concept of Kacser and Burns (1973) on the basis of the simplifying conditions as considered in Eq. (1b) (Pahlich, unpubl.). This simplification is acceptable because the flux through the sequence is strictly unidirectional, as was concluded from the highly exergonic energy profile of this sequence (Pahlich et al. 1981). A steep deviation from equilibrium in favor of proline synthesis exists under physiological conditions (Pahlich et al. 1982).

Scheme 1

$$\text{Glutamate} + \text{NADPH} + \text{ATP} \Rightarrow \text{glutamate semialdehyde} + \text{NADP} + \text{ADP} + \text{P}_i \tag{a}$$

$$\text{Glutamate semialdehyde} \Rightarrow \Delta^1\text{-pyrroline-5-carboxylate} + \text{H}_2\text{O} \tag{b}$$

$$\frac{\Delta^1\text{-Pyrroline-5-carboxylate} + \text{NADPH} \Rightarrow \text{proline} + \text{NADP}}{\Sigma\ \text{glutamate} + 2\text{NADPH} + \text{ATP} \Rightarrow \text{proline} + 2\text{NADP} + \text{ADP} + \text{P}_i + \text{H}_2\text{O}} \tag{c}$$

$$F = k \cdot \text{Glu} \cdot \text{NADPH}^2 \cdot \text{ATP} \cdot \left(1 - \frac{\dfrac{\text{Pro} \cdot \text{NADP}^2 \cdot \text{ADP} \cdot \text{P}_i}{\text{Glu} \cdot \text{NADPH}^2 \cdot \text{ATP}}}{K^1_{app} \cdot K^2_{eq} \cdot K^3_{app}} \right)$$

$$k = \left[\frac{Km_1}{Vmax_1} \cdot \text{NADPH} + \tau_2 \cdot \frac{\text{NADP}^+ \cdot \text{ADP} \cdot \text{P}_i}{K^1_{app}} \cdot \text{NADPH} + \frac{Km_3}{Vmax_3} \cdot \frac{\text{NADP}^+ \cdot \text{ADP} \cdot \text{P}_i}{K^1_{app} \cdot K^2_{app}} \right]^{-1}, \tag{5}$$

where F = stationary flux rate of the reaction sequence, Glu = concentration of glutamic acid, Pro = concentration of proline, and K_{app} = apparent equilibrium constants of the respective reactions.

In the numerator the thermodynamic driving force ΔS of the sequence is formulated which results from the concentrations of the starting substrates and the end products of the entire process. In this open reaction chain the intermediates do not contribute to this important term. This results from the summation in Scheme 1. The denominator – kinetic term k – considers the enzyme parameters of all reaction steps of the biosynthetic sequence. This means that any enzyme of the reaction chain contributes to the flux establishment. The kinetic constants are formulated as relaxation times, τ (= K_m/V_{max}) (Higgins 1963; Bücher and

Rüssmann 1963). (The spontaneous cyclization of the semialdehyde must be expected to be very small. Therefore the term containing τ_2 probably could be neglected in the summation.) The kinetic constants are modulated by the actual substrate, product, and coenzyme concentrations. This is an additional aspect which makes it difficult to interpret measured changes of single parameters or concentrations in terms of flux developments in situ. Again, only a rigorous quantitative treatment which includes all terms of Eq. (5) can give insight into the adaptive changes of a metabolism if the arguments are based on enzyme kinetic data and on pool-concentration determinations.

Some general conclusions concerning the character of metabolic reaction sequences and their possible adaptive changes can be drawn from these examples. Equation (5) formulates a "minimum set" of criteria which contribute to the establishment of the steady flux in the cell. This equation is certainly based on the Michaelis and Menten formalism, but represents "another dimension" of complexity than can be detected from Eq. (1). Intuitive extrapolations from kinetic data on metabolic functioning are ruled out with this insight, as also suggestions on the respective flux responses of adaptive metabolic processes which are based on enzyme and pool measurements must be derived from and governed by appropriate and hence complex rate laws.

Under certain circumstances, simplifications of rate laws are justified. For instance, if a reaction is shown to occur strictly unidirectionally in metabolism, then the denominator terms describing the reverse reaction may be neglected. In Eq. (3) this is the second part of the denominator which is multiplied by the term $1/Vm_r$. Also terms of the remaining sum can be neglected if they are very small as compared to the rest of the terms. The numerator term is unavoidable as long as products are present in the reaction system. If molecules of the mass action term have additional signal functions – which is often true for a feedback function of nucleotides – then further regulatory terms must be considered in the rate laws. The glucose-6-phosphate dehydrogenase might be taken as an example. The enzyme is coupled to the dehydrogenases of the proline biosynthetic sequence via the NADP/NADPH moiety and is regulated by a negative NADPH feedback loop (Pahlich 1992). A biphasic response curve of the stationary coenzyme concentrations emerges if the parameters of the dehydrogenase change differentially in comparison to those of the pyrroline carboxylate reductase. Without knowing this response curve with its sensitive and insensitive branches, it is impossible to make suggestions about the effect of environmentally induced coenzyme or enzyme parameter changes on the flux of this particular pathway or any other pathway in general.

b) Flux Control of Metabolic Sequences Under Stationary Flow

From Eq. (5) it becomes apparent that the rate of a metabolic reaction sequence arises from the multiplication of two terms: the thermodynamic driving force and the kinetic term. The kinetic term k in Eq. (5) defines the system's character of the metabolic sequences. This term combines all kinetic parameters of all reaction steps which are involved in the biosynthetic process. Therefore every reaction step obviously contributes to the development of a stationary flux. The quantitative portion of this contribution of any of the involved reaction steps can be elaborated with the help of control theory (Kacser and Burns 1973; Heinrich and Rapoport 1974). This theory also paves the way for an appropriate design of analytical approaches for metabolic system's analysis.

The entire concept cannot be summarized, and the reader is referred to the original literature (Kacser and Burns 1973, 1979; Kacser 1983; Kacser and Porteous 1987). Only a few arguments which are crucial for practicing the control concept will be mentioned here. (Useful experimental advise concerning the analysis of stationary metabolic states is found in the interesting paper by Bücher and Rüssmann 1963).

It has been shown theoretically that multienzyme reaction sequences show biphasic response curves if the changing concentration of a single enzyme is plotted against the system's flow (Fig. 4). The response curve in Fig. 4 has been calculated with a generalized rate equation (legend to Fig. 4) which can be derived from rate laws of the form of Eq. (5). The nonlinear, biphasic character of the response curve increases with the number of enzymes which are involved in the process (Kacser and Burns 1980).

This is an important finding in terms of adaptive potential in plants. For instance, if in different genetic lines of plants multiple gene copies of an enzyme exist, then an enzyme's position can be shifted from the sensitive part of the profile towards the plateau phase – the insensitive branch. A seemingly small difference in the genetic potential results in a major functional effect because environmentally induced parameter changes in such an enzyme are then of minor or even negligible influence on the flux of the reaction chain. This situation was mentioned already in Section 4.

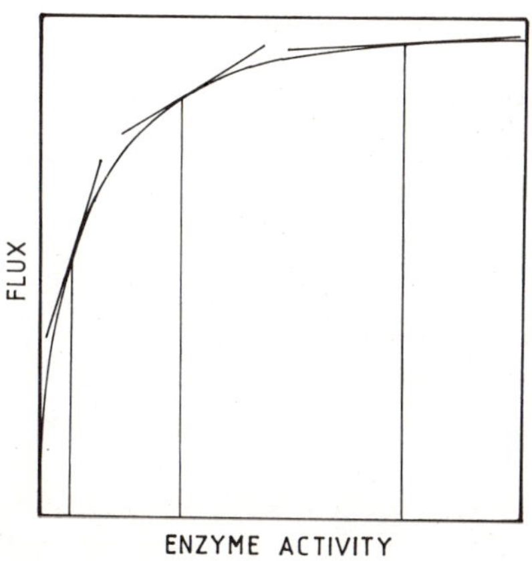

Fig. 4. Flux rate (F) and enzyme concentration (E) (activity). The flux rates in linear multistep reaction sequences are nonlinear functions of single enzyme concentrations. The sensitivity coefficient Z – which is a measure of the contribution to flux control of a single enzyme – is given by the slope of the tangent dF/dE at respective enzyme concentrations. (After Kacser and Burns 1979)

A quantitative measure of the flux regulating potential of an enzyme is its control coefficient Z [Eq. (6)] which compares the fractional flux change with the fractional change in the respective enzyme concentrations. (The original name "sensitivity coefficient" has been discarded meanwhile by recommendation of Burns et al. 1985). The control coefficient must be determined experimentally and is the slope of the tangent at different enzyme concentrations (see Fig. 4).

$$Z = (dF/F)/(dE/E). \tag{6}$$

Therefore if measurements of fluxes in vivo are possible and if additional data on enzyme concentration (or parameter) changes are available (for instance with immunochemical techniques), then sets of data result which are best suited for Z determinations. From Eq. (6) it is seen that experimental conditions resulting in rather small changes in fluxes and enzyme rates are preferable.

The control theory proposes that the sum of all control coefficients of a sequence add up to unity (summation theorem; Kacser and Burns 1973). The value of a control coefficient close to unity marks a "bottle neck" or "rate-limiting step", or a "dis-equilibrium reaction", which in reality can exist as a limiting situation but which probably is the exception rather than the rule. The more enzymes are involved in a metabolic network the smaller will be the individual control coefficients, and hence also the chance that single values of the coefficient approach unity. A very small value of this coefficient shows that this enzyme is of minor importance for flux control. It catalyzes an "equilibrium reaction". The summation theorem substantiates in particular the system's character of metabolic reaction sequences: any change in the value of the control coefficient of any step leads to partial changes in all the other coefficients so that they sum up to unity again. This effect is a part of the earlier mentioned self-organization process of metabolic networks.

In order to determine the control coefficients experimentally, two principles can be recommended which are described in Eqs. (7) and (8). Both methods allow the determination of the relative ratios of the control coefficients either from enzyme kinetic data or from measurements of the pool ratios. Both types of measurements are the "classical targets" of metabolic analysis.

$$Z_1 : Z_2 : Z_3 = Km_1/Vmax_1 : Km_2/Vmax_2{}^* Kapp_1 : Km_3/Vmax_3{}^* Kapp_1{}^* Kapp_2. \tag{7}$$

$$Z_1 : Z_2 : Z_3 = (1-\delta_1) : \delta_1(1-\delta_2) : \delta_1\delta_2(1-\delta_3)\ldots, \tag{8}$$

where δ = disequilibrium ratio Γ/Kapp, where Γ is the mass action ratio P/S. P and S symbolize the actual concentrations of the products and substrates of a reaction. Kapp = apparent equilibrium constant of the respective reaction. The other symbols have the earlier defined meanings.

Equation (7) describes the principle for Z determinations on the basis of enzyme kinetic data. Considered is an irreversible reaction sequence with one substrate-one product reactions. The combination of the enzyme parameters and chemical equlibrium constants of the single reaction steps according to Eq. (7) elucidates the Z ratios of the involved reaction steps and hence the system's character and its proposed changes during adaptive events.

The same result can be achieved [Eq. (8)] if the respective mass action ratios of the single reaction steps are determined and used for the calculation. Both equations also show how the "classical targets" of biochemical analysis are integrated into the recent concept of metabolic flux control, and how they should be used in order to make suggestions about metabolic fluxes, flux changes, and regulatory properties in reaction sequences.

The discussion concerning this very well-established "sound theory of control" (Porteous 1983) left open the important question as to how the boundary conditions are marked by metabolic stationary fluxes, whether fluxes can become instable, and what follows after passing such presumed instabilities. The answer to this question is gained from a rather complicated mathematical approach (Nicolis and Prigogine 1989); but the existence of different states or phases of stationary flows and their transients to new forms of stability is of major importance in environmentally induced metabolic adaptations. Therefore, this question will be discussed in a following section with the help of purely graphical methods.

6. External Environment and Metabolic Parameter Structure

With the steady-state concept as the basic principle of metabolic functioning, it is usually also assumed that the interior of a cell has more or less the character of a chemostat, i.e, a stable "normal physiological environment". This character of a chemostat could be reached only if the parameter structure, which is the result of a given gene expression, were a fixed entity of "constants". With the so-called stress responses of plants, however, it is well documented that the metabolism responds to environmental changes on short time scales and with intensive metabolic pool alterations. This environmental effect must somehow be related to cell internal milieu changes. Therefore, environmentally induced changes of an existing parameter pattern must be viewed as immediate or even primary stress effects. A phenomenological description of what happens if stress is applied to plants has been defined by Larcher (1987): "... stress can be described as a state in which increasing demands made upon a plant lead to an initial destabilization of functions, followed by normalization and improved resistance. If the limits of tolerance are exceeded and the adaptive capacity is overtaxed, permanent damage or even death result."

Larcher's concept considers a hierarchy of subsequent phases of stress responses which include states of stability, instability, and transitions to alternative states of stability and functionality. The transitions can approach states of general instability which lead to death; but which is the kinetic concept that can explain the hierarchy of all these metabolic transitions and states? In order to answer this question, it is necessary to take a closer look at the interactions between external environment, cell water, and parameter structure of a cell. (A shortened version of the concept has been discussed elsewhere: Pahlich 1993.)

7. Environmentally Induced Parameter Changes Which Are Cell Water-Mediated

The role of cell water for metabolic functioning is of paramount importance and has been investigated particularly in respect to specific physiological responses and misfunctions of plants under water and salt stress (Slatyer 1967; Paleg and Aspinall 1981; Somero et al. 1991). Water is the solvent in cells, it is the media for homogeneous catalysis, diffusion, and transport and – in respect to subcellular functioning – it is also the "shaping media" which guaranties the correct folding of the primary gene products, the proteins (Anfinsen 1973; Wolfenden et al. 1979). Furthermore, its "sorting out effect" forces lipids to assemble as membranes and to remain stable under these conditions (hydrophobic effect: Tanford 1978). Yet correctly folded proteins and appropriately associated membranes provide the parameters for catalysis and for transport, and hence the "hardware" for metabolic functioning. From this point of view, the problem arises as to whether or not environmentally stipulated changes in the water fractions of a cell, affect metabolic functioning.

It is known that the water in plant cells and tissues responds sensitively to environmental changes, which is evident from numerous water potential measurements. The factors of the physical environment which cause these changes are known to affect the properties and the structure of water itself (Franks 1972; Lewin 1974) and in mutual interactions also the properties of solutes like ions (Wiggins 1973), amino acids and oligopeptides (Schrier and Robinson 1971 and 1974; Pahlich ans Stadermann 1984) and also of proteins (Timasheff 1993) and membranes (Lewin 1974, Crowe et al. 1989). Therefore, the interaction of the external environment with the plant water phases must also be expected to provide primary information for restructuring of proteins and membranes and hence also for metabolic perturbations.

The solute properties of the "water compartments" of a plant (cytoplasm, vacuole, free space, and vascular tubes) and their changes in accordance with a water fraction are best described in the more general terminology of thermodynamics. The state of water in a cell – as in that of any other component in the solution – can be assigned by a free energy per mole of each species. This quantity is the chemical potential μ_j of the species j and is defined in Eq. (9a) (Nobel 1991).

$$\mu_j = \mu_{j'} + RT \ln a_j + V_jP + z_jFE + m_jgh, \tag{9a}$$

$$\mu_j* = \mu_{j'} + RT \ln a_j^* + V_jP^* + z_jFE^* + m_jgh^*, \tag{9b}$$

where μ_j and $\mu_{j'}$ = chemical potentials of a_j and of a reference state, a_j = thermodynamic activity of species j, a = thermodynamic activity of a ($\gamma \cdot$ concentration), V_j = partial molal volume of j, P = pressure, z_j = electric charge of j, F = Faraday constant, E = electrical potential, m_j = mole mass of j, g = gravitation constant, and h = height.

The chemical potential of the molecule species, j, is composed of several terms of a sum which are the thermodynamic activity (the "biochemically effective concentration" Nobel 1991) of a molecule, the pressure, the electrical potential, and the gravity. A change in any of these terms alters the chemical potential of a molecule. In biochemical treatises of reactions, all terms besides the thermodynamic activity (the concentration term) are usually neglected. Yet the differences in chemical potential provide the "driving forces" for reaction processes as well as for diffusion, and transmembrane transport, in the same way as a temperature gradient directs a heat flow. Therefore, it is unavoidable that any changes in the chemical potentials of all involved molecular species in reaction networks result in a general perturbation of the hitherto balanced system. A "demand" is put on the system, which is forced to move towards "alternative balanced states".

Besides the contributions resulting from several terms which constitute the chemical potential of a molecule, it is important to note that its value is additionally influenced by its "solute milieu". This "milieu effect" is also well known to enzymologists, who observe specific "buffer effects" on enzyme activities and on enzyme parameters. This "buffer effect" is usually hard to interpret because it results from a mutual change of any component of the solute. In a strict biochemical sense therefore, all concentrations of the components should be expressed as thermodynamic activities including the concentration of the enzyme itself and its complexes with substrates and products (Webb 1963). Exactly these "milieu effects" must be expected to occur also in cells the water milieu of which is changed. In terms of Eq. (9) this means that the chemical potential as defined in Eq. (9a) switches to the value as defined in Eq. (9b).

The mutual changes in the chemical potentials in a solution are quantitatively expressed in the Gibbs-Duhem relation (van Hippel and Schleich 1969, Moore and Hummel 1976) (the Gibbs-Duhem relation is formulated in Table 1). Any external change which influences the cell water or its chemical composition leads to unpredictable mutual changes of all chemical potentials of all solute components which always add up to zero. The concept of such a solvent perturbation has been outlined for a ternary water-protein-salt sytem by Klotz (1966) and Timasheff (1993). Water perturbations of cells will therefore disturb established steady states in the cell via small "concentration" changes of solutes and particularly via enzyme parameter changes, and of course also membrane parameter alterations (Blumenthal et al. 1970). This effect must be viewed as "increasing demand (which can lead to) initial "destabilization of functions" in terms of Larcher's definition of stress. What really happens depends on the particular solute conditions and cannot be predicted or generalized. (The analogy of these interdependencies with those resulting from the summation of the control coefficients should be noticed.) Besides the water potential itself, these perturbations, in a mutual and milieu-dependent way, alter concentrations, protein structures, membranes, and gradients.

Table 1. "Key-Lock" versus solvent perturbation signaling

Source of information	Signal (hormone, nucleotide, photon)	Change of the physical environment (temperature, drought, salt, pressure)
Primary event	Signal molecule-receptor interaction ("Key-lock" interaction) $R + S \leftrightarrow RS$	Solvent perturbation (mutual thermodynamic activity changes of the solvent water and the solutes) a_P ; $a_S \leftrightarrow a_W$ $\longleftrightarrow$ a'_P ; $a'_S \leftrightarrow a'_W$
Quantitative description	Mass action ratio $\Delta G^\circ = -RT \ln K_{eq}$	Gibbs-Duhem relation $N_W d\mu_W + N_S d\mu_S + N_P d\mu_P = 0$
Signal transduction	"Monocausal" sequence of events	Differential perturbation of the system's parameters
Result	Induction of a master process (differential gene activation)	Metabolic phase transition (switching between multistable stationary states, transition to oscillatory states)

The Gibbs-Duhem relation describes that changes of the external environment should be recognized differentially in cells and tissues which have reached different developmental stages and which therefore also have different chemical compositions. In this respect, meristem cells should respond in a different manner to a given environmental change than, for instance, highly specialized cells. Both might have identical water potentials yet entirely different solute milieus in the aqueous compartments and vice versa. Weatherly (1965) referred to this fact in an early paper on the water movement in the leaf. Therefore, details of these perturbation effects can hardly be detected and measured if total plant homogenates are analyzed. Tissue-specific recordings are necessary to elucidate the postulated differential effects of the environmental impact.

The widely accepted concept of the partitioning of the cell water into "bound" and "free" water fractions in the aqueous cytosol underlines the importance of the solute milieu changes for stress recognition. This partitioning into "vicinal" and "bulk" water fractions (Clegg 1981), which provide a "compartmentation without membranes", gives rise to very fragile and delicate distribution equilibria of ions, metabolites, and even enzymes (Clegg 1981). The distribution equilibria cannot be actively controlled, like membrane-reconciled intercompartimental gradients, by active transport or other processes. Immediate equilibration occurs after a perturbation of the water phases according to the new chemical potentials of the water itself and its solutes. The equilibration is a very sensitive response to any environmental change and can provoke a major metabolic perturbation in the aqueous cytoplasm without any gross change of enzyme contents. This results from a differential effect of the water activity changes (water potential changes) on individual enzymes (Corcuera et al. 1989). The effect of milieu perturbations on "amphitropic membrane proteins" (Burn 1988), on local diffusion reaction clusters (Masters et al. 1987), or on metabolons in general remains to be elucidated. Especially these latter aspects of metabolic functioning underline the potential of cell water perturbations for environmentally induced metabolic switching. Therefore the perturbation process itself may be viewed as the "primary signal" which can slightly increase or decrease stationary fluxes ("demand") or – under conditions which will be discussed later – can even shift them towards their boundaries of stability and hence switch between different metabolic states.

8. Water Perturbation-Mediated Metabolic Switching

Table 1 outlines the basic difference between the well-known "key-lock-switch" and the "solvent perturbation mediated switch" in metabolic systems. In the key-lock-switch, the information is encoded in the structure of a signal molecule "S" which might be a phytohormone. The signal information is realized through the interaction with a receptor R. The activated signal-receptor complex RS (which in plants might also be a photon-mediated change of a protein structure) typically initiates a vectorial sequence of events which, after passing the signal chain, finally results in the activation of a master process. The light-induced differential gene activation of the phenylpropane pathway is an example for such an event and describes a straight forward "monocausal" process of differential gene activation.

Perturbations of the water phases of a cell reach all solute components at the same time. Therefore a solvent perturbation mediated process of switching (right part of Table 1) most likely acts according to the Gibbs-Duhem relation. Thermodynamic activity changes of the water phase (or that of any solute) will affect all other components (membranes included) in mutual interactions ($a_p/a_s/a_w$ change to $a'_p/a'_s/a'_w$). All cell functions which hitherto were balanced in a thermodynamic sense are disturbed (see list in Sect. 9). In the metabolic reaction net a search for new stable states is initiated which is a process of self-organization. The "initial destabilization of functions (is) followed by normalization" if new stable states can be established. This transition usually is time-dependent (slow transitions of molecular structures) and its relaxation time is related to the complexity of the disturbed biological system. For proline accumulation it has been shown, for instance, that stressed suspension cultures reach new steady states in minutes, while it takes many hours in plant seedlings to establish the stationary state under stress (Pahlich 1990).

9. Primary Responses in Solute Perturbed Plant Cells

In agreement with the picture developed in the preceding section, immediate disturbing effects are expected to occur in solvent-perturbed cells which should result in the following responses which are molecule-species-specific, and thus milieu-dependent differential effects:

1. changed thermodynamic activities of solutes,
2. altered partitioning of solutes (ions, metabolites, proteins) between vicinal and bulk water phases,
3. reshaped protein structures, including denaturation,
4. changed subunit assembly,
5. disturbed substrate and ligand binding,
6. altered rate of enzyme catalysis,
7. effects on protein lipid interactions,
8. changed membrane fluidity and membrane transport.

Most of these disruptions are known from in-vitro studies from traditional biochemical research (Timasheff and Fasman 1969). It is a problem, however, to measure these multifaceted perturbations in vivo. Most of these effects are easily reverted in buffer homogenates because the solutes adjust their chemical potentials to the buffer milieu; but how then can information be gained about the in vivo situation of environmentally perturbed cells?

10. Uncovering Metabolic Perturbations

The "metabolic control theory" (Kacser and Burns 1973; Heinrich and Rapoport 1974) provides a useful tool which might help to tackle at least parts of this problem. As has been discussed already, the enzyme and transport parameters of cells – "the system's constants"–characterize the respective metabolic situation. In contrast to the enzyme parameters, which adjust to the buffer environment after homogenization (via flexible structure transitions), the concentrations of the metabolites (unfortunately not their thermodynamic activities) can be trapped in their in vivo ratios under appropriate homogenization conditions. These ratios therefore approximate also the system's character, and the parameter structure can be deduced from this "mirror image" of the appropriately determined metabolite concentration patterns.

[A kinetic analysis has shown (Pahlich, in progress) that in roots of wheat seedlings the parameter structure of the cytosolic dehydrogenases and transaminases remains nearly unchanged after applying water stress on the plant; but the accompanying metabolite patterns of the respective reactions in roots – the mass action ratios – are entirely different in these differently treated roots. The results indicate that under in vivo conditions the enzyme structures and/or their kinetic properties must have been distorted in such a way that the in vivo parameter patterns are different from those which are detected in the vitro assay].

"Appropriate metabolite determinations" are those which allow the calculation of the mass action ratios Γ (Γ = PQ/AB) of single reaction steps of metabolic units. This aspect is delineated in Table 2 with the metabolites of the glycolytic sequence of cultured plant cells (Kubota and Ashihara 1990). The mass action ratios Γ were calculated from their data with the formula given in Eq. (8). From the knowledge of δ, the ratios of the control coefficients ($Z_1 : Z_2 : Z_3 \ldots$) of the reaction steps become apparent. These Z ratios consider all regulatory properties (feedback loops, inhibitions, activations, partitions, thermodynamic activities, etc.) which were active in the cell in the moment of the harvest. It should be possible to detect metabolic alterations which are caused by environmental perturbations and which are triggered due to the Gibbs-Duhem relation via sets of appropriate metabolite concentrations. The data in Table 2 show the following:

A value of Z close to unity marks a bottle neck which seems to exist in this case for the hexose kinase reaction. Control coefficients far from unity identify reactions of minor importance for flux control. In the example given, this is true for the rest of the reaction steps of the glycolytic chain. Minor changes in their enzyme activities have to be considered ineffective in terms of rate control or even in terms of changes of the system's character.

Controllability may change, however, with the respective environment, and flux-controlling steps can be shifted to other parts of the sequence ("normalization of an initially destabilized metabolism"). For instance, different regulatory patterns of this sequence would arise if the hexose phosphate pools where not at rate limiting concentrations. Then reaction step one can be omitted from the calculation. The reaction sequence is then the steps 2 to 8. Two contrasting situations are considered in Table 2. If control is achieved in compartments containing the phosphofructokinase isoenzyme, then the glucose isomerase is most sensitive to flux control, followed by the phosphofructokinase itself. The glyceraldehyde phosphate dehydrogenase (GAPDH) reaction exerts a negligible role for flux control under these circumstances (Table 2, lane 3); but in compartments or cells which contain the PPi phosphofructokinase isoenzyme (lane 4) the former plays the significant role in flux control. The control sensitivity of the GAPDH reaction is comparable to that of the PPi phosphofructokinase. Therefore, the consequence of a measured change in any enzyme activity can hardly be interpreted unless its control coefficient is evaluated.

To my knowledge, this powerful analytical tool has never been used to elucidate the changes of environmentally controlled system characters in plants. Complete data sets which are of the type of those given in Table 2 are missing in plant adaptibility research.

11. Bifurcations: a Particular Type of Metabolic Network Switching

According to the described concept, environmentally induced metabolic changes are most likely initiated by a solvent perturbation of the cell's interior. As has been outlined, this perturbation can change the flux rates of established steady states to a certain extent; but what happens if the environmental impact (the stress factor) continues to increase or decrease? How are the boundaries of stationary states of fluxes defined and where are the limits of flux stability? This problem will be discussed and answered with the reactions of the glyceraldehyde phosphate dehydrogenase (GAPDH) and the malate dehydrogenase (MDH). Both enzymes have cytoplasmic isoforms which closely interact in the cell via the NAD/NADH couple (Atkinson 1986), which – in adult cells – is considered to be moiety conserved. Through the coenzyme couple, both reactions constitute an example for a simple metabolic net (reaction scheme in Fig. 5). The GAPDH reaction – as part of the central glycolytic sequence – must always be operative in growing roots, and therefore depends on a high steady-state concentration of its oxidized coenzyme NAD. NAD is recycled from NADH by the reverse reaction of the MDH. This recycling process thus interconnects two opposing pathways to a metabolic net. The entire metabolic network behind this scheme has been neglected here because the described "minimum reaction couple" exhibits the basic character of a metabolic phase transitions.

It has been shown by Blackwood and Miflin (1976) that the K_m values of the MDH for oxaloacetic acid and for NADH change drastically and differentially with the salt environment. This effect might be taken as a measure for the flexible parameter responses of the MDH under varying salt burdens – it is viewed as a solvent perturbation effect. Unfortunately, the urgently needed value for calculations of the second parameter V_{max} is not given in the paper. Therefore it

Table 2. Concentration patterns of the glycolytic metabolites and ratios of control coefficients

Glycolytic reaction steps		Mass action ratios (Γ)	Disequilibrium ratios ($\delta = \Gamma/K_{eq}$)	Ratios of the control coefficients (Z)	
1. Fructokinase EC 2.7.14	Fru ⇒ F6P	1.2×10^{-3}	3.2×10^{-6}	0.999	0.999
Glucose kinase EC 2.7.1.2	Gluc ⇒ G6P	7.1×10^{-3}	1.3×10^{-6}	0.999	0.999
2. P-Glucose-isomerase EC 5.3.1.9	G6P ⇒ F6P	0.18	0.38	0.62	0.62
3. P-Fructikinase EC 2.7.1.11	F6P ⇒ F1, 6P	0.21	1×10^{-4}	0.38	–
3a. PP_i-P-Fructokinase EC 2.7.1.90	F6P ⇒ F1, 6P	1.9	0.59	–	0.165
4. Aldolase EC 4.1.2.13	F1, 6P ⇒ GAP/DHAP	12.8×10^{-5}	0.98	7.6×10^{-7}	4.5×10^{-3}
5. Glyceroaldehyde-P-dehydrogenase EC 1.2.1.12 EC 2.7.2.3	GAP ⇒ 3PGA	7.9×10^{2}	0.53	1.1×10^{-5}	0.103
6. 3 PGA-Mutase EC 5.4.2.1	3PGA ⇒ 2PGA	0.8	0.40	1.2×10^{-5}	7×10^{-2}
7. Enolase EC 4.2.1.11	2PGA ⇒ PEP	1.4	0.30	5.5×10^{-6}	3.3×10^{-2}
8. Pyruvate-kinase EC 2.7.14	PEP ⇒ Pyr	5.0	2.5×10^{-4}	2.4×10^{-6}	1.4×10^{-2}

Ratio of Control Coefficients: $Z1 : Z2 : Z3 = (1 - \delta_1) : \delta_1(1 - \delta_2) : \delta_1 \delta_2(1 - \delta_3) \ldots$
$Z1 : Z2 : Z3 = K_{M1}/V_{max1} : K_{M2}/V_{max2} * K_{eq1} : K_{M3}/V_{max3} * Keq1Keq2 \cdot\cdot$

was assumed in this discussion that V_{max} is identical with the measured enzyme activity (which is not at all self-evident from a kinetic point of view). De facto the salt conveys a parameter change which usually is brought about by signals like inhibitors or activators. This fact is important in terms of the system's functioning, despite missing knowledge about details of the salt-enzyme interactions. There is no doubt that the K_m and V_{max} values are functions of the salt milieu, the details of which are not worked out experimentally. These functions are the equivalents of the well-established inhibitor and activator terms, which on a formal basis are multipliers of the K_m and/or V_{max} values (Wong 1975). Unfortunately, no sytematic investigations are available which allow the interpretation of solvent perturbation in terms of a quantitative effect on v/s profiles. (Some respective systematic approaches along the required way concerning temperature adaptation of enzymes are reviewed by Somero 1978).

The consequence for the dynamics of the reaction cycle of the salt-induced flexible parameter responses of the MDH are analyzed and interpreted in Fig. 5, which shows the v/s profiles of the MDH as functions of the mole fractions of NAD (solid lines). [The calculation of the MDH profiles were performed with the reduced rate law for an unidirectional bi-substrate reaction as discussed with Eq. (3). It should be noticed that under these conditions the actual reaction rates of the MDH are very low and far beyond the normalized V_{max} value of unity (y-axis). This is in accordance with recent views of metabolic flux rates which are considered to be far apart from their maximal rates. A 100-mM salt stress activates the enzyme (Fig. 5, upper curve) as compared to the rate in the salt free buffer environment (middle curve). High salt concentrations instead are inhibiting (lower curve). From the calculated ΔS values it becomes apparent that it is not possible to reverse the MDH reaction under the considered concentrations of substrates, products, and coenzymes. Malate oxidation (the reverse reaction) would be possible only with mole fractions of NAD much higher than 0.95, i.e., only under extreme oxidative NAD redox charges in a cell.

The network aspect between both reactions arises if the activity profile of GAPDH (dotted line in Fig. 5) is combined with that of the MDH. The inhibition curve was measured in extracts of wheat roots (Henrich 1993, unpubl.). The GAPDH is practically inactive if the NAD mole fraction approaches the value 0.3 (30% of the coenzyme moiety is NAD and 70% NADH). The intersection points of both rate profiles mark the possible stationary states between the opposing reactions. They are numbered I, II, and III and occur at quite different mole fractions of NAD (x-axis). The respective steady flux rates (y-axis) of course differ also.

The system character of the network becomes apparent from a secondary plot (inset in Fig. 5). This plot depicts the stationary values of NAD as functions of the salt environment. Increasingly low salt concentrations increase the stationary NAD values while high salt concentrations drastically decrease them. An upper branch of oxidative NAD redox charges is opposed by a lower branch of more reducing states. (The diagram is partially hypothetical because of lack of exper-

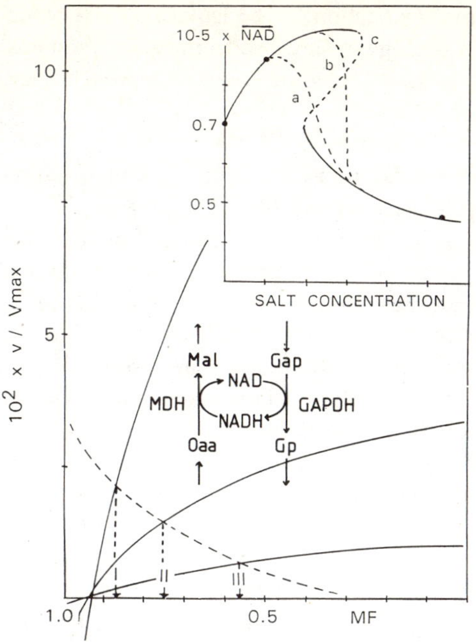

Fig. 5. Metabolic stationary states of coupled reaction systems. The malate dehydrogenase (*MDH*) is coupled to the glyceraldehyde phosphate dehydrogenase (*GAPDH*) via the NAD/NADH moiety. The functional properties are shown which arise from such coupling. The normalized activity (*v/Vmax*) of both enzymes is plotted against the variable catabolic redox charge (expressed as NAD mole fraction [NAD/(NAD+NADH)] = MF). The MDH profiles (*solid lines*) were calculated with a rate law for a reversible ordered bi-reaction [Eq. (3)] (Segel 1975). The K_m values for NADH and oxalacetic acid in salt media (100 to 500 mM NaCl) stem from Blackwood and Miflin (1976). The NADH-inhibited GAPDH activity curve (*dotted line*) was measured in wheat extracts and scaled according to the concentration range of coenzymes considered for the MDH reaction. *Inset* The steady-state concentrations of NAD (taken from the intersection points I, II, and III of the primary plot) are incorporated into a parameter portrait of the reaction couple. Salt-dependent high and low steady-state regimes of NAD become apparent (*solid lines*). The transition between both regimes can be a sliding down (*a*), an abrupt (*b*), or jump-like (*c*) process. The latter two transitions must be regarded as "metabolic triggers" which switch between metabolic states of different quality

imental data.) The upper branch of the profile might be termed the favorable metabolic state because it permits a high glycolytic flux in the growing root. At high salt concentrations a low NAD/NADH ratio is established instead. In terms of glycolytic support of a growing root, this marks a rather unfavorable situation. Both redox states are represented by the so-called deterministic stable thermodynamic branches of this reaction network and are system properties.

These properties cannot be evaluated from inspection of the isolated MDH or the GAPDH reactions, or from the fact that both genes are producing their products. The interesting question now arises as to how the upper branch approaches the lower branch at intermediate salt concentrations.

From the theoretical concepts of network dynamics (Reich and Sel'kov 1981) one can expect three different shapes of the transient (Fig. 5, dotted lines): (a) a sliding down, (b) an abrupt drop, or (c) a jump-like change.

Which possiblity is valid depends on the respective cell milieu and its parametric structure, which again is a function of the respective external physical environment and which could be included as additional z-axis into the figure. The transitional paths (b) and (c) represent solutions which give a clear demarcation line between two characteristic metabolic NAD redox states and which are functions of the salt concentration (x-axis). [Situation (a) is discussed in the last section of this chapter]. These transition points mark a change in the system's quality: they separate two regimes of stable states of the NAD redox charges. A hitherto stable NAD redox state has switched to an alternative stable state. The salt burden has changed the parameter pattern of the MDH in such a way that this change is "viewed" like a destabilizing signal. The switch is not simply caused by a mass action effect arising from the changed NAD mole fraction. Both regimes are characterized by deviating sets of parameters (the inhibiting parameters become apparent only under high salt conditions). The transition or threshold works like a switch; it can be viewed as a "metabolic trigger". The new metabolic quality following the switching event is established by self-organization and exhibits "normalization and improved resistance" (Larcher 1987) to the salt environment. The environmentally induced parameter change of the MDH has driven the stable thermodynamic upper branch towards its "instability point" and forced the system to change its character for purely kinetic reasons.

The salt-dependent approach of the "instability point" – the threshold of the upper stable thermodynamic branch – is called a "bifurcation" in terms of system dynamics. A bifurcation point is reached at "any value of a parameter at which the number or stability of steady states change" (Glass and Makey 1988). In this case, the changed parameter does not result from system internal signaling. It results from the earlier mentioned "salt function", which acts on the enzyme parameters like a destabilizing signal.

The inset in Fig. 5 also shows that a decrease of the external control parameter salt does not necessarily force the system back on the same track which was observed for its forward movement. Instead a hysteretic loop (Neumann 1973) arises (inset mode c). In order to reconstitute the initial metabolic situation, the salt concentration must drop far beyond that value which was the "trigger concentration" in the forward process. Then another "instability point" is reached, the passage of which triggers back the system to its original state. In the region of the loop, the system exerts two stable solutions at one salt concentration. It depends on the system's history (the movement on the respective branch) which state can exist and is established. It is said that such a system exerts a memory. The example clearly shows that the knowledge of stationary values of concentrations and activities is not sufficient to interpret a system's character or to foresee its future development. One has to gain insight into its positioning on respective thermodynamic branches which dictate the possible developmental movements.

The example also gives some clues concerning the problem how much salt must be added to a system until it changes its character or quality. As long as the stationary states are moving on a stable thermodynamic branch the system has its original quality. Only after passing the threshold does this quality change. The amount of salt which is necessary to trigger the switch obviously depends on the shape and expansion of the stable thermodynamic branches. This shape is primarily governed by the parametric structure of the enzymes, which might vary in genetic lines of plants. Because of differential flexible responses of enzymes towards their particular cell internal environment, this shape can also be quite different in cells and tissues of different states of differentiation. Switching is brought about by even small changes of the salt concentrations if the system parameters are already in the vicinity of the "threshold". In this case, a sensitive response would be observed. A parameter set which locates the "starting situation" on a part of the branch, which is apart from the transition point needs a major change of the salt concentration until it reaches the threshold. Therefore, sensitivity can arise from a particular parameter structure, but also from the instantaneous "milieu state" of a cell.

Metabolic switching as viewed here can occur without extra differential gene activation and is caused exclusively by the salt (or other environmental factors)-induced parameter changes of enzymes (and transport proteins). Up to now it is not known whether the observed osmotic regulation of the rates of enzyme production (Jones and Armstrong 1971), the osmotic gene regulation (Higgins et al. 1987), and generally the environmentally induced protein synthesis (Ho and Sachs 1989) – at least in part – can also be explained in terms of "metabolic switching". It could well be that the new metabolic quality after the passage of a transition point triggers also a new quality of protein biosynthesis rates, of protein degardation rates, or of secondary protein modification rates (recall facets of gene potential). This effect on the half-lives of proteins could provide an explanation for the rate enhancement (or decrease) of those environmentally induced biosynthetic sequences of metabolites, which are already operating at the moment of stress application [synthesis of proline, γ-amino butyric acid, polyhydrols (Yancey et al. 1982), etc.]. Of course, there is no doubt that a given metabolism is the consequence of the gene expression being active at the time; but the situation after metabolic switching can cause differential and altered rates of gene expression in a broad sense (transcription rate, translation rate, processing rate, rate of secondary modification, etc.). This again can increase (or decrease) the "milieu inductions" because new sets of parameters arise with the consequence that further changes in the cell internal milieu occur. A feedforward loop of regulation arises. A dynamic process of fine tuning of the metabolism is initiated, which finally leads to optimal performance under the changed environment. The search for "the least understood process ... the molecular mechanism underlying the perception of stress signals" (Ho and Sachs 1989) might prove not to exist in the form of the conventional type of the key-lock signal effect.

12. The Transport-Catalysis Switch

The combination of transport and/or diffusion and catalysis is another metabolic source for the generation of bi- or even multistability. This again will be demonstrated graphically with the GAPDH reaction. This enzyme is known to be substrate-inhibited by GAP (Orsi and Clealand 1972), the extent of which is dependent on the NAD mole fraction of the system (Henrich and Pahlich, unpubl.). If the supply of the cell with GAP is altered (changed GAP export from chloroplast, altered diffusion gradients between the bulk and vicinal water fractions) then the transport processes between the macro- and microenvironments of the enzyme are also changed. (The micro-environment refers to the situation at the active site of an enzyme.) This situation is considered in Fig. 6, which

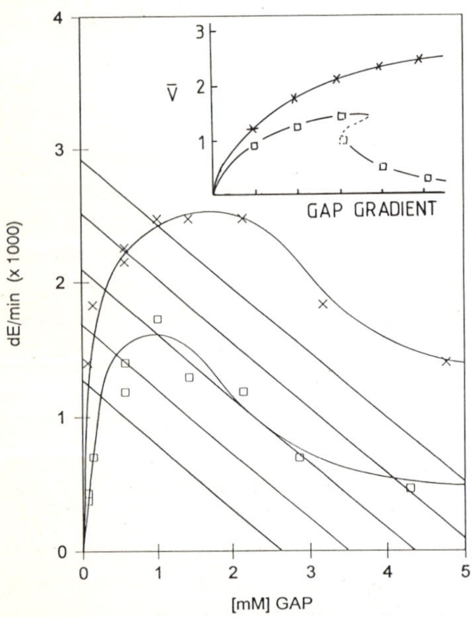

Fig. 6. Stationary states of a transport-catalysis-system. The substrate-inhibited activity profiles of wheat glyceraldehyde phosphate dehydrogenase (GAPDH) were measured at NAD mole fractions of 1 (×—×) and of 0.75 (□—□). Linear transport (or diffusion) fluxes with decreasing gradients of GAP superimpose the saturation profiles. The *intersections* of the transport and catalytic fluxes mark the stationary states of the combined system. The *inset* is a profile of the stationary flux states as functions of the GAP gradients. Under the conditions assumed stationary states are established at low GAP concentrations if solely NAD (and no NADH) is available in the reaction milieu. Under the physiologically more realistic NAD mole fraction of 0.75 instead two stable stationary branches arise – one at low and a second at high GAP concentrations. Both states stand for different metabolic qualities: the former state is controlled by the substrate saturation of the GAPDH, while the latter is transport controlled

shows the measured substrate inhibition curves of wheat GAPDH at high (×- - - - -×) and low (- - - - -) NAD redox charges (Henrich 1993). The straight lines represent the assumed diffusion or transport fluxes with identical transport coefficients (represented by the slope) but with different gradients of GAP between the enzyme and a GAP source.

The intersection points of the transport and reaction rates again mark the stationary states of these pairs of fluxes. Under the conditions considered stationary fluxes are established at high NAD redox charges (×—×) which occur only at low steady GAP concentrations; this holds for all five GAP gradients which are considered. De facto only one stable metabolic branch arises from this if the GAP gradients vary as indicated in the inset; but at reduced NAD mole fractions (□—□), a maximum of three steady intersections of different qualities are possible in the same range of gradient variation. If steep gradients exist which direct the transport flow, then stable stationary states are found only with the inhibited part of the reaction profile (intersections at the right). If less steep transport gradients exist, stable stationary states are established only on the initial part of the saturation profile (left side). A third set of intersections is found at intermediate gradient levels which mark unstable positions. In the secondary plot (inset in Fig. 6), again two thermodynamic branches of stability arise: a single one under high NAD redox charges (upper curve). With the inhibited enzyme under reduced NAD redox charges (lower curve), two stable branches, one at high and one at low transport rates are formed, which form a hysteretic loop. The former branch is characterized by low and the latter by high steady-state concentrations of GAP. The low and high GAP branches of the transport reaction system therefore represent also different metabolic qualities. At low GAP concentrations the system is kinetically controlled. It is the degree of saturation with GAP which determines the stationary flux; but at high transport levels, transport control is exerted. The transition again is jump-like because of the existence of "transition points". As in the previous example with the MDH, two different metabolic regimes also arise with different characters and qualities. Comments on their adaptive qualities could be only speculative at present; but there is no doubt that also this system has the potential for developing bifurcations, which is an adaptive potential.

An example for such a transport-catalysis system which also exhibits the described bistability has been measured with the NADH peroxidase by Degn et al. (1979). The interaction of the Calvin cycle with the phosphate transporter of the chloroplast membrane is also thought to produce bi-stable dynamic states (Pettersson and Ryde-Pettersson 1987).

13. The "Environmental Cusp"

The discussion in this chapter centered around the problem of how minor differences in the genetic potentials of plants – which are observed in clones of plants or in closely related lines of crop plants – can exert responses towards an unfavorable or even extreme environment which range from normal development up to complete failure to survive (Fig. 1). The principal of the outlined concept is summarized and generalized schematically in Fig. 7. On the y-axis a system's response is depicted as, for instance, a metabolic flux. (On a higher level of biological organization this could also be a developmental trait.) The x-axis shows a parameter change (which might be dependent on the salt concentration) and the z-axis may symbolize a parameter change which is water potential-dependent. In the x-y-plane (the water potential is close to zero) the nonlinear flux response is shown as a function of the salt concentration. The y-z-plane gives the system's response if the salt concentration is constant and the water potential is changing. The shapes of both response curves depend on the respective parameter sets of the system; they are expected to differ if genetic lines of plants are compared, and therefore their adaptive values will also differ. This insight can be gained if a concomitant change of both parameters occurs (x- and z-axis change). Then an interesting x-y-z structure can arise in regulated metabolic networks, which has the form of a fault or cusp. (For a mathematically based discussion of the character of a cusp, especially concerning the boundaries between different qualitative states, see Nicolis and Prigogine 1989.) The shape of this cusp, of course, differs in different genetic lines of plants. Three different regimes of responses are observed in this structure. At low values of the water potential (a sector parallel to the x-y-plane), the already mentioned nonlinear response regime is found;

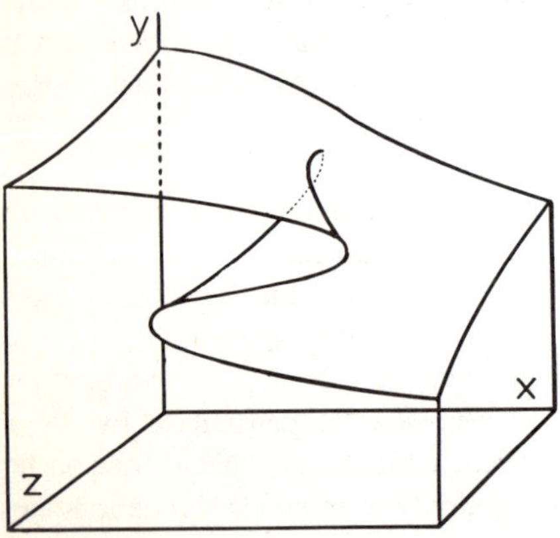

Fig. 7. The "environmental cusp". The scheme is a presentation of a portrait of the type which is discussed in the inset of Fig. 5. On the x-y-plane stationary values of a process (the y-axis might show stationary concentrations, stationary catalytic fluxes, stationary diffusion, or membrane transport rates) are plotted as functions of an environmental parameter such as an increasing salt burden, which shifts an internal enzyme parameter (x-axis). The z-axis carries another external control parameter (temperature, water potential) which also acts on an internal parameter. The fold which arises at a certain x-y combination separates an upper and a lower plateau of stable but qualitatively different metabolic regimes (for details see text)

but at higher values of the water potential, a discontinuity (bifurcation) emerges which originates from a point in which there is a unique dependence of the salt concentration-dependent parameter on the water potential-dependent parameter. From this point on, three dependencies of these parameters exist as is best seen from the s-shaped curve in the front plane of the figure. The two well-known stable upper and lower thermodynamic branches become apparent with the unstable branch in between; but the cusp shows that these branches in fact are parts of "upper" and "lower" plateaus in higher-order networks. The plateaus mark different metabolic or even developmental regimes in an environmental net. These regimes or phases (Eigen 1988a) represent different biological qualities which are reached via jump-like transitions. The ontogenetic development of plants and the stress induced metabolic transitions must be viewed as a sequence of passages over such thresholds: "... das Wesen der Lebenserscheinungen ... äußert sich in einer Abfolge von Phasensprüngen" (Eigen 1988a). This theoretically founded insight was intuitively approached with Larcher's definition of stress hierarchies. The transition phenomenon is found on all levels of biological processes, as in the origin of life, metabolism, development, and evolution.

Adaptability research therefore should distinguish and elucidate the environmentally forced transitions between these different phases. A measured change of a pool, of an enzyme activity, or of a transport rate can occur within any of the mentioned phases, but can also describe the situation after a transition to another phase. Both changes refer to entirely different situations. Discrimination between the two entirely different situations is not possible on the basis of two comparative measurements (especially if the experiments are designed in a way such that "big differences of the responses" can be expected). The former case represents a situation where the regime retains its earlier character or quality. The change is an adjustment of pool sizes, enzyme activities, or transport rates to an altered environment, and might be viewed as a metabolic acclimatization. Through acclimatization, the system, of course, can be pushed towards a transition point (the fold). Under such a movement, the system becomes more and more sensitive towards environmental changes the closer it approaches the discontinuity; but only the passage of the fault drives the system into a different metabolic phase and hence also quality. This new quality can exert a changed degree of tolerance or even resistance to the environment. From the "viewpoint of the plant" a classification of the respective regimes in terms of a state for "better" or "less good" performance can hardly be given. Both phases express situations where the catalysts are best fitted to perform under a given environment, which is an expression of adaptation and optimal performance.

The hypothetical scheme in Fig. 7 also elucidates which data sets are needed to analyze adaptive responses: they are data of parameter changes under the influence of a wide range of physical conditions which allow the construction or calculation of respective environmental cusps. Systematic determinations of the K_m, V_{max}, K_i and transport constants must be performed over a wide physiological range of temperatures under varying ionic strength, water potentials, or pH values, which give insight into the flexibility potentials of catalysts and transporters.

These comprehensive data sets are missing as yet. The control theory anticipates how these data sets have to be quantitatively arranged and interpreted.

For reasons of completeness a note must be added on additional dynamic states which have not been considered in this chapter. Beyond bistability, stable patterns of regular or irregular oscillations can arise which can run into deterministic choas. The whole set of these states has been measured with the already mentioned isolated NADH peroxidase (Degn et al. 1979) and in vitro and in vivo with yeast glycolysis (Hess and Markus 1987). These intuitively unordered states on a time scale prove to be highly ordered, even esthetic, forms if they are visualized as "strange attractors" (Hess and Markus 1987). In botanical research, this actual field of research is being initiated now and some interesting findings with the CO_2 metabolism of CAM plants have been published (Lüttge and Beck 1992). There is no doubt that these dynamic states are also part of the adaptive strategies of plants.

With this concept in mind, it is plausible that even minor differences in the gene potential can give rise to major differences because of the differential plastic capacity of the gene products and their "specific functions... in the larger process" (Veit et al. 1991) of dynamic structure development. The result for these differential responses arises from the ability to establish slightly modified "environmental cusps" with alternative stable thermodynamic branches or even additional dynamic states. Respective cusps of particular dynamic systems in genetic lines of plants have not yet been determined and remain to be uncovered. The analytical tools for doing this research are available. Knowledge of "environmental cusps" would also provide new knowledge of the characters of dynamic systems, which probably would enable molecular biologists to improve the adaptive potential of mesophytes and other plants of agronomical importance by genetic tailoring.

References

Anfinsen CB (1973) Angew Chem 24:1065–1073. **Aktinson DE** (1986) In: Damjanovich S, Keleti T, Tron L (eds) Dynamics of biochemical systems. Elsevier, Amsterdam, pp 129–143.

Blackwood GC, Miflin BJ (1976) Plant Sci Lett 7:435–446. **Blumenthal R**, Changeux J-P, Lefever R (1970) J Membr Biol 2:351–357. **Boiteux A**, Hess B, Sel'kov EE (1980) Curr Top Cell Regul 17:171–201. **Boyer JS** (1982) Science 218:443–448. **Braam J**, Davis R (1990) Cell 12:357–364. **Bradshaw AD** (1965) Adv Genet 13:115–155. **Bücher T**, Rüssmann W (1963) Angew Chem 19:881–893. **Burn P** (1988) TIBS 13:79–83. **Burns JA**, Cornish-Bowden A, Groen AK, Heinrich R, Kacser H, Porteous JW, Rapoport SM, Rapoport TA, Stucki JW, Tager JM, Wanders RJA, Westerhoff HV (1985) TIBS 97: 639–666.

Clausen J, Keck DD, Hieseg WM (1945) Carnegie Inst Wash Publ 564. **Corcuera LJ**, Hintz M, Pahlich E (1989) Phytochemistry 6:1569–1571. **Clegg JS** (1981) Collect Phenom 3:289–312. **Close TJ**, Bray EA (eds) (1993) Responses to cellular dehydration during environmental stress. Curr Top Plant Physiol. 10. **Crowe JH**, Hoekstra FA, Crowe LM (1989) Proc Natl Acad Sci USA 86:520–523.

de Mendoza D, Cronan JE Jr (1983) TIBS 8:49–52. **Degn H**, Olsen LF, Perram JW (1979) In: Gurel O, Rössler OE (eds) Bifurcation theory and applications in scientific disciplines. Ann NY Acad Sci 316:623–638.

Eigen M (1976) Ber Bunsen-Ges 80:1059–1081. **Eigen M** (1988a) Prespektiven der Wissenschaft. Deutsche Verlagsanstalt, Stuttgart. **Eigen M** (1988b) In: Hierholzer K, Wittmann H-G (eds) Phasensprünge und Stetigkeit in der natürlichen und kulturellen Welt. Wissenschaftliche Verlagsgesellschaft, Stuttgart, pp 113–149. **Ellis JJ**, van der Vlies SM (1991) Annu Rev Biochem 60:321–347.

Fischer EH (1993) Angew Chem 105:1181–1188. **Franks F** (1972) Water. A comprehensive treatise. Vol 1. Plenum Press, New York.

Glass I, Mackey MC (1988) From clocks to chaos. Princeton University Press. **Goldberg ME** (1985) TIBS 10:388–391. **Grime JP** (1979) Plant strategies and vegetation processes. John Wiley, Chichester.

Heinrich R, Rapoport TA (1974) Eur J Biochem 42:89–95. **Henrich S** (1993) Diplomarbeit, JLU Gießen. **Hess B**, Markus M (1987) In: Küppers B-O (ed) Ordnung aus dem Chaos. Piper, München, pp 157–174. **Higgins CF**, Cairney J, Stirling DA, Sutherland L, Booth IR (1987) TIBS 12:339–344. **Higgins J** (1963) Ann NY Acad Sci 108:339–344. **Ho T-H D**, Sachs MM (1989) In: Jones HG, Jones MB, Flowers TL (eds) Plants under stress. Cambridge University Press, Cambridge, pp 157–180.

Jennings DH, Trewavas AJ (eds) (1986) Plasticity in Plants. Symp Soc Exp Biol Nr XXXX Cambridge. **Jones HG**, Flowers TJ, Jones MB (eds) (1989) Plants under stress. Cambridge University Press, Cambridge. **Jones RL**, Armstrong JE (1971) Plant Physiol 48:137–141.

Kacser H (1960) Kinetic models of development and heredity. Symp Soc Exp Biol Vol XIV p 13. **Kacser H** (1983) Biochem Soc Trans 11:35–40. **Kacser H**, Burns JA (1973) Symp Soc Exp Bot Vol XXVII, pp 65–104. **Kacser H**, Burns JA (1979) Biochem Soc Trans 7:1149–1160. **Kacser H**, Burns JA (1980) Genetics 97:639–666. **Kacser H**, Porteous JW (1987) TIBS 12:5–12. **Katterman F** (ed) (1990) Environmental injury to plants. Academic Press, San Diego. **Klotz IM** (1966) Arch Biochem Biophys 116:92–96. Krebs EG (1993) Angew Chem 105:1173–1183. **Körner EH** (1993) In: Schulze ED, Mooney HA (eds) Biodiversity and Ecosystem Function. Springer, Berlin Heidelberg New York, pp 117–140. **Kubota K**, Ashihara H (1990) Biochem Biophys Acta 1036:138–142.

Laidler KJ (1958) The chemical kinetics of enzyme action. Oxford University Press, London. **Larcher W** (1987) Naturwissenschaften 74:158–167. **Levitzki A** (1978) Quantitative aspects of allosteric mechanisms. Springer, Berlin Heidelberg New York. **Lewin S** (ed) (1974) Displacement of water and its control of biochemical reactions. Academic Press, London. **Lüttge U**, Beck F (1992) Planta 188:28–38.

Masters CJ, Reid S, Don M (1987) Mol Cell Biochem 76:3–14. **Michaelis L**, Menten ML (1913) Biochem Z 49:333–369. **Moore WJ**, Hummel DO (1976) Physikalische Chemie. Walter de Gruyter, Berlin, 2.

Neumann E (1973) Angew Chem 10:430–444. **Nicoles G**, Prigogine I (1989) Exploring complexity. WH Freeman, New York. **Nobel PS** (1991) Physicochemical and environmental plant physiology. Academic Press, San Diego.

Orsi BA, Clealand WW (1972) Biochemistry. 11:102–109.

Pahlich E, Jäger H-J, Kaschel E (1981) Z Pflanzenphysiol 107:137–144. **Pahlich E**, Jäger H-J, Horz M (1982) Z Pflanzenphysiol 105:475–478. **Pahlich E**, Stadermann T (1984) J Plant Physiol 115:91–96. **Pahlich E** (1990) Bull Soc Bot Fr 137:3–11. **Pahlich E** (1992) In: Mengel K, Pilbeam DI (eds) Nitrogen metabolism of plants. Clarendon Press, Oxford, pp 187–200. **Pahlich E** (1993) Rev Bras Fisiol Veg 5:209–216. **Paleg LG**, Stewart GR, Bradbee JW (1984) Plant Physiol 75:974–978. **Paleg LG**, Aspinall D (eds) (1981) The physiology and biochemistry of drought resistence in plants. Academic Press, Sydney. **Pettersson G**, Ryde-Pettersson U (1987) Eur J Biochem 169:423–429. **Porteous JW** (1983) TIBS 8:5–12. **Purich DL**, Fromm HJ (1972) Curr Top Cell Reg 6:131–167.

Reich JG, Sel'kov EE (1981) Energy Metabolism of the cell. A theoretical treatise. Academic Press, London. **Richards RA** (1993) In: Close TJ, Bray EA (eds) Plant responses to cellular dehydration during environmental stress. Curr Top Plant Physiol 10:211–223. **Rosen R** (1970) Dynamical system theory in biology. John Wiley, New York.

Schrier EE, Robinson RA (1971) J Biol Chem 246:2870–2874. **Schrier EE**, Robinson RA (1974) J Solution Chem 3:493–501. **Segel IH** (1975) Enzyme Kinetics. John Wiley, New York. **Shinde U**, Inouye M (1993) TIBS 18:442–446. **Slatyer RO** (1967) Plant-water relationships. Academic Press, London. **Somero GN** (1978) Annu Rev Ecol Syst 9:1–29. **Somero GN**, Osmond CB, Bolis CL (eds) (1992) Water and life. Springer Berlin Heidelberg, New York.

Tanford C (1978) Science 200:1012–1018. **Thye E**, Pahlich E (1994) Eur J Agron (in press). **Timasheff SN** (1993) Annu Rev Biophys Biomol Struct 22:67–97. **Timasheff SN**, Fasman GD (eds) (1969) Structure and stability of biological macromolecules, vol 2. Marcel Dekker, New York.

Von Hippel PH, Schleich T (eds) (1969) In: Timasheff SN, Fasman GD (eds) (1969) Structure and stability of biological macromolecules. Marcel Dekker, New York, pp 417–469. **Veit B**, Greene B, Lowe B, Mathern J, Sinha N, Vollbert E, Walko R, Hake S (1991) Dev Suppl 1:105–111.

Weatherley PE (1965) The state and movement of water in the leaf. Symp Soc Exp Biol XIX:157–185. **Webb JL** (1963) Enzyme and metabolic inhibitors vol 1. Academic Press, New York. **Wiggins PM** (1973) Biophys J 13:385–398. **Wolfenden RV**, Cullis PM, Southgate CCF (1979) Science 206:575–577. **Wong JT-F** (1975) Kinetics of enzyme mechanisms. Academic Press, London.

Yancey PH, Clark ME, Hand SC, Bowlus RD, Somero GN (1982) Science 217:1214–1222.

Zaccai G, Eisenberg H (1990) TIBS 15:333–337

Prof. Dr. E. Pahlich
Institut für Allg. Botanik und
Pflanzenphysiologie der JLU
Heinrich Buff Ring 54–62
35392 Gießen, Germany

VIa. Phytochrome

By Bertold Hock

1. Introduction

Phytochrome appears to be the best characterized of the photoreceptors involved in plant morphogenesis and also in photoperiodism. It is used by plants for measuring light intensity, quality, direction, and duration. This red/far-red (r/fr) reversible chromoprotein exists as a dimer of two approximately 120-kDa polypeptides, each with a covalently attached linear tetrapyrrole chromophore. The characteristics have been thoroughly reviewed, e.g., recently by Smith and Whitelam (1990), Quail (1991), Thomas and Johnson (1991), Thompson and White (1991), Furuya (1993), and Vierstra (1993).

Major progress has been achieved during the past years in the study of the role of individual phytochromes for photomorphogenesis and signal transduction. Moreover, the paradigm that phytochrome is no enzyme has been refuted, at least for moss phytochrome, and a better understanding of the nature of phytochrome action in the green plant has been reached. This has broadened the concept of ecological functions of the phytochrome.

2. Photomorphogenesis and Photoperiodism

Photomorphogenesis refers to the control of development by light. In higher plants all formative processes such as seed germination, seedling development, vegetative growth, and flowering are concerned. In its most general sense, photomorphogenesis is the developmental strategy if and as long light is available, while skotomorphogenesis (from the Greek skotos = dark) is the developmental strategy of choice as long as light is not yet, or no longer available (Mohr and Shropshire Jr. 1983). In other words, photomorphogenesis can be considered as affluence strategy and skotomorphogenesis as survival strategy (Mohr and Schopfer 1992). Light and darkness trigger the transition from one pattern of gene expression to the other. Photoperiodism refers to the control by the photoperiod of flow-

Progress in Botany, Vol. 56

ering and other long-term qualitative changes during development such as bud dormancy, tuber formation, or the expression of male and female flowers.

Both photomorphogenesis and photoperiodism are controlled by phytochrome and/or the blue light/UV-A photoreceptor(s). They perceive the light signal, which is subsequently transduced into a biochemical change. Complex interactions between the signal transduction chains of different photoreceptor groups have been observed, the most intensively studied being the coaction of phytochrome and the blue/UV-A photoreceptor. The availability of an increasing number of photomorphogenic mutants has allowed us to dissect the function of individual photoreceptors and to analyze parts of the signal transduction pathways.

3. Ecological Significance of Phytochrome Responses

Plant ecology deals with the relations between plants and with interactions between plants and their environment. Both aspects depend upon phytochrome as a photoreceptor for the perception of the red (r) and far-red (fr) wavelength regions (ca. 600–800 nm). Phytochrome conveys information to plants not only on the presence or absence of light during early germination but also on fluctuations occurring naturally under three situations of ecological significance: at the beginning and end of the day (EOD); under water; and in the proximity of neighboring plants. This subject has been thoroughly reviewed by Smith (1992).

As germination of many wild plants, especially weeds, requires light, it is obvious that light can play an important photomorphogenic role beneath the soil surface. Therefore the spectral properties of soil or organic litter are of special importance. Light can penetrate several millimeters of soil, depending on soil type and moisture. This subject has been intensively discussed and further investigated by Mandoli et al. (1993). They showed that light penetration was significantly better through sands than through either sandy or silty loams. For sandy soils, the r/fr photon ratio dropped by more than 30% in the first few millimeters, decreasing the P_{fr}/P_{total} ratio by 10–15%. Similar spectral changes were obtained when sample depth for sandy soils was held constant, and particle size of that soil was decreased. Water saturation greatly increases the light transmittance of these soils and also the r/fr photon ratios. It should be noticed, however, that loams exhibit entirely different properties. Mandoli et al. (1993) discuss in detail the potential biological significance of the soil properties mentioned above. It is tempting to attribute a role in seed germination, root gravitropism, and other pre-emergent photomorphogenesis to soil depth, ped size, and water content.

Control of seed germination by light is observed in most of the wild plants growing in open habitats, including weeds in crop fields. As large-seed crop plants have been selected for light-independent germination, this difference has been successfully used by Hartmann and Nezadal (1990) in intensely cultured crop fields to significantly reduce the coverage by weeds. The concept is based on the fact that the emergence of weeds in a crop field is mainly due to repeated short light exposures that occur during the mechanical cultivation in daylight. The very low fluence response of phytochrome enables plants to react to low active phytochrome (P_{fr}) levels that even far-red light is sufficient to evoke a response. Most of the seed bank that has accumulated over the decades in the soil of intensely cultured crop fields is found in deeper layers, not reached by light, down to 15–40 cm. Since light penetrates only to about 1 cm into the soil, the following strategy has been used for weed control: (1) carrying out early cultivation (plowing and tillage) on bright days to provoke a large flush of emerging weeds in the subsequent 4 weeks; (2) destroying the developing weeds by harrowing during dark; and (3) sowing or planting the crops in the dark, i.e., between 1 h after sunset and 1 h before sunrise. This strategy combines reduction of the seed bank with the selective emergence of the crop plants.

Emerging seedlings and other plants not only receive direct sunlight or light filtered through a canopy but also upwardly reflected light from soil as well as radiation reflected from adjacent plants. Phytochrome conveys some kind of a primitive kind color vision to plants because it can sense shifts in the r/fr composition of the spectral environment as they occur under the conditions mentioned above and signal distance from the soil surface or competition by other plants. A thorough review of the morphological responses to these different environments and their practical implications has been given by Kasperbauer (1992).

Kasperbauer and Hunt (1992) showed that seedling morphology of cotton is greatly influenced by the amount of fr light reflected from soil, dead plant residue, or even artificially colored surfaces. An upwardly reflected r/fr ratio lower than the ratio of incoming sunlight resulted in larger shoots and a higher shoot/root biomass ratio. A reflected r/fr ratio higher than the ratio in incoming sunlight resulted in larger roots and a lower shoot/root biomass ratio.

A promising practical application is the use of colored mulch or plastic foils, which are usually applied for soil and water conservation as well as for weed control, if they reflect r/fr ratios lower than the ratio in incoming sunlight. Soybean seedlings developed longer stems over green and red surfaces, which reflected lower amounts of blue and lower r/fr ratios than were present in incoming sunlight. On the other hand, the largest root and the lowest shoot/root biomass ratio for root crop (i.e., radish) developed over orange mulch, which reflected a r/fr ratio higher than the ratio of the incoming light (Kasperbauer 1992).

It has long been known that light passing through a canopy exhibits not only reduced photon flux densities but also significantly reduced r/fr ratios (for review cf. Smith 1992) due to the effective absorption by leaves in the red (and blue) region. This "green" shading, that is quite different from neutral shade, allows shade-avoiding plants to allocate their resources towards more rapid extension growth, whereas shade-tolerant plants, which normally grow in a shaded environment, have acclimated photosynthetically to low light levels and show little or no alteration of their stem extension rate as they are exposed to different r/fr environments.

Another important source of irradiation is the reflection by neighboring plants, even before mutual shading occurs. Consequently, phytochrome is used for sensing the nearness of other plants. For instance, close-spaced tobacco began developing longer internodes and narrower leaves *before* shading effects became apparent (Kasperbauer 1973). This subject has been further investigated by Ballaré et al. (1990, 1991). When individual seedlings from *Datura ferox* and *Sinapis alba* were transferred to populations formed by plants of similar structure, they responded with an increase in the rate of stem elongation within 3 days after transplanting, i.e., before shading among neighbors became apparent. Blinding individual stem segments with annular cuvettes containing copper sulfate solutions, to fr radiation scattered by the surrounding seedlings, abolished this response. Here, the low r/fr ratios are received at the stem level. It appears to be crucial for success in the competition for light to acquire information about the relative position in a stand at a very early stage of canopy development.

Quantitative analyses were carried out by Ballaré et al. (1991) on the effect of light quality ("green shade") and light quantity ("neutral shade") on fully de-etiolated plants. In low canopy densities, where only the r/fr ratio is reduced (due to fr reflected from nearby leaves), internode elongation could be promoted by low r/fr ratios even if they were established during the central part of the photoperiod, when plants are exposed to very high fluence rates. On the other hand, in denser populations, where the photon fluence rates are also significantly reduced, elongation growth is promoted by the low r/fr ratio and the reduced fluence rate. The latter seems to be independently sensed by phytochrome and the blue-light receptor. Again, perception takes place at the stem level.

Light appears to specifically influence the pattern of elongation along the stem axis. Shinkle et al. (1992) noted that hypocotyls from dim red light-grown cucumber seedlings have an even distribution of growth whereas dark-grown seedlings show an apical to basal graded decline in growth rate. It converts to the nongradient pattern when the seedlings are transferred to dim red light. This strategy of growth is interpreted as an ecological advantage for seedlings germinating under natural light because a larger elongation zone means an increased capacity to respond to anticipated shading.

Ballaré et al. (1992) studied a further role of phytochrome, which can act as a phototropic sensor in de-etiolated plants. Continuous unilateral fr light provided against a background of white light induced bending away from the fr light source (negative curvature) in hypocotyls of cucumber seedlings. If they were grown in the field, they also showed negative phototropic responses to the proximity of neighboring plants that back-reflected fr radiation. The bending response was reduced in the long-hypocotyl (lh) mutant, which is deficient in phytochrome B. (cf. p. 206). The ecological significance must be seen in connection with the blue-light photoreceptor that mediates positive phototropic bending. In patchy canopy environments, young seedlings actively project new leaves into light gaps via stem bending responses, which are elicited by a blue-absorbing photoreceptor and phytochrome.

4. At Least Two Types of Phytochromes Regulate Plant Development

During the last decades, it has been deduced from physiological studies of the complex phytochrome responses that more than one type of phytochrome exists. Two physiologically distinct phytochrome pools were proposed by Hillman (1967). Rigorous proof was finally provided during the mid-1980s by immunochemical techniques.

The classically known light-labile phytochrome occurs predominantly in etiolated tissue because it accumulates in the dark, but decays in the light to very low levels. It has been named type I (cf. Furuya 1989). In addition, a relatively stable phytochrome regardless of light conditions was first revealed at much lower concentrations in green tissue and named type II.

Because of the difficulties in distinguishing the different roles of these photoreceptors in plant morphogenesis, new approaches have been recently taken, the analysis of phytochrome-deficient plants (photoreceptor mutants) and of phytochrome overproducers (transgenic plants). The subject has been reviewed by Kendrick and Nagatani (1991), Chory (1992), Reed et al. (1992), and Chory (1993). A seminal role was assigned to *Arabidopsis thaliana* because of its favorable features suitable for genetic studies: small genome size, accumulated genetic information, self-fertility, ease of cultivation and genetic analyses, short generation time, small body, and development of experimental procedures such as mutagenesis, transformation, and gene tagging (Okada and Shimura 1992). The presently available photomorphogenetic mutants have been listed by Chory (1993). In addition, similar mutants of other dicots, such as *Cucumis sativus, Pisum sativum, Brassica rapa, Lycopersicon esculentum,* and the monocots *Sorghum bicolor* and *Triticum aestivum*, as well as transgenic *Nicotiana tabacum* and *Avena sativa* have greatly extended the concept of different photosensory roles of the phytochrome types mentioned above.

Up to now, four groups of photomorphogenic mutants can be discriminated: (1) Chromophore mutants such as hy1, hy2, and hy6 of *Arabidopsis*. These are blocked to chromophore biosynthesis at steps prior to the formation at biliverdin IXa, a direct precursor of phytochromobilin (cf. below), the phytochrome chromophore. Mutations in the corresponding genes lack spectrophotometrically detectable phytochrome in etiolated seedlings and reduce or eliminate the function of presumably all the phytochromes. The phenotypes are characterized by long hypocotyls not only in the dark but also in the light. The mutants lack responsiveness to both continuous r and continuous fr. Normal photomorphogenesis can be restored by exogenously supplied tetrapyrroles such as biliverdin (Parks and Quail 1991). (2) Photoreceptor mutants with defects in gene coding for the phytochrome apoprotein. They are deficient in only one type of phytochrome and are therefore particularly suited for relating specific phytochromes to specific functions. (3) Photoreceptor mutants overexpressing one type of phytochrome. These transgenic plants express high levels of one foreign phytochrome species in addition to the native types and can be therefore monitored for an altered phenotype. (4) Signal transduction mutants. Here components of the phytochrome signal chain are defective. These mutants have greatly improved our understanding of how the phytochrome signal is transduced into cellular activity.

The most extensive genetic analyses have been carried out in *Arabidopsis thaliana*. They have revealed five different phytochrome genes (phyA, phyB, phyC, phyD, phyE), with phyA coding for phytochrome type I, and the others coding for phytochrome type II. Detailed analyses have been carried out for phyA and B.

This nomenclature has now been adopted also for other plants and will be used in this chapter, where appropriate.

5. Phytochrome B Is Involved in Most Photoresponses of Etiolated and De-etiolated Plants

Although there are several subpopulations of stable phytochrome from Type II which differ from that which is abundant in etiolated tissues, most data are available for phytochrome B functions. They have been mainly obtained from the functional analysis of phytochrome B-deficient mutants (*Arabidopsis* hy3, *Brassica* ein, *Cucumis* lh, *Sorghum* ma_3), and of phytochrome B overexpressing transgenic plants (*Arabidopsis* Box). It was shown by Wagner et al. (1991) in transgenic *Arabidopsis* seedlings (with a short hypocotyl phenotype) overexpressing phytochrome B that this type is synthesized in the P_r form and undergoes reversible conformational changes after r/fr-light irradiation. The spectral properties are very similar, however, to those of phytochrome A. This has been also confirmed for phytochrome B from tomato (Sharma et al. 1993).

Phytochrome from green leaves of *Avena sativa* has been purified on a large scale by Pratt et al. (1991). An apparent monomer size of 123 kDa was evaluated. The spectral analysis yielded no marked differences between phytochromes from green and etiolated oat shoots except for a small shift of the P_r minus P_{fr} difference spectrum for phytochrome from green oats to shorter wavelength (12 nm maximum and 4 nm minimum).

From the available data, it can be concluded that the following photomorphoses are controlled by phytochrome B: deetiolation in white light and continuous r, but not in continuous fr, as measured, for instance, by the suppression of hypocotyl, petiole, and floral stem growth, promotion of chloroplast number under the same conditions, suppression of root hair elongation in the light, mediation of the end-of-day far-red response (EOD-FR). Furthermore, phytochrome B is constitutively expressed (Sharrock and Quail 1989).

The EOD-FR growth promotion concerning the internode growth is perceived mainly by the leaves and triggered by low r/fr ratios, as is the case in the natural environment at the end of the photoperiod. The EOD-FR response is experimentally used to simulate shade responses, and it has become the standard assay for the action of light-stable phytochrome as growth is strongly dependent on the state of phytochrome at the end of the photoperiod. The *Arabidopsis* hy3 mutant and other B^- mutants, where phytochrome B is undetectable, either do not display or exhibit only reduced EOD-FR growth response. Furthermore, this mutant type, which is also represented by *Sorghum* ma^3, lacks photoperiodic control of flowering and exhibits defects in the light-induced anthocyanin synthesis and stimulation of leaf elongation. It has been recently proven that the hy3 mutant of *Arabidopsis* has a defect in the gene coding for phytochrome B (Reed et al. 1993).

Taken together, these data prove that the most important light-dependent steps in the life of a green plant are controlled by phytochrome B. Overexpression results in an enhanced responsiveness to continuous red, but not to far-red irradiation, as seen in *Arabidopsis* (McCormac et al. 1993). It should be kept in mind, however, that several phytochrome B-mediated responses can be mediated in addition by the blue/UV-A photoreceptors. The complex interactions have been studied recently by Oelmüller and Kendrick (1991).

Stewart et al. (1992) were able to prove with light-grown oat seedlings that end-of-day light quality in plants growing in otherwise natural light environment alters the level of at least one of the type II phytochromes predominating in light-grown tissues. The fluctuation of the amount of this photoreversible phytochrome extracted under the different conditions were approximately in the range of the fourfold.

It is usually taken for granted that P_r is the biologically inactive form while light-induced P_{fr} triggers the light-responsive switch to "on" or "off". Liscum and Hangarter (1993b) have now provided genetic evidence that the P_r form of phytochrome B in *Arabidopsis* regulates normal hypocotyl gravitropism in darkness, whereas depletion of the P_r form by conversion to P_{fr} attenuates hypocotyl gravitropism. Two independently isolated hy3 lines lacking phytochrome B showed randomized hypocotyl growth direction in the dark. On the other hand, Reed et al. (1993) did not notice any defect in root gravitropism in the phytochrome B mutants. It remains to be seen whether other responses can be found that are controlled by the P_r form.

Up to now, not much is known on responses controlled by other type II phytochromes. Liscum and Hangarter (1993c) have recently shown that apical-hook opening in *Arabidopsis* may be controlled by such a type II phytochrome being different from phytochrome B. However, it should be noticed that a blue-light photoreceptor independently mediates this reaction. Comparing growth of wild type and lh mutant seedlings of *Cucumis sativus* in white light supplemented with fr by conventional and transducer technology, Smith et al. (1992) found that the mutant seedlings responded to additional fr radiation by an increment in extension rate to that observed with the wild type seedlings. The authors conclude that the species of phytochrome deficient in the lh mutant (apparently phytochrome B), is not solely responsible for the shade-avoidance syndrome in light-grown plants.

6. Phytochrome A Mediates the HIR Responses

Type I phytochrome (= phytochrome A) represents the most abundant phytochrome species which has been biochemically and biophysically studied in great detail. This light-labile photoreceptor accumulates at a high level in the dark, about a 100 times higher than type II; but it is rapidly degraded upon illumination to a low level in the light. In light-grown tissue, both photoreceptor types are present at roughly comparable levels. The absence of phytochrome A does not seem to greatly impair plant development. Phytochrome A^--deficient mutants such as *Arabidopsis* hy8, where a single nucleotide change within the hy8 locus results in phytochrome A transcripts containing a stop codon, as well as fre1, fhy2, or the *Lycopersicon* mutant aurea (au) exhibit in general normal responses to EOD-FR treatments, continuous white or red light treatments, and appear normal in continuous white and red light.

However, there is direct evidence that endogenous phytochrome A is responsible for the far-red controlled high-intensity response (FR-HIR) of etiolated seedlings. This type of action, with maximum effectiveness at 716 nm (Hartmann 1966), was first reported for growth inhibition and anthocyanin synthesis in *Sinapis alba* seedlings (Mohr 1957) and in turnip seedlings (Siegelman and Hendricks 1957).

The degree of fr effectiveness depends on the energy. Hartmann (1966) found that an equal response could be obtained by simultaneous irradiation with 658- and 768-nm bands, both of which are ineffective when used alone. Both the single 716 band and the dual irradiance maintain about 3% in the P_{fr} form. It has been concluded that P_{fr} must be held at a relatively low level while repeatedly exciting P_r and P_{fr}. The low level of P_{fr} was considered as diminishing its destruction.

In *Arabidopsis* lack of phytochrome A seems to affect only the FR-HIR response, i.e., the corresponding mutants etiolate in continuous fr, but the absence does not affect the de-etiolation in response to prolonged red or white light (Parks and Quail 1993). The authors derive from the fact that the hy8 mutant develops normally in continuous white light from the seedling stage to mature adults that phytochrome A might not even be necessary for normal photomorphogenesis. The same conclusion has been drawn from work on the mutants fre1 (Nagatani et al. 1993b) and fhy2 (Whitelam et al. 1993).

According to this view, the function of phytochrome A might be highly specialized and restricted to certain phases of development. It is considered to play a crucial role in triggering the light-induced developmental switch from heterotrophic to autotrophic growth at early stages.

This implies that A and B obviously have transiently antagonistic roles in regulating early seedling morphogenesis. Parks and Quail (1993) point out the transient nature of the inhibitory effect of the FR-HIR in the elongation of growth. After the transition from heterotrophic to autotrophic development, when the phytochrome A levels have dropped, the shade-avoidance reaction assumes dominance and elongation growth accelerates in response to enhanced levels of FR.

Substantial work has been carried out in this direction with the aurea mutant of tomato (au) which was originally isolated as a nongerminating, gibberellin-responsive plant (Koornneef et al. 1985). It has long hypocotyl, yellow-green leaves, due to reduced chlorophyll accumulation (Ken-Dror and Horwitz 1990), reduced levels of anthocyanin, and defective and delayed development of chloroplasts. Sharma et al. (1993) have shown that this mutant is deficient in photoactive phytochrome A at the etiolated stage, when it contains a spectrally inactive phytochrome A. However, the mutant contains spectrally active phytochrome B in its green tissue as does the wild type.

Casal and Kendrick (1993) examined in greater detail the shade-avoidance reactions in wild-type and aurea-mutant seedlings of tomato. Compared to wild-type seedlings, the au-mutant seedlings grown under natural radiation show higher rates of internode extension growth as well as deficient sensitivity of internode extension growth responses to reduction in the r/fr ratio, as is observed in the presence of neighboring vegetation due to reflecting a significant amount of fr after receiving full sunlight or to the presence of a dense shade canopy. The ability to set the *degree* of response to r/fr signals according to the amount of phytochrome absorbable radiation (as observed in wild-type seedlings) is interpreted by Casal and Kendrick (1993) as a strategy to avoid strong responses both to very distant plants shading the plant under consideration only at the extremes of the photoperiods, and to the twilight drop in r/fr. Therefore, it is not surprising that light-grown au mutants respond to EOF-FR treatment with an enhancement in the rate of stem elongation and to high-irradiance supplementary fr light during the photoperiod, a strong shade signal.

Sharma et al. (1993) discuss the possibility that the au mutation, which does not affect the amount of translatable mRNA of phytochrome, affects multiple species of phytochrome, including phytochrome A, at the etiolated stage only. This may also explain the reduced level of protochlorophyll in etiolated au seedlings and contribute to their yellow color. On the other hand, a factor necessary for phytochrome stabilization (possibly a cheaper one) may be absent from this mutant (Smith et al. 1993).

Smith et al. (1993) examined in older tomato seedlings the photosynthetic acclimation to shade. Again, the aurea mutant showed a reduced capacity to respond to vegetation shade (characterized by a low r/fr ratio) with increased extension. This leads to a reduced maximum rate of net photosynthesis, although the au lesion has little effect on the capacity to exhibit photoadaption to vegetational shade when measured at the thylkoid level. The authors come to the conclusion that in wild-type plants phytochrome A may also be functional in de-etiolated plants.

The search for additional functions of phytochrome A is complicated by the fact that the almost normal photomorphogenesis of the aurea mutant in white light is primarily due to the light being absorbed by the blue-light/UV-A photoreceptor in the blue-light/UV-A region of the visible spectrum (Oelmüller and Kendrick 1991). In red light, the transcript levels for the light-harvesting chlorophyll a/b-binding (cab) protein of photosystem I and II are strongly reduced in the mutant.

Under these conditions, it eventually loses the ability to develop into a normal plant.

It is therefore interesting to study plants with elevated phytochrome A levels. The availability of full-length cDNA clones encoding phytochrome A has made it possible to express this phytochrome species in transgenic plants. This strategy has been successfully applied by several groups, e.g., Boylan and Quail (1989, 1991) for oat phytochrome A in transgenic tomato and in *Arabidopsis*, Kay et al. (1989) for rice phytochrome A in transgenic tobacco, and Keller et al. (1989) for oat phytochrome A in transgenic tobacco. Phytochrome A overexpressors exhibit continued persistence of growth inhibition by far-red light in an irradiance-dependent manner, as may be expected from an extended HIR response. Recently, Nagatani et al. (1993a) has proved that introduced rice phytochrome A into a transgenic tobacco line greatly inhibited hypocotyl elongation in the presence of continuous red light. However, overexpressed phytochrome A did not significantly affect the greening process, although the chloroplast in the guard cells of the overexpressors appeared to be more developed than those of wild-type plants. The lacking response in greening is in contrast to observations by Cherry et al. (1991) and McCormac et al. (1991). It might be due to the difference in the age of the plants examined and/or the host cultivars of tobacco used for the transformation.

Phytochrome A has been characterized by Heyer and Gatz (1992). Its mRNA levels varied in different organs and were modulated by white light. The highest levels were detected in the etiolated stage of seedlings and sprouts. In addition to the transcriptional control, phytochrome is also regulated at the level of translation, as is also known for other dicots. The DNA sequence of potato phytochrome A shows some peculiarities. The amino acid directly adjacent to cysteine 323, to which the chromophore is covalently attached, is not serine as in all other type A phytochrome genes, but tyrosine, whereas type B phytochrome sequences from monocots and dicots encode a glycine in this position.

7. Phytochrome Structure

Although phytochrome structure has been studied in great detail for a considerable time, major progress has only recently been achieved on the significance of individual domains of the apoprotein for phytochrome action. Cherry et al. (1992) used transgenic tobacco plants expressing oat phytochrome. Full-length oat phytochrome caused a number of phenotype alterations, including delayed leaf senescence, increased leaf chlorophyll content, decreased apical dominance, and decreased stem elongation, whereas none of these phenotypic alterations was exhibited by the transformants expressing a truncated phytochrome without the amino acids 7–69. The latter still exists as a homodimer in solution, is capable of covalently binding the chromophore, and is rapidly degraded in vivo after photoconversion to the fr-absorbing form.

Parker et al. (1992) studied synthetic polypeptide fragments corresponding to the 6-kDa and terminal region of oat phytochrome A and demonstrated the α-helix-forming ability in an amphiphilic environment (SDS micelles). This is consistent with earlier views that the N-terminal chain of phytochrome A interacts with a hydrophobic chromophore pocket that is exposed upon P_r to P_{fr} conversion in an amphiphilic α-helical manner. The N-terminal α-helices would then be stabilized by interacting with the hydrophobic surface. The authors further showed that the synthetic peptide formed a native-like complex with phytochrome A lacking the N-terminal region.

The quaternary structure of phytochrome is tripartite, consisting of two interacting carboxy domains that dimerize to form a base from which extend the two globular amino-terminal ends. The amino- and carboxy-terminal domains of each subunit are joined by a region defined as the hinge (Jones and Erickson 1989).

The fine mapping of the dimerization regions has now been carried out by Edgerton and Jones (1992). They used a novel assay for dimerization based on a protein fusion system with λ repressor. Short sequences in the carboxy-terminal domain proved to be involved in dimerization, one residing between residues V623 and S673 and the other between N1049 and Q1129. They are highly conserved between proteins belonging to the phytochrome A family. Embedded within these sequences are four segments that could potentially form

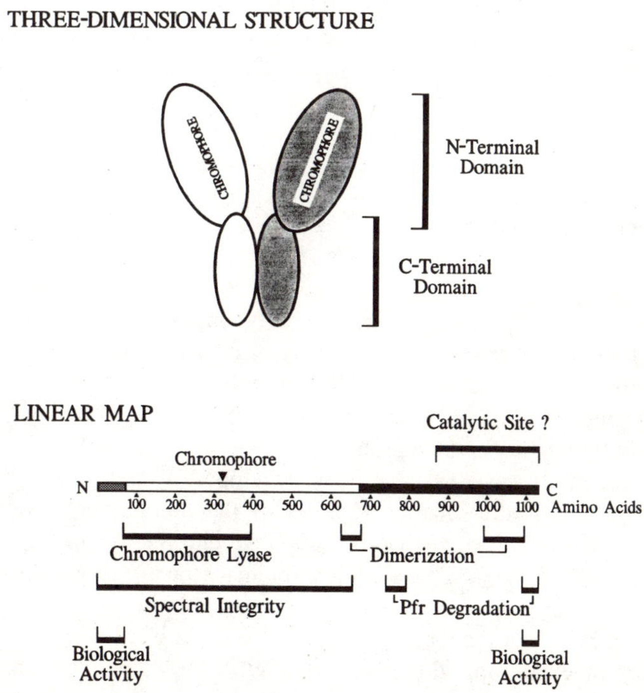

Fig. 1. Structure of phytochrome A. (Vierstra 1993)

amphipathic α-helices, two of them being conserved between phytochrome A, B, and C genes. This finding supports earlier hypotheses on the specific role of phytochrome heterodimers for the detection of wide range of fluence levels, magnitude, and nature of phytochrome responses. Figure 1 summarizes the present knowledge of phytochrome structure.

A fuller understanding of the structural changes associated with the P_r to P_{fr} transition would facilitate the search for the docking between phytochrome and its receptor(s). In this context, the conformational changes of the chromophore itself are of special importance. Although substantial evidence for a semicircular conformation in P_r exists, the structure of the chromophore in P_{fr} is less certain. The double bond at C15 has been proposed as the site of isomerization from Z to E configuration (e.g., Rüdiger et al. 1983), triggering conformational changes particularly of the N-terminal chromophore containing region of the protein.

Gabriel and Hoober (1991) examined through molecular modeling a 51-amino acid sequence that includes cysteine-321, the attachment site, of *Avena* phytochrome A, and forms a distinct pocket. The semicircular conformation of the P_r chromophore easily fits within this pocket. The authors compared models of Z and E isomers at the C-4 and C-15 double bonds. An isomerization at C-4 caused the chromophore to move partially out of the pocket and brought the side-chain carboxyl groups and D to the surface of the polypeptide. Whereas this model predicts a rotation of the chromophore of the magnitude determined experimentally, the models of C-15 isomerization predicted no significant rotation and no significant change of the chemical reactivity of the chromophore. The authors conclude that the change in orientation by an isomerization at C-4 is compatible with the observed interaction of P_{fr} with metal ions, which is possibly a component in the physiological activity of the photoreceptor.

8. Phytochrome Synthesis and Assembly

Phytochrome is synthesized in the P_r form, e.g., during early stages of seed germination. Synthesis of the holoprotein involves the convergence of two biosynthetic pathways, one for the chromophore and the other for the apoprotein. The assembly is autocatalytic, requiring only apophytochrome-free bilin with an ethylidene functionality in the A-ring (Lagarias and Lagarias 1989).

In vitro phytochrome assembly systems have been developed by the group of J.C. Lagarias and used for the further characterization of the holophytochrome synthesis. Wahleithner et al. (1991) expressed oat apophytochrome A_3 in both *Saccharomyces cerevisiae* and *E. coli*. The soluble protein extracts are capable of covalently attaching the chromophore analogue phycocyanobilin as well as phytochromobilin, the natural chromophore synthesized from biliverdin by cucumber etioplast preparations, yielding the spectrally active holophytochrome. Most interestingly, the phycocyanobilin-apophytochrome adduct proved also to be photoreactive. It undergoes a light-induced conformational change similar to the native holoprotein. However, the analogous phycoerythrobilin (PEP) adduct is locked in a photochemically inactive protein confôrmation that is similar to the P_r form of phytochrome (Li and Lagarias 1992). Because of the lack of the C15–C16 double bond (cf. Fig. 2) in the PEB adduct, the authors conclude that the C15 double bond that is necessary for photoreversibility and that the P_r to P_{fr} photoconversion involves the Z to E isomerization of the C15 double bond. The reconstitution from recombinant apophytochrome and different bilins opens the possibility for the further elucidation of the bilin-apophytochrome interaction and for preparing novel phytochromes with altered biological activities.

The present knowledge on the biosynthetic pathway of the chromophore is summarized by Fig. 2. Both 5-aminolevulinic acid and biliverdin IXα are precur-

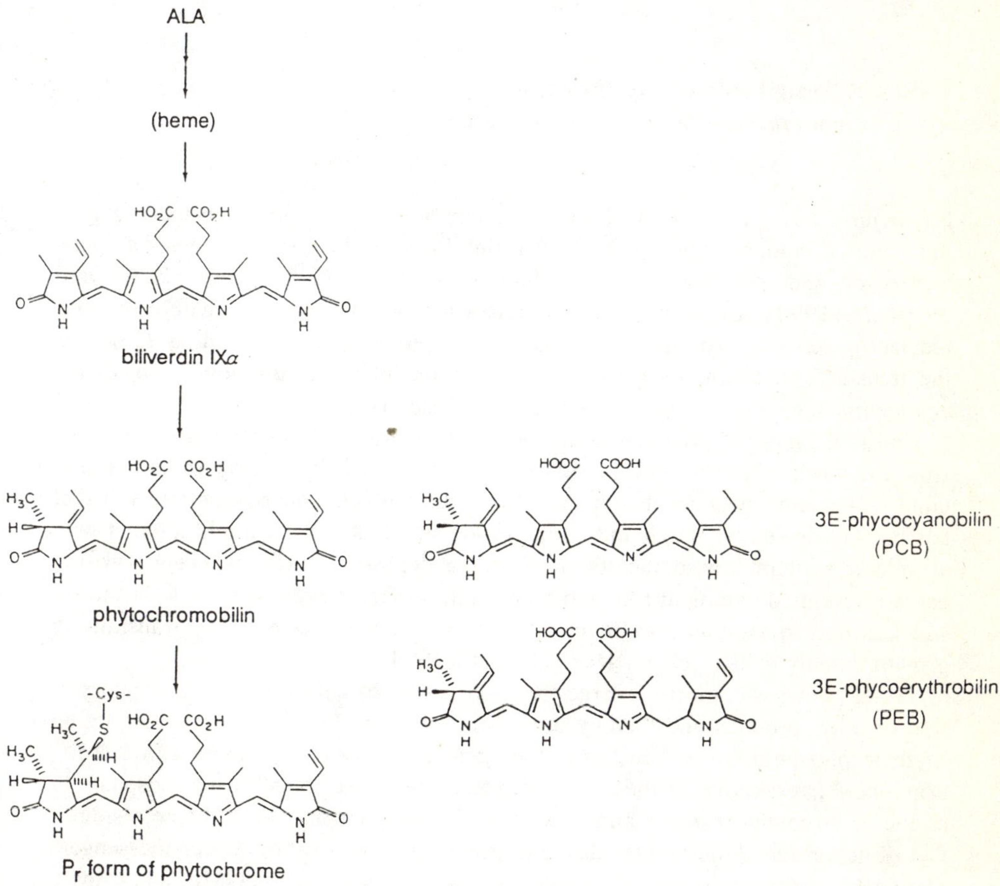

Fig. 2. Proposed biosynthetic pathway of the phytochrome chromophore. *ALA* 5-aminolevulinic acid. The analogues of phytochromobilin, phycocyanobilin, and phycoerythrobilin are indicated. (Combined after Wahleithner et al. 1991 and Li and Lagarias 1992)

sors. Terry and Lagarias (1991) have shown that isolated plastids from cucumber cotyledons convert biliverdin IXα to phytochromobilin in the presence of NADPH and ATP and the plastid-localized enzyme phytochromobilin synthase. The authors assume that all of the enzymes of the chromophore biosynthetic pathway reside in the plastid.

It is known from a variety of systems that successful folding of many proteins requires accessory proteins called molecular chaperones (Hemmingsen et al. 1988). Grimm et al. (1993) carried out renaturation experiments with phytochrome purified from etiolated oat seedlings and the bacterial GroEL molecular chaperone. In vitro folding experiments with denatured phytochrome revealed that the successful reconstitution of the photoreversible photoreceptor required the presence of GroEL and MgATP. GroEL related proteins have been identified in the cytosol of oat cells (Grimm et al. 1991).

9. Signal Transduction May Involve a Dephosphorylation-Phosphorylation Cascade

Phytochrome acts not only as a photoreceptor but also as phototransducer. Signal transduction comprises all processes that link P_{fr} with the terminal steps, i.e., gene expression and/or changes in membrane conductance. As has been put forward by Haupt (1990), a minimal model requires the interaction of P_{fr} with a primary reactant generating an internal signal. This step is termed coupling of P_{fr} to the transduction chain. In a second step, intracellular transmission is postulated connecting the internal signal with the final response.

The first hints for potential components of the signal transduction chain came from the work of Datta et al. (1985) on the stimulation by phytochrome and calcium of protein phosphorylation in isolated pea nuclei. Subsequent research that has been reviewed by Singh and Song (1990) suggests that P_{fr} might interact with a cellular protein kinase thereby initiating a dephosphorylation-phosphorylation cascade. Related events are known from quite different processes such as the visual cascade in vertebrates (Rayer et al. 1990) or the Ras pathway transmitting growth signals to the cell nucleus (cf. Marx 1993).

Protein kinases comprise a large family of enzymes catalyzing the phosphorylation of other proteins and thereby modulating their biological activity. According to their phosphorylation reactions, two general classes of kinases can be discriminated irrespective of their compartmentation, those transferring phosphate to serine or threonine residues and those transferring phosphate to tyrosine residues. Cytosolic protein kinases are also grouped according to the second messengers that act on them, such as Ca^{2+} and calmodulin, diacyglycerol, cGMP, and cAMP. In contrast, the membrane-bound receptor protein tyrosine kinases are activated by ligands binding to the extracellular domain, causing a conformational change in the receptor, and increasing the catalytic activity of the cytoplasmic protein kinase domain. Protein kinases, together with their complementary protein phosphatases (dephosphorylating enzymes), form a network of enzymes involved in signal amplification. This issue has been reviewed recently by Cohen (1992) and Ma (1993).

In the case of phytochrome, the physiologically active P_{fr} form is expected to trigger subsequent events by the exposure of the tetrapyrrole chromophore followed by conformational changes of the N-terminal domain of the phytochrome apoprotein. It should be kept in mind, however, that in higher plants phytochrome in vivo has always been detected in the cytosol rather than in association with membranes. Therefore a direct application of animal models, dealing with membrane spanning protein kinases, to plants may not be permissible.

The involvement of a dephosphorylation-phosphorylation cascade in phytochrome-mediated responses has come within reach by the finding that there might be a functional association of phytochrome and protein kinase activity which are co-purified (Grimm et al. 1989; Kim et al. 1989) in higher plant preparations. Hofmann et al. (1991) showed that phytochrome can serve as a substrate for protein kinases.

Although these data do not provide rigid proof for a direct in vivo interaction between activated phytochrome and protein kinase, an unexpected discovery has further substantiated this hypothesis. Schneider-Poetsch described in a series of seminal papers (Schneider-Poetsch et al. 1991a,b; Schneider-Poetsch and Braun 1991; Schneider-Poetsch 1992) a striking coincidence of phytochrome and bacterial sensor protein sequences. Homologies of angiosperm and fern phytochrome were found in their C-terminal domains to the cytosolic region of membrane-bound bacterial sensor proteins. Bacterial sensor proteins belong to a group of two-component sensors that receive environmental stimuli (e.g., nitrogen or phosphate limitation, presence of plant exudates) through a N-terminal sensor module and transmit the information via a series of conformational changes to a C-terminal transmitter module, which in turn is transitionally autophosphorylated at a conserved histidine, a remarkable property of all the bacterial sensor proteins mentioned above (Stock et al. 1990). From the histidine the phosphate is transferred to the receiver module of a regulator protein which eventually acts on transcription or enzyme activity.

A similar mode for signal transductions can be envisaged for phytochrome. Homologous sequences occur at coinciding positions in the C-terminal part of phytochromes (Schneider-Poetsch et al. 1991a), although the crucial histidine is not conserved in phytochromes. Instead, a conserved tyrosine is found spatially very close to the histidine position. It is tempting to ascribe kinase function to this domain of fern and angiosperm phytochromes.

Extensive analyses have been carried out with deletion mutants of oat phytochrome A in transgenic tobacco (Cherry et al. 1993) which allowed the identification of essential regions for biological activity in addition to chromophore attachment, photoreversibility, spectral integrity, and dimerization. It was clearly shown that the carboxy terminus contains sites necessary for biological activity beyond those needed to maintain the quaternary structure of the chromoprotein.

Removal of as little as 35 amino acids from the C terminus renders the molecule nonfunctional. Unfortunately this stretch, a not-well-conserved sequence when compared to other phytochromes, is distal to the specific residues proposed by Schneider-Poetsch (1992) to be directly involved in protein phosphorylation. Another region indispensible for the functional integrity of phytochrome A has been discovered by Debesh et al. (1993) in *Arabidopsis*. If glycine in position 727 is replaced by glutamate (mutant hy8–3), a photochemically active phytochrome without detectable regulatory activity is expressed. Again, no nonphytochrome proteins with significant sequence similarity to this polypeptide could be detected.

10. Moss Phytochrome Is a Light-Regulated Protein Kinase

A more favorable situation in terms of phytochrome functions is encountered with moss phytochrome. The major breakthrough was achieved by Thümmler et al. (1992) and Algarra et al. (1993) with their molecular studies on phytochrome of the moss *Ceratodon purpureus*. This lower plant phytochrome gene phyCer codes for a 145 kDa polypeptide with a COOH-terminal domain which is homologous to the catalytic domain of eukaryotic protein kinases, e.g., the protein-tyrosine kinases of *Dictyostelium discoideum* or the products of retroviral oncogenes belonging to the Raf-Mos serine/threonine kinase family. The situation is distinctly different from the situation of fern and spermatophyte phytochromes, which exhibit sequence similarities to the catalytic domains of the membrane-bound bacterial receptor-histidine kinases mentioned above, although no evidence for histidine kinase activity has as yet been provided for a higher plant phytochrome. This is in accordance with the fact that the phyCer gene exhibits high sequence homology to the N-terminal chromophoric domain of higher plant phytochrome genes, whereas its C-terminal part shows no sequence homology to any reported phytochrome genes (Thümmler et al. 1992).

Biochemical evidence has been given that the photoreceptor molecule phytochrome itself is an active, light-regulated protein kinase. Although phyCer represents a novel phytochrome gene, it is related to the light-stable type II phytochrome. The steady-state concentration of its mRNA is not affected by light as it is the case with type I mRNA. The photoreceptor is a very low abundant soluble protein. It does not seem to be an integral membrane protein and should therefore be involved in intracellular signal transduction rather than in *trans*-membrane signalling. The enzyme activity of the photoreceptor has been proven by in vitro phosphorylation experiments with protein extracts of protonemata. A 140-kDa protein as substrate was phosphorylated in a r/fr-dependent manner. It is obvious that the forthcoming analysis of the *Ceratodon* phytochrome action will shed light on the transduction pathway.

These investigations raise the question as to the properties of other lower plant phytochromes. Cryptogamic phytochromes, e.g., from algae, mosses, and ferns, organisms that do not etiolate, have been less frequently investigated than those from etiolating seed plants. Sequencing data on *Selaginella martensii* by Hanelt et al. (1992) yielded homologies with phytochromes of higher plants which range between 62 (phytochrome B of *Arabidopsis*) and 55 or 56% (phytochrome C of *Arabidopsis* or *Avena*, respectively). An unrooted phylogenetic tree based upon mutation distances includes all sequences from cryptogamic plants (*Mougeotia, Ceratodon, Selaginella, Psilotum*) together with those of phytochrome B and C from *Arabidopsis* on a branch, well separated from the branch formed by type I phytochromes from etiolating plants. The B-type phytochrome is interpreted to be closer to the hypothetical phytochrome ancestor than other phytochromes.

The phytochrome of the fern *Adianthum* appeared to be similar to the *Selaginella* phytochrome. It is most closely related to the B-type of seed plants (Okamoto et al. 1993). *Mougeotia* probably contains only one phytochrome gene, also of the B-type. However, phytochrome transcription is downregulated in the light, which on the other hand is typical for type I phytochromes.

It certainly will be rewarding to take a closer look in the structure-function relation of B-type phytochromes.

11. G-Proteins and Further Components of the Phytochrome Signal Transduction Chain Controlling Gene Expression

Signal transduction mutants of *Arabidopsis*, tomato, and pea have significantly contributed to the elucidation of further events following phytochrome activation. In these mutants, light signals are uncoupled from a number of light-dependent processes. They are characterized by normal phytochrome levels but most of them, such as cop1, det1, and det2 in *Arabidopsis*, lip1 and lw in pea, Tibet Dwarf in wheat, hp in tomato, differ from wild plants by de-etiolated phenotypes in the dark. They are constitutively morphogenic. Recent progress has been summarized by Chasan (1993) and Chory (1993).

The det (for de-etiolated) and cop (for constitutive morphogenic) mutations of *Arabidopsis* are recessive. This suggests that their wild-type gene products act to repress photomorphogenesis in the dark. Light activation by phytochrome is therefore expected to reverse the repression. For photomorphogenesis to occur, P_{fr} must therefore negate the repressive effects of the det/cop proteins in *Arabidopsis* or similar ones in other plants. The fact that the det and cop mutations have pleiotropic but not completely overlapping effects suggests that multiple, separable pathways may transduce the signal from P_{fr}.

Deng et al. (1992) have cloned the cop1 locus. The amino-terminal half of the endoded protein contains a putative DNA-binding motif, zinc finger domains, an array of tandem repeats of a zinc binding domain characterized by invariant pairs of cysteines and histidines. The carboxy-terminal half of cop1 contains a G protein-related domain which is homologous to the WD-40 repeat motif of the

β-subunit of the trimeric G proteins, but is also found in the yeast transcriptional repressor. The authors consider cop1 to be a regulatory protein with the capacity to function as a negative transcriptional regulator capable of direct interaction with components of the G protein signaling pathway.

Guanine nucleotide-binding proteins can be grouped into several different classes. They function as molecular switches in signaling events of cell growth and differentiation. With respect to the phytochrome signal transduction chain, the heterotrimeric G proteins deserve special attention, although the other members, among them protein synthesis elongation and initiation factors as well as the low molecular weight ras proteins, should not be neglected. For instance several animal systems are known where signals are transmitted from a receptor tyrosine kinase to ras through specific protein-protein interactions (McCormick 1993). Recent investigations by Clark et al. (1993) on small monomeric GTP-binding proteins (related to the ras family) in pea nuclei indicate a participation in phytochrome-controlled cellular changes. In passing, it should be mentioned that many human cancer cells contain mutated alleles of the ras gene coding for impaired ras proteins with defects in GTPase activity.

The structure and function of heterotrimeric GTP-binding and hydrolyzing proteins (G proteins) have been recently reviewed by Hepler and Gilman (1992) and Linder and Gilman (1992). The α-subunits have a single, high-affinity binding site for GDP or GTP. The GDP-form binds with high affinity to the tightly associated complex $\beta\gamma$, whereas the GTP-bound form of α dissociates from $\beta\gamma$. Both the GTP-activated α and $\beta\gamma$ serve as regulators of effector proteins such as mating factors in yeast. All α-subunits possess intrinsic GTPase activity, thereby inactivating the active complex. The GTPase activity can be viewed as an internal clock that controls an on/off switch.

The α unit is modified by several toxins, which have been extensively used in phytochrome research during the past year. Cholera toxin constitutively activates α by inhibiting its GTPase activity, whereas the modification by pertussis toxin prevents receptor-mediated activation of G proteins. The heterotrimer as well as $\beta\gamma$ and (in most cases) α are bound through lipid modifications to the plasma membrane.

The regulation of G proteins by phytochrome has been first described in etiolated *Avena* seedlings (Romero et al. 1991a,b), where the binding of GTP is regulated in vivo by r/fr light. Cholera toxin that covalently modifies GTP-binding proteins via ADP ribosylation, upregulated cab (chlorophyll a/b-binding protein) expression in the absence of a light signal. Studies on the phosphorylation/dephosphorylation of nuclear proteins provided evidence for their regulation by light. The phosphorylation pattern was again affected by cholera toxin providing stimulatory and the GDP analogue GDPβS providing inhibitory conditions, suggesting that G proteins are involved in the signal transducing cascade. Further analysis has been carried out by Romero and Lam (1993) in dark-adapted soybean cells. Already 10 ng/l cholera toxin were sufficient to stimulate cab expression after a few hours in the dark, indicating that the phytochrome-dependent gene expression can be uncoupled from the light stimulus. Moreover, the calmodulin (CaM) antagonist W-7, a naphthalenesulfonamide derivative, prevented the induction of the photoregulated gene by phytochrome or the bacterial toxin. The involvement in the transduction chain of a Ca^{2+}- or CaM-dependent protein further downstream from the G protein seems to be reasonable.

A bypass of the normal requirement for light induction by blocking the action of a repressor of photomorphogenesis can also be taken by heat-shock treatment. Kloppstech et al. (1991) and Otto et al. (1992) reported photomorphogenesis-like effects in dark-grown pea seedlings subjected to periodic temperature shifts or cyclic heat shock treatments. This phenomenon may indicate that the hypothetical negative regulator of photomorphogenesis is sensitive to heat in vivo. The phenotype shows a striking similarity to the dark-grown mutant of pea (Frances et al. 1992), which has several features in common with the *Arabidopsis* mutants det and cop.

The newly discovered cop9 mutant of *Arabidopsis*, leading to the constitutive morphogenetic phenotype similar to cop1, is supposed to code for protein that is closely associated with the cop1 protein (Wei and Deng 1992). A new quality of mutants is described for *Arabidopsis* cop2, cop3, and cop4 (Hou et al. 1993). Whereas the dark-grown seedlings exhibit the normal etiolation pattern with long hypocotyls, the cotyledons show the wild-type appearance with open and enlarged cotyledons, and no apical hooks as they would occur in the light. This suggests that light-regulated development of the cotyledons and the hypocotyl are separable processes. All three mutants have typical etioplasts in their darkgrown cotyledons, which implies that the cotyledon chloroplast development and cellular differentiation can be uncoupled during light-regulated seedling development. Since the cop4 mutant is defective in both root and shoot gravitropic responses, it can be deduced that the two processes share a common regulatory element.

The complex interplay of regulatory signals between chloroplasts and the nucleus has been investigated by Susek et al. (1993) by three new *Arabidopsis* mutants gun1, gun2, and gun3 (for genomes uncoupled). Whereas in the wild type the transcription of several nuclear genes such as cab and rbcS (small subunit of RuBisCo) in the light depends upon chloroplast development, the gun mutants with defects in chloroplast development allow the partial expression of cab and rbcS. Consequently, the gun genes, whose products are as yet unknown, coordinate nuclear cab and rbcS transcription with nuclear function. Susek and Chory (1992) provided first evidence that a chlorophyll precursor, heme, or other porphyrin could act as such a signal. It is very likely that the intracellular crosstalk also involves the mitochondria. Organellar development thus appears to be controlled by a complex intracellular regulatory network.

A highly innovative approach has been taken by Neuhaus et al. (1993) to dissect the phytochrome signal transduction pathway on the level of single cells in tomato hypocotyls. For this purpose, the tomato mutant aurea containing less than 5% of the phytochrome A pool was used for the microinjection of oat phytochrome A into subepidermal cells. Only in this case, did a complete restoration of anthocyanin synthesis, chloroplast development and gene photoregulation (as assayed using a plasmid with the cab-β-glucuronidase reporter gene construct coinjected into aurea hypocotyl cells) take place, although the yield (viable cells) was only in the range of 5–8% for injections done in the light. The effect was restricted to the injected cell.

The microinjection system of Neuhaus et al. (1993) allowed the introduction of the G protein antagonists GDPβS and pertussis toxin and of the G protein agonists GTPβS, cholera toxin, and GMP-PNP into single cells. The application of the antagonists resulted in the loss of a detectable phytochrome response in the light, suggesting the involvement of heterotrimeric G proteins rather than small G proteins (because of the selectivity of pertussis toxin) in phytochrome signal transduction. On the other hand, the agonists substituted for activated phytochrome A because they stimulated the responses, normally triggered by light, in the dark in the absence of phytochrome A. This means that light is not required for the execution of the phytochrome phototransduction processes downstream of the G protein.

The same strategy was used by Neuhaus et al. (1993) to address the question of whether Ca^{2+} and CaM participate in the phytochrome signal transduction pathways. It was clearly shown that light-activated anthocyanin synthesis does not require Ca^{2+} or CaM, whereas the expression of cab and chloroplast development depended upon these components. This suggests that phytochrome signaling involves at least two different pathways downstream of G protein activation. In this context, the existence of Ca^{2+}-regulated nuclear enzymes, that are phytochrome-dependent, is of general interest. Roux (1992) mentions in his review a CaM-regulated nucleoside triphosphatase and a CaM-independent but Ca^{2+}-dependent protein kinase. This microinjection approach also bears relevance on the question of an intercellular propagation of the phytochrome stimulus within a single cell. Because of the lack of expression of phytochrome A-regulated processes in neighboring cells that were not microinjected, the authors come to the conclusion that phytochrome signaling is cell-autonomous and not transmitted from cell to cell like the systemic acquired resistance signal.

However, intercellular signals have been recently discovered by Nick et al. (1993) leading to long-range inhibitions of anthocyanin synthesis. In this case, a microbeam illumination was used to stimulate different sub-regions of mustard cotyledons and combined with in situ hybridization of chalcone synthase mRNA with gene-specific probes. Irradiation with red light of only three cells from different cotyledonary areas in general resulted in a localized induction of anthocyanin, although for vein and intercostal cells the responses appeared to be locally confined to a lesser extent than the responses of tip and margin cells. Long-range suppression of anthocyanin formation in the margin was observed after microbeam irradiation of the cells adjacent to the vein. Different models were discussed, such as mutual competition between individual cells for limited resources or suppressive interactions due to actual inhibitors. In any case, these physiological signals are superimposeed to the prepatterns defining the symmetry axes and organ development.

12. Regulation of Photogene Transcription by DNA-Binding Proteins

The analysis of light-responsive gene regulation is currently concentrated on DNA elements responsible for light-responsive expression or repression, respectively, and on DNA binding proteins with specificity for light-responsive DNA elements. This issue has been reviewed by Gilmartin et al. (1990). Considering the fact that 30 years have passed since the gene activation hypothesis was put forward for important phytochrome responses (Hock and Mohr 1964), one may ask after the main difficulties preventing a complete understanding of the phytochrome signal transduction chain leading to changes at the DNA level. These obstacles are not only due to the complexity of the regulation of transcription per se including *cis*- and *trans*-acting elements, but also to the fact that details on eukaryotic transcription factors have come up only recently.

Regulation of transcription in eukaryotes by DNA-binding has been reviewed by Krajewska (1992), transcription factors such as TFIIβ and TBP by Hahn (1993) and Struhl (1994). An overview on *cis*-acting elements, which provide binding sites for sequence-specific transcription factors (the *trans*-acting factors) has been given by Gruissem (1990). *Cis*-acting elements are parts of promoters of protein-encoding genes. External and other stimuli modulate in many cases gene expression at the level of promoter activity. Up to now, the following common motifs have been identified in promoters of light-responsive genes: (1) the G-box, originally defined in rbcS promoters from different plant species (Giuliano et al. 1988), (2) the GATA box occurring in the regulatory regions of cab, cauliflauer mosaic virus and rbcS (e.g., Lam and Chua 1989), (3) an AT-rich, factor-binding sequence in promoters of rbcS genes (e.g., Datta and Cashmore 1989), and (4) several GT elements in a pea rbcS gene (Green et al. 1988) and the rice phytochrome A promoter (Debesh et al. 1990).

Another DNA-binding protein recognizing the oat phytochrome A3 promoter has been characterized in rice by Nieto-Sotelo et al. (1994). This is of special interest, as phytochrome downregulates the expression of its own phytochrome A gene, one of the fastest light-induced effects on transcription in plants. The binding protein acts as a nuclear transcription factor (positive factor = PF1) by binding to the AT-rich PEl element, which together with the positive element PE3 is part of the phytochrome A3 promoter and supports high levels of transcription in the absence of light. The 213 amino acids long polypeptide shows binding characteristics similar to the mammalian high mobility group proteins (HMG I-Y proteins), important DNA-binding proteins, that are involved in cell division and in the maintenance of the undifferentiated state, and to histone H1 domains. The rice PF1 is considered as a member of a relatively new class of DNA-binding proteins that have the potential for transcriptional activation.

Okubara et al. (1993) studied in *Lemna gibba* promoter elements involved in negative regulation by phytochrome. A striking similarity between the deduced amino acid sequences and those of a number of ABA-induced or seed-specific mRNAs was noted, raising the possibility that these gene products have a role in desiccation or water stress. However, these *cis*-elements are different from the negative phytochrome-dependent response element RE1 discussed before. It seems likely that the *Lemna* genes studied by Okubara et al. (1993) respond to multiple regulatory signals, like other phytochrome-regulated genes.

Orozco and Ogren (1993) analyzed deletions in the spinach RuBisCo activase promoter in transgenic tobacco to define the regions necessary for conferring light-inducible and tissue-specific expression. A light-responsive element acting in leaves was found between −150 and −78 relative to the transcription start site. It included a putative G box responsible for gene expression in the leaves and a GT motif. Deletion of the latter caused a 10- to 100-fold increase in expression in roots. Interactions between multiple *cis* elements are therefore required for light-inducible and tissue-specific gene expression.

Important progress has been achieved in the work on plant mutants. It enabled the identification of genes coding for transcriptional repressors such as the already mentioned copl protein (cf. Deng et al. 1992). This function is supported by its similarities to a subunit of the *Drosophila* TFIIB (Dynlacht et al. 1993). The DNA-binding protein GT-1 has been characterized by Gilmartin et al. (1992) in tobacco. The nuclear protein interacts with box II, a regulatory component of a light-responsive element present within the upstream region of pea rbcS. The 43 kDa protein exhibits a novel helix-helix-turn-helix (HHTF) binding motif. It is constitutively expressed; but its function in light-responsive transcription requires further components. Accessory negative regulators blocking transcriptional activation in the dark, and positive accessory proteins activating transcription in the light are discussed in addition to the direct photoreversible modification of GT-1.

Better knowledge of key components of the phytochrome signaling pathway, such as protein kinases, G proteins, and DNA binding proteins, may serve to identify the missing links by comparing other systems known in more detail, such as the RAS pathway in yeast (Broach 1991) for mating signal transduction or in mammals (Satoh et al. 1992) for growth factor signal transduction (for review of Kaziro et al. 1991). Comparative studies are also expected to throw light on poorly understood phenomena such as additive, synergistic or antagonistic effects of different photoreceptors. Signal integration is a well-known mechanism at the level of protein kinases and protein phosphatases (Cohen 1992). Hormone receptor physiology will also certainly profit from these approaches.

13. Photomodulations

It has long been known that there are many phytochrome-controlled photoresponses which cannot be explained by concepts involving gene expression. These are found especially among photomodulations such as light-dependent leaf movements, chloroplast migrations, or activation of ion channels. These readily reversible responses are usually attributed to light effects on membranes. The subject has been reviewed by Racusen and Galston (1983).

The green alga *Mougeotia* is frequently used as a model system for phytochrome-mediated photomodulations. Research has concentrated on the effects of light upon the positioning of the single large chloroplast which orients its face to low-intensity light or its profile to high-intensity light (for review cf. Haupt 1982, 1991, 1993).

Photoresponses of *Mougeotia* are ideally suited for the analysis of the signal transduction chain because they can be studied here at the cell level without interference from other cells. In addition, the whole chain of events requires a time range between a few minutes and an hour. It is independent of the nucleus, and consequently does not require new products of gene expression. In *Mougeotia*, phytochrome is localized close to the cell surface. Undirectional light transforms more phytochrome to the P_{fr} form at the proximal and distal surface than at the flanks. This gradient stores the information about the light direction, necessary for the reorientation movement. The motive force is generated by the interaction of actin and myosin with the active filaments being localized in the peripheral cytoplasm close to the chloroplast's edge.

Research during the 1980s has specified several elements of the transduction chain, including calcium, calmodulin, and actin activity. Progress has been thoroughly reviewed by Tretyn et al. (1991) and Wada et al. (1993).

A major issue during the last years has been the question of the role of Ca^{2+} in the signal transduction chain of photomodulations. Two possibilities were considered: (1) the phytochrome gradient is translated directly into a Ca^{2+} gradient to guide chloroplast positioning. This could be achieved by P_{fr}-mediated opening of Ca^{2+} channels in the plasma membrane (vectorial Ca^{2+} effect). (2) The alternative, a scalar Ca^{2+} effect, assumes a uniform distribution of Ca^{2+} in the cytoplasm. The sole function of Ca^{2+} would then be to establish cytoskeletal competence for the performance of chloroplast movement. In this case, P_{fr} is thought to act on anchorage sites to actin on the plasma membrane (vectorial anchorage effect, cf. Grolig and Wagner 1988).

As has been recently summarized by Tretyn et al. (1991), it seems unlikely that Ca^{2+} is part of the signaling process itself as the application of Ca^{2+} entry blockers or the incubation in Ca^{2+}-depleted media (e.g., Schönbohm et al. 1990; Russ et al. 1991) failed to influence the final chloroplast position with respect to light. It is rather envisaged that Ca^{2+} establishes the cytoskeletal competence in the thin cortex layer (0.2–0.3 μm), enabling rapid response when a signal is dispatched by an incipient beam of light (Wagner et al. 1992). The observed changes in cytosolic viscosity in response to modifications of cytosolic Ca^{2+} concentrations are compatible with changes in the polymerization/depolymerization behavior of microtubuli. In *Mougeotia* interphase cells, they are located just beneath and in close contact to the plasma membrane. Most interestingly, Ca^{2+}-binding globules are found most abundantly in the area where the chloroplast is in contact with the cortical cytoplasm. They are thought to act as Ca^{2+} exchangers providing local increases of Ca^{2+} concentrations.

The situation in *Mougeotia* is complicated by the fact that in addition to phytochrome the blue-light/UV-A photoreceptor effects chloroplast orientation. Blue light irradiation leads to a rapid and prominent rise of cytosolic Ca^{2+} concentrations in contrast to red light treatment.

The participation of Ca^{2+} in fast phytochrome actions of higher plants has been the subject of intensive investigations. The existence of direct responses as observed in leaf movements and of indirect responses in cytoskeleton modifications (cf. above) has been substantiated during the last years.

Phytochrome-controlled swelling of protoplasts from etiolated cereal leaves has been discovered by Blakeley et al. (1983). The photoreversible effect shows an absolute Ca^{2+} requirement. This swelling response is believed to be involved in the phytochrome-regulated leaf growth and unrolling. It has been ruled out that this reaction may be directly mediated by plasma membrane H^+-ATPase (Terry et al. 1992).

Chae et al. (1990) succeeded in loading the Ca^{2+} probe quin2 by a simple incubation method into oat protoplasts in order to determine cytosolic Ca^{2+} concentrations. Red-light irradiation for 5 min caused an increase from 30 to 193 nM Ca^{2+}. A subsequent fr irradiation for 5 min decreased it by about 48 nM. Even when extracellular Ca^{2+} was completely removed, red light increased the cytosolic Ca^{2+} concentrations by about 51 nM. Therefore P_{fr} is involved in the mobilization process of Ca^{2+} from intracellular Ca^{2+} pools in addition to the regulation of Ca^{2+} influence.

The effect of phytochrome on the subcellular localization of Ca^{2+} in etiolated oat coleoptile cells was studied by Tretyn et al. (1992) using an antimonate staining procedure and X-ray microanalysis. Ca^{2+} was mainly observed in the plasma membrane zone on both sides of the membrane of dark control cells being most likely the result of a passive Ca^{2+} influx into the cell and active pumping out of the cell. After exposure to red light, a change of intracellular Ca^{2+} localization occurs. Ca^{2+} is removed from the plasma membrane zone within the cytosol and accumulates in the ER cisternae. Far-red light, after red light, stimulates transport of Ca^{2+} to the outside of the cell and its accumulation into the cell wall, as well as transport into the vacuole, leading to a decrease of cytosolic Ca^{2+}. Obviously, the ER and the tonoplast play a major role in the regulation of Ca^{2+} concentrations.

A causal relation between changes of cytosolic Ca^{2+} concentrations and phytochrome-mediated swelling of leaf protoplasts has been finally proven by Shacklock et al. (1992). The red light-induced swelling of etiolated wheat leaf protoplasts is preceded by a transient increase in free Ca^{2+} in the cytosol, followed by a decrease to below the resting. These events take place within ca. 2 min after a 3-min red-light pulse. When the red light treatment was followed immediately by fr light, no changes of Ca^{2+} levels were observed. Cytosolic Ca^{2+} was imaged in Fluo-3 loaded protoplasts using a laser scanning confocal microscope equipment with a transmission detector and a laser providing fluorescence exitation. The artificial transient elevation of Ca^{2+} concentrations by releasing Ca^{2+} from caged probes (Nitr-5/AM) in the presence of UV lead to an even greater volume increase as compared to the phytochrome effect. The use of caged probes for dissecting the molecular pathway between phytochrome and red light-induced response is expected to clarify further details on the transduction pathway of these fast reactions.

Volotovski et al. (1993) recorded the time characteristics of the changes in cytosolic Ca^{2+} concentration of oat protoplasts caused by the action of r and fr light in the presence of 2 mM $CaCl_2$ in the medium. In this case, $[Ca^{2+}]$ was determined using the fluorescence probe quin2 loaded into the cells. As a reflection of the Ca^{2+} uptake into the protoplasts, an increase of the fluorescence intensity of quin2 is observed immediately after the addition of Ca^{2+} in the dark. Red light induces oscillations of the fluorescence signal which appear about 12 s after switching on the light. When the light is switched off, the oscillations disappear. Similar observations are observed after fr irradiation. The authors come to the conclusion that the Ca^{2+} oscillations may be associated with Ca^{2+} stores located in the ER or vacuoles, and that phytochrome is likely to control the Ca^{2+} permeability of these membranes. As a mediator inositol-1,4,5,-triphosphate is discussed being known to induce Ca^{2+} oscillations in animal cells when injected into them (Berridge 1990).

The role of plant phosphoinositides in intracellular signalling has been extensively reviewed by Drøbak (1993). Guron et al. (1992) showed with etiolated corn leaves that as little as 15 s of red light immediately increased the level of phosphatidylinositol bisphosphate three to six, fold compared to the control. The effect can be reversed if fr follows the r irradiation. Photocontrol is discussed to act at the phosphatidylinositol phosphate kinase and phospholipase level.

The approaches reported so far were dealing with Ca^{2+} as second messenger in fast phytochrome reactions. However, this does not explain how the turgor changes during swelling or shrinking of protoplasts are brought about. The monkey pad tree *Samanea saman* with its circadian rhythm in leaflets provides an elegant system to address this question because the phase of the rhythm can be altered by phytochrome (and the blue light receptor) and because ion fluxes can be studied in protoplasts isolated from opposite sides of the pulvinus (Kim et al. 1993). The authors proved that exposure to red light for 1 min followed by darkness, which favors leaflet closing, caused K^+ channels to open in flexor and close in extensor protoplasts. This effect is reversed by fr in the case of flexor protoplasts, whereas the extensor protoplasts are exclusively regulated by the blue light photoreceptor.

Takagi et al. (1992) during their studies on the mechanical properties of cytoplasm in mesophyll cells of *Vallisneria gigantea*, discovered one of the fastest phytochrome responses reported so far. A centrifuge microscope with stroboscopic illumination was used to analyze the passive movement of chloroplasts upon irradiation at low centrifugal force (approximately 5 g). After a brief irradiation with red light (1–2 s) the rate of passive gliding increased two- to threefold after an apparent latent period of 15–30 s. This effect was abolished by fr. Obviously, the cytosol becomes more fluid in the presence of P_{fr} and more rigid when phytochrome is converted back to the P_r form. Irradiation with actinic light also induces the release of Ca^{2+} from the mesophyll protoplasts. The decrease in Ca^{2+} concentration is supposed to activate the microfilament system to bring about the generation of the motive force for the streaming.

14. Intracellular Localization of Phytochrome

Further progress in the analysis of phytochrome signal transduction chains depends upon a better definition of the subcellular localization of phytochrome species. This would significantly facilitate the search for phytochrome receptors transmitting the stimulus downstream the transduction pathway.

Early work on phytochrome localization disclosed the pelletabililty after red light irradiation (e.g., Quail et al. 1973). It was followed by immunolocalization studies using phytochrome-specific antibodies (e.g., McCurdy and Pratt 1986; Speth et al. 1986). The pelletable phytochrome P-730 is confined to sequestered areas in contrast to P-660 that is diffusively distributed thoughout the cytosol. The sequestered areas of phytochrome also contain ubiquitin and are therefore the places of phytochrome destruction. Hofmann et al. (1991) purified the sequestered particles of phytochrome from oat seedlings. Although protein kinase activities were not directly associated with these particles, phytochrome purified from these particles served as a substrate for protein kinase activity. However, it never showed any autophosphorylation activity, as would be expected if it were a kinase itself.

Sequestered areas of phytochrome are most probably not the subcellular sites of phytochrome action. Several approaches have been directed during the past towards the identification of membrane-associated phytochrome. Although this type of binding would fit current ideas of the signal transduction chain, especially with respect to the membrane-bound trimeric G-proteins, it is hard to match this hypothesis with the available data for the immunolocalization of phytochrome, which show no membrane association. However, antibodies would fail to detect phytochrome in the bound form, if the epitopes recognized by existing antibodies were concealed in the membrane-bound form, e.g., by binding a third partner.

Recent evidence for membrane-associated phytochrome has been obtained for a variety of plant sources by Lamparter et al. (1992). It could be partly solubilized with detergents and separated by gel filtration. A distinct, phytochrome-positive, high molecular weight fraction could be identified, which is assumed to contain phytochrome associated with a hydrophobic protein. This association is different from sequestered phytochrome. The identity of the membrane(s) to which phytochrome binds, is not yet clear. Neither the plasma membrane nor mitochondria or "normal" ER are favored as predominant sites of phytochrome associations.

Immunolocalizations have also been carried out in the green alga *Mougeotia* (Hanstein et al. 1992a,b). Phytochrome was recognized by a monoclonal antibody raised against phytochrome I from corn. Immunoblotting yielded a monomer size of the polypeptide of approximately 110 kDa. *Mougeotia* phytochrome also proved to be to a large extent a soluble protein, abundant throughout the cytosol. It is assumed to be in equilibrium with a reactant-bound P_{fr}-pool at the plasmalemma.

15. Phytochrome Interactions with the Blue-Light/UV-A Photoreceptor

a) Characteristics of the Blue-Light Photoreceptor (Cryptochrome)

Among the blue-light (BL)/near-ultraviolet (UV-A) mediated responses, the increase of stomatal aperture, changes of the expression of several nuclear- and chloroplast-coded genes coding for leaf proteins, suppression of stem elongation and phototropic curvature of the stem have been studied most intensively. A summary on the transduction of BL signals has been given by Kaufman (1993). Phototropic responses in fungi are discussed in detail in Ootaki (1993).

It is not yet clear how many photoreceptors are involved in BL/UV-A-mediated processes, nor is the question of the chromophores settled. Most likely candidates are flavins, but also pterins (Galland and Senger 1988), and the carotinoid zeaxanthin (Quinones et al. 1993). The main difficulty is related to the fact that a spectroscopic assay, which was available in the case of phytochrome, is lacking. For simplicity, the receptor is referred to in the following as BL receptor.

Work on *Arabidopsis* mutants with reduced responses to BL has been summarized by Chory (1993). Two types were revealed, those that affect the inhibition of hypercotyl growth rate, and those that affect phototropic curvature but show normal hypocotyl elongation in BL. The hy4 mutant isolated by Koornneef et al. (1980) belongs to the former type. The defect was located by Amad and Cashmore (1993) in a gene that encodes a protein with significant sequence identity to prokaryotic DNA photolyases, flavoproteins that repair DNA damaged by UV light. Although the chromophore-binding domain of HY4 shares homology to photolyase, the HY4 protein most likely does not repair DNA, and its mode of action is unknown.

In spite of many unsolved questions concerning BL receptors, it is generally agreed that they are plasmalemma-bound chromoproteins. Similarly to the phytochrome signal transduction chain, heterotrimeric G proteins, associated with the plasma membrane, protein kinases, and gene activation, are discussed as essential elements for signal transduction in BL-mediated photomorphoses. Short et al. (1993) enriched pea plasma membrane fractions exhibiting the specific BL-mediated, rapid phosphorylation of a 120-kDa plasma membrane protein, described before (Gallagher et al. 1988) in connection with phototropism. They obtained evidence that the crucial components, the photoreceptor, a kinase, and a 120-kDa substrate protein, are localized together in the plasma membrane.

The signal transduction pathway in the BL response of stomatal guard cells appears to be very short. It involves the activation of an H^+ ATPase and H^+

extrusion leading to plasma membrane hyperpolarization (Assmann et al. 1985). The membrane potential induces the opening of voltage-gated K^+ channels in the plasma membrane, and K^+ ions are taken up electrophoretically into the cells. Shimazaki et al. (1993) used specific inhibitors to dissect the signaling pathway from BL perception to the activation of the H^+ ATPase. It was shown that the calmodulin antagonist W-7 and the myosin light chain kinase (MLCK) inhibitor ML-7 interfere with the BL signaling pathway without directly affecting the H^+ ATPase. This indicates that MLCK or a MLCK-like protein are involved. Striking similarities to BL-dependent pH changes in guard cells have been found by Schmid and Dring (1993) at the outer surface of brown algae and some green algae, but not in red algae. The acidification caused by BL affects the availability of CO_2 for photosynthesis by the release from bicarbonate that together with carbonate is the main source of inorganic carbon of seawater at the normal pH of ca. 8.2 but is not readily taken up.

Weisshaar et al. (1991) studied light-mediated expression of chalcone synthase in *Petroselinum crispum*. They identified *cis*-acting promoter elements and trans-acting protein factors binding to these elements. It appears that the signal transduction chains of the BL/UV-A receptors and phytochrome use similar elements.

Flower formation in *Arabidopsis* is induced by BL. Kaldenhoff et al. (1993) searched for genes that are activated by BL and initiate the morphogenetic programme leading to flower formation. A single-copy gene was further characterized. It contained several regulatory elements including GA and ABA sequence motifs. The inducibility by either BL or ABA (in the dark) in relation to flower formation indicates that both factors share common elements in the signal transduction pathway. As the putative gene product a transmembrane channel protein has been predicted.

b) Interactions of the BL Receptor and Phytochrome in Seed Plants

Considerable work has been carried out during the past years on BL receptor-phytochrome interactions. It strengthens the view that the different signal transduction chains merge either before or at the gene level.

Earlier work has been summarized by Mohr (1986). This review concentrates on a few examples taken from recent literature on plant responses at the organ level and then proceeds to the biochemical analysis that allows dissection of the interaction between the BL receptor and phytochrome.

One of the examples is apical hook opening and cotyledon unfolding in etiolated *Arabidopsis thaliana* seedlings, which are controlled by the activities of multiple photosensory systems, phytochrome (including the high-irradiance phytochrome system) and a high irradiance BL photoreceptor (Liscum and Hangarter 1993a). The inductive low fluence BL effect is fr-reversible, indicating the role of phytochrome in this response. Other well-studied responses controlled by phytochrome and the BL receptor are the expansion of dicotyledonous leaves (for review, see Dale 1988), as has been measured by Blum et al. (1992) in discs, excised from primary leaves of *Phaseolus vulgaris*, and the inhibition of stem elongation (e.g., Gaba and Black 1979).

Photomodulations such as the leaflet movements of legumes also provide interesting examples for responses triggered by phytochrome and the BL receptor. Roblin et al. (1990) showed in *Cassia fasciculata* that phytochrome mediates the phytochrome-induced closure of leaflets in a Ca^{2+}-dependent manner that can be inhibited by Ca^{2+} channel blockers and promoted by Ca^{2+} channel activators. In contrast, the BL receptor triggers the light-induced opening of leaflets and is not affected by these Ca^{2+} channel modulators although this movement is dependent upon the presence of Ca^{2+}, which may be mobilized from internal stores.

The following part is restricted to ultrastructural and biochemical aspects of control by both photoreceptors. Zandomeni and Schopfer (1993) investigated the orientation of cortical microtubules in the outer epidermal wall of maize coleoptiles that are related to the direction of microfibril deposition at the inner surface of the cell wall and consequently to elongation growth. Transverse microtubules are correlated with a stimulation, longitudinal ones with an inhibition of elongation growth. It is controlled by both the BL receptor and the phytochrome. Dark-grown coleoptiles exhibited an irregular pattern of parallel microtubules with a random distribution of orientations. A 1-h irradiation with red light caused a uniformly transverse alignment (with respect to the long axis), which shifted after 1–2 h continued irradiation into a longitudinal orientation. However, BL induced within 10 min a stable longitudinal alignment. Auxin caused a transverse microtubule arrangement, even in BL. Zandomeni and Schopfer (1993) interpret the role of microtubules not as a casual involvement in cell growth but rather as a response to vectorial changes that may result from mechanical alterations in the cell wall. In this way, they could transmit directional information to the cytoplasm.

The group of H. Mohr intensively studied the coaction of phytochrome and the BL/UV-A receptor on enzymes of the glutamine synthetase/glutamate synthase pathway in Scots pine seedlings (e.g., Elmlinger and Mohr 1991; Elmlinger et al. 1994). These studies led to the conclusion that P_{fr} is the effector proper that acts on gene expression in photomorphogenesis while BL/UV-A light acts to control the effectiveness of P_{fr}. The plastidic ferredoxin-dependent glutamate synthase (Fd-GOGAT) in the cotyledonary whorl was controlled by red and by blue light. Whereas red could replace BL up to 10 days after sowing, thereafter BL was required for a further increase of the (Fd-GOGAT) activity. However, it was also shown that BL does not enable enzyme appearance if the P_{fr} level is low. This means that P_{fr} is still operating in BL. Similarly, thc strong specific effect of BL on the plastidic glutamine synthetase (GS) in the later stage of germination can be attributed to an increase in the responsiveness to P_{fr}. It was shown that light-mediated changes in the transcript level were exclusively caused by phytochrome action, whereas BL specifically increased the responsiveness of the transcriptional process to P_{fr}. If the P_{fr} level is kept low, BL does not affect transcription. Most interestingly, no correlation between the transcript level and the rate of protein synthesis has been observed; but again, BL operates at the level of·GS synthesis, that totally depends upon BL, only by establishing their responsiveness to P_{fr}.

Work of Oelmüller and Kendrick (1991) on four nuclear genes coding for plastidic proteins in the dark-grown aurea mutant tomato seedlings, which is deficient in phytochrome A, has shown that BL is required for survival. Conversely, in wild-type plants, development appears to be normal in red light, which only activates phytochrome. From these data it is concluded that BL sensitizes the mutant to the residual phytochrome which allows normal gene expression and survival of the mutant under daylight conditions.

An entirely different situation is encountered in the red and BL-mediated stomatal opening in *Vicia faba* (Talbott and Beiger 1993). Low-intensity BL (below threshold for guard cell photosynthesis) induces a fast, but transitory malate accumulation, most likely acting as counter ion for K^+ uptake, and a steady continued increase of sucrose. There is also a large increase in the levels of maltose and maltotriose, which together with sucrose appear to be formed from products of the hydrolytic pathway of starch degradation. On the other hand, red light-stimulated opening is DCMU-sensitive and therefore directly dependent on photosynthesis. Red light primarily induces the accumulation of sucrose derived from photosynthetic CO_2 fixation in guard cells, which show no decrease in starch content and no accumulation of K^+ under these conditions. The latter is obviously used primarily during the initial, rapid phase of BL-stimulated opening, whereas longer-term responses, dependent upon both red and blue light, involve two different types of carbohydrate metabolism.

c) Interactions of the BL Receptor and Phytochrome in Lower Plants

Whereas light-oriented chloroplast movement is mediated by BL receptors in many species ranging from angiosperms to algae, there are some notable exceptions where phytochrome is responsible, in addition, to the BL receptor. This is the case in the Zygnematales *Mesotaenium* and *Mougeotia*, which, because of their simple wall structure, are especially suited to study the BL receptor-phytochrome interactions.

The unicellular alga *Mesotaenium*, with one large ribbon-shaped chloroplast, shows a strong potentiation of phytochrome effects by BL (Kraml and Herrmann 1991). The fast movement response of the chloroplast from profile position to face position is only achieved if a red pulse (R) is followed by 10 min BL irradiation whereas single treatments are rather ineffective. The authors showed that the R-BL sequence is much more effective than BL-R. However, the memory of R before BL holds only for some minutes. The transient signal mediated by P_{fr} can be transformed to a more stable internal signal by the interaction with BL. This interaction takes place in the transduction behind the photoperception because P_{fr} can be removed by P_r before the onset of BL. However, in *Mougeotia* the preferred pigment interaction is the irradiation sequence BL-R.

An obstacle in the investigation of BL receptor/phytochrome interactions is the fact that not only the BL receptor but also P_{fr} absorbs BL due to its blue-absorption band. Hayami et al. (1992) succeeded in separating the phytochrome- from the BL receptor-mediated BL response in the fern *Adiantum capillus-veneris* by combining the techniques of cell centrifugation and of microbeam irradiation with linearly polarized light of different qualities. Both receptors appear to be localized in the ectoplasm and/or the plasma membrane, the cell structure whose position remains unchanged after basipetal centrifugation. The transition moments of their blue-absorbing band are nearly parallel to the surface.

Both photoreceptors are also involved in *Dryopteris sparsa* protonemata. This fern behaves in many respects similarly to *Mesotaenium*, mentioned above, except that the lifetime of P_{fr} is much longer whereas the BL-induced chloroplast arrangement began to decay immediately. Further analysis of the chloroplast movement showed that obviously the P_{fr} gradient determines the direction of chloroplast movement and P_{fr} holds the chloroplast in its vicinity. For continuous movement of chloroplasts towards a high P_{fr} area, a continuous supply of newly formed P_{fr} is required.

References

Ahmad M, Cashmore AR (1993) Nature 366:162–166. **Algarra P**, Linder S, Thümmler F (1993) FEBS Lett 315:69–73. **Assmann SN**, Simonici L, Schroeder JI (1985) Nature 318:285–287.

Ballaré CL, Scopel AL, Radosevich SR, Kendrick RE (1992) Plant Physiol 100:170–177. **Ballaré CL**, Scopel AL, Sánchez RA (1990) Science 247:329–332. **Ballaré CL**, Scopel AL, Sánchez RA (1991) Plant Cell Environ 14:57–65. **Berridge MJ** (1990) J Biol Chem 265:9583–9586. **Blakeley SD**, Thomas B, Hall JL, Vince–Prue D (1983) Planta 158:416–421. **Blum DE**, Elzenga TM, Linnemeyer PA, Van Volkenburgh E (1992) Plant Physiol 100:1968–1975. **Boylan MT**, Quail P (1991) Proc Natl Acad Sci USA 88:10806–10810. **Boylan MT**, Quail PH (1989) Plant Cell 1:765–773. **Broach JR** (1991) Trends Genet 7:28–33.

Casal JJ, Kendrick RE (1993) Plant Cell Environ 16:703–710. **Chae Q**, Park HJ, Hong SD (1990) Biochim Biophys Acta 1051:115–222. **Chasan R** (1993) Plant Cell 5:137–140. **Cherry JR**, Hershey HP, Vierstra RD (1991) Plant Physiol 96:775–785. **Cherry JR**, Hondred D, Walker JM, Vierstra RD (1992) Proc Natl Acad Sci USA 89:5039–5043. **Cherry JR**, Hondred D, Walker JM, Keller JM, Hershey HP, Vierstra RD (1993) Plant Cell 5:565–575. **Chory J** (1992) Development 115:337–354. **Chory J** (1993) TIG 9:167–172. **Clark GB**, Memon AR, Tong C-G, Thompson GA, Roux SJ (1993) Plant J 4:399–402. **Cohen P** (1992) TIBS 17:408–413.

Dale JE (1988) Annu Rev Plant Physiol Plant Mol Biol 39:267–295. **Datta N**, Cashmore AR (1989) Plant Cell 1:1069–1977. **Datta N**, Chen YR, Roux SJ (1985) Biochem Biophys Res Commun 128:1403–1408. **Debesh K**, Bruce WB, Quail PH (1990) Science 250:1397–1399. **Debesh K**, Franci C, Parks BM, Seeley KA, Short TW, Tepperman JM, Quail PH (1993) Plant Cell 5:1081–1088. **Deng X-W**, Matsui M, Wei N, Wagner D, Chu

AM, Feldmann KA, Quail P-H (1992) Cell 7:791–801. **Drøbak BK** (1993) Plant Physiol 102:705–709. **Dynlacht BD**, Weinzierl ROJ, Admon A, Tijan R (1993) Nature 363: 176–179.

Edgerton MD, Jones AM (1992) Plant Cell 4:161–171. **Elmlinger MW**, Mohr H (1991) Planta 183:374–380. **Elmlinger MW**, Bolle C, Batschauer A, Oelmüller R, Mohr H (1994) Planta 192:189–194.

Frances S, White MJ, Edgerton MD, Jones AM, Elliott RC, Thompson WF (1992) Plant Cell 4:1519–1530. **Furuya M** (1989) Adv Biophys 25:133–167. **Furuya M** (1993) Annu Rev Plant Physiol Plant Mol Biol 44:617–645.

Gaba V, Black M (1979) Nature 278:51–54. **Gabriel JL**, Hoober JK (1991) J Theor Biol 151:541–556. **Gallagher S**, Short TW, Ray PM, Pratt LH, Briggs WR (1988) Proc Natl Acad Sci USA 85:8003–8007. **Galland P**, Senger H (1988) Photochem Photobiol 48:811–820. **Gilmartin PM**, Sarokin L, Memelink J, Chua NH (1990) Plant Cell 2:369–378. **Gilmartin PM**, Memelink J, Hiratsuka K, Kay SA, Chua N-H (1992) Plant Cell 4:839–849. **Giuliano G**, Pichersky E, Malik VS, Timko MP, Scolnik PA, Cashmore AR (1988) Proc Natl Acad Sci USA 85:7089–7093. **Green PJ**, Yong M-H, Cuozzo M, Kano-Murakami Y, Silverstein P, Chua N-H (1988) EMBO J 7:4035–4044. **Grimm R**, Gast D, Rüdiger W (1989) Planta 178:199–206. **Grimm R**, Speth V, Gatenby AA, Schäfer E (1991) FEBS Lett 286:155–158. **Grimm R**, Donaldson GK, van der Vies SM, Schäfer E, Gatenby AA (1993) J Biol Chem 268:5220–5226. **Grolig F**, Wagner G (1988) Bot Acta 101:2–6. **Gruissem W** (1990) Plant Cell 2:827–828. **Guron K**, Chandok MR, Sopory SK (1992) Photochem Photobiol 56:961–995.

Hahn S (1993) Nature 363:672–673. **Hanelt S**, Braun B, Marx S, Schneider-Poetsch HAW (1992) Photochem Photobiol 56:751–758. **Hanstein C**, Lange U, Schneider-Poetsch HAW, Grolig F, Wagner G (1992a) Acta Histochem Suppl XLI:223–230. **Hanstein C**, Grolig F, Wagner G (1992b) Bot Acta 105:55–62. **Hartmann KM** (1966) Photochem Photobiol 5:349–366. **Hartmann KM**, Nezadal W (1990) Naturwissenschaften 77:158–163. **Haupt W** (1982) Annu Rev Plant Physiol 33:205–233. **Haupt W** (1990) Photochem Photobiol 52:261–263. **Haupt W** (1991) Introduction to photosensory transduction chains. In: Louci F et al. (eds) Biophysics of photoreceptors and photomovements in microorganisms. Plenum Press, New York. **Haupt W** (1993) Naturwissenschaften 80:116–124. **Hayami J**, Kadota A, Wada M (1992) Photochem Photobiol 56:661–666. **Hemmingsen SM**, Woolford C, van der Vies SM, Tilly K, Dennis DT, Georgopoulos CP, Hendrix RW, Ellis RJ (1988) Nature 333:330–334. **Hepler J**, Gilman AG (1992) TIBS 17:383–387. **Heyer A**, Gatz C (1992) Plant Mol Biol 18:535–544. **Hillman WS** (1967) Annu Rev Plant Physiol 18:301–324. **Hock B**, Mohr H (1964) Planta 61:209–228. **Hofmann E**, Grimm R, Harter K, Speth V, Schäfer E (1991) Planta 183:265–273. **Hou Y**, von Arnim AG, Deng X-W (1993) Plant Cell 5:329–339.

Jones AM, Erickson HP (1989) Photochem Photobiol 49:479-483.

Kaldenhoff R, Kölling A, Richter G (1993) Plant Mol Biol 23:1187–1198. **Kasperbauer MJ** (1973) Agron J 65:447–450. **Kasperbauer MJ** (1992) Photochem Photobiol 56:823–832. **Kasperbauer MJ**, Hunt PG (1992) Photochem Photobiol 56:579–584. **Kaufman LS** (1993) Plant Physiol 102:333–337. **Kay SA**, Nagatani A, Keith B, Deak M, Furuya M, Chua N-H (1989) Plant Cell 1:775–782. **Kaziro Y**, Ihoh H, Kozaka T, Nakafuku M, Satoh T (1991) Annu Rev Biochem 60:349–400. **Keller JM**, Shanklin J, Vierstra RD, Hershey HP (1989) EMBO J 8:1005–1012. **Kendrick RE**, Nagatani A (1991) Plant J 1:133–139. **Ken-Dror S**, Horwitz BA (1990) Plant Physiol 92:1004–1008. **Kim HY**, Coté GG, Crain

RC (1993) Science 260:960–962. **Kim I-S**, Bai U, Song P-S (1989) Photochem Photobiol 49:319–323. **Kloppstech K**, Otto B, Sierratta W (1991) Mol Gen Genet 225:468–473. **Koornneef M**, Rolff E, Spruit Z (1980) Z Pflanzenphysiol 100:147–160. **Koornneef M**, Cone JW, Dekens RG, O'Herne-Robers EGO, Spruit CJP, Kendrick RE (1985) J Plant Physiol 120:153–165. **Krajewska WM** (1992) Int J Biochem 24:1885–1898. **Kraml M**, Herrmann H (1991) Photochem Photobiol 53:255–259.

Lagarias JC, Lagerias DM (1989) Proc Natl Acad Sci USA 86:5778–5780. **Lam E**, Chua N-H (1989) Plant Cell 1:1147–1156. **Lamparter T**, Lutterbüse P, Schneider-Poetsch HAW, Hertel R (1992) Photochem Photobiol 56:697–707. **Li L**, Lagarias JC (1992) J Biol Chem 267:19204–19210. **Linder ME**, Gilman AG (1992) Sci Am 267(1):36–43. **Liscum E**, Hangarter RP (1993a) Plant Physiol 101:567–572. **Liscum E**, Hangarter RP (1993b) Plant Physiol 103:15–19. **Liscum E**, Hangarter RP (1993c) Planta 191:214–221.

Ma H (1993) TIG 9:228–230. **Mandoli DF**, Ford GA, Waldron LJ, Nemson JA, Briggs WR (1993) Plant Cell Environ 13:287–294. **Marx J** (1993) Science 260:1588–1590. **McCormac AG**, Cherry JR, Hershey HP, Vierstra RD, Smith H (1991) Planta 185:162–170. **McCormac AC**, Wagner D, Boylan MT, Quail PH, Smith H, Whitelam GC (1993) Plant J 4:19–27. **McCormick F** (1993) Nature 363:15–16. **McCurdy DW**, Pratt LH (1986) J Cell Biol 103:2541–2550. **Mohr H** (1957) Planta 49:389–405. **Mohr H** (1986) Coaction between pigment systems. In: Kendrick RE, Kronenberg GMH (eds.) Photomorphogenesis in plants. M Nijhoff, Dordrecht, pp 547–564. **Mohr H**, Schopfer P (1992) Pflanzenphysiologie, 4th edn. Springer, Berlin Heidelberg New York. **Mohr H**, Shropshire W Jr (1983) An introduction to photomorphogenesis for the general reader. In: Shropshire W Jr. Mohr H (eds) Photomorphogenesis. Encyclopedia of plant physiology; new series, vol 16A, Springer, Berlin Heidelberg New York, pp 24–38.

Nagatani A, Nishizawa NK, Mori S, Kay SA, Chua N-H, Furuya M (1993a) Plant Cell Physiol 34:825–833. **Nagatani A**, Reed JW, Chory J (1993b) Plant Physiol 102:269–277. **Neuhaus G**, Bowler C, Kern R, Chua N-H (1993) Cell 73:937–952. **Nick P**, Ehmann B, Furuya M, Schäfer E (1993) Plant Cell 5:541–552. **Nieto-Sotelo J**, Ichida A, Quail PH (1994) Plant Cell 6:287–301.

Oelmüller R, Kendrick RE (1991) Plant Mol Biol 16:293–299. **Okada K**, Shimura Y (1992) Cell 70:369–372. **Okamoto H**, Hirano Y, Abe H, Tomizawa K-i, Furuya M, Wada M (1993) Plant Cell Physiol 34:1329–1334. **Okubara PA**, Williams SA, Doxsee RA, Tobin EM (1993) Plant Physiol 101:915–924. **O'Neill S** (1992) Photochem Photobiol 56:789–801. **Ootaki T** (ed) (1993) Growth and behavioral responses of fungi to environmental conditions. IKS Series No 4. Institute of Genetic Ecology, Tohoku University, Sendai, Japan. **Orozco BM**, Ogren WL (1993) Plant Mol Biol 23:1129–1138. **Otto B**, Ohad I, Kloppstech K (1992) Plant Mol Biol 18:887–896.

Parker W, Partis M, Song P-S (1992) Biochemistry 31:9413–9420. **Parks BM**, Quail PH (1991) Plant Cell 3:1177–1186. **Parks BM**, Quail PH (1993) Plant Cell 5:39–48. **Pratt LH**, Shimazaki Y, Steward SJ, Cordonnier M-M (1991) Planta 184:81–86.

Quail PH (1991) Annu Rev Genet 25:389–409. **Quail PH**, Marmé D, Schäfer E (1973) Nature 245:189–191. **Quinones MA**, Lu Z, Zeiger E (1993) Plant Physiol 102:S-15.

Racusen RH, Galston AW (1983) Developmental significance of light-mediated electrical responses in plant tissue. In: Shropshire W Jr, Mohr H (eds) Photomorphogenesis. Encyclopedia of plant physiology; new series, vol 16B. Springer, Berlin Heidelberg New York, pp 687–703. **Rayer B**, Naynert M, Stieve H (1990) J Photochem Photobiol B 7:107–148.

Reed JW, Nagpal P, Chory J (1992) Photochem Photobiol 56:833–838. **Reed JW**, Nagpal P, Poole DS, Furuya M, Chory J (1993) Plant Cell 5:147–157. **Roblin G**, Fleurat-Lessard P, Everat-Bourbouloux A, Bonmort J, Moyen C (1990) Photochem Photobiol 52:192–202. **Romero LC**, Lam E (1993) Proc Natl Acad Sci USA 90:1465–1469. **Romero LC**, Sommer D, Gotor C, Song P-S (1991a) FEBS Lett 282:341–348. **Romero LC**, Biswal B, Song P-S (1992b) FEBS Lett 282:347–350. **Roux SJ** (1992) Photochem Photobiol 56:811–814. **Rüdiger W**, Thümmler F, Cmiel E, Schneider S (1983) Proc Natl Acad Sci USA 80:6244–6248. **Russ U**, Grolig F, Wagner G (1991) Planta 184:105–112.

Satoh R, Nakafuku M, Kaziro Y (1992) J Biol Chem 267:24149–24152. **Schmid R**, Dring MJ (1993) Plant Physiol 101:907–913. **Schneider-Poetsch HAW**, Braun B (1991) J Plant Physiol 137:576–580. **Schneider-Poetsch HAW**, Braun B, Marx S, Schaunburg A (1991a) FEBS Lett 281:245–249. **Schneider-Poetsch HAW**, Sensen C, Hanelt S (1991b) Z Naturforsch 46c:750–758. **Schneider-Poetsch HAW** (1992) Photochem Photobiol 56:839–846. **Schönbohm E**, Meyer-Wegener J, Schönbohm E (1990) J Photochem Photobiol 5:331–341. **Shacklock PS**, Read ND, Trewavas AJ (1992) Nature 358:753–755. **Sharma R**, López-Juez E, Nagatani A, Furuya M (1993) Plant J 4:1035–1042. **Sharrock RA**, Quail PH (1989) Genes Dev 3:1745–1757. **Shimazaki K-i**, Omasa K, Kinoshita T, Nishimura M (1993) Plant Cell Physiol 34:1321–1327. **Shinkle JR**, Sooudi SK, Jones RL (1992) Plant Physiol 99:808–811. **Short TW**, Reymond PR, Briggs WR (1993) Plant Physiol 101:647–655. **Siegelman HW**, Hendricks SB (1957) Plant Physiol 32:393–398. **Singh BR**, Song RS (1990) Photochem Photobiol 52:249–254. **Smith H** (1992) Photochem Photobiol 56:815–822. **Smith H**, Whitelam GC (1990) Plant Cell Environ 13:695–707. **Smith H**, Turnbull M, Kendrick RE (1992) Photochem Photobiol 56:607–610. **Smith S**, Samson G, Fork DC (1993) Plant Cell Environ 16:929–937. **Speth V**, Otto V, Schäfer E (1986) Planta 168:299–304. **Steward SJ**, Pratt LH, Cordonnier-Pratt M-M (1992) Plant Physiol 99:1708–1710. **Stock JB**, Stock AM, Mottonen JM (1990) Nature 344:395–400. **Struhl K** (1994) Science 263:1103–1104. **Susek R**, Chory J (1992) J Plant Physiol 19:387–399. **Susek RE**, Ausubel FM, Chory J (1993) Cell 74:787–799.

Takagi S, Kamitsubo E, Nagai R (1992) Protoplasma 168:153–158. **Talbott LD**, Zeiger E (1993) Plant Physiol 102:1163–1169. **Terry MJ**, Lagarias JC (1991) J Biol Chem 266:22215–22221. **Terry MJ**, Thomas B, Hall JL (1992) J Plant Physiol 140:761–764. **Thomas B**, Johnson CB (eds) (1991) Phytochrome properties and biological action. Springer, Berlin Heidelberg New York. **Thompson WF**, White MJ (1991) Annu Rev Plant Physiol Plant Mol Biol 42:423–466. **Thümmler F**, Dufner M, Kreisl P, Dittrich P (1992) Plant Mol Biol 20:1003–1017. **Tretyn A**, Kendrick RE, Wagner G (1991) Photochem Photobiol 54:1135–1155. **Tretyn A**, Kendrick RE, Kopcewicz J (1992) J Exp Bot 43:439–448.

Vierstra RD (1993) Plant Physiol 103:679–684. **Volotovski ID**, Sokolovsky SG, Nikiforov EL, Zinchenko VP (1993) J Photochem Photobiol B: Biol 20:95–100.

Wada M, Grolig F, Haupt W (1993) J Photochem Photobiol B 17:3–25. **Wagner D**, Tepperman JM, Quail PH (1991) Plant Cell 3:1275–1288. **Wagner G**, Russ U, Quader H (1992) Calcium, a regulator of cytoskeletal activity and cellular competence. In: Menzel D (ed): The cytoskeleton of the algae. CRC Press, Boca Raton. **Wahleithner BA**, Li L, Lagarias JC (1991) Proc Natl Acad Sci USA 88:10387–10391. **Wei N**, Deng X-W (1992) Plant Cell 4:1507–1518. **Weisshaar B**, Armstrong GA, Block A, da Costa e Silva O, Hahlbrock K (1991) EMBO J 10:1777–1786. **Whitelam GC**, Johnson E, Peng J, Carol P, Anderson ML, Cowl JS, Harberd NP (1993) Plant Cell 5:757–768. **Winands A**, Wagner G, Marx S, Schneider-Poetsch HAW (1992) Photochem Photobiol 56:765–770.

Yatsuhashi H, Kobayashi H (1993) J Photochem Photobiol B 19:25–31.

Zandomeni K, Schopfer P (1993) Protoplasma 173:103–112.

Prof. Dr. Bertold Hock
Technische Universitat München
(Weihenstephan)
Lehrstuhl für Botanik
85350 Freising, Germany

VIb. Movement of Pulvinated Leaves

By W.-E. Mayer and R. Hampp

1. Introduction (Phenomenology, Terminology)

Leaf movements of many plant species are mediated by curvatures of pulvini, located at the base of petioles and laminae. In pulvini, the central vascular core is surrounded by parenchymous cortex cells. The curvatures of pulvini are brought about by the coordinated and simultaneous swelling and shrinking of the cortex cells (the motor cells) in two opposing halves of the pulvinus. Motor cells that are expanded axially in the horizontal day (open, unfolded) position of leaves or leaflets are designated extensor cells, those that are contracted axially in the respective position flexor cells. Depending on whether the leaves (pulvini) move to their vertical night position (closed, folded position) downwards (e.g., secondary pulvini of *Phaseolus* and *Samanea*), or upwards (e.g., secondary pulvini of *Cassia*), the extensor cells are located in the lower (abaxial) or upper (adaxial) half of the pulvinus.

Leaf movement of pulvinated leaves is controlled by endogenous and exogenous signals. Under constant conditions the leaves show circadian (period length about 24 h) and/or ultradian (period length some minutes or several hours) movements. These are modified by external signals. Photonastic movements are induced by dark/light transition, scotonastic (nyctinastic) movements by light/dark transition, and phototropic movements by unilateral light. Phototropic movements in response to the position of the sun are called heliotropic (suntracking). Two types can be distinguished: diaheliotropic and paraheliotropic. In the former, leaf laminae remain perpendicular to the incident light, whereas in the latter, leaf laminae can obtain any angle between perpendicular and parallel. In addition, other stimuli such as temperature, gravity, touch, or shaking can affect rhythmic and light-driven movements.

For analyzing the mechanisms of swelling and shrinking of pulvinar motor cells, mainly the biophysical and biochemical events that are associated with the rhythmic, the light-driven, and the seismonastic movements have been investigated.

Since the last review on leaf movements in this periodical (Hensel 1987), the recent knowledge (up to 1990) regarding the physiology of movements of pulvinated leaves has been summarized by several experts in a booklet on the

Progress in Botany, Vol. 56

pulvinus, edited by Satter et al. (1990). Light-driven leaf movements are reviewed by Koller (1990). The present chapter will try to give a condensed overview on the most important results described in these reviews, supplemented by papers published afterwards. Excluded from this report are the seismonastic movements of *Mimosa*, and gravitropic movements. The reader is referred to the reviews by Fromm and Eschrich (1990), Roblin (1990), Fromm (1991), and Sibaoka (1991).

2. Mechanisms of Motor Cell Swelling and Shrinking

It is now generally accepted that the mechanisms of swelling and shrinking are similar in stomatal guard cells and in pulvinar motor cells (Gorton 1990). For pulvinar motor cells these mechanisms are depicted in almost identical models by Moran (1990), Satter and Morse (1990), Crain (1990), and Lee (1990). They comprise modifications of the older one by Satter and Galston (1981). The general concept is as follows.

Pulvinar motor cells swell when protons are pumped by the plasma membrane H^+-ATPase, thus creating a proton gradient and a shift of the membrane potential to more negative values. The hyperpolarization opens inward K^+ channels (K_H channels) and drives K^+ entry through these channels down the electrical gradient. Cl^- influx occurs presumably through H^+/Cl^- cotransport, driven by the H^+ concentration gradient. Though the models do not depict the transport of these ions across the tonoplast, it is evident that ions entering the cell have to be stored in the vacuole. With the increase of the concentrations of intracellular osmotica, the motor cells take up water and start to swell.

Shrinkage of pulvinar motor cells is not the reversal of swelling. Membrane depolarization, which may be due to Ca^{2+} uptake [through Ca^{2+} channels or stretch-activated (SA) channels] and/or Cl^- efflux (through Cl^- channels or anion permeable SA channels), opens outward K^+ channels (K_D channels) and provides the driving force for K^+ efflux. Loss of KCl through the open K_D and Cl^- channels leads to water release and cell shrinking.

This model is supported now by strong experimental evidence. Particularly, the H^+ and K^+ transport across the plasma membrane of motor cells of some species and the properties of the K_D and K_H channels in the plasma membrane of *Samanea* pulvinar motor cell protoplasts have been investigated thoroughly.

a) Hydrogen Ion Transport

Motor cells secrete H^+ when they swell and take up H^+ when they shrink. This was found for motor cells in intact pulvini with H^+-sensitive microelectrodes, for swelling protoplasts, and for excised pulvinar tissues.

In the *Samanea* pulvinus the pH of the water-free space (WFS) of the apoplast changed upon darkening from 6.2 to 7.2 in the shrinking extensor, and from 5.5 to 4.9 in the swelling flexor. Subsequent white light, which induces reopening of the pulvinus (swelling in extensor cells and shrinking in flexor cells), changed the pH of the WFS of the extensor apoplast back to 6.2 and that of the flexor apoplast to 6.3 (Lee and Satter 1989).

In *Phaseolus,* the circadian movements and the circadian changes in the pH of the WFS of the extensor apoplast were measured simultaneously in the same pulvinus. Downward movement (shrinking of extensor cells) was correlated with an increase of the pH in the WFS of the extensor apoplast from between 5–6 to 6–7, upward movement (swelling of extensor cells) with reversed pH changes (Starrach 1988; Starrach and Mayer 1989; Mayer 1990).

Protoplasts of extensor cells of the *Phaseolus* pulvinus exhibit K^+-induced swelling (Erath et al. 1988). Addition of 10 mM KCl or K^+-iminodiacetate (an impermeable anion) resulted in a volume increase of approx. 50%. This swelling of protoplasts was accompanied by an acidification of the external medium.

In intact pulvini of *Phaseolus*, the K^+, Cl^- and malate content increases and the ψ_{tissue} decreases in the extensor half of the pulvinus when the leaf is irradiated with blue light (BL) or far-red light (FR) for 6 h at the beginning of the photoperiod (in a 12-h light/12-h dark period). Respective changes in the ion content and ψ_{tissue} occurred in the flexor when irradiated with red light (R). These results indicate that in this phase of the light/dark cycle BL and FR irradiations are swelling signals for the extensor and R for the flexor (Bialczyk and Lechowsky 1990a). Excised extensor tissue really acidified the bathing medium when irradiated with BL and FR, excised flexor tissue when irradiated with R (Bialczyk and Lechowsky 1992).

Older measurements with excised pulvinar tissues of *Samanea* are in accordance with these results. Excised strips of the pulvini acidified the medium in their swelling phase and this acidification was blocked by metabolic inhibitors or inhibitors of the plasma membrane H^+-ATPase (Iglesias and Satter 1983; Lee and Satter 1987; Satter et al. 1987).

Apoplastic pH changes in the abaxial part of the *Desmodium motorium* pulvinus (Antkowiak 1992) appear to contradict the relation of H^+ extrusion and motor cell swelling or H^+ uptake and motor cell shrinking. As measured with H^+ sensitive double-barrelled microelectrodes in the intact pulvinus, the pH of the WFS of the apoplast oscillated rhythmically between approx. 6 and 6.5. The period length (about 3 to 4 min) of the apoplastic pH rhythm was identical with the period lengths of the rhythms of leaflet movement, of apoplastic K^+ concentration, of extracellular electric potential, and of membrane potential. However, the phase correlation between the leaflet movement rhythm and the pH oscillation was such that during cell shrinking, the H^+ concentration in the WFS was highest while it decreased during cell swelling. Furthermore, the rhythmic changes in H^+ and K^+ concentrations in the apoplast were nearly in phase, e.g., the apoplastic H^+ concentration was high when the apoplastic K^+ concentration was high and decreased when the apoplastic K^+

concentration decreased. The physiological significance of these phase correlations in the *Desmodium* pulvinus is not clear. However, the finding that in the *Phaseolus* pulvinus extensor cell shrinking and swelling during the slow circadian pulvinar movements were correlated with an increase respectively decrease in the WFS-pH, whereas during the quicker ultradian movements (period length approx. 150 min) this correlation was not found (Starrach and Mayer 1989) indicates that during short-term pH changes cation exchange processes in the cell wall might obscure the phase relation between H^+ extrusion and cell swelling or H^+ uptake and cell shrinking.

b) K^+ Transport

For three species it has been shown that the K^+ concentration in the WFS of the motor cell apoplast of intact pulvini changes in accordance with the model.

Using K^+-sensitive microelectrodes, Lowen and Satter (1989) showed that in open pulvini of *Samanea* (extensor cells swollen), the K^+ concentration in the WFS of the extensor apoplast was about 15 mM. It increased rapidly to 80–100 mM upon turning off the light (a shrinking signal for the extensor cells in this leaf position). In the flexor the apoplastic K^+ concentration was between 10–35 mM in the open pulvinus (flexor cells shrunken) and decreased to about half the amount upon darkening (a swelling signal for flexor cells in this leaf position).

Similar results were obtained for *Phaseolus* pulvini (Starrach and Mayer 1989). Downward movement of the lamina (light-off-induced or during a circadian cycle) caused an increase of the K^+ concentration in the WFS of the extensor apoplast from approx. 10 mM to approx. 50 mM. These changes corresponded well with those of the amount of apoplastic K^+ of tangential extensor sections. When extensor sections were cut during the subjective night position of the leaf, the apoplastic K^+ content was about twice of that of the subjective day position (0.45 and 0.23 mmol g^{-1} DW_{tissue} respectively). In the flexor, the apoplastic K^+ content changed in a reversed manner to that of the extensor and thus behaved in accordance with the concept that shrinking cells lose K^+ to the apoplast, while swelling cells take it up from the apoplast (Freudling et al. 1988).

From the changes of the apoplastic and the total K^+ contents of tangentially cut pulvinar sections, it was concluded that 30–40% of the K^+ gained or lost by the cells during cell swelling and shrinking was shuttled between the protoplasts and their surrounding apoplast, the remaining 60–70% being transported between different pulvinus parts. The high concentration and large amount of fixed negative charges in the walls of motor cells (Starrach et al. 1985; Starrach and Mayer 1986; Mayer 1990) allows for both a considerable accumulation of K^+ and the binding of large amounts of H^+. Since depletion of the apoplastic K^+ reservoir is coupled to acidification and filling to alkalization of the apoplast (Starrach and Mayer 1989), it is likely that the H^+/K^+ exchange within the cell wall is most important for its function as K^+ reservoir.

Desmodium is another example where motor cell swelling and shrinking was found to be correlated with respective changes of the apoplastic K^+ concentration. The apoplastic K^+ concentration in the abaxial part of intact but mechanically fixed pulvini oscillated between 10 to 20 mM with a stable phase relationship to the simultaneously measured ultradian rhythm of the extracellular electric potential (Antkowiak and Engelmann 1989; Antkowiak 1992). Only seconds after the rapid hyperpolarization of the extracellular potential, the apoplastic K^+ concentration increased and quickly reached the highest value. As in moving leaflets the extracellular hyperpolarization in the abaxial motor tissue always preceded the downward movement of the leaflet (Antkowiak et al. 1991), it is evident that cell shrinking is coupled with a sudden increase of the apoplastic K^+ concentration (up to 20–30 mM), cell swelling with a respective decrease (down to 5–15 mM).

c) Cl^- and Malate

It is well documented that K^+ and Cl^- are the main osmotica which are accumulated in swelling and depleted in shrinking parts of pulvini (Satter and Galston 1981; Satter et al. 1988; Mayer et al. 1987). In contrast, little is known about the Cl^- transport.

Excised flexor tissue of the *Samanea* pulvinus acidified the bathing medium upon darkening (flexor cell swelling) much more when the impermeable anion iminodiacetate replaced permeable anions. Thus the uptake of permeable anions such as Cl^- may be coupled to the uptake of H^+ via a H^+/anion cotransport (Satter et al. 1987). Since the amount of malate increased in the flexor in the absence of permeable anions, it was concluded that malate synthesis could substitute for H^+/Cl^- uptake in charge balance (Satter et al. 1987).

In 14-day-old soil-grown plants of *Phaseolus coccineus*, malate was shown to be involved in the osmotic motor. In a 12-h light/12-h dark cycle, the malate content changed rhythmically in the extensor and flexor, being lower in the shrunken than in the swollen pulvini parts (Bialczyk and Lechowski 1986). Fusicoccin, which stimulates H^+ extrusion and promotes opening of closed leaves, increased the malate content, whereas ABA, which induces closing of leaves, reduced it (Bialczyk and Lechowski 1987). Metabolic and ATPase inhibitors (CCCP, DES, DCCD) and the K^+ channel blocker TEA reduced the malate and the K^+ content in the extensor and flexor during both light and dark phase of a light/dark cycle (Bialczyk and Lechowski 1990b). Furthermore, 6-h treatments with wavelengths known to induce swelling of extensor (BL or FR) and flexor (R) cells, increased the content of malate, K^+, and Cl^-. Treatments with wavelengths known to be without effect on the volume of extensor (R) and flexor (BL or FR) cells, did not affect the respective ion pools. These results indicate that light as well as fusicoccin induce malate synthesis and ion uptake in swelling motor cells (Bialczyk and Lechowski 1990a).

When isolated leaves of *Phaseolus coccineus* were fed for 12h via the transpiration stream with K^+-iminodiacetate (an impermeable anion), the malate content of both flexor and extensor was increased, whereas it was decreased upon incubation with KCl. The K^+ content increased in both cases. This provides evidence that malate is synthesized and used in the osmotic motor when Cl^- is too low to compensate for the electric charges carried by K^+ (Bialczyk and Lechowski 1989). Results obtained for older (3-to 4-week-old) *Phaseolus coccineus* plants support this conclusion (Mayer et al. 1987). In soil-grown plants of this age, the malate content was low and did not change rhythmically with the extensor and flexor cell swelling and shrinking, whereas in Cl^--deprived plants of the same age, the malate content was considerably higher and changed (together with citrate) rhythmically, being higher in the swollen and lower in the shrunken pulvini parts.

d) Ion Channels

Application of the patch-clamp technique to plasma membranes of pulvinar motor cells added a new dimension to the understanding of the ion transport in pulvinar cells.

In *Samanea* motor cell protoplasts, several channel types were found: depolarization-activated K^+ channels (K_D channels), hyperpolarization-activated K^+ channels (K_H channels), Cl^- channels, stretch-activated channels (SA channels), and Ca^{2+} permeable, nonselective, depolarization-activated channels (Moran 1990). Of all these channels, the best described so far by the patch-clamp technique is the K_D channel.

The K_D channel of extensor and flexor protoplasts is characterized as follows: it opens upon depolarization and its conductance ranges from 15 to 40 pS at external K^+ concentrations between 5 and 125 mM, and more than 100 mM K^+ on the cytoplasmic side. It is selective for K^+ over other monovalent ions with the relative permeability sequence $K^+ > Rb^+ > Na^+ \simeq Cs^+ \simeq Li^+$ (Moran et al. 1990), it can be blocked by TEA, quinine, Cs^+, Ba^{2+}, Cd^{2+}, and Gd^{3+} (Moran et al. 1988; 1990), and its activity is regulated by phosphorylation (Moran 1993). The K_D channel appears to be the most accessible channel in the plasma membrane and it occurs in numbers sufficient to permit rates of K^+ efflux which are necessary to account for the K^+ loss in shrinking pulvinar cells.

As analyzed by the patch-clamp technique, it is known that the K_H channel is activated by membrane hyperpolarization and has a conductance of about 25 pS (Moran and Satter 1989; Moran 1990). However, properties and regulation of K^+ channels, presumably the K_H channels which mediate K^+ influx, were investigated recently in more detail by measuring the K^+-induced changes in membrane potential with the membrane potential sensitive fluorescent dye 3,3′-dipropylthiadicarbocyanine iodide (Kim et al. 1992, 1993). With this method, the open state of K^+ channels is indicated when the membrane potential depolarizes (indicated by fluorescence increase) after the extracellular K^+ concentration is increased from 20 to 200 mM; the closed state is indicated when no change of the membrane potential occurs upon this treatment. Mediation by K^+ channels of the measured depolarization is indicated by the findings that the K^+ channel blockers TEA and Al^{3+} inhibit the K^+-induced depolarization, and that other cations (Na^+, Li^+, and Ca^{2+}) have no or little effect on the membrane potential.

The state of activation (opening) of the inward K^+ channels changed rhythmically in flexor and extensor protoplasts of the *Samanea* pulvinus, and the rhythms were phase-shifted by 180°. As expected from the K^+ fluxes from leaflet movement of intact plants in constant darkness, the K^+ channels were open in flexor and closed in extensor protoplasts during the first 8 h in darkness (intact leaflets are closed), whereas at hours 16–21 after onset of darkness, these K^+ channels were open in the extensor and closed in the flexor (intact leaflets are open). Opening and closing of these K^+ channels is also regulated by light. However, the light-induced reactions of the channels depend on the state of activation and therefore on the phases of the circadian oscillator. Light effects on the K^+ channels were investigated at two phases of the circadian oscillator which was synchronized to a 8-h dark/16-h light cycle: at hours 4 to 8 of the dark period (Kim et al. 1992) and at hours 10 to 12 of the light period (Kim et al. 1993). During the dark period, when the K^+ channels were open in the flexor and closed in the extensor protoplasts, white light (WL) and BL pulses (but not R pulses) induced transient membrane hyperpolarization in both extensor and flexor protoplasts and opened the inward K^+ channels in the extensor but closed them in the flexor protoplasts. Vanadate inhibited hyperpolarization in response to the light pulses, but did not affect the opening of the K^+ channels. This indicates that hyperpolarization resulted from activation of the plasma membrane H^+-ATPase, but that this hyperpolarization is not the sole factor controlling opening of these channels. During the light period when the K^+ channels were open in the extensor and closed in the flexor protoplasts, darkening closed the channels in the extensor, but did not open them in the flexor protoplasts. R pulses (30–60 s) given before darkening had no effect on the closing reaction of extensor protoplasts, but opened the inward K^+ channels in the flexor protoplasts. In the flexor, the R-induced opening was reversed by FR.

Little is known about the properties of three other types of channels (Moran 1990). One of them, designated Cl^- channel, was observed in flexor protoplasts when the cytoplasmic surface of the membrane was exposed to a high Ca^{2+} concentration. Another, detected in extensor protoplasts, was permeable to Ca^{2+} and K^+, but was not affected by the K^+ channel blockers TEA and Ba^{2+}. It was activated at membrane potentials depolarized by more than −160 mV. Finally, by applying negative pressures (−10 to −60 mm Hg) several voltage-independent SA channels were found. Judging by amplitude, five channel types could be distinguished (3, 9, 14, 27, 50 pS). Judging by their selectivities, two types were observed: one (9 and 50 pS channels) with an appreciable permeability ratio of $P_{Cl}/P_K > 20$, and the other one (3, 27, and 14 pS channels) poorly discriminating (Moran 1990, 1991). Moran supposes that opening of the non-selective SA channels (or the depolarization-activated Ca^{2+} channels) could provide the rise in the cytosolic Ca^{2+} which is presumably important for the opening of the Cl^- channels, and that the opening of the Cl^--selective SA channels could provide both the depolarizing signal for opening of the K_D channels and a pathway for Cl^- efflux during cell shrinking.

e) Membrane Potential

K^+ uptake through open K^+ channels requires a membrane potential more negative than the equilibrium potential for K^+. For the efflux of K^+ and Cl^- through open K^+ and Cl^- channels a membrane potential is needed that is depolarized to values more positive than the equilibrium potential for K^+, and more negative than that for Cl^-. Thus membrane hyperpolarization should be found in swelling, membrane depolarization in shrinking motor cells. Such a correlation was found in recent investigations (Antkowiak et al. 1991; Antkowiak 1992; Nishizaki 1987, 1988, 1989, 1990, 1992), whereas results of older measurements are contradictory (Racusen and Satter 1975; Racusen and Galston 1977; Scott et al. 1977; Freudling et al. 1980). There are, however, indications that in the older reports the measured electric potentials do not represent the membrane potential of the cells in situ. Freudling et al. (1980) showed for extensor cells of the *Phaseolus* pulvinus that in freshly excised tissue slices the cells were hyperpolarized when they were shrunken, and depolarized when swollen. However, the membrane potential of the extensor cells in the freshly cut slices did not change upon illumination whereas the extensor cells in aged slices displayed large light-on-induced depolarization and light-off-induced hyperpolarizaton, similar to that found for extensor cells in the intact pulvinus in air (Starrach et al. 1984). This indicates that the measured electric potential in freshly excised slices did not reflect the membrane potential of the extensor cells in situ. Disturbance of H^+ extrusion by the H^+ pump is probably the cause for the atypical membrane potential in freshly excised pulvinar tissue, since the membrane potential in the aged slices was rather negative (about $-133\,mV$) as compared to that in freshly cut slices (about $-40\,mV$), and the CN^--induced depolarization was much larger than in the freshly cut slices (Freudling et al. 1980).

Similarly, the circadian changes of the membrane potential in flexor and extensor cells of *Trifolium repens* (Scott et al. 1977) and *Samanea* (Racusen and Satter 1975; Racusen and Galston 1977) are in contrast to the predicted correlation between cell volume and membrane potential changes. Cutting and/or submersion of pulvini may well be the cause for membrane potentials which do not reflect those of the cells in situ.

The membrane potential of pulvinar cells of *Desmodium* oscillates with an ultradian period, and in a manner that indicates that membrane depolarization drives and controls K^+ efflux (cell shrinking), whereas membrane hyperpolarization is connected to K^+ influx (cell swelling) via potassium channels (Antkowiak et al. 1991; Antkowiak 1992). In the abaxial part of the intact pulvinus in air, the extra- and the intracellular electric potential differences against the bathing solution of the petiole were measured with two microelectrodes simultaneously. From the differences between the intra- and extracellular potentials the membrane potential in situ (the electric potential difference across the plasma membrane) was estimated and shown to oscillate between −136 and −36 mV. Oscillation was in phase with the intracellular but inverse to the extracellular potential. The phase relationship between the leaflet movement rhythm and the membrane potential was as follows: downward movement (shrinking of abaxial motor cells) was preceded and accompanied by a strong depolarization, upward movement (swelling of abaxial motor cells) by hyperpolarization. Assuming a K^+ concentration in the cytosol of 150 mM and an apoplastic K^+ concentration of 5 to 30 mM (see above), the K^+ equilibrium potentials can be estimated to be −87 and −41 mV. Values found for the hyperpolarized or depolarized membrane potential in situ were more positive or more negative than the potassium equilibrium potentials and thus in accordance with the predicted mechanisms of the K^+ transport.

Light-induced changes of membrane potential and leaf movement were investigated in the intact pulvini in air of *Desmodium* and *Phaseolus vulgaris*.

When unilateral WL (50 W m^{-2}) was directed onto the abaxial part of the *Desmodium* pulvinus, the leaflet was forced to the lowered position and oscillated in this position with small amplitudes. When the light beam was directed towards the adaxial part, it was forced to the upper position and oscillated there with small amplitudes. Thus, unidirectional light induced shrinking in the directly irradiated motor cells. Most likely, this light-induced cell shrinking is due to an inhibition of the plasma membrane H^+-ATPase by the light, causing depolarization of the membrane potential and efflux of K^+. The following data provide evidence for this conclusion. (1) Unilateral light directed to the abaxial part of the pulvinus, but also vanadate applied via the transpiration stream, reduced the amplitude of the oscillating membrane potential of abaxial motor cells by decreasing the negative peak values during the spontaneous hyperpolarization. Vanadate, finally, arrested the membrane potential in the depolarized state. (2) Vanadate or anoxia conditions suppressed leaflet movement and arrested leaflets in the lowered position (Antkowiak et al. 1992).

These results are in accordance with similar investigations using pulvini of *Phaseolus vulgaris* (Nishizaki 1986, 1987, 1988, 1989, 1990, 1992). By inserting two microelectrodes into one part of the intact pulvinus in air, Nishizaki (1987) recognized that: (1) when a microelectrode is inserted into the pulvinus in air and the electric potential difference is measured against the bathing solution of the petiole, this potential mostly represents an extracellular potential; (2) in an early report (Nishizaki 1986) this potential was erroneously designated as membrane potential; (3) the intracellular potential (electric potential difference between a microelectrode inserted into a motor cell and the reference electrode in the bathing solution of the petiole) is more negative than the extracellular one. It changed upon illumination by unilateral BL and subsequent darkening in a mirror-image relationship to the extracellular potential, i.e., BL-on depolarized the intracellular potential and hyperpolarized the extracellular potential. BL-off caused a reversion.

Analysis of the BL- and R-effects on the intracellular potential and on leaf movement resulted in the following findings and conclusions. The unilateral BL induces a positive phototropic response: the pulvinus curved always towards the direction of the incident light, i.e., upwards, upon illumination of flexor, and downwards, when the extensor was irradiated. These movements were always preceded by membrane depolarizations of the motor cells in both pulvinus parts, and the extent of depolarization was always larger in the cells of the irradiated side. The BL-induced depolarizations are caused by inhibition of the plasma membrane H^+-ATPase. However, membrane depolarization via inhibition of the H^+-pump by BL occurred only when the pump was active before illumination. The activity of the plasma membrane H^+-ATPase, and thus the state of hyperpolarization of the respective membranes, depends on the supply of energy from respiration and/or photophosphorylation. When the plasma membrane of the motor cell is not in the fully hyperpolarized state, the H^+-ATPase can be activated by directional R via a supply of ATP provided by photophosphorylation. Under anoxia, R, directed on the flexor, caused downward movement, which could be changed to upward movement by superposition of BL. Data in support of these conclusions are: (1) in the fully hyperpolarized state of the membrane directional BL (or BL pulses) induced the largest degree of (transient) depolarization, whereas R was without effect. The extent to which BL induced depolarization was proportional to the magnitude of the intracellular potential at the time at which the pulse was applied; (2) in the depolarized state of the membrane (induced by anoxia, NaN_3, or CCCP), BL did not affect the intracellular potential. R induced hyperpolarization under anoxia, but not, when depolarization was induced by NaN_3 or CCCP. (3) DCMU, an inhibitor of light-dependent electron transport, completely inhibited the hyperpolarization of the membrane by R.

These results show, that unilateral irradiation with WL or BL (after darkening the leaves during the photoperiod of a light/dark cycle) induces a positive phototropic movement reaction in the pulvini of *Phaseolus vulgaris* and *Desmodium*. In irradiated cells, light inhibits the plasma membrane H^+-ATPase, depolarizes the membrane and induces cell shrinking, presumably by causing an efflux of K^+, Cl^-, and water. In shaded cells, light also induces depolarization via inhibition of the plasma membrane H^+-ATPase. However, the pulvinar movement reaction indicates that these cells swell. Nishizaki (1992) explains this movement response with the quantitatively different effects in the irradiated and shaded motor cells. The larger depolarization in the irradiated cells should induce a larger turgor decrease than in the shaded cells. The resulting turgor difference should cause shrinking of the irradiated cells and swelling of the shaded cells (movement towards the light direction).

The photonastic response occurring after irradiation of pulvini with diffuse light in the dark period, or at the beginning of the light period in a light/dark cycle, is an opening movement which is mediated by extensor cell swelling and flexor cell shrinking. The light effects on the membrane potential in this photonastic leaf movement response were different from those found for the phototropic responses.

Protoplasts of extensor and flexor cells of the *Samanea* pulvinus respond with hyperpolarization upon WL and BL pulses between 4 to 8 h after beginning of the dark phase of a 16-h light/8-h dark cycle. The inhibitory effect of vanadate and DCCD indicates that this hyperpolarization is caused by the WL- and BL-induced activation of the plasma membrane H^+-ATPase (Kim et al. 1992, see above). These results are consistent with the finding that protoplasts of extensor cells of the *Phaseolus coccineus* pulvinus swell upon transition from dark to light in a 10-h light/14-h dark cycle (Mayer, unpubl.). Indirect evidence for the light-induced activation of the H^+-ATPase is provided by the fact that the apoplastic pH and the K^+ concentration decrease in the extensor region of the *Samanea* pulvinus after WL and BL irradiation (Lee and Satter 1989; Lowen and Satter 1989; see above). Thus, the causal steps in the photonastic response of the leaves are contrary to those in the phototropic response. In extensor cells most likely light-induced membrane hyperpolarization via activation of the plasma membrane H^+-ATPase causes opening of K_H channels, K^+ (and Cl^-) influx, and finally cell swelling. This chain of events is also known for the light-induced swelling reaction of guard cells (Kearns and Assmann 1993). In flexor cells, however, the sequence of steps leading to the photonastic reaction is not yet clear; in flexor *protoplasts* WL and BL induce transient hyperpolarization via an activation of the H^+-pump and the closing of K^+ influx channels (Kim et al. 1992), while an increase of the apoplastic pH was measured in the flexor apoplast of *intact pulvini* (Lee and Satter 1989).

3. Biochemistry of Leaf Movement

a) Basic Carbon Pathways

Availability of data on motor cell metabolism is still very limited, which is partly due to difficulties in the analytical distinction between motor and other cells. One approach to overcome these limitations is the application of methods of quantitative histochemistry (Hampp et al. 1990). This way it has been shown that, e.g., with regards to the occurrence of basic carbon pathways, motor cells appear to have some peculiarities. Protein-related activities of ribulose bisphosphate carboxylase/oxygenase (Rubisco; Calvin cycle) and of hydroxypyruvate reductase (HP-R; photorespiration pathway) were low in pulvinar tissues from *Phaseolus coccineus*, while those of 6-phosphogluconate dehydrogenase (6-PGDH; ox. pentose posphate pathway), fumarase (TCA cycle), and NAD-dependent triose phosphate dehydrogenase (NAD-GAPDH; glycolysis) exceeded laminar activities severalfold (Rieger and Hampp 1991).

Determination of the longitudinal and transversal compartmentation of the activities of fumarase, NAD-GAPDH, and 6-PGDH showed that the levels of activity of all three enzymes increased towards the middle part of the pulvinus. Here, expressed on a dry weight basis, the analysis of cross sections showed highest activities in the outer parts of the extensor.

As far as Rubisco and HP-R are concerned, their activities were low in comparison to laminar tissue when related to protein or dry weight. They were, however, comparable on a chlorophyll basis which is due to the low number of chloroplasts contained in pulvinus cells (Erath et al. 1988). In this respect, pulvinus cells differ from guard cells which, per chlorophyll, contain levels of activity of both enzymes that should not be of metabolic significance (Hampp et al. 1982; Outlaw 1989). The conclusion from this finding is that, although pulvinus motor cells contain only a few, but nevertheless oxygen-evolving chloroplasts (Erath et al. 1988), these should also be functional with regards to the reductive pentose-phosphate pathway. They could thus contribute to some extent to the net supply to organic counterions for the uptake of K^+ during active swelling. This conclusion is in agreement with results reported by Vanhinsberg and Horton (1990) who showed that pulvini of *Phaseolus vulgaris* have a limited ability to fix externally supplied carbon dioxide. In this investigation the activity of Rubisco was approx. ten times lower in the distal part of the pulvinus when compared with the lamina. Fixation rates are also limited by CO_2 availability as pulvini (*Phaseolus vulgaris*) lack stomates (Vanhinsberg and Horton 1990).

The considerable demand for metabolic energy in the motor organ may be reflected by the levels of apparent activity of enzymes which are involved in catabolic pathways such as glycolysis, oxidative pentose-phosphate pathway, or citrate cycle. Fumarase, GAPDH, and 6-PGDH, all showed increased levels of activity in cells directly involved in turgor-mediated movement, such as extensor and flexor, or which are suggested to constitute potassium reservoirs (flanks).

All levels of activity investigated exhibited no differences between day and night positions. This is in support of the notion that cells of motor tissues have generally evolved a high capacity for heterotrophic energy-converting processes, and is in accordance with the dependence of leaf movement on mitochondrial respiration (Mayer 1981; Saeedi and Roblin 1986; Fleurat-Lessard et al. 1988).

b) Metabolite Pools

The distribution of products and substrates of these pathways should somehow reflect metabolic compartmentation of pulvinus cells. Again, applying techniques of quantitative histochemistry, Rieger et al. (1992) analyzed the distribution of soluble carbohydrates, starch, and malate within pulvini of *Phaseolus coccineus* in comparison to other leaf tissues. In a series of cross sections along individual pulvini, only small differences in total carbohydrate levels were detected between day and night positions. In contrast, the level of malate changed during diurnal leaf movement, especially in the central part of the pulvinus. Glucose and fructose increased towards the transition zones between pulvinus and lamina, and pulvinus and petiole, and this trend was even more pronounced for starch. By contrast, sucrose levels were highest in the pulvinus proper. The transverse compartmentation of metabolites was studied in distinct, approximately 0.5-mm-thick tissue slices from the central part of the pulvinus. These were dissected further into up to 14 distinct subsamples, representing bundle, bundle sheath, motor tissues, and flanks. Irrespective of the position of the leaf (day or night), the central vascular core and the surrounding bundle sheath had high levels of sucrose and were low in glucose and fructose, while in the cortex the situation was reversed. In the night position, the level of sucrose decreased by approx. 30% in the bundle sheath and the central vascular core but not in the other sections. From this study it can be concluded that due to the relatively small diurnal changes in their cortical pools, soluble sugars are obviously not involved in the osmotic processes resulting in leaf movement.

In contrast to pulvini from older leaves (26-day-old plants), those of younger ones (14-day-old plants) exhibited an opposite diurnal change in pools of starch and malate in the outermost layer of the extensor. Here, starch increased at night while the malate pool was lowered nearly stoichiometrically. Inverse pool sizes were found in the day position of the respective leaves. Although less significant, the opposite diurnal variation occurred in samples taken from the flexor region. Owing to the tissue-specific differences, areas of different carbohydrate activities in the laminar pulvinus of *Phaseolus coccineus* could be located. The central vascular core, including the bundle sheath, is obviously involved in temporary storage of photoassimilates, while the cortical regions are responsible for osmotically driven leaf movement.

c) Carbohydrates and Leaf Movement

As pulvinus cells possess only limited abilities for photoassimilation (see above) import of photoassimilates is a prerequisite for their function. As there are no other reports in the recent literature, we rely here mainly on data obtained with *Phaseolus* pulvini. A study on organ-specific pools of carbohydrates during growth of *Phaseolus* plants showed that, until 80 days after sowing, leaves were the major contributors to the nonstructural carbohydrate content in the rachis and the pulvinus (Pena-Valdivia and Ortega-Delgado 1991). Sucrose, the major transport vehicle for assimilated carbon (Ziegler 1975; Giaquinta 1983), is in this context of prime interest. Studies on its transverse distribution in *Phaseolus* pulvini indicate its selective accumulation in the bundle sheath (Rieger et al. 1992). Motor tissues contained only 15 to 25% of the sucrose contained in the bundle sheath (Rieger et al. 1992). This largely rules out a direct role of sucrose in the osmotic processes responsible for leaf movement. The selective accumulation of sucrose in the bundle sheath, as well as diurnal fluctuations therein, provide strong evidence for its origin from adjacent sieve tubes. This assumption is supported by tracer studies. $^{14}CO_2$ fixation by tips of leaflets of *Mimosa pudica* resulted in an accumulation in tertiary pulvini of labeled photoassimilates after stimulation. Microautoradiography of unstimulated tertiary pulvini showed labeled assimilates only in the phloem. In the stimulated state, label occurred also in the xylem and apoplast of collenchymatous tissue and beyond, into the extensor (Fromm and Eschrich 1990). Stimulation of seismonastic movement obviously triggers unloading of sugars out of the sieve tubes, creating an increased osmotic potential in the apoplast, and thereby an increased turgor loss of the extensor cells. During recovery (turgor increase of motor cells), label migrated into the outer extensor cells (Fromm and Eschrich 1990).

Uptake of photoassimilates could be via H^+-sucrose cotransport, as suggested for *Samanea saman* leaf pulvini (Satter et al. 1988). The increased levels of sucrose in bundle-sheath cells of *Phaseolus coccineus* are in accordance with the assumption that the vascular system also takes part in the temporary storage of photosynthate, thereby moderating the effect of changing activities of either source or sink tissues (Wardlaw 1990). Sucrose is obviously also in part converted to starch within the bundle sheath, and, after transportation, within extensor cells, because starch levels are temporarily increased in both cell types (Rieger et al. 1992). Tissue-specific pools of glucose and fructose, reported by these authors, indicate that pathways of sucrose hydrolysis could be different in bundle-sheath and cortical cells. The former exhibited a glucose/fructose ratio of close to unity which is indicative for sucrose breakdown by (apoplastic) invertase. This assumption is supported by a report which shows that phloem unloading of sucrose in the stem of *Phaseolus vulgaris* operates by passive leakage (Minchin and Thorpe 1987). Cortical cells, by contrast, exhibited glucose/fructose ratios <1 which could be the result of sucrose breakdown by cytosolic sucrose synthase, yielding UDP-glucose and fructose.

With regards to turgor changes of motor cells, a comparison of pools of malate and starch are of interest. In pulvinus cells of *Phaseolus*, levels of malate were significantly higher when leaves were in the day position, while the opposite was found for starch (Rieger et al. 1992). As the changes were nearly stoichiometrical they could well be linked to turgor changes within motor cells as reported for, e.g., the elongation zone of growing barley leaves (Hampp et al. 1987). Thus, starch could act as a buffer for diurnal malate turnover. Malate formed in this way could compensate in part for the positive charges resulting from potassium uptake during swelling. This implies that pulvinus starch is not a reservoir for energy supply of movement.

Instead, hydrolysis of sucrose in the cortex (ratio of glucose/fructose $= 1$) should yield immediate substrates for glycolysis. Thus, energy demands for leaf movement should be covered by sucrose supply, and there are reports in favor of this assumption. Satter et al. (1976) found, e.g., that in leaf pulvini of *Albizzia julibrissin* rhythmic movement occurred during a long dark period without damping as long as sucrose was available. Sucrose was even able to re-establish a damped movement of *Albizzia* leaf pulvini which were kept in darkness. Supply of carbohydrates by laminar photosynthesis as well as increased consumption during movement were also demonstrated by analyzing glucose pools in lower halves of secondary *Phaseolus* pulvini (Degli Agosti et al. 1990). In this study interruption of illumination of the corresponding leaves led to a decay of the glucose content which depended on the duration of dark treatment. The decay was slow in the steady state, but rapid during movement.

The decrease in sucrose pools from the bundle sheath towards the cortex, as reported by Rieger et al. (1992), is paralleled by increasing levels of extractable activity of enzymes such as fumarase or NAD-GAPDH which represent catabolic pathways (Rieger and Hampp 1991). This is in agreement with the role of sucrose supplying substrate for respiration.

d) Energy Metabolism

Ion fluxes associated with leaf movement require metabolic energy. The energy status of a given tissue can be described by terms such as energy charge, phosphorylation potential, and ATP/ADP ratio. In *Samanea saman* highest rates of oxygen consumption were measured during the opening phase of leaflets (Satter et al. 1979). Determination of adenylates in the pulvinus of *Mimosa pudica* revealed very high levels of ATP compared to other tissues (Fromm and Eschrich 1988). The ATP/ADP ratio was highest in unstimulated secondary pulvini (> 9) and dropped to 1.4 upon stimulation (Fromm and Eschrich 1990).

Again with *Phaseolus coccineus*, microdissection of freeze-dried pulvini was used to assay the amounts of adenine nucleotides contained in cell clusters (about five to ten cells) which were obtained from different areas of a pulvinus cross section (Hampp et al. 1990). The total amount (ATP+ADP+AMP) was highest in the flexor and flank regions, with values of up to 3.25 mmol/kg dry weight. Dry weight-related data depend, however, largely on the cell wall thickness which can vary considerably among the different functional cells of a pulvinus. Physiologically more significant is therefore the ratio of ATP/ADP which is a measure of the energy status of a cell and which is independent of a given reference. Highest ATP/ADP ratios were detected in flexor and adjacent flank cells as well as in the flank/extensor interspace (Hampp et al. 1990). This can be taken as an indication for an increased catabolic turnover in these regions, most possibly due to high rates of ion (K^+) transport between apoplast and symplast of flexor, extensor, and flanks. The tissue-specific distribution of enzyme activities and metabolite pools described above supports this view.

4. Phototransduction

Light affects leaf movement in a complex manner. Several types of responses are described. Light entrains leaf movement to various light/dark cycles, or changes the period of the circadian movements by acting on the circadian oscillator which controls leaf movement. Light directly affects swelling and shrinking of motor cells, and thus leaf movement (phototropic, photonastic, scotonastic reactions).

The responses, however, occur only at permitted phases of the circadian clock. Furthermore, the same light signal induces opposite reactions in the pulvinus, namely, motor cell swelling and motor cell shrinking in opposing pulvinus parts. Taken into account the complexity of light effects on leaf movement, several phototransduction pathways can be expected. To date, only the light-accelerated phosphatidylinositol turnover has been proposed as a transduction pathway for light signals in pulvini (Crain 1990). However, cAMP could also be involved in phototransduction (Roblin et al. 1990; Bourboulux et al. 1992).

a) Photoreceptors

Phytochrome is clearly implicated in the scotonastic folding response (Koller 1990; Satter and Morse 1990; Roblin et al. 1989) as well as in the phase shifting response of the circadian clock (Satter and Galston 1981). BL and FR, absorbed by the unidentified BL-absorbing pigment, are effective in the photonastic unfolding response occuring after dark/light transition (Satter and Galston 1981; Roblin et al. 1989; Koller 1990; Bialczyk and Lechowski 1990a). BL, absorbed presumably by the same pigment which is involved in the photonastic unfolding reaction, is effective in both the phototropic responses (Koller 1990; Nishizaki 1992) and in phase-shifting reactions of the circadian clock (Satter and Galston 1981). R, absorbed presumably by chlorophyll, may also be involved in the phototropic reactions (Nishizaki 1992; see Sect. 2e).

Kim et al. (1992; 1993; see Sect. 2d) provide evidence that light of a specific spectral range influences the membrane transporters of flexor and extensor protoplasts of the *Samanea* pulvinus only at permitted phases of the circadian clock. In flexor protoplasts BL, presumably absorbed by the BL-absorbing pigment which transmits the photonastic reaction in intact leaflets, activates the H^+-ATPase at the end of the dark period (thus causing hyperpolarization of the protoplasts and closing of the inward K^+ channels), whereas at the end of the light period, R that was absorbed by phytochrome and followed by dark, opens the inward K^+ channels. In the extensor protoplasts, photoexcitation of the BL-absorbing pigment brings about H^+-ATPase activation and opening of inward K^+ channels at the end of the dark period. However, at the end of the light period, extensor protoplasts are not responsive to R, and closing of inward K^+ channels occurs upon darkening (removal of BL), which indicates that the K^+ channels are regulated by the blue light photoreceptor rather than by phytochrome.

The effects of light quality of 6-h irradiations at the beginning of a 12-h light/12-h dark cycle on malate synthesis and ion uptake of motor cells of 14-day-old *Phaseolus* plants differ from those of the *Samanea* protoplasts. In the extensor tissue obviously both the specific BL-absorbing pigment and phytochrome, in the flexor phytochrome only perceives the effective light. This is deduced from the action spectrum, the effectiveness of the nonspecific quencher of triplet excited flavins, phenylacetic acid, and from R/FR reversion experiments (Bialczyk and Lechowsky 1990a).

b) Light-Stimulated Phosphatidylinositol Turnover

In whole *Samanea* pulvini, a brief (5–30 s) exposure to WL causes an increase in the concentration of inositol phosphates (inositol-1,4-bisphosphate, IP_2; inositol-1,4,5-trisphosphate, IP_3) and diacylglycerol (DAG). This is accompanied by a decrease in phosphatidylinositol-4-phosphate and phosphatidylinositol-4,5-bisphosphate (Morse et al. 1987a,b; 1989a,b; 1990; Coté et al. 1989; Crain 1990), and is consistent with a stimulation of the phosphatidylinositol (PI) cycle, known to act as signal transduction pathway in animals. However, light was given to whole pulvini and at the end of the dark period when BL promotes leaflet opening (extensor cell swelling and flexor cell shrinking) and R phase shifts the movement rhythm, so that the results do not indicate whether the response was specific for extensor cell swelling or flexor cell shrinking, or to BL or R. Nevertheless, it is assumed (Crain 1990; Coté and Crain 1993) that further steps in the stimulus/response coupling include an IP_3-stimulated increase in cytosolic free Ca^{2+} which is accompanied by the closure of inward K^+ channels, induction of a depolarizing current through anion permeable channels, and the opening of K_D channels. In addition, the DAG-stimulated increase of the activity of protein kinases should modulate, by phosphorylation, the activity of ion channels and/or the activity of the H^+-ATPase. These assumptions are based on the paradigma of the PI cycle in animals, on data demonstrating phosphoinositide-mediated signal transduction in guard cells (Kearns and Assmann 1993; Coté and Crain 1993), and on the observations that Ca^{2+}-agonists and -antagonists strongly influence leaf movement.

c) Ca^{2+}

For pulvinar cells, there is no direct evidence in support of the hypothesis that upon illumination the concentration of free Ca^{2+} increases in the cytosol, due to IP_3-mediated Ca^{2+} release from internal stores. There is, however, evidence that Ca^{2+} plays a role in light-induced leaf movements.

In *Albizzia lophanta* external Ca^{2+} and the Ca^{2+} ionophore A23187 mimic and increase the R-induced (phytochrome-mediated) scotonastic folding movement, while the Ca^{2+} chelator EGTA and the intracellular Ca^{2+} antagonist TMB-8 inhibit this movement (Moysset and Simon 1989, 1990).

In *Cassia fasciculata*, Roblin et al. (1989; 1990) found indications that external Ca^{2+} and the Ca^{2+} transport through Ca^{2+} channels in the plasma membrane of motor cells are involved in phytochrome-mediated scotonastic folding, whereas mobilization of Ca^{2+} from internal stores could be important for the (blue pigment-mediated) photonastic unfolding. Inhibition by the Ca^{2+} channel blockers verapamil and nifidipine of scotonastic folding movement, but not of the photonastic unfolding, supports this concept. Accordingly, the Ca^{2+} channel activator Bay K 8644 increased the rate of the scotonastic folding movement, but had no effect on the photonastic unfolding. Inhibition of calmodulin action by trifluorperazine, R 2451, and W-7, and increasing the intracellular cAMP concentration via application of db-cAMP, prostaglandins, or forscolin (adenylate cyclase activators), or ICI 58301 (phosphodiesterase inhibitor) promoted both the opening and closing movements. In contrast, a treatment that should decrease the cAMP concentration (by activating the phosphodiesterase by imidazole) inhibited both movements. These data were taken as evidence for Ca^{2+}-calmodulin acting on ion transport in pulvinar cells via cAMP metabolism.

In conflict with the above hypothesis is the inhibition of both the light- and dark-induced movements (Roblin et al. 1991) by the phosphodiesterase inhibitors theophylline, aminophylline, and caffeine. The partial reversibility of the theophylline effect by $CaCl_2$ indicates that methylxanthines affect leaf movements not by changing the intracellular cAMP, but by affecting the intracellular Ca^{2+} concentration. This has been proposed for the phase-shifting and period-changing effects of theophylline and caffeine on the circadian leaf movement rhythms of *Trifolium* and *Phaseolus* (Mayer and Scherer 1975; Mayer et al. 1975; Bollig et al. 1978).

An involvement of Ca^{2+} in the IAA-induced leaflet unfolding was shown for *Cassia fasciculata*. The phytohormones IAA, GA_3 and 6-BAP inhibited rate and amplitude of the scotonastic movements and promoted that of the photonastic opening movements, probably by exerting an effect on ionic fluxes in pulvinar cells (Everat-Bourbouloux et al. 1990). IAA alone induced unfolding of the leaflets in the middle of the dark period. The inhibition of this movement by EGTA, $LaCl_3$, and by the intracellular Ca^{2+} antagonist TMB-8, and the enhancement by the Ca^{2+} ionophore A23187 indicates a functional involvement of Ca^{2+} in the IAA-induced opening movement. Because this movement was not affected by the Ca^{2+} channel blockers verapamil and nifedipine, but hindered by LiCl (which is known to block the phosphatidylinositol cycle), Bourbouloux et al. (1992) concluded that the IAA-effect is mediated through the phosphatidylinositol cycle, and the increase of the concentration of cytosolic Ca^{2+} through depletion of intracellular Ca^{2+} stores. In addition, Ca^{2+}-calmodulin is possibly involved in this reaction, since inhibitors of calmodulin (compound 48/80, W-7) inhibit IAA-induced opening.

Interpretation of effects of various substances on photonastic or scotonastic movements of whole pulvini are difficult as in both movements not only cell swelling but also cell shrinking is involved. Thus it is still open whether the substances affect the movements via their effects on cell swelling or/and on cell shrinking.

Data from excised extensor and flexor tissues from *Phaseolus coccineus* pulvini (Bialczyk and Lechowsky 1992) provide evidence that changes in the cytosolic Ca^{2+} concentration, achieved via transport of Ca^{2+} through Ca^{2+} channels in the plasma membrane, are an important step in the transduction of light signals to the inward K^+ channels. In the extensor tissue, the BL- and FR-induced (BL-absorbing pigment-mediated) influx of Ca^{2+} and K^+ (which is accompanied by acidification of the external medium), was inhibited by the Ca^{2+} channel blockers verapamil and nifedipine. Taking into account the BL effects on protoplasts from extensor cells of *Samanea* (H^+-pump activation, hyperpolarization, opening of inward K^+ channels which occurs also when hyperpolarization is inhibited by vanadate; Kim et al. 1992) these results support the conclusion that an increase of the cytosolic Ca^{2+} concentration by Ca^{2+} uptake is necessary for the opening of the inward K^+ channels in the early phases of the photoperiod. In the flexor tissue, the R-induced (phytochrome-mediated) influx of Ca^{2+} and K^+ (accompanied by acidification of the external medium), was also inhibited by the two Ca^{2+} channel inhibitors.

5. Leaf Movements as Hands of Biological Clocks

Movements of pulvinated leaves are under the control of endogenous oscillators, i.e. under constant conditions they move with circadian (period length about 24 h) and/or with ultradian periods (period lengths from several minutes to several hours). Often the circadian movements are superimposed by ultradian movements as described for example for *Glycine, Phaseolus*, and *Oxalis* (Guillaume and Koukkari 1987; Millet et al. 1988; Johnsson and Skrove 1988; Starrach and Mayer 1989). For *Oxalis* it was shown that ultradian rhythms often occur after the circadian movements have faded out. Low irradiance or lowering the coupling between pulvini by physical cuts along the petiole increase the occurrence of ultradian movements (Pedersen et al. 1993). In *Desmodium motorium*, the lateral leaflets oscillate with an ultradian (3 to 4 min), the terminal leaflets with a circadian period. When administered in an appropriate manner, external signals synchronize, shift the phase, or change the period of the endogenous oscillators and thus the leaf movements (Bünning 1973). However, due to the lack of knowledge about the molecular mechanisms of the circadian and ultradian oscillators, the signal transduction chains from the oscillators to the effectors of cell swelling or shrinking, as well as the signal transduction from the environment to the clock are largely unknown. If leaf movement is controlled by endogenous clocks, effects on the phase or period of the clock should be reflected in respective changes of the leaf movement, whereas effects on the movement process per se should not alter the phase or period of the oscillators (Bünning 1973; Johnsson et al. 1990).

a) Circadian Rhythms in Protoplasts of Pulvinar Motor Cells

There is now strong evidence that each motor cell of the pulvinus behaves as a circadian oscillator. In protoplasts of extensor and flexor cells of the *Samanea* pulvinus, the state of opening of inward K^+ channels changes in constant darkness in a circadian manner, and the rhythms are phase-shifted by 180° as expected from the movement of intact leaflets (Kim et al. 1993). The different phase-dependent light responses of these flexor and extensor protoplasts in a given light/dark cycle (see above) are evidence for different properties of the oscillators of flexor and extensor cells with respect to synchronizing or phase-shifting light signals.

The circadian volume changes of protoplasts from extensor, flexor, and flank cells of the *Phaseolus* pulvinus (Mayer and Fischer 1994) correspond well with the circadian rhythms of the opening state of the K^+ channels in the *Samanea* pulvinus. They further indicate that the activity of the plasma membrane H^+-ATPase also changes in a circadian manner, since this is a prerequisite for circadian ion fluxes. The period length of the protoplasts volume rhythm is similar to that of the pulvinar cells in situ and synchronisation by a 14-h light/10-h dark cycle of the volume rhythm of extensor protoplasts occurred as expected from the behaviour of the extensor cells in situ. Taking into account the results on the light effects on the protoplasts of the extensor cells of *Samanea,* it is likely that the light-dependent synchronization pathway of the volume rhythm of the extensor protoplasts is initiated by the light-on induced activation of the plasma membrane H^+-ATPase (causing hyperpolarization and opening of inward K^+ channels) and by the light-off induced closing of these channels at the end of the light period.

The results about the circadian rhythms in individual protoplasts from pulvinar motor cells show that, in a given phase of the circadian cycle, a pulvinus consists of populations of circadian oscillators with different properties with respect to light signals. Thus light-induced phase response curves of intact leaves do not reflect the response of one population of oscillators with similar properties (as in other circadian systems) but the overall response of at least two oscillator populations with differing properties.

b) Involvement of K^+ Channels and of Plasma Membrane H^+-ATPases in the Control and Generation of the Biological Clocks

The question of whether the H^+-ATPase and the K^+ channels of the plasma membrane of pulvinar motor cells are involved in the functioning of the biological clocks, or are only controlled by them (merely as hands of the clocks), has been studied for the ultradian movement of *Desmodium* and the circadian movement of *Phaseolus* and *Oxalis*.

Inhibition of the plasma membrane H^+-ATPase by vanadate, or by reducing the ATP supply through anoxia (leaflets in N_2 instead of air), lengthens the period of the ultradian leaflet movement rhythm of *Desmodium* and, finally, arrests the leaflet in the lowered position (Antkowiak et al. 1992; Weber et al. 1992). This indicates that the plasma membrane H^+-ATPase controls a physiological parameter involved in determining the period length of the ultradian clock, and that the H^+-pump is an essential component of the mechanism of this clock. The same conclusion holds for K^+ transport through K^+ channels, because the K^+ channel blocker TEA also lengthens the period of the leaflet movement rhythm (Weber et al. 1992). The slight period-shortening effects of other alkali ions and the strong period-lengthening caused by LiCl are not in accordance with the permeability sequence found for the K_D channels in the *Samanea* motor cell protoplasts. Thus, these ions affect the clock not by their interaction with these channels, but via other means.

In contrast to *Desmodium*, the period of the circadian leaf movement rhythm in *Phaseolus coccineus* is shortened when substances which should inhibit the plasma membrane H^+-ATPase directly, or by means of a decreased ATP supply are applied via the transpiration stream (Mayer et al. 1994). This holds also for the K^+ channel blocker TEA. Application of these inhibitors for a limited time (4-h pulses) induces phase shifts of the circadian leaf movement rhythm. The shape of the phase response curves towards such 4-h pulses of vanadate, erythrosin B, azide, cyanide, low temperature, and TEA are all similar. The pulses are less effective and produce no advances (or only small ones) during the physiological day position of the leaf, but induce largest advances in the physiological night position. This indicates that the H^+-ATPase and the K^+ channels have at least a key function in coupling external signals to the circadian clock. They could also represent essential components of the causal loop which generates circadian rhythmicity as proposed by the membrane models for the mechanism of the circadian clock (Engelmann and Schrempf 1980; Edmunds 1988; Rensing and Hardeland 1990).

In *Oxalis*, 4-h pulses of vanadate induce phase shifts, but no changes in the period length (Pedersen 1994).

c) Treatments Affecting Clock-Mediated Leaf Movements

Direct current pulses (below 100 μA) sent through the leaflet, pulvinus, and petiole of *Desmodium* induced phase shifts, period lengthening, and reduction of the amplitude, or stopped the ultradian movement rhythm for up to approx. 90 min. It is not yet clear whether these effects on the clock are brought about via disturbance of the membrane potential and subsequent changes of the ion transport of the pulvinar motor cells, or via affecting the electric coupling between these cells (Johnsson et al. 1993)

Pulses of electromagnetic radio-frequency radiation (homogeneous 27.12 MHz or non-homogeneous 27 MHz) induced temporary changes in amplitude, period, and phase of the ultradian leaf movement rhythm of *Desmodium* (Ellingsrud and Johnsson 1993). Again, these clock effects could be due to field effects on the membrane potential, or on the electric coupling of the motor cells.

Increasing the leaf weight by positioning different weights on the leaflet surface was shown to lengthen the period and reduce the amplitude of the ultradian leaflet movement rhythm of *Desmodium*. In *Oxalis*, loading had no effect on the period but lowered the amplitude of the circadian movement. This indicates different dependencies of the period of the ultradian and the circadian clocks on the water potential of the motor cells (Pedersen et al. 1990).

In *Oxalis*, Pedersen et al. (1992) found that spot irradiation of only one leaflet (4-h red light) induced phase shifts of the same direction and magnitude also in the shielded leaflets. This indicates strong coupling between the oscillator populations in the three pulvini.

6. Ecological Implementations of Phototropic Movements

As outlined in the introduction, a variety of plants can adjust the position of the leaf lamina in response to the position of the sun, and this type of movement is named "heliotropic".

A diaphototropic response (leaf laminae remain perpendicular to the incident light) maximizes light perception of the respective leaf. For example in *Lupinus*, the lamina tracks the sun within a period of 2h after sunrise until 2h before sunset (Wainwright 1977). A heliotropic response can be modulated by environmental factors such as water or nutrient availability, temperature, and irradiance. This results in an adjustment of the angle at which the leave perceives solar radiation (= paraheliotropic movement; for overviews see Koller 1986, Werker et al. 1991).

a) Water Availability

The heliotropic response increases the water use efficiency of photosynthesis. Its modulation by stress (paraheliotropic response) is of importance for plants not able to tolerate water stress (Koller 1990). For soybean plants it was shown that, as the water potential decreases, leaves move closer to parallel to the direct beam of the light, thus intercepting less direct beam radiation (Kao and Forseth 1991, 1993). The authors argue that without this adjustment, plants would suffer more from photodamage of the photosynthetic apparatus.

b) Nitrogen Availability

Many paraheliotropic N-fixing legumes are found at sites which are low in soil nitrogen (Prichard and Forseth 1988). Under such conditions, acquisition of N causes large carbon costs. Plants that are able to adapt/maximize their rates of photosynthesis in relation to carbon loss/cost of nitrogenase activity should thus be more competitive (Bloom et al. 1985; Kao and Forseth 1993). Noninoculated soybean plants grown under low soil N availabilities (lower growth rate, lower rates of light-saturated photosynthesis) showed increased paraheliotropism (= light avoidance) relative to plants grown under high soil N levels (Kao and Forseth 1991), i.e., low NO_3 plants had more vertical leaf angles and received lower solar irradiances at midday than high NO_3 plants (Kao and Forseth

1993). The authors termed this an enhanced photosynthetic carbon return on N-investment in photosynthesis. There was also a high degree of correspondence between gas exchange and heliotropic response to soil nitrogen in soybean inoculated with *Bradyrhizobium*.

c) Leaf Temperature

In *Phaseolus vulgaris* air temperature modified laminar orientation in light through its effect on tissue temperature in the pulvinus region, not that of the lamina or petiole (Fu and Ehleringer 1989). Temperature effects were not expressed under darkness. Movement resulted in leaf (lamina) temperatures which stayed close to the thermal optimum of photosynthesis. However, even when the pulvinus temperature was kept constant, leaflets progressively oriented away from a direct light beam in response to progressive photon flux density, i.e., changes in light intensity alone, can control paraheliotropic leaf movement in *Phaseolus vulgaris*. Photosynthesis played no role in perception of vectorial excitation in heliotropic movements (Fu and Ehleringer 1989).

d) Organismic Interaction

Mycorrhization has been shown to also exert some influence on heliotropic movements. Legumes with VAM associations displayed less paraheliotropism than did plants without (*Leucaena leucocephala,* Huang et al. 1985). This was credited to a possibly improved water status.

In summary, tropic leaf responses to vectorial light can be modified by pulvinus temperature, N and water availability. Reduced water availability modulates the tropic response such that, as the leaf water potential becomes more negative, light interception by the lamina decreases, too. Restricted N availability can increase this effect. This will help to cope with an increased susceptibility of the photosynthetic apparatus in some plants to environmentally induced damage under reduced N availability (Ogren 1988).

Acknowledgement. We are indebted to Prof. Engelmann for critical reading of the manuscript. As far as own results are presented, financial support by the Deutsche Forschungsgemeinschaft is gratefully acknowledged.

Abbreviations

ABA, abscisic acid; 6-BAP, benzylaminopurine; BL, blue light; cAMP, cyclic 3′,5′-adenosine monophosphate; CCCP, carbonylcyanide-m-chlorophenylhydrazone; DAG, diacylglycerol; db-cAMP; N^6, 2′-O-dibutyryladenosine 3′,5′-cyclic monophosphate; DCCD, dicyclohexylcarbodiimide; DCMU, dichlorphenyldimethylurea; DES, diethylammoniumchloride; EGTA, ethylene glycol-bis (amino ethylether) N,N,N′,N′-tetraacetic acid; FR, far-red light; GA_3, gibberellic acid; GAPDH, glyceraldehyde-phosphate dehydrogenase; HP-R, hydroxypyruvate reductase; IAA, indole-3-acetic acid; IP_2, inositol-1,4-bisphosphate; IP_3, inositol-1,4,5-trisphosphate; 6-PGDH, 6-phosphogluconate dehydrogenase; PI, phosphatidylinositol; R, red light; Rubisco, ribulose-1,5-bisphosphate carboxylase; TEA, tetraethylammoniumchloride; TMB-8, 3,4,5,-trimethoxybenzoic acid 8-(diethylamino)-octyl ester hydrochloride; VAM, vesicular-arbuscular mycorrhiza; W-7, N-(6-aminohexyl)-5-chloro-1-naphtalenesulfonamide; WFS, water free space; WL, white light.

References

Antkowiak B (1992) PhD Thesis. University of Tübingen, Tübingen. **Antkowiak B**, Engelmann W (1989) J Interdiscip Cycle Res 20:164–165. **Antkowiak B**, Mayer W-E, Engelmann W (1991) J Exp Bot 42:901–910. **Antkowiak B**, Engelmann W, Herbjørnsen R, Johnsson A (1992) Physiol Plant 86:551–558.

Bialczyk J, Lechowski Z (1986) Plant Cell Physiol 27:981–987. **Bialczyk J**, Lechowski Z (1987) New Phytol 105:469–475. **Bialczyk J**, Lechowski Z (1989) Biochem Physiol Pflanz 184:79–86. **Bialczyk J**, Lechowski Z (1990a) Plant Physiol Biochem 28:315–320. **Bialczyk J**, Lechowski Z (1990b) New Phytol 115:595–601. **Bialczyk J**, Lechowski Z (1992) Acta Soc Bot Pol 61:241–252. **Bloom AJ**, Chapin FS III, Mooney HA (1985) Annu Rev Ecol Syst 16:363–392. **Bollig I**, Mayer K, Mayer W-E, Engelmann W (1978) Planta 141:225–230. **Bourbouloux A**, Roblin G, Fleurat-Lessard P (1992) J Exp Bot 43:63–71. **Bünning E** (1973) The physiological clock. Circadian rhythms and biological chronometry. Springer, Berlin Heidelberg New York.

Coté GG, DePass AL, Quarmby LM, Tate BF, Morse MJ, Satter RL, Crain RC (1989) Plant Physiol 90:1422–1428. **Coté GG**, Crain RC (1993) Annu Rev Plant Physiol Plant Mol Biol 44:333–356. **Crain RC** (1990) In: Satter RL, Gorton HL, Vogelmann TC (eds) The pulvinus: motor organ for leaf movement. Vol 3. American Society of Plant Physiologists, Bethesda, pp 175–188.

Degli Agosti R, Naidet C, Millet B (1990) Botanica Helvetica 100:249–256.

Edmunds LN Jr (1988) Cellular and molecular bases of biological clocks: models and mechanisms for circadian timekeeping. Springer, Berlin Heidelberg New York. **Ellingsrud S**, Johnsson A (1993) Bioelectromagnetics 14:257–271. **Engelmann W**, Schrempf M (1980) Photochem Photobiol Rev 5:49–86. **Erath F**, Ruge WA, Mayer W-E, Hampp R (1988) Planta 173:447–452. **Everat-Bourbouloux A**, Fleurat-Lessard P, Roblin G (1990) J Exp Bot 224:315–324.

Fleurat-Lessard P, Bonmort J, Roblin G (1988) Plant Physiol (Suppl) 86:147. **Freudling C**, Mayer W-E, Gradmann D (1980) Plant Physiol 65:966–968. **Freudling C**, Starrach N, Flach D, Gradmann D, Mayer W-E (1988) Planta 175:193–203. **Fromm J** (1991) Physiol

Plant 83:529–533. **Fromm J**, Eschrich W (1988) Trees 2:18–24. **Fromm J**, Eschrich W (1990) In: Satter RL, Gorton HL, Vogelmann TC (eds) The pulvinus: motor organ for leaf movement. Vol 3. American Society of Plant Physiologists, Bethesda, pp 25–43. **Fu QA**, Ehleringer JR (1989) Plant Physiol 91:1162–1167.

Giaquinta RT (1983) Annu Rev Plant Physiol 34:347–387. **Gorton HL** (1990) In: Satter RL, Gorton HL, Vogelmann TC (eds) The pulvinus: motor organ for leaf movement. Vol 3. American Society of Plant Physiologists, Bethesda, pp 223–237. **Guillaume FM**, Koukkari WL (1987) In: Pauly JE, Scheving LE (eds) Advances in chronobiology, Part A. Alan R Liss, New York, pp 47–57.

Hampp R, Outlaw WH Jr, Tarczynski MC (1982) Plant Physiol 70:1582–1585. **Hampp R**, Outlaw WH Jr, Ziegler (1987) Z Naturforsch 42c:1092–1096. **Hampp R**, Rieger A, Outlaw WH Jr (1990) In: Linskens HF, Jackson JF (eds) Physical methods in plant sciences. Springer, Berlin Heidelberg New York, pp 124–147. **Hensel W** (1987) Prog Bot 49:171–180. Iglesias A, Satter RL (1983) Plant Physiol 72:564–569. **Huang RS**, Smith WK, Yost RS (1985) New Phytol 99:229–243.

Johnsson A, Skrove D (1988) Plant Physiol Life Sci Adv 7:1–5. **Johnsson A**, Engelmann W, Antkowiak B (1990) In: Satter RL, Gorton HL, Vogelmann TC (eds) The pulvinus: motor organ for leaf movement. Vol 3. American Society of Plant Physiologists, Bethesda, pp 79–100. **Johnsson A**, Bostrøm A-C, Pedersen M (1993) J Interdiscip Cycle Res 24: 17–32.

Kao WY, Forseth IN (1991) Plant Cell Environ 14:287–293. **Kao WY**, Forseth IN (1993) Am J Bot 80:886–891. **Kearns EV**, Assmann M (1993) Plant Physiol 102:711–715. **Kim HJ**, Coté GG, Crain RC (1992) Plant Physiol 99:1532–1539. **Kim HJ**, Coté GG, Crain RC (1993) Science 260:960–962. **Koller D** (1986) Photochem Photobiol 44:819–826. **Koller D** (1990) Plant Cell Environ 13:615–632.

Lee Y (1990) In: Satter RL, Gorton HL, Vogelmann TC (eds) The pulvinus: motor organ for leaf movement. Vol 3. American Society of Plant Physiologists, Bethesda, pp 130–141. **Lee Y**, Satter RL (1987) Plant Physiol 83:856–862. **Lee Y**, Satter RL (1989) Planta 178:31–40. **Lowen CZ**, Satter RL (1989) Planta 179:421–427.

Mayer W-E (1981) Planta 152:292–301. **Mayer W-E** (1990) In: Satter RL, Gorton HL, Vogelmann TC (eds) The pulvinus: motor organ for leaf movement. Vol 3. American Society of Plant Physiologists, Bethesda, pp 160–174. **Mayer W-E**, Fischer C (1994) Chronobiol Int 11:156–164. **Mayer W-E**, Scherer I (1975) Z Naturforsch 30c:855–856. **Mayer W-E**, Gruner R, Strubel H (1975) Planta 125:141–148. **Mayer W-E**, Ruge WA, Starrach N, Hampp R (1987) Z Naturforsch 42c:553–558. **Mayer W-E**, Betz S, Schöffel S (1994) Biological Rhythm Res 25:301–314. **Millet B**, Botton A-M, Hayoum C, Koukkari WL (1988) Chronobiol Int 5:187–193. **Minchin PEH**, Thorpe MR (1987) J Exp Bot 38:211–220. **Moran N** (1990) In: Satter RL, Gorton HL, Vogelmann TC (eds) The pulvinus: motor organ for leaf movement. Vol 3. American Society of Plant Physiologists, Bethesda, pp 142–159. **Moran N** (1991) Biophys J 59:456a. **Moran N** (1993) Biophys J 64:387a. **Moran N**, Satter RL (1989) In: Dainty J et al. (eds) Plant Membrane Transport. Elsevier, Amsterdam. **Moran N**, Ehrenstein G, Iwasa K, Mischke C, Bare C, Satter RL (1988) Plant Physiol 88:643–648. **Moran N**, Fox D, Satter RL (1990) Plant Physiol 94:424–431. **Morse MJ**, Crain RC, Satter RL (1987a) Plant Physiol 83:640–644. **Morse MJ**, Crain RC, Satter RL (1987b) Proc Natl Acad Sci USA 84:7075–7078. **Morse MJ**, Crain RC, Coté GG, Satter RL (1989a) Plant Physiol 89:724–729. **Morse MJ**, Satter RL,

Crain RC, Coté GG (1989b) Physiol Plant 76:118–121. **Morse MJ**, Crain RC, Coté GG, Satter RL (1990) In: Morré DJ, Boss WF, Loewus F (eds) Plant inositide metabolism. Alan R Liss, New York, pp 201–215. **Moysset L**, Simon E (1989) Plant Physiol 90: 1108–1114. **Moysset L**, Simon E (1990) Plant Cell Physiol 31:187–193.

Nishizaki Y (1986) Plant Cell Physiol 27:155–162. **Nishizaki Y** (1987) Plant Cell Physiol 28:1163–1166. **Nishizaki Y** (1988) Plant Cell Physiol 29:1041–1046. **Nishizaki Y** (1989) In: Tazawa M et al. (eds) Plant water relations and growth under stress Yamada Science Foundation, Tokyo, pp 425–427. **Nishizaki Y** (1990) Plant Cell Physiol 31:591–596. **Nishizaki Y** (1992) Plant Cell Physiol 33:1073–1078.

Ogren E (1988) Photosynth Res 18:263–275. **Outlaw WH Jr** (1989) Physiol Plant 77:275–281.

Perdersen M (1994) Physiol Plant (in press). **Pedersen M**, Johnsson A, Herbjørnsen R (1990) Z Naturforsch 45c:859–862. **Pedersen M**, Eidesmo T, Johnsson A (1992) Physiol Plant 84:514–520. **Pedersen M**, Johnsson A, Mähle J, Dalløkken R (1993) Physiol Plant 89:277–284. **Pena-Valdivia CB**, Ortega-Delgado ML (1991) J Sci Food Agric 55:563–578. **Prichard JM**, Forseth IN (1988) Am J Bot 75:1201–1211.

Racusen R, Galston AW (1977) Planta 135:57–62. **Racusen R**, Satter RL (1975) Nature 255:408–410. **Rensing L**, Hardeland R (1990) Chronobiol Int 7:353–370. **Rieger A**, Hampp R (1991) Planta 184:415–421 **Rieger A**, Lutz A, Hampp R (1992) Planta 187:95–102. **Roblin G** (1990) In: Satter RL, Gorton HL, Vogelmann TC (eds) The pulvinus: motor organ for leaf movement. Vol 3. American Society of Plant Physiologists, Bethesda, pp 44–58. **Roblin G**, Fleurat-Lessard P, Bonmort J (1989) Plant Physiol 90:697–701. **Roblin G**, Fleurat-Lessard P, Everat-Bourbouloux A, Bonmort J, Moyen C (1990) Photochem Photobiol 52:197–202. **Roblin G**, Fleurat-Lessard P, Bonmort J (1991) J Plant Physiol 139:166–170.

Saeedi S, Roblin G (1986) Plant Physiol 82:270–273. **Satter RL**, Galston AW (1981) Annu Rev Plant Physiol 32:83–110. **Satter RL**, Morse MJ (1990) In: Satter RL, Gorton HL, Vogelmann TC (eds) The pulvinus: motor organ for leaf movement. Vol 3. American Society of Plant Physiologists, Bethesda, pp 10–24. **Satter RL**, Applewhite PB, Chaudri J, Galston AW (1976) Photochem Photobiol 23:107–112. **Satter RL**, Hatch AM, Gill MK (1979) Plant Physiol 64:379–381. **Satter RL**, Xu Y, DePass A (1987) Plant Physiol 85:850–855. **Satter RL**, Morse MJ, Lee Y, Crain RC, Coté GG, Moran N (1988) Bot Acta 101:205–213. **Satter RL**, Gorton HL, Vogelmann TC (eds) (1990) The pulvinus: motor organ for leaf movement. Vol 3. American Society of Plant Physiologists, Bethesda. **Scott BIH**, Gulline HF, Robinson GR (1977) Aust J Plant Physiol 4:193–206. **Sibaoka T** (1991) Bot Mag Tokyo 104:73–95. **Starrach N** (1988) PhD Thesis. University of Tübingen, Tübingen. **Starrach N**, Mayer W-E (1986) J Plant Physiol 126:213–222. **Starrach N**, Mayer W-E (1989) J Exp Bot 40:865–873. **Starrach N**, Freudling C, Mayer W-E, Gradmann D (1984) Planta 160:88–90. **Starrach N**, Flach D, Mayer W-E (1985) J Plant Physiol 120:441–455.

Vanhinsberg N, Horton RF (1990) Biochem Physiol Pflanz 186:37–42.

Wainwright CM (1977) Am J Bot 64:1032–1041. **Wardlaw IF** (1990) New Phytol 116:341–381. **Weber U**, Engelmann W, Mayer W-E (1992) Chronobiol Int 9:269–277. **Werker E**, Shak T, Koller D (1991) Bot Acta 104:144–156.

Ziegler H (1975) In: Zimmermann MH, Milburn JA (eds) Encyclopaedia of Plant Physiology, vol 1. Springer, Berlin Heidelberg New York, pp 59–100.

Prof. Dr. R. Hampp
Dr. W.-E. Mayer
Universität Tübingen
Physiologische Ökologie der Pflanzen
Auf der Morgenstelle 1
72076 Tübingen, Germany

C. Genetics

I. Replication: The Role of *Escherichia coli* Heat Shock Proteins in DNA Replication

By Ellen L. Zechner

1. Introduction

Our understanding of how bacterial cells replicate their DNA originates from intensive studies of the modes of DNA replication used by bacteriophages and plasmids. The patterns of DNA replication and their control observed in bacteria have in turn been used as models to guide the investigation of DNA synthesis in more complex eukaryotic systems. What has emerged over the years is a remarkably detailed picture; one, however, that is by no means complete. The study of DNA replication continues to receive significant attention. Studies devoted to the relatively simple bacteriophage and plasmid models continue to provide new insights that are relevant to our understanding of prokaryotes and eukaryotes alike. The topic of this chapter is an example of such a recent contribution. Members of a large group of proteins, common to prokaryotes and eukaryotes, called molecular chaperones, have been found to interact with replication proteins important to bacteriophage and plasmids such that the replication activity is regulated. The specifics of those interactions will be described, as well as the relevance of this finding to the broader theme of cellular DNA replication.

a) Molecular Chaperones and the Heat Shock Response

Molecular chaperones are found in a great diversity of organisms. Nearly all chaperones, whether from plants, prokaryotes, or mammals, belong to three highly conserved protein families, the Stress-70, Stress-60, and Stress-90 families. Members of these families are found in bacteria, and in mitochondria, chloroplasts, cytoplasm, and the lumen of the endoplasmic reticulum of eukaryotic cells. Molecular chaperones perform many functions in vivo due to their ability to recognize and modulate the state of folding of polypeptides in cells. A typical function of a molecular chaperone is to bind transiently to newly synthesized polypeptides, either to promote the correct folding of the nascent protein, or to stabilize the polypeptide in an unfolded conformation. The latter function is necessary when the nascent protein must be maintained in an unfolded state until it reaches a specific subcellular location, is translocated across a membrane, or until the

Progress in Botany, Vol. 56

protein can productively associate with additional polypeptides to form an oligomeric complex. Once the proper context is attained, folding of the polypeptide to its final conformation is promoted by the associated chaperone, which then dissociates. Some chaperones also interact with native protein molecules to modulate protein-protein interactions in oligomeric structures. It is important to emphasize that molecular chaperones do not remain as components of the functional assembled structures, hence the name "chaperones".

Many protein chaperones are essential to cell viability under conditions of normal growth and are produced constitutively and abundantly. In response to conditions adverse to the cell where proteins are easily denatured or misfolded, such as a sudden increase in temperature, the rate of synthesis of a subset of chaperones, the stress or heat shock proteins (hsps), undergoes a large and transient increase. In *E. coli*, about 20 proteins are synthesized at a higher rate when the cells are grown at elevated temperatures. After induction, the abundance of hsps can reach 20% of the total cell protein. Heat-inducible chaperones are thought to interact with unfolded conformations of polypeptides at high temperatures preventing undesirable side reactions such as aggregation and speeding refolding and assembly. The stress inducible chaperones exist, therefore, in addition to constitutively expressed chaperones to aid in protein repair when an excess need is present.

Excellent reviews describing molecular chaperones, stress-inducible proteins, and the physiology of the stress response have been published in recent years (Pelham 1986; Neidhardt and VanBogelen 1987; Lindquist and Craig 1988; Rothman 1989; Ellis and van der Vies 1991; Gething and Sambrook 1992).

b) Heat Shock Proteins Are Involved in DNA Replication

In recent years, the ability of *E. coli* chaperones to assemble and disassemble oligomeric polypeptides, multienzyme complexes, and macromolecular structures has been increasingly linked with the process of DNA replication. A role for the heat-inducible chaperones DnaK, DnaJ, and GrpE is now established for replication of some plasmid and bacteriophage genomes, and is suggested in the initiation stage of *E. coli* chromosomal replication.

The requirement for hsps in bacteriophage λ replication was first identified by the isolation of *E. coli* mutants, *gro*, that failed to support growth of λ. Some of these mutants were suppressed by alteration of λp, a gene required for DNA replication at the λ origin (*ori*λ). *gro* classes of this type include mutations in the *dnaB*, *dnaK*, *dnaJ*, and *grpE* genes. In concert with λP's known function at *ori*λ, suppression by λp mutations connected these gene products with the initiation stage of λ replication. Some years later came the surprising revelation that the products of *dnaJ*, *dnaK*, and *grpE* expression were direct participants in the stress response of *E. coli*. *dnaJ*, *dnaK*, and *grpE* are essential for normal growth. The *dnaK* and *dnaJ* genes form an operon with the order promoter-*dnaK*-*dnaJ*. Transcription of these genes, as well as other genes involved in the heat shock response, is positively regulated by *rpoH*. *rpoH* encodes a 32-kDa sigma factor (σ^{32})

which, when present in the holoenzyme, directs RNA polymerase to the promoters of heat shock genes.

This chapter focuses on the role of the *E. coli* hsps in (1) activating RepA and RepE, crucial replication initiation factors encoded, respectively, by the plasmid prophage P1 and the prototypic fertility factor, F, and (2) sequential protein assembly and disassembly at the bacteriophage λ origin (*ori*λ) prior to the initiation of DNA replication. Additionally, the role of the heat-inducible *rpoH* gene product, σ^{32}, in controlling replication of the F plasmid is described. Lastly, recent findings linking the heat-inducible chaperones to activation of the *E. coli* chromosomal initiator protein, DnaA, are presented. Readers are directed specifically to an additional article that addresses these topics (Zylicz 1993).

2. Regulation of the Initiation of Plasmid Replication

Plasmids are extrachromosomal, autonomously replicating, double-stranded circular DNA elements found in most bacteria. In order to replicate, the majority of plasmids utilize proteins provided by the host except for a specific plasmid-encoded initiator protein. Unlike the genomes of bacteriophage, plasmids are maintained at a fixed number of copies per host chromosome that is regulated carefully, chiefly at the level of the initiation of DNA replication. Control of plasmid copy number is achieved by two general mechanisms of negative regulation. Plasmids conforming to the first regulatory motif, the ColE1 family, the drug resistance plasmid R1, the IncB plasmids, and others have evolved elaborate mechanisms utilizing antisense RNA regulators to limit production of a component of the initiation reaction. The limiting factor can be either the initiator protein itself or the first primer where DNA synthesis begins. In a second family of replicons, direct repeats of the initiator protein-DNA binding site (iterons) are multiply reiterated not only within but also sometimes outside of the origin region, and then compete for the available initiator protein and thereby limit initiation frequency. As illustrated in Fig. 1, very similar or identical iterons are frequently located in the promoter region of the initiator gene. Binding of the initiator protein to these promoter/operator sites results in autorepression of initiator synthesis. The affinity of the initiator protein for the promoter/operator DNA is frequently higher than for the origin iterons. Thus the general effect of iteron-based regulation is a reduction in the frequency of initiation of DNA replication from the origin based on preferential binding of the initiator to the operator versus origin DNA.

The complete process of iteron-based regulation is significantly more complex. A comprehensive description of the interplay between iteron-regulated replicons and their respective initiators that controls replication frequency and copy number is beyond the scope of this chapter. Detailed discussion of this topic appears elsewhere (Trawick and Kline 1985; McEachern et al. 1986; Womble and Rownd 1987; Chattoraj et al. 1988; Pal and Chattoraj 1988; Nordström 1990; Kittell and Helinski 1991).

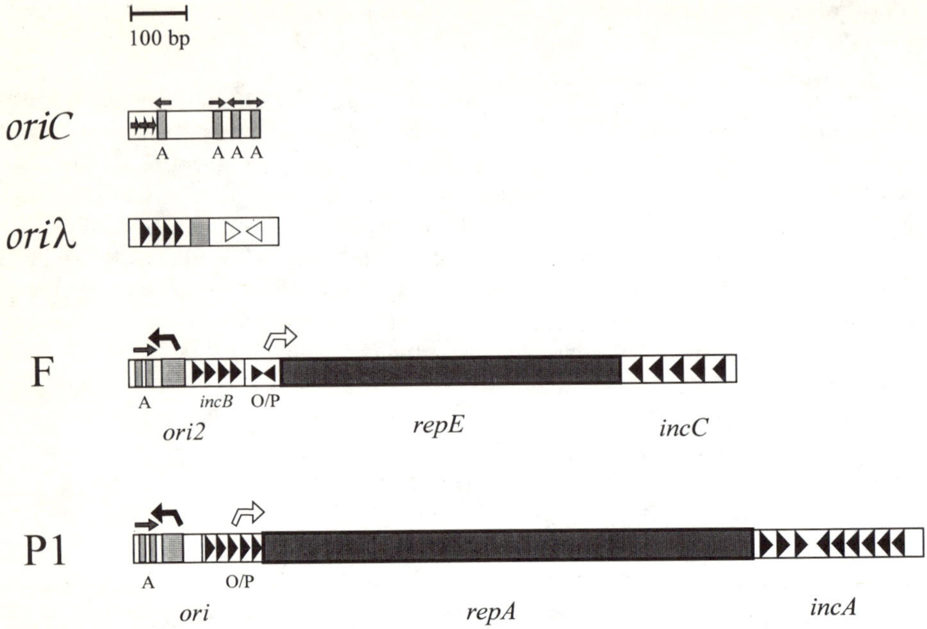

Fig. 1. Structural organization of the replication origins. *Small gray filled boxes* marked below with an *A* represent 9-bp sequences bound by DnaA. *Gray filled arrows above* indicate directionality of the DnaA repeat sequence. Location and directionality of repeat sequences bound by the initiator protein specific to each origin, λO in *oriλ*, RepE in F, and RepA in P1 are indicated by *black triangles*. O/P designates repeat sequences constituting the operator/promoter sites of the initiator genes for plasmids F and P1, which are depicted by the *large gray boxes*. *White arrows above* indicate the start and direction of transcription of the initiator genes. A+T rich region of the minimal *oriC* is located at the *extreme left*. *Stippled arrows* in this region designate three 13-bp repeat sequences. A+T-rich regions in the other origins are illustrated by a *stippled box*. *Black arrows above the A+T-rich regions* indicate the direction of unidirectional DNA replication for plasmids F and P1. Remaining features are described in the text

Examples of plasmids organized according to the iteron regulatory motif include pSC101, F, R6K, RK2, and the plasmid prophage P1. Initiation at an iteron-type origin normally involves both the action of the plasmid-specified initiator protein and the binding of the *E. coli* chromosomal initiator protein, the *dnaA* gene product, to its consensus binding site, which is also present in multiple copies in the origin region. Occupation of these sites by their respective initiator proteins results in a partial denaturation of the duplex DNA that is stabilized by the subsequent association of additional replication factors leading to the assembly of replication forks which then commence replication of the plasmid DNA (for detailed descriptions, see Kornberg and Baker 1992; Marians 1992).

3. Heat Shock Proteins Activate Replication of the Prophage P1 and F Plasmid

a) Bacteriophage P1

Bacteriophage P1, a temperate coliphage, exists as a plasmid in its lysogenic state. The copy number of the prophage (Prentki et al. 1977) and of miniplasmids derived from the prophage (Sternberg and Austin 1981, 1983) is stringently controlled at one to two per bacterial chromosome. Mini-P1 plasmids contain 1.5-kb of the 90-kb prophage which enables the miniplasmids to maintain the essential replication characteristics of the P1 replicon. This 1.5-kb region includes the gene *repA* encoding the 32-kDa initiator protein, a replication origin containing five 19-bp repeat sequences which bind the RepA protein, and a control locus containing nine additional RepA-binding iterons, *incA* (Abeles et al. 1984; Chattoraj et al. 1984; Austin et al. 1985) (Fig. 1).

P1 replication in vivo requires the host *dnaA* (Hansen and Yarmolinsky 1986), *dnaC*, *dnaB*, and *dnaG* genes (Scott and Vapnek 1980) encoding initiation factors, a DNA helicase, and a DNA primase, respectively, in addition to the plasmid *repA* (Abeles et al. 1984). Additional requirements for DNA gyrase, RNA polymerase, and DNA polymerase III have been demonstrated in vitro (Wickner and Chattoraj 1987; Wickner 1990).

Interestingly, mini-P1 and mini-F plasmids were found to be unstable in *dnaJ*, *dnaK*, and *grpE* mutant strains, implying hsp involvement in replication of these plasmids (Tilly and Yarmolinsky 1989; Bukau and Walker 1989). Experiments using purified proteins in an in vitro replication system then demonstrated a direct role for DnaJ, DnaK, and GrpE in P1 plasmid replication (Wickner 1990). Subsequent in vivo and in vitro experiments indicated that the binding of the RepA protein to the origin could be defective in the absence of any one of the three hsps and that this most likely contributed to the replication defect (Wickner 1990; Tilly et al. 1990). It was originally proposed that DnaK and DnaJ acted to dissociate dimeric RepA formed at high protein concentrations (Swack et al. 1987; Wickner et al. 1991a), which was inactive for DNA binding, to the monomeric form of RepA, which readily bound DNA (Wickner et al. 1991b).

DnaJ was shown to form a stable binary complex with RepA. The presence of DnaJ in the complex apparently facilitated recognition of the target protein, RepA, by DnaK. The activation of RepA for P1 DNA synthesis required ATP hydrolysis (Wickner et al. 1991b). The hsps themselves were not bound to the DNA nor were they required at later steps in ori-P1 DNA replication (Wickner et al. 1992; Sozhamannan and Chattoraj 1993). Subsequent experiments demonstrated an absolute requirement for the GrpE protein in this activation step in vitro as well (Skowyra and Wickner 1993).

The monomerization hypothesis was recently disputed, however, when the equilibrium constants for RepA-RepA and RepA-DNA interactions were determined (DasGupta et al. 1993). These values taken together with the calculated in vivo concentration of RepA protein indicated that RepA exists primarily in the monomeric form in vivo (> 80%). These, plus other recent data, indicate rather that the hsps activate the RepA fraction by renaturing a monomeric but misfolded form of the protein. Biochemical characterization of mutant RepA proteins, which are efficient initiators in the absence of the hsps, should help determine the actual mechanism of action (see Sozhamannan and Chattoraj 1993).

b) Plasmid F

Similarly detailed enzymological analysis of the role of DnaK, DnaJ, GrpE in the process of initiation of DNA replication at the mini-F origin is not yet available (Bukau and Walker 1989; Ezaki et al. 1989; Kawasaki et al. 1990). Mini-F plasmids, derived from F, exhibit all the essential characteristics of the original F replicon including stringent control at one to two copies per *E. coli* chromosome (Timmis et al. 1975). The minimal region required for replication of the plasmid and regulation of copy number (Fig. 1) includes the *repE* gene, encoding the 29-kDa initiator protein, an origin of replication (*ori2*) and the region responsible for negative regulation of copy number, *incC* (Kline 1985). *ori2* contains four direct repeats of the 19-bp RepE-binding sequence (Murotsu et al. 1981), an A+T rich region, and two binding sites for the chromosomal initiator protein, DnaA, which is also required for mini-F replication (Hansen and Yarmolinsky 1986; Murakami et al. 1987). As this structural arrangement is very similar to the P1 origin, the mechanism of activation of replication by the hsps is expected to be highly analogous. Binding of RepE to its iterons located in *ori2*, *incC*, and to the promoter/operator region of *repE* is essential to the initiation of DNA replication, control of copy number and repression of its own synthesis at the transcriptional level (Kline 1985). Although the mechanism of activation remains obscure, recent studies indicate that in vivo DnaK, DnaJ, and GrpE serve to activate RepE either before it binds the DNA or following formation of the nucleoprotein complex yet prior to the initiation of DNA synthesis (Ishiai et al. 1992).

Characterization of mini-F plasmid mutations that suppressed the replication defect in *dnaK*, *dnaJ*, *grpE*, or *rpoH* (see below) mutant strains revealed either promoter mutations in *repE* that resulted in increased expression of RepE or missense mutations resulting in RepE hyperactive for initiation and poorly active in autorepression (Kawasaki et al. 1991; Ishiai et al. 1992). Experiments carried out with purified components (Kawasaki et al. 1992) indicated that the hyperactive mutant RepE bound *ori2* DNA with a higher affinity than wild-type RepE. No clear role for DnaK or GrpE was observed in vitro despite their requirement in vivo. DnaJ, however, was shown to markedly enhance binding of the wild-type RepE to the operator but less markedly to *ori2*. Notably, it is also DnaJ that interacts directly with the initiator protein in the P1 system. Hsp-activated RepA bound origin iterons and operator iterons with comparable affinity, however. The differential effect of DnaJ on RepE binding to origin versus operator sites contributes to the hypothesis that the nature of the DNA-initiator protein complexes involved in initiator activities and autorepression are qualitatively different in mini-F but are identical in P1 (Sozhamannan and Chattoraj 1993).

An additional interesting facet to the regulation of F plasmid replication is that transcription of the *repE* gene is mediated primarily by the sigma factor (σ^{32}) (Wada et al. 1987) which also initiates the heat shock response (Gross et al. 1990). The requirement for the host *rpoH* gene, which encodes σ^{32}, for mini-F plasmid to replicate (Wada et al. 1986) is therefore manifest not only in RepE synthesis but also to elicit expression of *dnaJ*, *dnaK*, and *grpE*, the products of which in turn activate the RepE protein.

4. Initiation of DNA Replication at *ori*λ and *oriC*

The process of initiation at *ori*λ and at *oriC* (Fig. 1) has been studied extensively using purified components to reconstitute specific initiation in vitro (Ogawa et al. 1985; van der Ende et al. 1985; Funnell et al. 1986, 1987; Sekimizu et al. 1987, 1988; McMacken et al. 1987, 1988; Mensa-Wilmot et al. 1989; Zylicz et al. 1989). Basic aspects of the biochemical pathways involved in initiation in both systems are quite similar. In each system the process begins when multiple copies of the initiator protein, λO or DnaA bind to iterons within the cognate origin, *ori*λ or *oriC*, respectively, to form a nucleosome-like structure, i.e., DNA wrapped around the aggregate of initiator protein (Dodson et al. 1985; Fuller and Kornberg 1983; Tsurimoto and Matsubara 1981; Fuller et al. 1984; Echols 1990). This structure serves to direct the ordered assembly of additional replication factors at the origin resulting in localized unwinding of the adjacent A+T rich region of the duplex (Baker et al. 1986; Bramhill and Kornberg 1988a, 1988b; Dodson et al. 1986; Schnos et al. 1988). The enzymes involved in propagation of replication forks, which are the same for both *ori*λ and *oriC* templates, are then able to gain access to the DNA through this opening of the duplex.

The actual loading of the *E. coli* DnaB helicase to the origin occurs somewhat differently in each system in this second stage of the reaction. In *oriC* DnaB is delivered from a preformed complex of DnaB and DnaC (Wickner and Hurwitz 1975) to the DnaA-bound *oriC* to form a prepriming complex with helicase activity. Once DnaB is bound to *oriC*, DnaC is released from the complex (Kornberg and Baker 1992; Funnell et al. 1987). In the λ reaction the phage counterpart to DnaC, λP protein, displaces the bacterial protein from its complex with DnaB forming a second, more tightly associated complex with DnaB (Mallory et al. 1990). The affinity of λP for λO protein then directs the λP-DnaB complex to the initiator-bound *ori*λ to form a *ori*λ-O-P-DnaB nucleoprotein structure (McMacken et al. 1988; Dodson et al. 1985, 1986; Yamamoto et al. 1987). In contrast to the *oriC* system, however, the enzyme activities of DnaB in the λ prepriming complex are suppressed, apparently by its continuing association with λP protein. A third stage in the initiation reaction is necessary in the λ system, therefore, to activate the helicase activity resident in the prepriming complex. This stage requires the hsps DnaK, DnaJ and GrpE (Georgopoulos et al. 1990).

Investigation of the activation step in vitro (Zylicz et al. 1989; Alfano and McMacken 1989a, 1989b; Dodson et al. 1989; Liberek et al. 1988) indicated that DnaJ and DnaK bind the λ prepriming complex. DnaJ stably binds the prepriming intermediate by means of its interaction with λP, λO, and DnaB proteins. Association of DnaK with this nucleoprotein complex is apparently much more transient, as intermediate structures containing DnaK have not been isolated. The combined effect of DnaJ and DnaK in the presence of ATP is the partial disassembly of the complex whereby λP and DnaJ are largely removed. Separation of λP from DnaB activates the helicase activity such that localized unwinding of the DNA duplex is achieved preparing the template for DNA replication. Because DnaK is not a tightly held component of the complex, it may act to bind λP as it is released, preventing λP from reassociating with DnaB. The role of GrpE in the disassembly reaction is less well understood. GrpE may also function to sequester released λP protein (Zylicz et al. 1988). Presence of GrpE in the initiation reaction lowers by tenfold the amount of DnaK required to mediate disassembly (Zylicz et al. 1989).

Detailed analysis of the interactions occurring between these components suggest that GrpE helps specify DnaK's interaction with its target protein such that molecules of λP associated with DnaB and DnaJ can be discerned from the additional λP positioned elsewhere in the large, multiprotein structure. In this way, a lower concentration of DnaK can mediate the gradual removal of λP in a functionally relevant manner that is keyed by the presence of DnaJ (Osipiuk et al. 1993). Additionally, GrpE together with DnaJ, can stimulate the ATPase activity of DnaK 50-fold (Liberek et al. 1991a). DnaJ alone stimulates the rate of hydrolysis of DnaK-bound ATP, while GrpE alone increases the rate of release of bound ATP or ADP without affecting the rate of hydrolysis. Presumably then these proteins can act jointly to stimulate the release of λP from DnaK by enhancing the turnover of ATP.

A clearly defined role for the hsps in replication of the *E. coli* chromosome has not been established. The results of genetic analyses done over a range of temperatures are equivocal (Sakakibara 1988; Ohki and Smith 1989). Perhaps in vivo DnaK functions to promote the initiation by mediating nucleoprotein complex assembly at *oriC* in a manner similar to its function at *oriλ*. Alternatively, DnaK could simply be required to stabilize a replication factor that is otherwise unstable at 42 °C. In this capacity, DnaK is known to protect the *E. coli* RNA polymerase from heat inactivation in vitro (Skowyra et al. 1990). DnaK, as well as phospholipase A_2, acts in vitro to convert large aggregates of the DnaA protein, which are inert in replication, to active monomers of DnaA (Hwang et al. 1990). GrpE protein does not participate in the disaggregation reaction. DnaK has also been shown to activate some mutant forms of DnaA in vitro (Hwang and Kaguni 1991; Hupp and Kaguni 1993a, 1993b, 1993c). The mechanism of activation of the mutant DnaA proteins by DnaK is apparently distinct from a disaggregation process, as the GrpE protein is absolutely required in addition to DnaK (Hupp and Kaguni 1993b). ATP hydrolysis is also necessary and is presumably catalyzed by DnaK. The reaction is independent of DnaJ. Productive interaction between the hsps and the mutant DnaA was shown to occur prior to the onset of DNA replication and prior to the formation of a prepriming complex at *oriC*. It is conceivable that DnaA normally acts to facilitate the release of DnaC from its association with DnaB bound to the origin as a prerequisite to helicase function. In that case, a defect in that activity of a mutant DnaA may be compensated by the hsps affecting release of DnaC. Further study is required to reveal whether wild-type DnaA interacts with GrpE and DnaK in vivo at some point prior to the initiation of DNA replication to regulate the activity of DnaA in initiation.

5. Conclusions

In summary, the role of the *E. coli* head shock proteins (hsps) in DNA replication is to potentiate the activity of proteins essential to the initiation stage of DNA replication. In the bacteriophage P1 and in the F plasmid of *E. coli*, this activation step involves a direct interaction between the hsps and a single initiation protein, RepA or RepE. Modulation of RepA by the hsps occurs apparently in solution. RepE is affected either in solution prior to DNA binding or as part of a nucleoprotein complex at the origin. Activation of initiation of bacteriophage λ replication by the hsps requires the rearrangement of a multienzyme complex assembled at *ori*λ. In this system, the hsps shuffle the components of the preinitiation complex such that an inactivating contact between λP and DnaB proteins is disrupted. As a result, λP is either exported from the complex or sequestered by a continuing interaction with one or more hsps in a different position within the complex which is removed from DnaB.

In each of these systems, it is the DnaJ protein that first interacts with the target protein. In λ and in the P1 system, the presence of DnaJ apparently enables DnaK to recognize the complex. GrpE may act in an auxiliary capacity at this point to strengthen or modify in some way the interaction between DnaK and the target protein. At *ori*λ the presence of GrpE enables DnaK to associate preferentially with the complex of λP-DnaJ as opposed to free λP. DnaJ and GrpE also function jointly to stimulate the weak ATPase activity of DnaK. Hydrolysis of ATP (Liberek et al. 1991b; Banecki et al. 1992) and perhaps the presence of a Mg^{2+} cofactor (Skowyra and Wickner 1993) induce conformational changes in DnaK that most likely modify the conformation of the associated initiator protein or trigger rearrangement and partial disassembly of the multiprotein structure, resulting in activation of initiation and recycling of the hsps.

To date, only DnaJ has been shown to associate with RepE. The requirement for all three heat shock genes – *dnaJ*, *dnaK* and *grpE* – for mini-F replication in vivo, however, suggests that, through future biochemical analyses, evidence of complex formation between RepE and the other hsps will most likely emerge.

Lastly, we are left with initiation at *oriC*. Are the *E. coli* hsps involved in the initiation of DNA replication from the chromosomal origin? The reports of

Hupp and Kaguni (1993a–c), showing that the DnaK and GrpE proteins can make a mutated form of DnaA function in initiation at *oriC*, are intriguing. DnaK's auxiliary proteins, DnaJ and GrpE, were identified because they enable bacteriophage λ to grow. Additional factors may well be encoded by the bacterial chromosome that function in conjunction with DnaK to help regulate the activity of DnaA at the origin. Genetic screens aimed at identifying loci that improve *E. coli* growth may yet uncover such factors. Lastly, as new factors that contribute to the activation or inhibition of initiation of DNA replication from *oriC* continue to be discovered, (see, for example, Hwang and Kornberg 1990; Skarstad et al. 1993) a role to play for the *E. coli* hsps in the complex interplay of protein components assembling and disassembling at this critical locus may yet be found.

In closing, it is worthwhile to reiterate that molecular chaperones are highly conserved proteins universally present in prokaryotes, as well as eukaryotes. It is reasonable to expect, therefore, that functions performed by these proteins in bacteria will be highly analogous to their functions in fungi, plants, and animal cells as well.

Acknowledgement. The author would like to thank Drs. G. Högenauer, R. Zechner, and H. Bergler for helpful discussions and critical reading of the manuscript. E.L.Z. is supported by the Fonds zur Förderung der wissenschaftlichen Forschung.

References

Abeles A, Snyder K, Chattoraj D (1984) J Mol Biol 173:307–324. **Alfano C**, McMacken R (1989a) J Biol Chem 264:10699–10708. **Alfano C**, McMacken R (1989b) J Biol Chem 264:10709–10718. **Austin S**, Mural R, Chattoraj D et al. (1985) J Mol Biol 183:195–202.

Baker T, Sekimizu K, Funnell B et al. (1986) Cell 45:53–64. **Banecki B**, Zylicz M, Bertoli E et al. (1992) J Biol Chem 267:25051–25058. **Bramhill D**, Kornberg A (1988a) Cell 52:743–755. **Bramhill D**, Kornberg A (1988b) Cell 54:915–918. **Bukau B**, Walker G (1989) J Bacteriol 171:6030–6038.

Chattoraj D, Cordes K, Abeles A (1984) Proc Natl Acad Sci USA 81:6456–6460. **Chattoraj D**, Mason R, Wickner S (1988) Cell 52:551–557.

DasGupta S, Mukhopadhyay G, Papp P et al. (1993) J Mol Biol 232:23–34. **Dodson M**, Roberts J, McMacken R et al. (1985) Proc Natl Acad Sci USA 82:4678–4682. **Dodson M**, Echols H, Wickner S et al. (1986) Proc Natl Acad Sci USA 83:7638–7642. **Dodson M**, McMacken R, Echols H (1989) J Biol Chem 264:10719–10725.

Echols H (1990) J Biol Chem 265:14697–14700. **Ellis RJ**, van der Vies S (1991) Annu Rev Biochem 60:321–347. **Ezaki B**, Ogura T, Mori H et al. (1989) Mol Gen Genet 218:183–189.

Fuller R, Kornberg A (1983) Proc Natl Acad Sci USA 80:5817–5821. **Fuller R**, Funnell B, Kornberg A (1984) Cell 38:889–900. **Funnell B**, Baker T, Kornberg A (1986) J Biol Chem 261:5616–5624. **Funnell B**, Baker T, Kornberg A (1987) J Biol Chem 262:10327–10334.

Georgopoulos C, Ang D, Liberek K et al. (1990) In: Morimoto R, Tissieres A, Georgopoulos C (eds) Stress proteins in biology and medicine. Cold Spring Harbor, New York, pp 191–221. **Gething M**, Sambrook J (1992) Nature 355:33–45. **Gross C**, Straus D, Erickson J et al. (1990) In: Morimoto R, Tissieres A, Georgopoulos C (eds) Stress proteins in biology and medicine. Cold Spring Harbor, New York, pp 167–189.

Hansen E, Yarmolinsky M (1986) Proc Natl Acad Sci USA 83:4423–4427. **Hupp T**, Kaguni J (1993a) J Biol Chem 268:13128–13136. **Hupp T**, Kaguni J (1993b) J Biol Chem 268:13137–13142. **Hupp T**, Kaguni J (1993c) J Biol Chem 268:13143–13150. **Hwang D**, Kaguni J (1991) J Biol Chem 266:7537–7541. **Hwang D**, Kornberg A (1990) Cell 63:325–331. **Hwang D**, Crooke E, Kornberg A (1990) J Biol Chem 265:19244–19248.

Ishiai M, Wada C, Kawasaki Y et al. (1992) J Bacteriol 174:5597–5603.

Kawasaki Y, Wada C, Yura T (1990) Mol Gen Genet 220:277–282. **Kawasaki Y**, Wada C, Yura T (1991) J Bacteriol 173:1064–1072. **Kawasaki Y**, Wada C, Yura T (1992) J Biol Chem 267:11520–11524. **Kittell B**, Helinski D (1991) Proc Natl Acad Sci USA 88:1389–1393. **Kline B** (1985) Plasmid 14:1–16. **Kornberg A**, Baker T (1992) In: DNA replication, 2nd edn. Freeman, New York, 931 pp.

Liberek K, Georgopoulos C, Zylicz M (1988) Proc Natl Acad Sci USA 85:6632–6636. **Liberek K**, Marszalek J, Ang D et al. (1991a) Proc Natl Acad Sci USA 88:2874–2878. **Liberek K**, Skowyra D, Zylicz M et al. (1991b) J Biol Chem 266:14491–14496. **Lindquist S**, Craig E (1988) Annu Rev Genet 22:631–677.

Mallory J, Alfano C, McMacken R (1990) J Biol Chem 265:13297–13307. **Marians KJ** (1992) Annu Rev Biochem **61**:673–719. **McEachern M**, Bott M, Tooker P et al. (1986) Proc Natl Acad Sci USA 86:7942–7946. **McMacken R**, Alfano C, Gomes B et al. (1987) In: McMacken R, Kelly T (eds) Mechanisms of DNA replication and recombination, UCLA Symp Mol Cell Biol Ser, vol 47. Alan R Liss, New York, pp 227–246. **McMacken R**, Mensa-Wilmot K, Alfano C et al. (1988) In: Kelly T, Stillman B (eds) Cancer cells, vol 6. Cold Spring Harbor, New York, pp 25-34. **Mensa-Wilmot K**, Seaby R, Alfano C et al. (1989) J Biol Chem 264:2853–2861. **Morimoto R**, Tissieres A, Georgopoulos C (eds) Stress proteins in biology and medicine. Cold Spring Harbor, New York, 450 pp. **Murakami Y**, Ohmori H, Yura T et al. (1987) J Bacteriol 169:1724–1730. **Murotsu T**, Matsubara K, Sugisaki H et al. (1981) Gene 15:257–271.

Neidhardt F, VanBogelen R (1987) In: Neidhardt F (ed) *Escherichia coli and Salmonella typhimurium*: cellular and molecular biology. Am Soc Microbiol, Washington, DC, pp 1334–1345. **Nordström K** (1990) Cell 63:1121–1124.

Ogawa T, Baker T, van der Ende A et al. (1985) Proc Natl Acad Sci USA 82:3562–3566. **Ohki M**, Smith C (1989) Nucleic Acids Res 17:3479–3490. **Osipiuk J**, Georgopoulos C, Zylicz M (1993) J Biol Chem 268:4821–4827.

Pal S, Chattoraj D (1988) J Bacteriol 170:3554–3560. **Pelham H** (1986) Cell 46:959–961. **Prentki P**, Chandler M, Caro L (1977) Mol Gen Genet 152:71–76.

Rothman JE (1989) Cell 59:591–601.

Sakakibara Y (1988) J Bacteriol 170:972–979. **Schnos M**, Zahn K, Inman R et al. (1988) Cell 52:385–395. **Scott J**, Vapnek D (1980) In: Alberts B (ed) ICN-UCLA symposium on mechanistic studies of DNA replication and genetic recombination. Academic Press, New York, pp 335–345. **Sekimizu K**, Bramhill D, Kornberg A (1987) Cell 50:259–265. **Sekimizu K**, Bramhill D, Kornberg A (1988) J Biol Chem 263:7124–7130. **Skarstad K**, Thöny B, Hwang D et al. (1993) J Biol Chem 268:5365–5370. **Skowyra D**, Wickner S (1993) J Biol Chem 268:25296–25301. **Skowyra D**, Georgopoulos C, Zylicz M (1990) Cell 62:939–944. **Sozhamannan S**, Chattoraj D (1993) J Bacteriol 175:3546–3555. **Sternberg N**, Austin S (1981) Plasmid 5:20–31. **Sternberg N**, Austin S (1983) J Bacteriol 153:800–812. **Swack J**, Pal S, Mason R et al. (1987) J Bacteriol 169:3737–3742.

Tilly K, Yarmolinsky M (1989) J Bacteriol 171:6025–6029. **Tilly K**, Sozhamannan S, Yarmolinsky M (1990) New Biol 2:812–817. **Timmis K**, Cabello F, Cohen S (1975) Proc Natl Acad Sci USA 72:2242–2246. **Trawick J**, Kline B (1985) Plasmid 13:59–69. **Tsurimoto T**, Matsubara K (1981) Nucleic Acids Res 9: 1789–1799.

van der Ende A, Baker T, Ogawa T et al. (1985) Proc Natl Acad Sci USA 82:3954–3958.

Wada C, Akiyama Y, Ito K et al. (1986) Mol Gen Genet 203:208–213. **Wada C**, Imai M, Yura T (1987) Proc Natl Acad Sci USA 84:8849–8853. **Wickner S** (1990) Proc Natl Acad Sci USA 87:2690–2694. **Wickner S**, Chattoraj D (1987) Proc Natl Acad Sci USA 84:3668–3672. **Wickner S**, Hurwitz J (1975) Proc Natl Acad Sci USA 72:921–925. **Wickner S**, Hoskins J, McKenney K (1991a) Nature 350:165–167. **Wickner S**, Hoskins J, McKenney K (1991b) Proc Natl Acad Sci USA 88:7903–7907. **Wickner S**, Skowyra D, Hoskins J et al. (1992) Proc Natl Acad Sci USA 89:10345–10349. **Womble D**, Rownd R (1987) J Mol Biol 195:99–113.

Yamamoto T, McIntyre J, Sell S et al. (1987) J Biol Chem 262:7996–7999.

Zylicz M (1993) Philos Trans R Soc Lond Biol 339:271–278. **Zylicz M**, Ang D, Liberek K et al. (1988) Biochem Biophys Acta 951:344–350. **Zylicz M**, Ang D, Liberek K et al. (1989) EMBO J 8:1601–1608.

Dr. Ellen L. Zechner
Institut für Mikrobiologie
Karl-Franzens-Universität Graz
Universitätsplatz 2
A-8010 Graz, Austria

II. Recombination: Sexual Reproduction in Plants: Self-Incompatibility as a Mechanism Promoting Outbreeding and Gene Flow

By Simon J. Hiscock, Ursula Kües, and Ulf Stahl

1. Introduction

The evolution of meiosis introduced an alternation between haploid and diploid phases into the life cycle of living organisms. With meiosis came the need to reestablish the diploid condition through fusion of two genetically distinct gametes (syngamy) each containing half of the parental genome scrambled during the recombinational processes leading up to meiotic division. The fusion of nuclei (karyogamy), following syngamy, can generate a basis for variation which natural selection can act upon only if the gametes involved originate from different individuals (Maynard Smith 1978). It is not surprising, therefore, that the majority of organisms show adaptations which favor fusion of gametes from different genetic backgrounds and, consequently, outbreeding (Jain 1976; Charlesworth and Charlesworth 1987; Jarne and Charlesworth 1993).

Many organisms, especially animals and some plants, avoid selfing by having individuals of two different sexes (dioecy), each of which produces one distinct type of gamete (anisogamy). Dioecy is the most effective means of enforcing outbreeding since all chance of selfing is eliminated. For dioecy to be truly effective, however, there must be free mixing of the sexes and/or their gametes because only 50% of individuals are able to produce offspring (Maynard-Smith 1978). Sessile or semi-sessile anisogamous organisms with restricted potential for mixing therefore often tend to be hermaphrodite, with both male and female gamete-producing organs on the same individual (Jain 1976; Ghiselin 1969; 1974). The hermaphrodite habit is widespread in plants, being almost universal in flowering plants (Richards 1986), and has also been reported in 21 of the 32 animal phyla (Jarne and Charlesworth 1993).

Hermaphroditism, whilst ensuring that all individuals have the capacity to produce offspring, brings with it the potential problem of selfing. Hermaphrodite organisms have therefore evolved an array of ingenious mechanisms which avoid, or reduce selfing; these include: (1) the desynchronization of gamete production or release (dichogamy) (Jarne and Charlesworth 1993), (2) the positioning of male and female organs such that selfing is rendered virtually impossible (herkogamy) (Faegri and van der Pijl 1979), (3) self-incompatibility (de Nettancourt 1977; Sabbadin 1982, Grosberg 1988; Bull and Pease 1989) and in

Progress in Botany, Vol. 56

certain angiosperms, (4) symbiotic and "deceptive" relationships with pollinators (Bergström 1978; Faegri and van der Pijl 1979; Paulus and Gack 1990).

The selection forces which have led to the evolution of outbreeding are far from certain (Maynard Smith 1978; Charlesworth and Charlesworth 1979; Michod and Levine 1988; Wilmsen Thornhill 1993). Evidence from computer-assisted modeling of breeding system behavior (Lande and Schemske 1985; Schemske and Lande 1985; Charlesworth and Charlesworth 1987; 1990; Charlesworth et al. 1990) suggests that the major selective force promoting outcrossing is inbreeding depression, which results from the expression of deleterious and lethal recessive mutations in the increasingly homozygous progeny generated by selfing (Charlesworth and Charlesworth 1979; 1987). The negative effects of inbreeding depression are thus thought to be more important in maintaining outbreeding than the positive effects of genetic recombination such as the maintenance and proliferation of beneficial mutations (Peck 1993) and heterosis (Williams 1959; Wright 1977).

While strong inbreeding depression occurs in outcrossing species, after selfing, it is virtually absent in species which normally self-fertilize (Lande and Schemske 1985). Habital self-fertilization is fairly common in plants (Jain 1976; Schemske and Lande 1985) and in certain animals (Ghiselin 1969, 1974; Jarne and Charlesworth 1993) and has significant advantages to an individual: (1) avoidance of the "costs of outbreeding" (Grant 1981; Lande and Schemske 1985), (2) swift elimination ("purging") of deleterious recessive alleles (Holsinger 1988; Charlesworth and Charlesworth 1990) and (3) the preservation of well-adapted genotypes from recombination (Maynard-Smith 1978; Grant 1981).

On the basis of this argument, it might be expected that organisms will be either outbreeders or inbreeders and that mixed breeding systems will be rare. Certainly, in plants there is much evidence in favor of this prediction (Lande and Schemske 1985), but there is also increasing evidence for the widespread occurrence of mixed breeding systems (Allard et al. 1977; Brown 1990; Milton 1992). Even some habitual selfers show low levels of outbreeding (Briggs and Walters 1984; Richards 1986), suggesting that the long-term evolutionary advantages of outbreeding (Lewis 1979a) outweigh the short-term advantages of selfing (Jain 1976; Williams 1975).

This chapter will focus on self-incompatibility as an important mechanism promoting outbreeding. Self-incompatibility (SI) has been described in a number of hermaphrodite animals (such as certain Tunicates: Sabbadin 1982; Grosberg 1988, and Ciliates: Bull and Pease 1989), but has been studied most extensively in plants (de Nettancourt 1977; Lewis 1979a; Shivanna and Johri 1985). SI is by far the most widespread mode of outbreeding within the angiosperms (East 1940; Darlington and Mather 1949; Whitehouse 1950; Crowe 1964; Richards 1986). There is only limited evidence to suggest that SI also exists in pteridophytes (Wilkie 1956) and the gymnosperms (Piatnitsky 1934). In many isogamous algae (Hoekstra 1982; Bull and Pease 1989) and fungi, which show no sexual differentiation, self-incompatibility is once more the most important means of outbreeding (Lewis 1979a,b). Self-incompatibility in fungi, mediated through mating type factors, was reviewed in a previous volume of this series (Kües and Stahl 1992). Here we will concentrate on the SI systems found in angiosperms and refer to fungal systems only for comparison.

2. Self-Incompatibility Systems in Angiosperms

Self-incompatibility (SI) defined as the "inability of a plant producing functional gametes to set seed upon self-pollination" (Brewbaker 1957) is present in approximately half of all angiosperms (Darlington and Mather 1949; Brewbaker 1959; Richards 1986). SI is primarily a prefertilization barrier (Lewis 1979a), with pollen tubes being inhibited on or within the pistil.

From studies of tropical species there is increasing evidence for the existence of "late acting" forms of SI, with ovarian or ovular, or even post-zygotic inhibition mechanisms (Bawa et al. 1985; Seavey and Bawa 1986).

On the basis of the genetic control of the incompatibility phenotype of the pollen (see Table 1), angiosperm SI systems can generally be divided into two groups (de Nettancourt 1977). In "gametophytic" systems (GSI), the haploid pollen (gametophyte) determines its own incompatibility phenotype (East and Mangelsdorf 1925) (Fig. 1), whereas in "sporophytic" systems (SSI), the incompatibility phenotype of the pollen is determined by its parent plant (sporophyte) (Bateman 1952, 1954, 1955; Crowe 1954; Hughes and Babcock 1950) (Fig. 2). Species with heteromorphic SI where differences in floral morphology (notably the formation of pistils of two or three different lengths – di- and tristyly, respectively) are linked with the SI phenotype, all exhibit sporophytic control of SI (Lewis 1979a; Gibbs 1986; Barrett and Richards 1990) (Fig. 2B). Species which show no such morphological differences between incompatibility genotypes are often referred to as "homomorphic", i.e., all other species with SSI and species exhibiting GSI.

A number of morphological features, however, have been correlated with the possession of either homomorphic SSI or GSI (Table 1): Species exhibiting homomorphic SSI (i.e., Brassicaceae and Asteraceae) have tricellular pollen and "dry" cuticularized stigmas, whereas species with GSI usually have bicellular pollen and noncuticularized, "wet" stigmas with a distinct surface secretion, (Heslop-Harrison and Shivanna 1977). Also, as a general rule, in species with homomorphic SSI, inhibition of incompatible pollen occurs at the stigma surface, while in species with GSI (i.e., Solanaceae and Rosaceae) pollen tubes are inhibited within the style (Ockendon 1972; Heslop-Harrison 1975), although there are many exceptions (de Nettancout 1977; Shivanna and Johri 1985).

Table 1. Self-incompatibility systems in angiosperms. (After Lewis 1979a; Brewbaker 1957; Seavey and Bawa 1986)

SI system	Loci	Alleles	Sib/general compatibility response[a]	Type of pollen	Site of incompatibility	No. of families	Examples	Reference
Gametophytic	1	Many	0.25–0.5	Bicellular	Style	14–17	Solanaceae Rosaceae Leguminosae Scrophulariaceae	Charlesworth (1988)
	1	Many	0.25–0.5	Bicellular	Stigma	2	Onagraceae Papaveraceae[b]	Dickinson and Lawson (1975) Lawrence (1975)
	2	Many	0.72	Tricellular	Stigma	1	Gramineae	Lundquist (1954, 1955)
	4	Many	0.92	Bicelluar	Style	2	Ranunculaceae Chenopodiaceae	Lundquist et.al. (1973)
Sporphytic Homomorphic	1	Many	0.25–0.5	Tricellular	Stigma	6	Brassicaceae Asteraceae	Charlesworth (1988)
	3–4	Many	0.75–0.97	Tricellular	Stigma	1	Brassicaceae	Verma et al. (1977)

Heteromorphic Distyly	1	2	1.00	Bicellular/ Tricellular	Stigma/style	22	Primulaceae Linaceae Rubiaceae	Kohn and Barrett (1992)
Tristyly	2	2	0.84	Bicellular	Style	3	Lythraceae Oxalidaceae Pontederiacea	Barrett and Richards (1990)
Gametophytic/ Sporophytic	1	Many	0.25–0.5	Bicellular	Ovule[c]	1	Sterculiaceae	Cope (1962a,b)
Late-acting systems (may be gametophytic)	Many	?	up to 1.00?	?	Ovule	1	Boraginaceae	Crowe (1971) Seavey and Bawa (1986)
	?	?	?	Bicellular/ Tricellular	Ovary/ovule	14	Leguminosae Hamamelidaceae Winteraceae Saxifragaceae	Seavey and Bawa (1986)

[a] From Lewis (1979b) – the ratio of sibling compatibility to general compatibility in the population is a measure of the control of sib-matings (Sect. 4).
[b] No style.
[c] Hollow style.

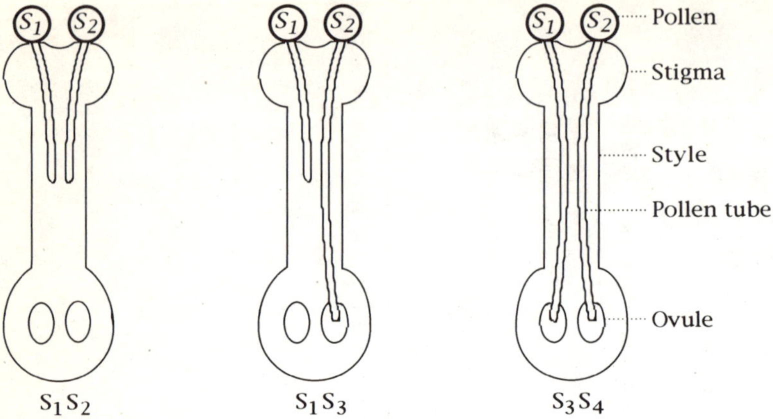

Fig. 1. Genetic control of gametophytic self-incompatibility. The self-incompatibility phenotype of the pollen is determined by its own haploid genome. When the *S*-allele in a pollen grain is matched with one of the two *S*-alleles in the pistil, the pollen is rendered incompatible. Pollen from an S_1S_2 parent is inhibited in its own style or that of another S_1S_2 individual; 50% of the pollen derived from an S_1S_3 plant is inhibited within the S_1S_2 style and all pollen from an S_3S_4 individual is compatible with the S_1S_2 style. S_1 and S_2 shown in *italics* in the pollen grain denotes the genotype of the pollen, $\mathrm{S_1S_2}$, $\mathrm{S_1S_3}$, and $\mathrm{S_3S_4}$ (in *plain letters*) underneath the pistil denotes its phenotype

3. The Genetic Basis of Self-Incompatibility

In the majority of homomorphic SSI and GSI systems, the SI response is controlled by a single polyallelic locus – the *S*-locus (de Nettancourt 1977). If *S*-alleles are matched in pollen and pistil the pollen is rendered incompatible (Figs. 1 and 2). Thus, in GSI *S*-alleles act in a codominant manner when expressed in female tissue, but in SSI where the pollen is in effect "diploid" for its incompatibility phenotype, dominance relationships between *S*-alleles do exist depending on whether they are expressed in male or female tissue (Thompson and Taylor 1966).

In species with homomorphic SSI, the two *S*-allele products which define the SI phenotype of the pollen are thought to be located in the pollen coating or tryphine which is added to the maturing pollen grain by the sporophytic tissue of the anther tapetum (Dickinson and Lewis 1973b; Heslop-Harrison et al. 1974; see Sect. 5b).

Some homomorphic SI systems are controlled by more than one locus. In the Gramineae, GSI is determined by two loci, *S* and *Z*, both of which are polyallelic (Lundquist 1954, 1955, 1956). In *Beta vulgaris* and *Ranunculus acris*, four loci control GSI (Lundquist et al. 1973; Østerbye 1975). Similarly, in *Eruca sativa* (Brassicaceae), SSI is controlled by three or four loci (Verma et al. 1977; Lewis 1977). Interestingly, it has been suggested that some homomorphic SI systems (*Theobroma cacao, Raphanus sativus,* and *Brassica campestris*) may be under dual sporophytic and gametophytic control (Cope 1962a,b; Lewis et al. 1988; Zuberi and Lewis 1988).

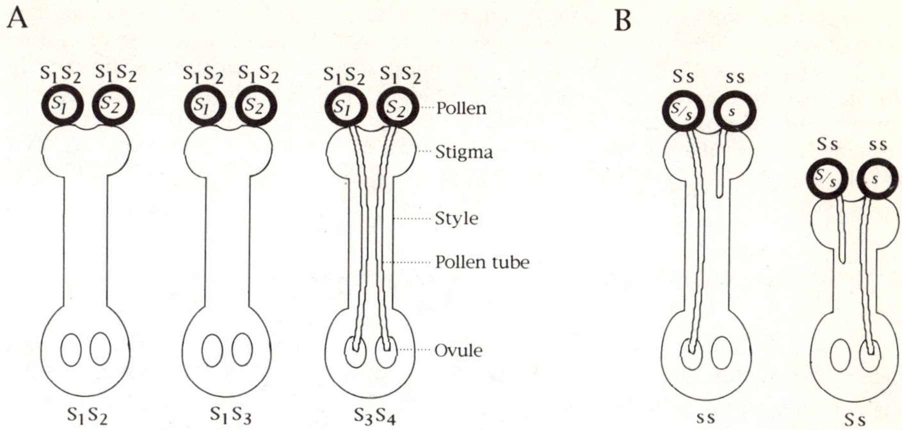

Fig. 2A,B. Genetic control of sporophytic self-incompatibility. The self-incompatibility phenotype of the pollen is determined by the diploid genome of its parent plant. **A** Homomorphic single-locus multiallelic sporophytic self-incompatibility (SSI). When either of the *S*-alleles of the pollen parent matches one of the two *S*-alleles present in the pistil, the pollen is rendered incompatible. Pollen from the S_1S_2 parent is inhibited on its own or another S_1S_2 stigma. All pollen derived from the S_1S_3 parent is functionally diploid for its SI phenotype and is therefore inhibited on the S_1S_2 stigma. Pollen derived from the S_3S_4 parent is compatible with the S_1S_2 stigma. S_1 and S_2 shown in *italics* in the pollen grain denote the genotype of the pollen, *bold circles* and $\mathrm{S_1S_2}$ (in *plain letters*) denote the pollen phenotype, and $\mathrm{S_1S_2}$, $\mathrm{S_1S_3}$, and $\mathrm{S_3S_4}$ (in *plain letters*) underneath denote the phenotype of the pistil. **B** Heteromorphic single-locus diallelic sporophytic self-incompatibility (SSI), (distyly). Only two self-incompatibility genotypes/phenotypes are possible (*Ss* and *ss*). Pollen from the short-styled (*Ss*) plant is compatible with the pistil of a long-styled (*ss*) plant, but incompatible with itself or another short-styled plant. The reciprocal is true for pollen from the long-styled (*ss*) plant. *S*/*s* and *s* shown in *italics* in the pollen grain denote the alternative or only possible genotype of the pollen, ss and SS (in *plain letters*) underneath denote the phenotype of the pistil

In heteromorphic sporophytic systems, the single *S*-locus controls not only the physiological incompatibility system but also many morphological features of the flower. For this reason the *S*-locus is often referred to as a supergene (Lewis 1979a; Barrett and Richards 1990). In distylous species, the *S*-locus has two alleles, *S* being dominant over *s*. An *Ss* genotype produces a short-styled sporophyte with Ss pollen, while an *ss* genotype produces a long-styled sporophyte with ss pollen (Fig. 2b). A similar genetic situation is found in tristylous species where short-, mid- and long-styled sporophytes exist. The tristylous condition is controlled by two diallelic loci – *S* and *M*; *M* being an epistatic modifier of *S* (de Nettancourt 1977, Lewis 1979a). The long-styled morph is a homozygous recessive (*mmss*), the mid-styled morph is *Mmss* and the short-styled morph is *mmSs* or *MmSs*.

The genetic basis of "late-acting" forms of SI is far from clear and it may be that most postzygotic forms of inhibition are simply a manifestation of inbreeding depression (Seavey and Bawa 1986). Genetic and cytological studies of such systems are rare. In *Theobroma cacao* (prefertilization ovular inhibition), dual sporophytic and gametophytic control of SI is postulated (Cope 1962a,b) while in *Borago officinalis* (thought to be postfertilization ovular inhibition), polygenic control of SI has been demonstrated (Crowe 1971).

4. Self-Incompatibility and Its Effect on Gene Flow

Gene flow, within and between populations, is an important force maintaining the genetic integration and homogeneity of a species (Mayr 1963) and counteracts the diversifying effects of genetic drift and weak selection (Devlin and Ellstrand 1990). As flowering plants are sessile, the extent of gene flow depends principally on the efficiency of seed and pollen dispersal (Grant 1981). Direct studies of the dispersal of seeds and pollen in natural and experimental populations have revealed their dispersal to follow a leptokurtic distribution – the majority of seeds and pollen falling just a few meters from the parent plant and only very few seeds and pollen grains achieving dispersal over long distances (Levin and Kerster 1974; Levin 1978). These findings led to the belief that gene flow in plant populations is low (Levin and Kerster 1974; Levin 1981,1984; Grant 1981; Hamrick 1982). This view is gradually being challenged as new methods for determining gene flow are applied to natural populations (Ellstrand 1992).

Indeed, studies in which gene flow is inferred from the distribution of genetic markers such as allozymes or RFLPs (Restriction Fragment Length Polymorphism) within populations have revealed higher levels of gene flow than was previously thought possible (Ellstrand and Marshall 1985; Ellstrand et al. 1989; Slatkin 1987; Devlin and Ellstrand 1990; Boshier et al. 1994), indicating the importance of long-distance dispersal of seeds and primarily of pollen in promoting gene flow (Campbell 1991; Phillipp et al. 1992; Boshier et al. 1994).

It is obvious that gene flow in species with adaptations for outbreeding should be more extensive than it is in selfing species (Grant 1981; Govindaraju 1988; Hamrick 1987; Ellstrand 1992). SI systems contribute to the promotion of gene flow by eliminating the possibility of selfing and by reducing the frequency of sib-matings (Bateman 1952).

Such systems are, however, only effective if floral adaptions are present which maximize pollen dispersal (Richards 1986); such adaptations include colorful showy corollas, attractive scent and nectar to attract insect or animal pollinators, or copious production of light pollen grains by anemophilous species (Faegri and van der Pijl 1979).

The leptokurtic pattern of seed and pollen dispersal in flowering plants means that often the nearest neighbors of any individual within a population will be its close relatives (parents or siblings). This leads to the occurrence of local inbreeding "neighborhoods" within many plant populations (Levin and Kerster 1974; Levin 1981; Brotschol et al. 1986; Moran and Hopper 1987; Lughadha and Parnell 1989), where most matings will be sib-matings. Sib-mating is not such a drastic form of inbreeding as selfing, but over longer periods can lead to the same effect – inbreeding depression (Lewis 1979a,b). Any mechanism which can reduce sib-matings will increase the efficiency of outbreeding in a population (Bateman 1952; Fisher 1963).

The relative effects of the various forms of SI on reducing sib-matings were assessed by Lewis (1979a,b) using ratios of sibling compatibility to general compatibility in a population (see Table 1). Distyly is ineffectual at reducing sib-matings and is, therefore, like dioecy, primarily an adaptation to avoid selfing. Tristyly gives a slight reduction in sib-mating (0.85% sib/general compatibility) but this advantage is offset by its wastage of pollen (within a population only 33.3% of the pollen can effect pollination).

In species with distyly 50% of siblings have long styles and 50% have short styles and their pollen is cross-compatible. The same situation is given in the general population and therefore sib/general compatibility must be 1. In species with tristyly, 25% have long styles, 25% have medium styles, and 50% have short styles. The percentage sibling compatibility will depend on the genotypes of the parents such that a *mmSs* × *mmss* (short × long) cross will give only short- and long-styled progeny (a two-group family) whereas a *Mmss* × *MmSs* (medium × short) cross will give short-, long-, and medium-styled progeny (a three-group family). Two group families have 25% compatible pollinations, whereas the three-group families have 33% compatible pollinations. There are seven possible two group families and five possible three-group families which gives a total sibling compatibility of 28.4%. When this sibling compatibility is compared with the general compatibility of 33%, the ratio of sib/general compatibility is 0.85, which indicates a slight reduction in the potential for sib-mating.

By contrast, the single locus polyallelic SSI and GSI system are extremely efficient mechanisms for reducing sib-matings, reducing them by as much as 50–75%. This is because there are large numbers of *S*-alleles leading to thousands of different pairwise combinations in diploid plants, which means that the general compatibility will approach 100%. In comparison, the multigene SSI and GSI systems are less efficient at reducing sib-matings because the alleles of two different loci must be matched (Lewis 1979a).

It is difficult to test such estimates of sib-mating reduction in natural populations, and no studies have attempted to do so, but the high numbers of *S*-alleles (20–60) estimated for natural populations in single polyallelic SSI and GSI (Emerson 1939; Ockendon 1974; Nou et al. 1991; Lawrence et al. 1993) indicate that sib-mating avoidance will be an important means of enforcing outbreeding and gene flow and will tend to act against the establishment of local inbreeding neighborhoods within these populations.

5. The Molecular Basis of Self-Incompatibility

Over the past 12 years, considerable progress has been made in the identification of molecules associated with SI, primarily in economically important species from the Solanaceae (GSI) (for reviews see Ebert et al. 1989; Haring et al. 1990; Singh and Kao 1992; Newbigin 1993; Dickinson 1994a) and Brassicaceae (homomorphic SSI) (for reviews see Nasrallah et al. 1991; Dzelzkalns et al. 1992; Trick and Heizman 1992; Dickinson et al. 1992; Nasrallah and Nasrallah 1993).

With the notable exceptions of *Papaver* (Franklin-Tong and Franklin 1992, 1993), *Pyrus* (Sassa et al. 1993), and *Antirrhinum* (Y. Xue, H. Dickinson, and E. Coen, pers. commun.), very limited molecular data exist for species outside of the Solanaceae and Brassicaceae. The majority of these molecular data tend to support a view that the male and female *S*-locus products are different. This is a departure from the long-held view, based on classical genetics, that the *S*-locus products from pollen and pistil will be the same (Lewis 1960).

Lewis (1960, 1965) proposed a "dimer hypothesis" as a unifying mechanism for all forms of SI. This hypothesis states that identical *S*-locus products in pollen and pistil recognize each other and dimerize to form an active "regulator" which inhibits pollen development. Many subsequent models for SI had this archetypal model as their basis (Ascher 1966; Linskens 1967; Nasrallah et al. 1970; Heslop-Harrison 1975; van der Donk 1975; Ferrari and Wallace 1977), but conclusive support for the model from molecular studies has never been forthcoming (Dickinson et al. 1992; Singh and Kao 1992; Thompson and Kirch 1992).

a) The Molecular Basis of Gametophytic Self-Incompatibility

Glycoproteins associated with specific *S*-alleles have been identified in the styles of members of the Solanaceae: *Nicotiana alata, Lycopersicon peruvianum, Petunia hybrida, Petunia inflata, Solanum chacoense* and *Solanum tuberosum* (Bredemeijer and Blass 1981; Mao et al. 1986; Kamboj and Jackson 1986; Ai et al. 1990; Xu et al. 1990; Kirch et al. 1989), Rosaceae: *Pyrus serotina* (Nakanishi et al. 1992; Sassa et al. 1993), and Scrophulariaceae: *Antirrhinum* sp. (Y. Xue, H. Dickinson and E. Coen, pers. commun.). All are basic proteins with molecular weights ranging from 23–34 kDa and are localized predominantly in stigmatic tissue and in the transmitting tissue of the style (Cornish et al. 1987). cDNA clones corresponding to a number of these glycoproteins have been characterized (Anderson et al. 1986; Clarke et al. 1990; Kheyr-Pour et al. 1990; Ai et al. 1990; Xu et al. 1990; Kaufmann et al. 1991; Tsai et al. 1992) and shown to be products of the *S*-locus (Newbigin et al. 1993). The amino acid sequences derived from these clones have revealed that the proteins contain eight conserved cysteine residues, two conserved histidine residues, five short conserved regions and a number of variable and hypervariable regions (Tsai et al. 1992). The overall amino acid sequence identify between S-glycoproteins from the different species varies from 38.7 to 93.5% (Singh and Kao 1992).

Sequence analysis of three different S-glycoproteins from *N. alata* revealed that some of the conserved regions of the molecules showed strong homology to amino acid sequences surrounding and including the catalytic domains of fungal ribonucleases (McClure et al. 1989). RNase activity was subsequently demonstrated *in vitro* for these three S-glycoproteins (McClure et al. 1989) and later for S-glycoproteins from other members of the Solanaceae (Broothaerts et al. 1991; Kaufmann et al. 1991; Ai et al. 1992), Rosaceae (Sassa et al. 1993) and Scrophulariaceae (Y. Xue, H. Dickinson and E. Coen, Pers. comm.). Attempts were made to determine whether S-RNases could specifically degrade RNA in self-pollen tubes (McClure et al. 1990; Gray et al. 1991) but the findings proved inconclusive (Singh and Kao 1992). Recently, however, stylar S-RNases have been shown conclusively to be required for the operation of GSI *in vivo* (Lee et al. 1994; Murfett et al. 1994).

Lee et al. (1994) transformed self-incompatible *P. inflata* with an antisense construct of its S-RNase gene and showed that the consequent reduction in S-RNase synthesis in the style was accompanied by a switch to a self-compatible phenotype. Murfett et al. (1994) transformed a self-compatible hybrid of *Nicotiana langsdorfii* × *N. alata* with an S-RNase gene from self-incompatible hybrid of *N. alata*; the resulting transformants synthesized large amounts of the S-RNase in the style and became incompatible with pollen carrying the same allele.

These data support an hypothesis for the mechanism of GSI based on specific degradation of pollen RNA by S-RNases during incompatible pollinations (McClure et al. 1989, 1990; Haring et al. 1990; Thompson and Kirch 1992; Newbigin et al. 1993; Fig. 3). The problem now is to determine how S-specificity is achieved in GSI; this requires identification of the pollen *S*-gene product(s). Even though low levels of S-RNase transcription have been detected in the pollen of *Nicotiana* (Dodds et al. 1993) there is a growing consensus of opinion amongst workers in the field of GSI that the male and female *S*-gene products are different (Singh and Kao 1992; Thompson and Kirch 1992; Franklin-Tong and Franklin 1993; Dickinson 1994). In the current models for GSI the *S*-locus product in the pollen is envisaged to be either a membrane-bound selective S-RNase transporter in the case of an incompatible stylar-pollen interaction or a highly specific cytoplasmic inhibitor of S-RNase activity in the case of compatible stylar-pollen interactions (Thompson and Kirch 1992; Singh and Kao 1992).

While the central role of S-RNases in the operation of GSI in Solanaceous species is now overwhelming, evidence from *Papaver rhoeas* suggests that its GSI does not operate through S-RNases (Franklin-Tong and Franklin 1993). S-proteins from *Papaver* have neither RNase activity nor do they share any homology with the amino acid sequences of fungal RNases (Franklin-Tong and Franklin 1992). Indeed, GSI in *Papaver* has been shown to operate via a Ca^{2+} signaling pathway in the pollen (Franklin-Tong et al. 1993; Foote et al. 1994) which appears to be linked with specific protein phosphorylation and dephosphorylation events

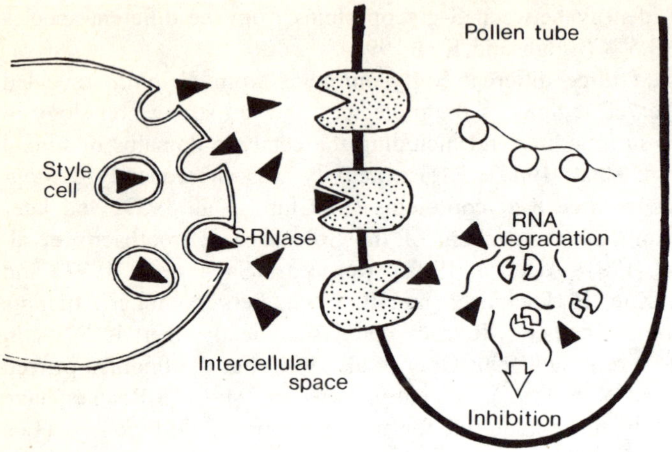

Fig. 3. Proposed mechanism for gametophytic self-incompatibility (GSI) in *Nicotiana* and *Petunia*. (After Lee et al. 1994; Murfett et al. 1994). The stylar *S*-allele product is an RNase (*black triangles*) and the *S*-allele product in the pollen is thought to be a receptor/transporter (*stippled*) of S-RNases. During an incompatible pollen-pistil interaction S-RNase molecules are specifically taken up by the S-RNase receptor/transporter and transported into the pollen tube where they degrade RNA (mRNA and ribosomal RNA) and thereby cause inhibition of further pollen tube development

(Franklin-Tong et al. 1992) and to the transcription of specific "response" genes in the pollen (Franklin-Tong and Franklin 1992) (Fig. 4).

b) The Molecular Basis of Sporophytic Self-Incompatibility

Stigma-specific glycoproteins which segregate with defined *S*-alleles have been identified in *Brassica oleraceae* and *B. campestris* (Nasrallah et al. 1970, 1972; Nishio and Hinata 1977; Hinata and Nishio 1978; Roberts et al. 1979; Nasrallah and Nasrallah 1984). These *S*-locus glycoproteins (SLGs) are all basic proteins with molecular weights of 55–65 kDa and exhibit allelic heterogeneity in molecular weight and charge (Nasrallah and Nasrallah 1984). SLG synthesis is developmentally regulated with maximal synthesis occuring in the papillar cells of the stigma just prior to anthesis coincident with the onset of SI (Nasrallah et al. 1985a). Once synthesized, SLGs accumulate in the papillar cell wall (Kandasamy et al. 1989).

SLG expression has been localized, by in situ hybridization using cloned cDNAs, to the papillar (epidermal) cells of the stigma (Nasrallah et al. 1988), where it is undetectable 5 days before anthesis but then increases to reach a maximum 1 day before anthesis when SLG protein synthesis is maximal and stigmas become fully self-incompatible (Nasrallah et al. 1985a).

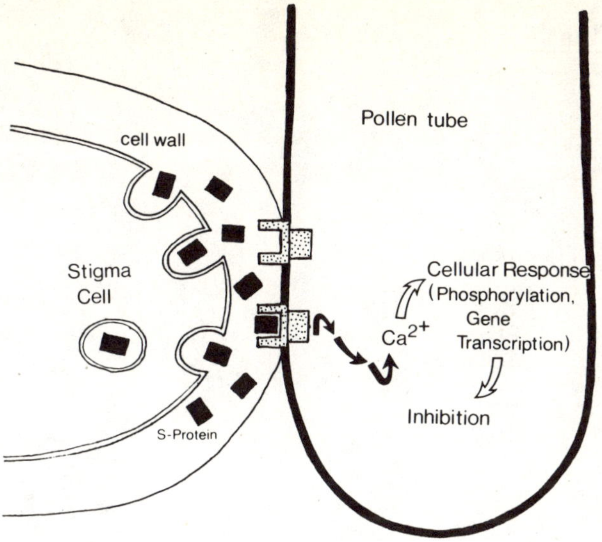

Fig. 4. Model for the mechanism of gametophytic self-incompatibility (GSI) in *Papaver*. (After Franklin-Tong and Franklin 1993). During an incompatible pollen-stigma interaction, the stigmatic S-protein (*black rectangles*) is thought to bind to a receptor (*stippled*) on the surface of the pollen tube. This initiates a second message which stimulates the release of Ca^{2+} from the endoplasmatic reticulum. The Ca^{2+} then causes inhibition of pollen tube development possibly through protein phosphorylation, leading to transcription of specific "pollen-response" genes

cDNAs encoding a number of SLGs have been cloned and their predicted amino acid sequences determined (Nasrallah et al. 1985b; Nasrallah et al. 1987; Takayama et al. 1987; Trick and Flavell 1989; Chen and Nasrallah 1990; Dwyer et al. 1991). These sequences reveal several, variable, N-glycosylation sites, 12 conserved cysteine residues, and a number of conserved, variable and hypervariable regions (Nasrallah et al. 1991; Trick and Heizmann 1992) (Fig. 5). Unlike the deduced sequences of *S*-associated glycoproteins from GSI in the Solanaceae, the amino acid sequences of SLGs do not predict a function (Dickinson et al. 1992).

Comparative studies of SLG sequences have defined two distinct classes of *S*-allele products which correlate well with their position in the *S*-allele dominance hierarchy (Nasrallah et al. 1991): SLGs of the dominant class I alleles share greater than 80% amino acid sequence similarity, while SLGs from the pollen recessive class II alleles share less than 70% amino acid similarity with class I alleles (Chen and Nasrallah 1990).

Furthermore, class II allele SLGs are not recognized by a monoclonal antibody (MAbH8) raised to a distinct glycoprotein (SLG_{13}), which recognizes all class I allele SLGs (Kandasamy et al. 1989).

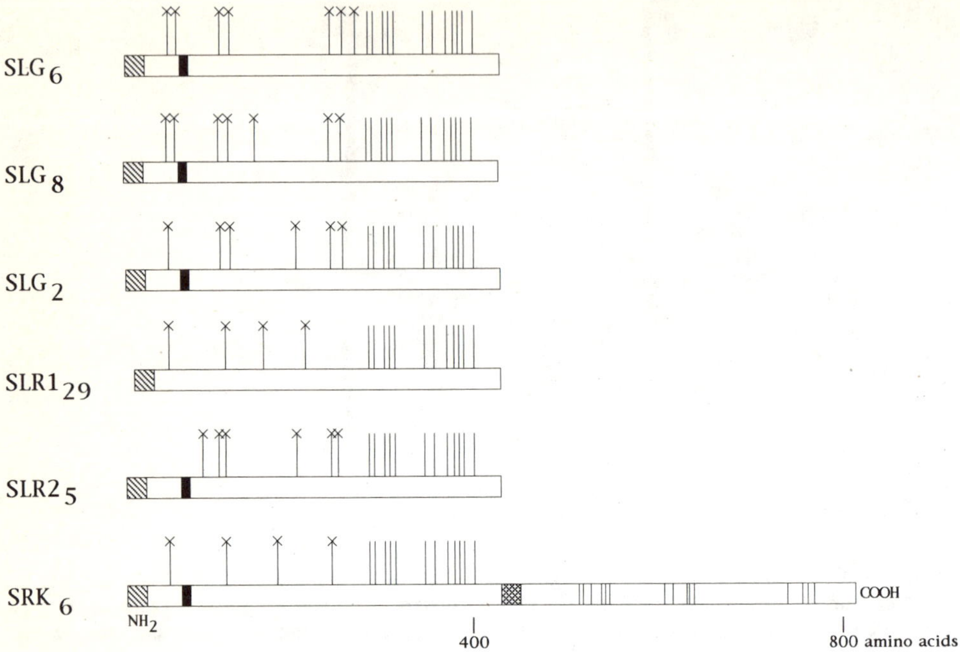

Fig. 5. Schematic representation of members of the *Brassica* S-glycoprotein family. *SLG* and *SRK* genes are linked to the *S*-locus but *SLR1* and *SLR2* genes are not. (Diagrams are drawn on the basis of data from Nasrallah et al. 1987; Takayama et al. 1987; Trick and Flavell 1989; Scutt et al. 1990; Stein et al. 1991). The *shaded box* at the N-terminus denotes a hydrophobic signal peptide; the *black box* represents an insertion/deletion difference which separates SLR1 glycoproteins from the others. *Vertical lines with crosses* denote N-glycosylation sites and *vertical lines without crosses* indicate conserved cysteine residues. For SRK_6 the *central shaded box* represents the hydrophobic transmembrane domain while the *vertical bars* towards the C-terminus denote amino acid residues which are conserved among serine-threonine kinases

As well as *SLG*, another, larger gene also resides at the *Brassica S*-locus: The S-receptor kinase (*SRK*) gene encodes a putative transmembrane receptor serine/threonine protein kinase which like *SLG* is expressed primarily in stigmatic tissue (Stein et al. 1991). The putative receptor domain at the N-terminus of SRK shares 90% amino acid similarity with the SLG encoded within the same *S*-allele (Stein et al. 1991; Dzelzkalns et al. 1992).

In vitro expression of the kinase domain of two SRKs has shown that this region does indeed encode a functional serine/threonine kinase (Goring and Rothstein 1992; Stein and Nasrallah 1993) but so far the nature of its substrate(s) (apart from SRK itself) and its mode of activation are unknown.

Recently it has been shown that certain self-compatible mutants of *Brassica oleracea* lack a functional *SRK* gene (Nasrallah et al. 1994), indicating that SRK is essential for the normal working of SSI. It has been proposed that during a self-incompatible response the SLG binds to the extracellular receptor domain of SRK. Binding stimulates the kinase domain to initiate a phosphorylation cascade in the papillar cell which culminates in the inhibition of pollen development (Dzelzkalns et al. 1992; Nasrallah and Nasrallah 1993; Fig. 6). The S-specificity of the proposed reaction must be mediated in some way by the pollen but the molecule or molecules involved have yet to be identified (Dickinson 1994a).

The model proposed for SSI has notable parallels with certain mammalian and avian recognition systems which are mediated by transmembrane receptor tyrosine kinases in-conjunction with secreted or truncated forms of their extracellular domain which lack the kinase domain (Dzelzkalns et al. 1992; Flickinger et al. 1992; DiStefano and Johnston 1988; Baumbach et al. 1989; Fantl et al. 1993). In the case of the epidermal growth factor receptor, it has been suggested that its soluble truncated form acts as a "linker/adapter" protein which binds to the extracellular domain of the receptor and thereby modifies the activity of the kinase (Fantl et al. 1993). Unlike SLG and SRK, however, all tyrosine receptor kinases and their truncated forms are encoded by the same gene and are generated by alternate splicing of the primary receptor kinase transcript or proteolytic cleavage of the translation product (Dzelzkalns et al. 1992).

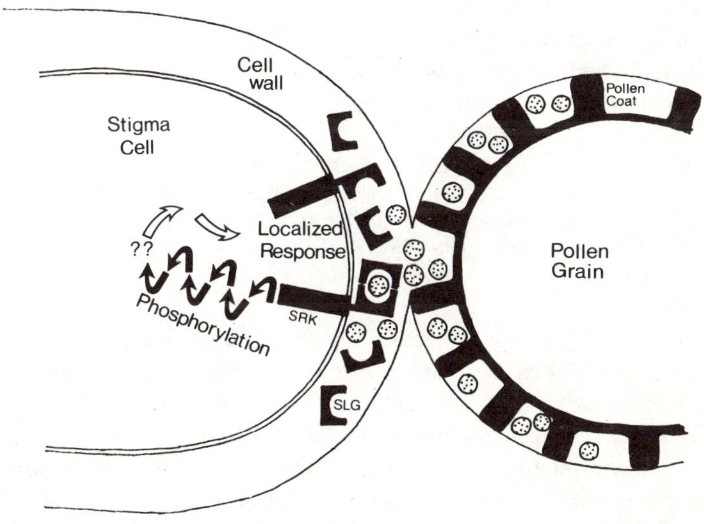

Fig. 6. Model for the mechanism of sporophytic self-incompatibility in *Brassica*. Inhibition of pollen development is on the stigma surface and is thought to be mediated by a highly localized response within the epithelial (papilla) cells of the stigma. An S-component (ligand?) from the pollen coating (*stippled circles*) is presumed to mediate the binding of the secreted *S*-locus glycoprotein (SLG) (*black* [) to the "S" receptor domain of the S-receptor kinase (SRK) (see text). This interaction is presumed to activate the kinase domain of SRK, which initiates a phosphorylation cascade within an epithelial cell of the stigma, the consequence of which is the inhibition of pollen development on that epithelial cell only

The male determinant of SSI, like the male determinant of GSI, has proved extremely elusive. Sporophytic determination of the pollen SI phenotype (Bateman 1954) dictates that if the *S*-allele product(s) reside within the pollen grain these alleles would have to be expressed before meiosis (Pandey 1970). An alternative hypothesis is that the *S*-alleles are expressed in the diploid tapetal cells of the anther during pollen development and are added to the mature pollen grain as part of the pollen coating (Dickinson and Lewis 1973 a,b; Heslop-Harrison et al. 1974). There is considerable indirect evidence that the male determinant of SSI does reside in the pollen coating because when incompatible pollen coating extracts are applied to appropriate stigmas they induce callose formation in the stigma (Dickinson and Lewis 1973b, 1975; Heslop-Harrison et al. 1975; Doughty et al. 1992). This stigmatic "callose response" is a characteristic of an incompatible pollen-stigma interaction in *Brassica* (Ellemann et al. 1992).

Furthermore, wiping *Brassica* stigmas with incompatible pollen coating in micromanipulation experiments prevents the development of compatible pollen, but only on those stigmatic papillae where incompatible coating had been applied (Dickinson 1994b). This indicates that the incompatibility response of the stigma is extremely focused – such extreme localization of the response is advantageous in nature where a simultaneous attachment of compatible and incompatible pollen on the same stigma can be expected.

The majority of experimental data available suggest that the S-specific determinant in the pollen is not an SLG (Dickinson et al. 1992) and it is very likely that the male and female *S*-allele products are different as appear to be the case in gametophytic systems (Newbigin 1993).

Despite the detection of low levels of *SLG* transcription in anthers (Guilluy et al. 1991) and activity of the *SLG* promoter in tapetal cells of the anther (Sato et al. 1991) extensive analyses of the proteins contained in the pollen (Gaude et al. 1988) and pollen coating of *Brassica* (Doughty et al. 1993, Hiscock 1993) have not revealed it to contain SLG proteins.

Recently a group of peptides with molecular weights of approximately 7 kDa have been identified in the pollen coating of *Brassica oleraceae* and *B. napus* which are able to interact with stigmatic SLGs *in vitro* (Doughty et al. 1993; Hiscock et al. 1994). These peptides, termed PCPs (pollen coat peptides) are highly polymorphic and are members of a large family of cysteine-rich thionin-like peptides (Doughty 1994). As yet no S-specificity has been demonstrated for interactions between PCPs and SLGs but there appears to be enough variability between these peptides to make them suitable candidates for signaling moelcules. Work is in progress using cloned *PCP* cDNAs to establish whether any *PCP* genes are linked to the *S*-locus (J. Doughty, pers comm.).

The fact that PCPs can interact with SLGs suggests that they may also interact with the receptor "S" domain of SRKs. If this can be proven to be so, then specific PCPs might function as the S-specific pollen ligand(s) envisaged to mediate SLG-SRK binding in the current models for SSI (Nasrallah and Nasrallah 1993; Dickinson 1994a; Fig. 6).

In addition to the *S*-locus-specific *SLG* and *SRK* genes, the *Brassica* genome contains a large number of *S*-related sequences which are not linked to the *S*-locus (Trick and Heizmann 1992). The first of these sequences to be characterized were termed *S*-locus-related (SLR) genes (Lalonde et al. 1989). So far two classes of such *SLR* genes have been identified – *SLR1*s (Lalonde et al. 1989; Trick and Flavell 1989; Trick 1990; Isogai et al. 1991) and *SLR2*s (Boyes et al. 1991; Trick and Heizmann 1992) – and their sequences predict polypeptides with a very similar structure to SLGs (Fig. 4).

*SLR1*s share approximately 65% nucleotide similarity with both *SLG*s and *SLR2*s whereas *SLR2* shares 70% nucleotide similarity with *SLG*s of class I alleles and 90% with *SLG*s of class II alleles (Nasrallah et al. 1991; Trick and Heizmann 1992). *SLR1* and *SLR2* are thus more like *SLG*s than they are like each other.

The expression pattern of *SLR1* is very similar to that of *SLG*s – the optimal expression occurring in stigmatic papillar cells at the onset on SI (Lalonde et al. 1989). Once synthezised, SLR1s, like SLGs, accumulate in the cell walls of papillar cells (Umbach et al. 1990). Unlike *SLG*s, however, *SLR1* and *SLR2* show very little polymorphism between alleles (Boyes et al. 1991). Furthermore, SLR1 proteins have been detected in many species within the Brassicaceae in both self-incompatible and self-compatible species (Umbach et al. 1990; Dwyer et al. 1992; Watanabe et al. 1992). The function of SLRs is unknown, but their widespread distribution and expression in stigmatic papillar cells suggest that they may play a fundamental role in pollination (Dickinson 1990). This assumption now seems more likely with the recent finding that SLR1 proteins are able to interact *in vitro* with members of the PCP family of pollen coat peptides (Hiscock et al. 1994).

Interestingly, *SRK*-like and *SLG*-like genes have also been found outside of the *Brassica* genus. *SRK*-like sequences have been detected in *Arabidopsis* (Tobias et al. 1992; Walker 1993) and in *Zea* (Walker and Zhang 1990) while SLG-like sequences have been detected in *Arabidopsis* (Tobias et al. 1992) and in *Daucus* (carrot) (van Engelen et al. 1993). All of these genes are expressed predominantly in vegetative tissues, indicating that their functions are unrelated to reproduction. Thus a picture is gradually emerging of a superfamily of *S*-genes with similar structure but different functions.

Much speculation has focused on the possible functions of *SRK*-like and *SLG*-like genes, and it has been suggested that they might be involved in defence responses against pathogens (Dickinson et al. 1992; Dickinson 1994a). If this is so, it is possible that the SSI system evolved from a preexisting pathogen defense mechanism which could explain why many aspects of the physiology of SSI and certain host-pathogen interactions (spore/pollen hydration dynamics, callose production) are so similar (Hodgkin et al. 1988).

6. Self-Incompatibility Systems in Angiosperms and Mating Type Factors in Fungi: Different Means to the Same End

Sexual dimorphism with specific male and female organs is rarely found in fungi. Cell fusion in most species of fungi occurs between undifferentiated cells and it is the mating type of the cells which determines whether cells are compatible or incompatible (Esser and Kuenen 1965; Raper 1966; Kües and Casselton 1992a; Kües and Stahl 1992; Elliott 1994).

For example, in basidiomycete fungi, sexual reproduction involves fusion of two compatible monokaryotic (haploid) hyphae from two mycelial colonies followed by nuclear migration to form a fertile dikaryon on which fruitbodies can develop and where, in specialized cells, karyogamy and meiosis occur (Raper 1966; Casselton 1978, 1994).

For fungal cells to be compatible, the mating types must be different. Some fungi have only two different mating types which are encoded at the same locus, e.g., the ascomycetes *Saccharomyces cerevisiae* (Herskowitz 1988, 1989; Dolan and Fields 1991) and *Neurospora crassa* (Glass and Lorimer 1991). Others have one or two different mating type loci with two or up to hundreds of different cross-compatible alleles. For example, in some *Coprinus* species mating is controlled by a single mating type locus – *A* – which is multiallelic (the bipolar system) (Kemp 1980), while in *Coprinus cinereus* and *Schizophyllum commune*, mating is controlled by two unlinked multiallelic mating type loci called the *A* and *B* factors (the tetrapolar system) (Casselton 1978; Raper 1966).

For a mating to be compatible *A* factors (bipolar) or *A* and *B* factors (tetrapolar) must be different (for reviews see Esser and Kuenen 1965; Raper 1966; Kües and Casselton 1992a,b; Casselton and Kües 1994). In the tetrapolar *Ustilago maydis* there is one locus, *a*, with only two different alleles and a multiallelic *b* locus (Banuett 1992; Saville and Leong 1992). In the related *Ustilago hordei, a* and *b* were found to be so closely linked that they always remain together, suggesting a bipolar system (Bakkeren et al. 1992; Bakkeren and Kronstad 1993). Molecular analysis may reveal that this is also true for other bipolar systems, such as in some *Coprinus* species.

In genetical terms, the bipolar mating type system may be thought of as being analogous to the angiosperm single locus GSI and SSI systems and the tetrapolar mating type system would be analogous with the two locus GSI system found in the Gramineae. Theoretically, the bipolar and tetrapolar systems of basidiomycetes with two alleles are only 50 and 75% efficient at preventing selfing, respectively (Lewis 1979a,b). However, with increasing numbers of alleles and efficient spore dispersal, the probability of a spore landing and germinating near a mycelium carrying different mating type factors will be extremely high, which means that in natural populations the ability of these mating type factor incompatibility systems to reduce selfing will approach or equal the 100% value achieved by the SI systems in angiosperms (Esser 1971; Koltin et al. 1972; Stamberg and Koltin 1973; Lewis 1979a,b).

As described earlier in this series (Kües and Stahl 1992), in *C. cinereus* and *S. commune* compatibility is achieved by complementation between different *A* genes and in *Ustilago* between different *b* genes. Molecular analysis has revealed that there are two types of genes present at these loci, termed *HD1* and *HD2*, which encode putative transcription factors with distinct homeodomain motifs. The products of the two different classes of genes are thought to dimerize to form a functional transcription factor complex which regulates the expression of genes involved in sexual development. Products of the two classes of genes encoded at the same locus are "incompatible", i.e., nonfunctional in eliciting a sexual reaction whereas gene products from allelic loci (derived from different mates) are compatible and complement each other to elicit a sexual reaction (Fig. 7). Functional and nonfunctional combinations can be distinguished because the gene and the protein sequences of different allelic forms are highly unique (for references see Gillissen et al. 1992; Herskowitz 1992; Kües et al. 1992, 1994 a,b,c; Kües and Casselton 1992c, 1993; Specht et al. 1992, 1994; Stankis et al. 1992; Saville and Leong 1992; Tymon et al. 1992; Bakkeren and Kronstad 1993; Challen et al. 1993; Richardson et al. 1993; Yee and Kronstad 1993; Gieser and May 1994; Luo et al. 1994; and those in Kües and Stahl 1992).

The two forms of the *U. maydis a* mating type locus encode putative pheromones and their corresponding pheromone receptors. Complementation should occur between the a1 pheromone and the receptor encoded in the *a2* locus and between the a2 pheromone and the receptor encoded in the *a1* locus. Distinction between the pheromones and between the receptors is possible because their genes have very different sequences (Froeliger and Leong 1991; Bölker et al. 1992; Bölker and Kahmann 1993; Spellig et al. 1994). A similar situation of complementation between pheromones and pheromone receptors is now expected for the *B* locus products of the higher basidiomycetes. Sequence data for one of the multiple *B* factors of *S. commune* has revealed the existence of genes encoding a putative pheromone and a putative pheromone receptor (Wendland and Kothe 1994; E. Kothe, pers. commun.).

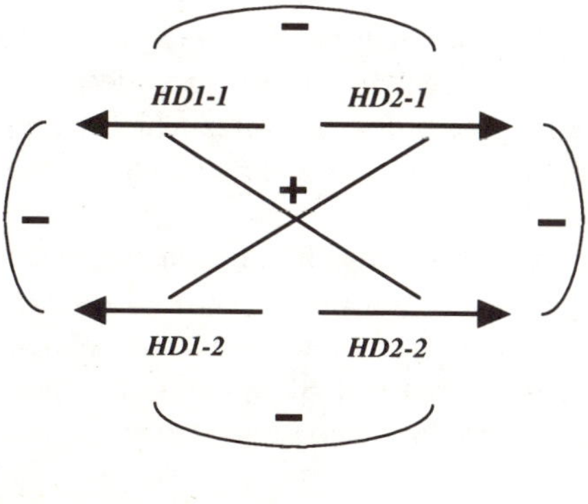

Fig. 7. Schematic representation of the possible HD1-HD2 gene interactions following mating in basidiomycete fungi (further details in Kües and Casselton 1993; Casselton and Kües 1994). HD1 and HD2 genes are divergently transcribed in pairs which encode transcription factors with distinct homeodomain motifs. *HD1-1* and *HD1-2*, *HD2-1* and *HD2-2* refer to different alleles of genes. + denotes a compatible interaction; – denotes an incompatible interaction

The situation described in this chapter for angiosperms is fundamentally different from that in fungi. In fungi complementation between mating type genes (or more correctly their products) leads to compatibility, whereas in angiosperms it is opposition between identical *S*-genes (though not identical *S*-gene products) which leads to incompatibility. This fundamental difference between two, highly efficient strategies for avoiding self-fertilization is a natural consequence of the different life-cycles of angiosperms and fungi and the stage of the life-cycle at which self-recognition takes place. In fungi there is a haploid/haploid interaction during dikaryon formation when a haploid nucleus enters a haploid (monokaryotic) cell, while in angiosperms there is a haploid/diploid or (functionally) diploid/diploid interaction between the pollen and the diploid sporophytic tissue of the pistil. For the haploid/diploid interaction, complementation between different mating type alleles is more efficient as a barrier to selfing than opposition between the same mating type alleles, whereas for a haploid/haploid interaction the presence of a diploid genome means that opposition between the same *S*-alleles will be more efficient at preventing selfing than complementation between different *S*-alleles (Lewis 1979a,b).

The fundamental differences between the life cycles of angiosperms and fungi have put different constraints on the evolution of their self-incompatibility mechanisms, and the data presented here and earlier (Kües and Stahl 1992) show that at the molecular level there may be several different solutions to the problem of avoiding self-fertilization. Fungi have evolved pheromone recognition and transcription regulator systems, while angiosperms have evolved a specific RNA degrading system and specific signaling pathways based on phosphorylation and/or Ca^{2+}.

7. Conspectus

The evolution of outbreeding systems is inextricably linked with the evolution of sex, which itself is a highly contentious issue amongst evolutionists and one which will not easily be resolved (Maynard-Smith 1978; Ghiselin 1974; Felsenstein 1988; Bernstein and Bernstein 1991; Wilmsen Thornhill 1993). The fact that self-incompatibility systems, as outbreeding mechanisms, are so widespread in plants and fungi is testimony to the very strong selection pressures which maintain outbreeding – particularly avoidance of the detrimental effects of inbreeding depression (Charlesworth and Charlesworth 1979, 1987).

With respect to the evolution of SI in angiosperms, it has often been suggested that a form of SI was present in the first angiosperms and may have been responsible for the explosive radiation of these plants during the late Cretaceous period (Whitehouse 1950; Crowe 1964). However, the variety of SI systems and their scattered distribution among so many angiosperm families suggest that SI has probably arisen many times during angiosperm evolution and so is more likely to be polyphyletic (Gibbs 1986; Charlesworth 1988).

We have reviewed the recent advances, made through molecular biology, in our understanding of the mechanisms of single-locus GSI and SSI. So far all of this data comes from the female side of the interaction – stylar S-RNases have been shown to be essential for GSI in the Solanaceae (Lee et al. 1994; Murfett et al. 1994) and a stigmatic S-receptor kinase is thought to mediate SSI in *Brassica* (Nasrallah and Nasrallah 1993). Nothing, however, is known about the S-molecule(s) present in pollen which denote "self" and "non-self". The identification and characterization of these male S molecules is now a major goal for those working on self-incompatibility in higher plants.

Despite their general genetical similarity, the self-incompatibility systems found in angiosperms and fungi are quite different at the molecular level. This great diversity of molecular mechanisms for self-incompatibility is a consequence of the tremendous selective advantages that can be accrued by a species through optimizing the beneficial effects of recombination through outbreeding.

Acknowledgement. The authors thank Dr. David Boshier, Professor Hugh Dickinson, Dr. David Haig and Dr. Lorna Casselton for helpful discussion and critical reading of the manuscript. Special thanks are due to Profs. Enrico Coen and Hugh Dickinson and to Drs. Yong Xue, James Doughty, and Erika Kothe for allowing us to discuss unpublished data. S.J.H. was funded by a grant from the Dunston trust (Dept. of Plant Sciences, Oxford) and U.K. was funded by a Violette and Samuel Glasstone Fellowship from the University of Oxford.

References

Ai Y, Singh A, Coleman CE, Lorger TR, Kheyr-Pour A, Kao TH (1990) Plant Reprod 3:130–138. **Ai Y**, Tsai D-S, Kao T-H (1992) Plant Mol Biol 19:523–528. **Allard RW**, Kahler AL, Clegg MT (1977) In: Christiansen FB, Fenchel TM (eds) Measuring selection in natural populations. Springer, Berlin Heidelberg New York, pp 1–19. **Anderson MA**, Cornish EC, Mau SL, Williams EG, Hoggart R, Atkinson A, Bonig I, Grego B, Simpson R, Roche PJ, Haley JD, Penschow JD, Niall HD, Tregear GW, Coghian JP, Crawford RJ, Clarke AE (1986) Nature 321:38–44. **Ascher PD** (1966) Euphytica 15:179–183.

Bakkeren G, Gibbard B, Yee A, Froeliger E, Leong S, Kronstad J (1992) Mol Plant-Microbe Interact 5:347–355. **Bakkeren G**, Kronstad JW (1993) Plant Cell 5:123–136. **Banuett F** (1992) Trends Genet 8:174–180. **Barrett SCH**, Richards JH (1990) Mem N Y Bot Gard 55:35–61. **Bateman AJ** (1952) Heredity 6:285–310. **Bateman AJ** (1954) Heredity 8:305–332. **Bateman AJ** (1955) Heredity 9:52–68. **Baumbach WR**, Horner DL, Logan JS (1989) Genes Dev 3:1199–1205. **Bawa KS**, Perry DR, Beach JH (1985) Am J Bot 72:331–345. **Bergström G** (1978) In: Harborne JB (ed) Biochemical aspects of plant and animal co-evoultion. Academic Press, London, pp 207–231. **Bernstein C**, Bernstein H (1991) Aging, sex and DNA repair. Academic Press, San Diego. **Bölker M**, Kahmann R (1993) Plant Cell 5:1461–1469. **Bölker M**, Urban M, Kahmann R (1992) Cell 68:441–450. **Boshier DH**, Chase MR, Bawa KS (1994) Am J Bot (in press). **Boyes DC**, Chen CH,

Tantikanjana T, Esch JJ, Nasrallah JB (1991) Genetics 127:221–228. **Bredemeijer GM**, Blass J (1981) Theor Appl Genet 59:185–190. **Brewbaker JL** (1957) J Hered 48:271–277. **Brewbaker JL** (1959) Ind J Genet Plant Breed 19:121–133. **Briggs D**, Walters SM (1984) Plant variation and evolution. Cambridge University Press, Cambridge. **Broothaerts WJ**, Decock B, Damme JV, Vendrig JC (1991) Sex Plant Reprod 4:258–266. **Brotschol JV**, Roberds JH, Namkoong G (1986) Silvae Genet 35:131–138. **Brown AHD** (1990) In: Brown AHD, Clegg MT, Kahler AL, Weir BS (eds) Plant population genetics, breeding and genetic resources. Sinauer, Sunderland. **Bull JJ**, Pease CM (1989) Evolution 43: 667–671.

Campbell DR (1991) Evolution 45:1965–1968. **Casselton LA** (1978) In: Smith JE, Berry DR (eds), The filamentous fungi, vol 3. Edward Arnold, London, pp 275–297. **Casselton LA** (1994) In: Esser K, Lemke P (eds) The mycota, vol II. Springer, Berlin Heidelberg New York (in press). **Casselton LA**, Kües U (1994) In: Esser K, Lemke P (eds) The mycota, vol I. Springer, Berlin Heidelberg New York, pp 307–321. **Challen MP**, Elliott TJ, Kües U, Casselton LA (1993) Mol Gen Genet 241:474–478. **Charlesworth D** (1988) Heredity 60:445–453. **Charlesworth D**, Charlesworth B (1979) Proc R Soc Lond Ser B 205:513–530. **Charlesworth D**, Charlesworth B (1987) Annu Rev Ecol Syst 18:237–268. **Charlesworth D**, Charlesworth B (1990) Evolution 44:870–888. **Charlesworth D**, Morgan MT, Charlesworth B (1990) Evolution 44:1469–1489. **Chen CH**, Nasrallah JB (1990) Mol Gen Genet 222:241–248. **Clark KR**, Okuley JJ, Colins PD, Sims TL (1990) Plant Cell 2:815–826. **Cope FW** (1962a) Heredity 17:157–182. **Cope FW** (1962b) Heredity 17:183–195. **Cornish EC**, Pettitt JM, Bonig I, Clarke AE (1987) Nature 326:99–102. **Crowe LK** (1954) Heredity 8:1–11. **Crowe LK** (1964) Heredity 19:435–457. **Crowe LK** (1971) Heredity 27:111–118.

Darlington CD, Mather K (1949) The elements of genetics. Allen and Unwin, London. **de Nettancourt D** (1977) Incompatibility in angiosperms – Monographs on Theoretical and Applied Genetics 3. Springer, Berlin Heidelberg New York. **Devlin B**, Ellstrand NC (1990) Evolution 44:248–259. **Dickinson HG** (1990) Bioessays 12:155–161. **Dickinson HG** (1994a) Nature 367:517–518. **Dickinson HG** (1994b) Sex Plant Reprod (in press) **Dickinson HG**, Lawson J (1975) J Cell Sci 18:519–532. **Dickinson HG**, Lewis D (1973a) Proc R Soc Lond Ser B 183:21–38. **Dickinson HG**, Lewis D (1973b) Proc R Soc Lond Ser B 184:149–165. **Dickinson HG**, Lewis D (1975) In: Duckett JC, Racey PA (eds) The biology of the male gamete. Suppl Biol J Linnean Soc 7:165–175. **Dickinson HG**, Crabbe MJC, Gaude T (1992) Int Rev Cytol 140:525–561. **Di Stefano PS**, Johnston ME (1988) Proc Natl Acad Sci USA 85:270–274. **Dodds PN**, Bönig I, Du H, Rödin J, Anderson MA, Newbigin E, Clarke AE (1993) Plant Cell 5:1771–1782. **Dolan JW**, Fields S (1991) Biochim Biophys Acta 1088:155–169. **Doughty J** (1994) PhD Thesis, University of Reading, UK. **Doughty J**, McCubbin A, Hedderson F, Elleman CJ, Dickinson HG (1992) In: Mulcahy DJ, Bergamini-Mulcahy G, Ottaviano E (eds), Angiosperm pollen and ovules: basic and applied aspects. Springer, Berlin Heidelberg New York, pp 1–14. **Doughty J**, Hedderson F, McCubbin A, Dickinson HG (1993) Proc Natl Acad Sci USA 90:467–471. **Dwyer KG**, Balent MA, Nasrallah JB, Nasrallah ME (1991) Plant Mol Biol 16:481–486. **Dwyer KG**, Lalonde BA, Nasrallah JB, Nasrallah ME (1992) Mol Gen Genet 231:442–448. **Dzelzkalns VA**, Nasrallah JB, Nasrallah ME (1992) Dev Biol 153:70–82.

East EM (1940) Proc Am Philos Soc 82:499–518. **East EM**, Mangelsdorf AJ (1925) Proc Natl Acad Sci USA 2:166–183. **Ebert PR**, Anderson MA, Bernatzky R, Altschuler M, Clarke AE (1989) Cell 56:255–262. **Ellemann CJ**, Franklin-Tong VE, Dickinson HG (1992) New Phytol 121:413–424. **Elliott CG** (1994) Reproduction in fungi: genetical and physiological aspects. Chapman & Hall, London. **Ellstrand NC** (1992) New Forests 6:241–256. **Ellstrand NC**, Marshall DL (1985) Am Nat 126:606–616. **Ellstrand NC**, Devlin B, Marshall DL (1989) Proc Natl Acad Sci USA 86:9044–9047. **Emerson S** (1939) Genetics

24:924–537. **Esser K** (1971) Mol Gen Genet 110:86–100. **Esser K**, Kuenen R (1965) Genetics of fungi. Springer, Berlin Heidelberg New York.

Faegri K, van der Pijl L (1979) The principles of pollination ecology, 3rd edn. Pergamon Press, Oxford. **Fantl WJ**, Johnson DE, Williams LT (1993) Annu Rev Biochem 62:453–481. **Felsenstein J** (1988) In: Michod RE, Levin BR (eds) The evolution of sex. Sinauer, Sunderland, pp 74–86. **Ferrari TE**, Wallace DH (1977) Theor Appl Genet 50:211–225. **Fisher RA** (1963) In: The theory of inbreeding. Oliver and Boyd, London, pp 139–145. **Flickinger WJ**, Maihle NJ, Kung H-J (1992) Mol Cell Biol 12:883–893. **Foote HCC**, Ride JP, Franklin-Tong VE, Walker Ea, Lawrence MJ, Franklin CH (1994) Proc Natl Acad Sci USA 91:2265–2269. **Franklin-Tong VE**, Franklin FCH (1992) Sex Plant Rep 5 (1):1–8. **Franklin-Tong VE**, Franklin FCH (1993) Trends Cell Biol 3:340–345. **Franklin-Tong VE**, Thorlby GJ, Lawrence MJ, Franklin FCH (1992) In: Mulcahy DJ, Bergamini-Mulcahy G, Ottaviano E (eds) Angiosperms pollen and ovules: basic and applied aspects. Springer, Berlin Heidelberg New York, pp 84–93. **Franklin-Tong VE**, Ride JP, Read DR, Trewavas AJ, Franklin FCH (1993) Plant J 4:163–177. **Froeliger EH**, Leong SA (1991) Gene 100:113–122.

Gaude T, Nasrallah ME, Dumas C (1988) Heredity 61:317–318. **Ghiselin MT** (1969) Q Rev Biol 44:189–208. **Ghiselin MT** (1974) The economy of nature and the evolution of sex. University of California Press, Berkeley. **Gibbs PE** (1986) Plant Syst Evol 154: 285–323. **Gieser PT**, May G (1994) Gene 146:167–176. **Gillissen B**, Bergemann J, Sandmann C, Schroer B, Bölker M, Kahmann R (1992) Cell 68:647–657. **Glass NL**, Lorimer IAJ (1991) In: Bennett JW, Lasure LL (eds) More gene manipulations in fungi. Academic Press, San Diego, pp 194–216. **Goring DR**, Rothstein SJ (1992) Plant Cell 4:1273–1281. **Govindaraju DR** (1988) Heredity 61:401–404. **Grant V** (1981) Plant speciation. 2nd edn. Columbia University Press, New York. **Gray LE**, McClure BA, Bonig I, Anderson MA, Clarke AE (1991) Plant Cell 3:271–283. **Grosberg RK** (1988) Q Rev Biol 63:377–412. **Guilluy C-M**, Trick M, Heizmann P, Dumas C (1991) Theor Appl Genet 82:466–472.

Hamrick JL (1982) Am J Bot 69:1685–1693. **Hamrick JL** (1987) In: Urbanska K (ed) Differentiation patterns in higher plants. Academic Press, London, pp 53–66. **Haring V**, Gray JE, McClure BA, Anderson MA, Clarke AE (1990) Science 250:937–941. **Herskowitz I** (1988) Microbiol Rev 52:536–553. **Herskowitz I** (1989) Nature 342:749–757. **Herskowitz I** (1992) Nature 357:190–191. **Heslop-Harrison J** (1975) Proc R Soc Lond Ser B 190:275–299. **Heslop-Harrison J**, Heslop-Harrison Y (1981) Acta Bot Neerl 30:289–307. **Heslop-Harrison Y**, Shivanna KR (1977) Ann Bot 41:1233–1258. **Heslop-Harrison J**, Knox RB, Heslop-Harrison Y (1974) Theor Appl Genet 44:133–177. **Heslop-Harrison J**, Knox RB, Heslop-Harrison Y, Mattsson O (1975) In: Duckett JG, Racey PA (eds). Biology of the male gamete. Linn Soc Lond Academic Press, London, pp 165–202. **Hinata K**, Nishio T (1978) Heredity 41:93–100. **Hiscock SJ** (1993) DPhil Thesis, Bodleian library, University of Oxford, UK. **Hiscock S**, Doughty J, Dickinson HG (1994) Planta (in press). **Hoekstra RF** (1982) J Theor Biol 98:427–451. **Hodgkin T**, Lyon GD, Dickinson HG (1980) New Phytol 110:557–569. **Holsinger KE** (1988) Evolution 42:1235. **Hughes MR**, Babcock EB (1950) Genetics 35:570–588.

Isogai A, Takayama S, Shiozawa H, Tsukamoto C, Kanbara T, Hinata K, Okazaki K, Suzuki A (1988) Plant Cell Physiol 29:1331–1336. **Isogai A**, Yamakawa S, Shiozawa H, Takayama S, Tanaka H, Kono T, Watanabe M, Hinata K, Suzuki A (1991) Plant Mol Biol 17:269–271.

Jain SK (1976) Annu Rev Ecol Syst 7:469–495. **Jarne P**, Charlesworth D (1993) Annu Rev Ecol Syst 24:441–466.

Kamboj RK, Jackson JF (1986) Theor Appl Genet 226:457–466. **Kandasamy MK**, Paolillo DJ, Faraday CD, Nasrallah JB, Nasrallah ME (1989) Dev Biol 134:462–472. **Kaufmann H**, Salamini F, Thompson RD (1991) Mol Gen Genet 226:457–466. **Kemp RFO** (1980) Trans Br Mycol Soc 74:355–360. **Kheyr-Pour A**, Bintrim SB, Lorger TR, Remy R, Hammond SA, Kao T-H (1990) Sex Plant Reprod 3:88–97. **Kirch HH**, Uhrig H, Lottspeich F, Salamini F, Thompson RD (1989) Theor Appl Genet 78:581–588. **Klein R**, Conway D, Parada LF, Barbacid M (1990) Cell 61:647–656. **Kohn JR**, Barrett SCH (1992) Evolution 46:43–55. **Koltin Y**, Stamberg J, Lemke PA (1972) Bacteriol Rev 36:156–171. **Kües U**, Casselton LA (1992a) Mycol Res 96:993–1006. **Kües U**, Casselton LA (1992b) In: Broda PMA, Oliver SG, Sims PFG (eds) The eukaryotic genome – organisation and regulation. Society for General Microbiology Symp 50, Cambridge University Press, Cambridge, pp 185–210. **Kües U**, Casselton LA (1992c) Trends Genet 8:154–155. **Kües U**, Casselton LA (1993) J Cell Sci 104:227–230. **Kües U**, Stahl U (1992) Prog Bot 54:277–294. **Kües U**, Richardson WVJ, Tymon AM, Mutasa ES, Göttgens B, Gaubatz S, Gregoriades A, Casselton LA (1992) Genes Dev 6:568–577. **Kües U**, Asante-Owusu RN, Mutasa ES, Tymon AM, Pardo EH, O'Shea SF, Göttgens B, Casselton LA (1994a) Plant Cell 6: 1467–1475. **Kües U**, Göttgens B, Stratmann R, Richardson WVJ, O'Shea SF, Casselton LA (1994b) EMBO J 13:4054–4059. **Kües U**, Tymon AM, Richardson WVJ, May G, Gieser PT, Casselton LA (1994c) Mol Gen Genet (in press).

Lalonde BA, Nasrallah ME, Dwyer KG, Chen CH, Barlow B, Nasrallah JB (1989) Plant Cell 1:249–258. **Lande R**, Schemske DW (1985) Evolution 39:24–40. **Lawrence MJ** (1975) Proc R Soc Lond Ser B 188:275–285. **Lawrence MJ**, Lane MD, O'Donnell S, Franklin-Tong VE (1993) Heredity 71:581–590. **Lee H-S**, Huang S, Kao T-H (1994) Nature 367:560–563. **Leung DW**, Spencer SA, Cachianes G, Hammonds RG, Collins C, Henzel WJ, Barnard R, Waters MJ, Wood WI (1987) Nature 330:537–543. **Levin DA** (1978) In: Richards AJ (ed) Ecological genetics: the interface. Springer, Berlin Heidelberg New York, pp 189–212. **Levin DA** (1981) Ann Mo Bot Gard 68:233–253. **Levin DA** (1984) In: Dirzo R, Sarkhan J (eds) Perspectives on plant population ecology. Sinauer Sunderland, pp 242–260. **Levin DA**, Kerster HW (1974) Evol Biol 7:139–220. **Lewis D** (1960) Proc R Soc Lond Ser B 151:468–477. **Lewis D** (1965) In: Geerts SJ (ed) Genetics today. Proc XI Int Congress Genet 1963. Pergamon Press, Oxford, pp 657–663. **Lewis D** (1977) Proc R Soc Lond Ser B 196:161–170. **Lewis D** (1979a) Sexual incompatibility in plants. Edward Arnold, London. **Lewis D** (1979b) N Z J Bot 17:637–644. **Lewis D**, Verma SC, Zuberi MI (1988) Heredity 61:355–366. **Linskens HF** (1967) In: Ruhland W (ed) Encyclopedia of plant physiology vol XXVIII. Springer, Berlin Heidelberg New York. **Lughadha EMN**, Parnell JAN (1989) Bot J Linn Soc 100:337–354. **Lundquist A** (1954) Hereditas 40:278–294. **Lundquist A** (1955) Hereditas 40:278–294. **Lundquist A** (1956) Hereditas 42:293–348. **Lundquist A**, Østerbye V, Larsen K, Linde-Laursen I (1973) Hereditas 74:161–168. **Luo Y**, Ullrich RC, Novotny CP (1994) Mol Gen Genet (in press).

Mau S-L, Williams EG, Atkinson A, Anderson MA, Cornish EC, Grego B, Simpson RJ, Kheyr-Pour A, Clarke AE (1986) Planta 156:505–516. **Maynard Smith J** (1978) The evolution of sex. Cambridge University Press, Cambridge. **Mayr E** (1963) Animal species and evolution. Harvard University Press, Cambridge. **McClure BA**, Haring V, Ebert PR, Anderson MA, Simpson RJ, Sakiyama F, Clarke AE (1989) Nature 342:955–957. **McClure BA**, Gray JE, Anderson MA, Clarke AE (1990) Nature 347:757–760. **Michod RE**, Levin BR (eds) (1988) The evolution of sex. An examination of current ideas. Sinauer, Sunderland. **Milton JB** (1992) New Forests 6:197–216. **Moran GF**, Hopper SD (1987) In: Saunders DA, Arnold GW, Burbidge AA, Hopkins AJM (eds) Surrey Beatty Sidney, pp 151–162. **Murfett J**, Atherton TL, Mou B, Gasser CS, McClure BA (1994) Nature 367:563–566.

Nakanishi T, Yamazaki T, Funadera K, Tomonaga H, Ozaki T, Kawai Y, Ichii T, Sato Y, Kurihara A (1992) J Jpn Soc Hortic Sci 61:239–248. **Nasrallah JB**, Nasrallah ME (1984) Experientia 40:279–281. **Nasrallah JB**, Nasrallah ME (1993) Plant Cell 5:1325–1335. **Nasrallah JB**, Doney RC, Nasrallah ME (1985a) Planta 165:100–107. **Nasrallah JB**, Kao T-H, Goldberg ML, Nasrallah ME (1985b) Nature 318:263–267. **Nasrallah JB**, Kao TH, Chen CH, Goldberg ML, Nasrallah ME (1987) Nature 326:617–619. **Nasrallah JB**, Yu SM, Nasrallah ME (1988) Proc Natl Acad Sci USA 85:5551–5555. **Nasrallah JB**, Nishio T, Nasrallah ME (1991) Annu Rev Plant Physiol Plant Mol Biol 42:3933–422. **Nasrallah JB**, Rundle SJ, Nasrallah ME (1994) Plant J 5:373–384. **Nasrallah ME**, Barber JT, Wallace DH (1970) Heredity 25:23–27. **Nasrallah ME**, Wallace DH, Savo RM (1972) Genet Res 20:151–160. **Newbigin E**, Anderson MA, Clarke AE (1993) Plant Cell 5:1315–1324. **Nishio T**, Hinata K (1977) Heredity 38:391–396. **Nou IS**, Watanabe M, Isogai A, Shiozawa H, Suzuki A, Hinata K (1991) Jpn J Genet 66:227–239.

Ockendon DJ (1972) New Phytol 71:519–522. **Ockendon DJ** (1974) Heredity 33: 159–171. **Østerbye U** (1975) Hereditas 80:91–112.

Pandey KK (1970) Euphytica 19:364–372. **Paulus HF**, Gack C (1990) Israel J Bot 39: 43–79. **Peck JR** (1993) Proc R Soc Lond Ser B 125:87–92. **Philipp M**, Madsen HES, Siegismund HR (1992) Heredity 69:32–42. **Piatnitsky SS** (1934) US For Serv Translation 290.

Raper JR (1966) Genetics of sexuality in higher fungi. Ronald, New York. **Richards AJ** (1986) Plant breeding systems. George Allen and Unwin, London. **Richardson WVJ**, Kües U, Casselton LA (1993) Mol Gen Genet 238:304–307. **Roberts IN**, Stead AD, Ockendon DJ, Dickinson HG (1979) Planta 146:179–183.

Sabbadin A (1982) Am Zool 22:765–773. **Sassa H**, Hirano H, Ikehashi H (1993) Mol Gen Genet 241:17–25. **Sato T**, Thorsness MK, Kandasamy MK, Nishio T, Hirai M, Nasrallah JB, Nasrallah ME (1991) Plant Cell 3:867–876. **Saville BJ**, Leong SA (1992) In: Setlow JK (ed) Genetic engineering, principles and methods, vol 14. Plenum Press, New York, pp 139–162. **Schemske DW**, Lande R (1985) Evolution 39:41–52. **Scutt CP**, Gates PJ, Gatehouse JA, Boulter D, Croy RRD (1990) Mol Gen Genet 220:409–413. **Sears ER**, Genetics 22:130–181. **Seavey SR**, Bawa KS (1986) Bot Rev 52:195–219. **Shivanna KR**, Johri BM (1985) The angiosperm pollen: structure and function. Wiley Eastern, Dehli. **Singh A**, Kao T-H (1992) Int Rev Cytol 140:449–483. **Slatkin M** (1987) Science 234:787–792. **Specht CA**, Stankis MM, Giasson L, Novotny CP, Ullrich RC (1992) Proc Natl Acad Sci USA 89:7174–7178. **Specht CA**, Stankis MM, Novotny CP, Ullrich RC (1994) Genetics 137:709–714. **Spellig T**, Bölker M, Lottspeich F, Frank RW, Kahmann R (1994) EMBO J 13:1620–1627. **Stamberg J**, Koltin Y (1973) Heredity 30:15–26. **Stankis MM**, Specht CA, Yang H, Giasson L, Ullrich RC, Novotny CP (1992) Proc Natl Acad Sci USA 89:7169–7173. **Stein JC**, Nasrallah JB (1993) Plant Physiol 101:1103–1106. **Stein JC**, Howlett B, Boyes DC, Nasrallah ME, Nasrallah JB (1991) Proc Natl Acad Sci USA 88:9916–8820.

Takayama S, Isogai A, Tsukamoto C, Ueda Y, Hinata K, Okazaki K, Suzuki A (1987) Nature 326:102–105. **Thompson KF**, Taylor JP (1966) Heredity 21:345–364. **Thompson RD**, Kirch H-H (1992) Trends Genet 8:381–387. **Tobias CM**, Howlett B, Nasrallah JB (1992) Plant Physiol 99:284–290. **Trick M** (1990) Plant Mol Biol 15:203–205. **Trick M**, Flavell RB (1989) Mol Gen Genet 218:212–217. **Trick M**, Heizman P (1992) Int Rev Cytol 140:485–523. **Tsai D-S**, Lee HS, Post LC, Kreiling KM, Kao TH (1992) Sex Plant Reprod 5:256–263. **Tymon AM**, Kües U, Richardson WVJ, Casselton LA (1992) EMBO J 11:1805–1813.

Umbach AL, Lalonde BA, Kandasamy MK, Nasrallah JB, Nasrallah ME (1990) Plant Physiol 93:739–747.

van der Donk JAWM (1975) Mol Gen Genet 141:305–317. **van Engelen FA**, Hartog MV, Terry LT, Taylor B, Sturm A, van Kammen A, deVries SC (1993) Plant J 4:855–862. **Verma SC**, Renuka M, Indra D (1977) Proc R Soc Lond Ser B 196:131–157.

Walker JC (1993) Plant J 3:451–456. **Walker JC**, Zhang R (1990) Nature 345:743–746. **Watanabe M**, Nou SI, Takayama S, Yamakawa S, Isogai A, Suzuki A, Takeuchi T, Hinata K (1992) Plant Cell Physiol 33:343–351. **Wendland J**, Kothe E (1994) Bio-Engineering 10(2):64. **Whitehouse HLK** (1950) Ann Bot 14:199–216. **Wilkie D** (1956) Heredity 10:247–256. **Williams GC** (1975) Sex and evolution. Princeton University Press, Princeton. **Williams W** (1959) Nature 184:527. **Wilmsen Thornhill N** (ed) (1993) The natural history of inbreeding and outbreeding. University of Chicago Press, Chicago. **Wright S** (1977) Evolution and the genetics of population vol III. Experimental results and evolutionary deductions. University of Chicago Press, Chicago.

Xu B, Grun P, Kheyr-Pour A, Kao T-H (1990) Sex Plant Reprod 3:54–60.

Yee AR, Kronstad JW (1993) Proc Natl Acad Sci USA 90:664–668.

Zuberi MI, Lewis D (1988) Heredity 61:367–377.

Dr. Simon J. Hiscock
Dr. Ursula Kües
Department of Plant Sciences
University of Oxford
South Parks Road
Oxford OX1 3RB, UK

Prof. Dipl.-Ing Dr. Ulf Stahl
Institut für Biotechnologie
Fachgebiet Mikrobiologie und Genetik
Technische Universität
Gustav-Meyer-Allee 25
13355 Berlin, Germany

III. Mutation: Modified Starch Metabolism in Mutant and Transgenic Plants

By P. Sathish, Chuanxin Sun, Andres Lönneborg and Christer Jansson

1. Introduction

Unravelling the various steps in a metabolic pathway is a challenging task. The partitioning of metabolites and enzymes between different cellular compartments, such as mitochondria, plastids, vacuoles and endoplasmic reticulum, in eukaryotic cells, adds an additional level of complexity to the work. The use of mutants, and more recently also transgenic organisms, is often a very powerful approach to establish the correlation between a genetic locus and a metabolic activity and to elucidate the sequential steps in a metabolic pathway.

The employment of mutants in the study of starch biosynthesis has long traditions. Already in the late 1850s Gregor Mendel used the garden pea (*Pisum sativum*) as a basis for his laws of inheritance (Mendel 1865). As a result, starch mutants often have a scoreable phenotype, thus the screening process is straightforward. Mendel concentrated on characteristics such as differences in flower colour and seed shape. The wrinkled pea mutation *rugosus* is caused by a transposon insertion in one of the central genes involved in starch synthesis (see Sect. 4.1; Bhattacharyya et al. 1990). The wrinkled phenotype has been used to identify starch mutations in the seed, while iodine staining has been used to detect mutants in green tissues. Today, mutants for many of the important enzymes and transporters in the biosynthetic pathway of starch have been detected. In this review we will describe some of these mutants and their importance in the study of starch synthesis.

2. Starch Structures

Starch is a mixture of amylose and amylopectin, both glucose polymers (Fig. 1). Amylose is a mostly linear polymer of 200–2000 α-1,4 bonded glucose moieties with rare α-1,6 branch points. Amylopectin, on the other hand, is highly α-1,6 branched, with a complex structure of 10^6–10^8 mol. wt. and up to 3×10^6 glucose subunits, making it one of the largest biological molecules in nature (see Morrison

Progress in Botany, Vol. 56

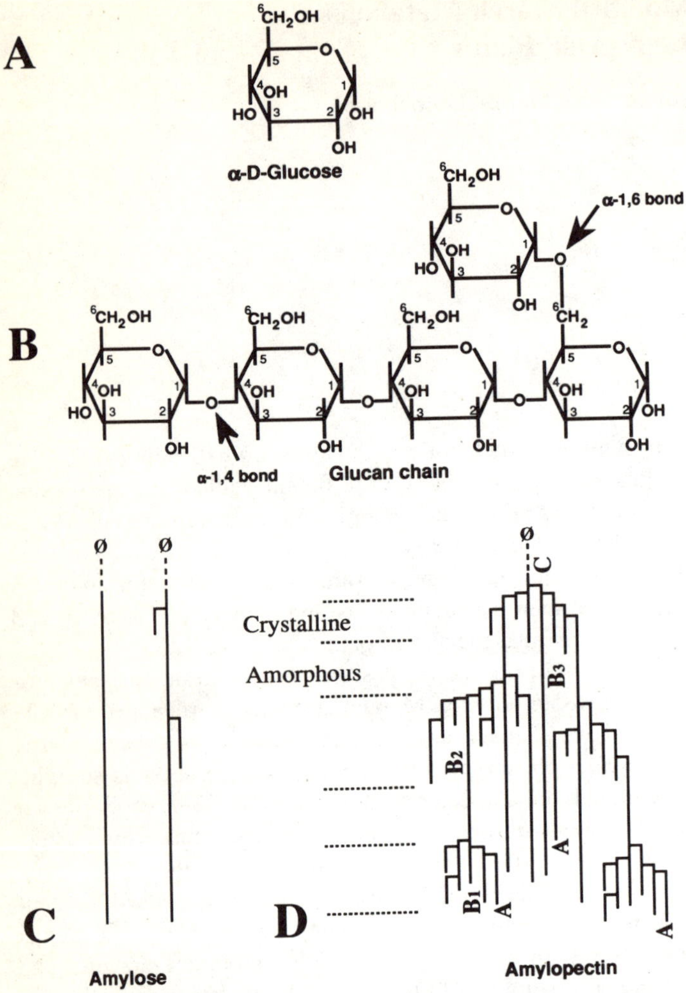

Fig. 1. Structure of starch and its components. **A** Glucose; **B** a glucan chain showing the α-1,4 and α-1,6 bonds; **C** amylose; **D** amylopectin. The crystalline branch clusters and the amorphous regions between the clusters are indicated. The A-chains are unbranched, whereas B-chains and the single C-chain with the reducing end (∅) are branched. The B-chains are further divided according to length from short B1-chains to long B3-chains (see Morrison and Karkalas 1990 for references)

and Karkalas 1990 for a detailed description of the amylose and amylopectin molecules). In the plant, starch is deposited as starch granules in chloroplasts of photosynthetic tissues or in amyloplasts of endosperm (seeds), tubers and roots.

In most plants, starch consists of 20–30% amylose and 70–80% amylopectin. The structure of the amylose and amylopectin molecules, the amylose/amylopectin ratio, the degree of substitution, and the association of lipids and proteins are responsible for the functional qualities of starch, and thereby affect industrially important properties such as gelatinization, retrogradation, viscosity, fermentability, behaviour as granules, and behaviour as pure and mixed polymer sheets. The size of the starch granules is heterogeneous, with a diameter of 0.5–100 μm. In some cereals, such as barley, starch is composed of one fraction with large granules (A-starch) and another fraction with small granules (B-starch). In oats, on the other hand, the starch granules are more homogeneous in size and correspond approximately to B-starch. Of the various forms of modified starches that occur in nature only phosphorylation of potato starch is understood to some extent.

3. Metabolic Pathways for Starch Synthesis

Excess photosynthate that is not used directly in cell metabolism is either transported as sucrose to sink tissues or temporarily accumulated as "transitory" starch in the chloroplasts (for recent reviews on starch metabolism in leaves, see Visser and Jacobsen 1993; Preiss 1991). Carbon enters starch via the gluconeogenesis pathway where glycerate-3-phosphate formed from CO_2 fixation is converted to glucose-1-phosphate in the chloroplast stroma. Glucose-1-phosphate is then activated to ADP-glucose. This reaction, which is catalyzed by the enzyme ADP-glucose pyrophosphorylase (ADP-G-PP), is the first committed step in starch synthesis. The enzyme ADP-G-PP is tightly regulated by the ratio of triose-phosphates (triose-P) to P_i. In this way, ADP-G-PP senses the energy status in the cell and can either promote starch synthesis under high energy conditions (high triose-P/P_i), or divert triose phosphates to the cytosol when energy is required (low triose-P/P_i). This is the main reason why ADP-G-PP is considered the key regulatory enzyme in starch synthesis.

In plants that store reserve starch in seeds, tubers or roots, the picture of starch synthesis may be somewhat different from that found in green tissues. Figure 2 shows a schematic representation of the metabolic pathways for starch synthesis in amyloplasts. Carbohydrates are transported from photosynthetic tissues as sucrose. There are different pathways possible for the conversion and transport of sucrose. In one pathway, sucrose synthase converts sucrose into glucose and fructose. Fructose and glucose can then be converted to glucose-1-phosphate in the gluconeogenesis pathway, or to the triose-P dihydroxyacetone phosphate and glyceraldehyde-3-phosphate, in glycolysis. Based on the results from photosynthetic tissues, it has long been assumed that carbohydrates are transferred to the amyloplast mainly in the form of triose-P, using the triose-P/P_i translocator. However, the results of Keeling et al. (1988) and Tyson and Rees (1988) demonstrated that glucose-1-phosphate can be transported across the envelope membrane of wheat. It appears that transport mechanisms in the amyloplast membrane can differ between plant species. In some plants, the triose-P/P_i translocator seems to be the major transport system, while a system involving glucose-1-phosphate or other hexoses dominates in others. As can be seen in Fig. 2, a transport system involving hexoses requires a lower energy input than the triose-P/P_i translocator. In addition to these two transport mechanisms, it has recently been shown that ADP–glucose can also be transported into amyloplasts

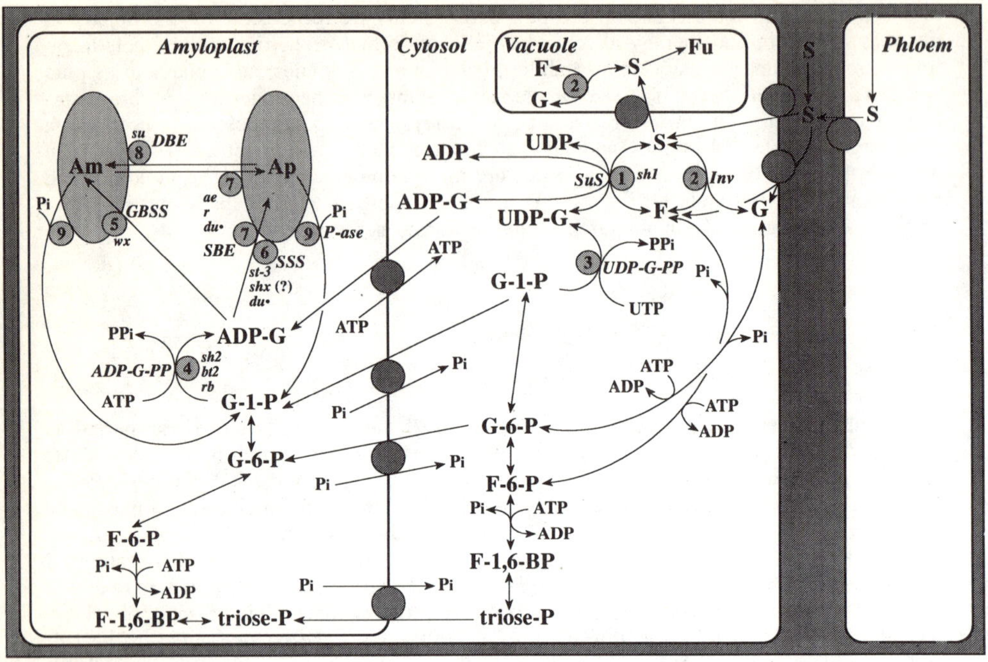

Fig. 2. A schematic representation of starch synthesis in sink tissues. The import of sucrose from the phloem is shown as an apoplastic unloading (rather than symplastic) which seems to be the dominating pathway in storage and reproductive tissues (Taiz and Zeiger 1991). Various transport mechanisms for import of carbon into the amyloplast are depicted (see text for details). Abbreviations, metabolites: *S* sucrose; *G* glucose; *F* fructose; *Fu* fructans; *G-1-P* glucose-1-phosphate; *G-6-P* glucose-6-phosphate; *F-6-P* fructose-6-phosphate; *triose-P* triose-phosphates; *Am* amylose; *Ap* amylopectin; *UDP* uridine diphosphate; *ADP* adenosine diphosphate; *UTP* uridine triphosphate; *ATP* adenosine triphosphate; *PPi* pyrophosphate; *Pi* orthophosphate; *UDP-G* UDP-glucose; *ADP-G* ADP-glucose. Abbreviations, enzymes: Selected enzymes involved in starch synthesis are numbered from *1–9*. SuS (*1*), sucrose synthase; Inv (*2*), sucrose invertase; UDP-G-PP (*3*), UDP-glucose pyrophosphorylase; ADP-G-PP (*4*), ADP-glucose pyrophosphorylase; GBSS (*5*), granule-bound starch synthase; SSS (*6*), soluble starch synthase; SBE (*7*), starch branching enzyme; DBE (*8*), debranching enzyme; P-ase (*9*), phosphorylase. Abbreviations, mutations: Mutations affecting enzymes or transporters in starch metabolism are listed by respective protein. Mutations affecting regulatory rather than metabolic genes are indicated (•)

of cultured sycamore (*Acer pseudoplatanus*) cells and into spinach (*Spinacia oleracea* L.) chloroplasts (Pozueta-Romero et al. 1991a,b). Thus, it is possible that carbon can enter also the amyloplast through an ADP-G transporter. Production of cytosolic ADP-glucose in developing rice endosperm was shown to be catalyzed by an ADP-specific isoform of sucrose synthase (SuS) (Lim et al. 1992). The dramatic effects on starch accumulation that are seen in ADP-G-PP mutants (see Sect. 4.a–c below) are not easily understood if ADP-sucrose synthase should contribute significantly to the level of ADP-glucose. An obvious explanation to the disparate results from work with different cereals is that the activities of sucrose-metabolizing enzymes in the amyloplast vary between different plants and between photosynthetic and non-photosynthetic tissues of the same plant.

In photosynthetic and non-photosynthetic tissues alike, the glucose moiety of ADP-glucose is incorporated in the growing starch polymer with the help of starch synthases. Granule-bound starch synthase (GBSS) is believed to exclusively catalyze the synthesis of amylose, while soluble starch synthase (SSS) appears to be responsible for the synthesis of the linear α-1,4 structure of amylopectin (Smith and Denyer 1992). In addition, a role for plastidial starch phosphorylase (P-ase) in starch synthesis rather than degradation cannot be ruled out (Smith and Denyer 1992). The formation of α-1,6 linkages in amylopectin is catalyzed by starch branching enzymes (SBE). Evidence from studies with the *sugary* (*su*) mutant of maize indicates that a starch debranching enzyme (DBE) might also be required for the synthesis of the mature amylopectin molecule.

Most (if not all) plants contain multiple forms of both GBSS, SSS and SBE in the amyloplasts. In general, two forms of GBSS (GBSSI and GBSSII) and SSS (SSSI and SSSII) have been reported (Preiss 1991). It has recently been suggested, however, that, at least in pea, GBSSII and SSSII are identical (Smith and Denyer 1992; Denyer et al. 1993). The arguments supporting this notion are compelling. First, SSSI and SSSII in both pea embryos and maize kernels are kinetically and antigenetically dissimilar and of different molecular weights, as are GBSSI and GBSSII (Preiss 1991; Smith and Denyer 1992; Denyer et al. 1993). Second, SSSII and GBSSII in pea embryos are antigenetically related, have similar molecular weights, exhibit a similar pattern upon peptide mapping, and obtained amino acid sequences of SSSII are identical to those deduced for GBSSII (Denyer et al. 1993). Also, the kinetic properties of the pea SSSII and GBSSII are similar (Smith and Denyer 1992). Third, SSSII in maize is primer-dependent in vitro in the presence of 0.5 M citrate, whereas SSSI is primer-independent (Preiss 1991). For SBE, two or more forms have been identified in a large number of plants and tissues (Preiss 1991). The reasons for the multiple forms of starch synthases and SBE are not clear. It is conceivable that various isoforms could be recruited during different stages of plant development or during different phases of starch synthesis.

It should be clear from the discussion above that there is a large number of aspects that need to be addressed in order to satisfactorily understand and exploit starch metabolism in plants. These include the identity of the transport mechanism that provides carbon for starch synthesis in the amyloplast, the function of the various isoforms of GBSS, SSS and SBE, and the interaction between GBSS, SSS and SBE during amylose and amylopectin synthesis. The use of novel and available mutants and transgenic plants has the potential to shed light on these and other questions.

4. Mutants

The mutants described in this section are categorized on the basis of their effect on the enzymes involved in starch metabolism. Their role in altering the activity of the enzymes involved in starch metabolism is not fully understood. A compilation of starch mutants, their primary lesions, and phenotypes is shown in Table 1.

a) Shrunken-2

Shrunken-2 (*sh-2*) is one of the two mutations in two unlinked loci of *Zea mays* that is known to affect the activity of ADP-G-PP in the endosperm. Maize *sh-2* mutants exhibit poorly developed, shrunken endosperms. The kernels have a high concentration of sucrose and very low levels of starch. The ADP-G-PP activity was dosage-dependent on *sh-2* (Tsai and Nelson 1966), indicating that *sh-2* is a structural gene for ADP-G-PP. Mutational alterations at this locus that result in approximately 10% wild-type ADP-G-PP activity levels have an almost wild-type phenotype, while those with less than 2% of the wild-type activity appear to be lethal (Hannah et al. 1980).

Examination of the deduced amino acid sequence of the maize *sh-2* allele reveals close similarity to known ADP-G-PP subunits. However, at the DNA level, there is little

Table 1. Plant mutant genes and enzymes of starch synthesis

Plant	Mutant[a]	Gene (ISPMB No.)/ gene product (E.C. No.)	Phenotype	Reference
Amaranthus hypochondriacus	*wx*	*gss1*(251)/GBSS I (2.2.4.1.21:1)	Negligible levels of amylose	Konishi et al. (1985); Okuno and Sakaguchi (1982)
Arabidopsis thaliana	*adg-1*	*agp1* (246), *agp2* (247)/ADP-G-PP (2.2.7.7.27:1, 2.2.7.7.27:2)	Negligible levels of starch	Lin et al. (1988a)
	adg-2	*agp2* (247)/ADP-G-PP (2.2.7.7.27:2)	Low levels of starch	Lin et al. (1988a)
Chlamydomonas reinhardtii	*wx*	*gss1* (251)/GBSS I (2.2.4.1.21:1)	Negligible levels of amylose	Delrue et al. (1992)
	st-3	*sss2*/SSS II	Low levels of starch, high levels of amylose, and modified form of amylopectin	Fontaine et al. (1993)

Table 1. (Continued)

Plant	Mutant[a]	Gene (ISPMB No.)/ gene product (E.C. No.)	Phenotype	Reference
Hordeum vulgare	*wx*	*gss1* (251)/GBSS I (2.2.4.1.21:1)	Negligible levels of amylose	Rohde et al. (1988)
	shx	*sss1*(253)/SSS I (2.2.4.1.21:2)	Altered starch granule size	Schulman and Ahokas (1990)
	Risø	Not characterized	Lowered starch levels and altered granule size	Kreis (1986); Tester et al. (1993)
Oryza sativa	*wx*	*gss1* (251)/GBSS I (2.2.4.1.21:1)	Negligible levels of amylose	Sano (1984)
	ae	*sbe3*/SBE III (2.4.1.18:3)	High levels of amylose and abnormal amylopectin	Mizuno et al. (1993); Kaushik and Khush (1991)
	du	Not characterized	Lowered starch levels and higher amylose levels	Kaushik and Kush (1991)
	flo	Not characterized	High in amylose	Yano et al. (1985); Satoh and Omura (1981)
	white core	Not characterized	Not characterized	Satoh and Omura (1981)
Pisum sativum	*rb*	*agp*/ADP-G-PP	Lowered levels of starch and high levels of sucrose	Wang and Hedley (1991); Kooistra (1962)
	r	*sbe1*/SBE I (2.4.1.18:1)	Lowered levels of starch, higher levels of amylose, sucrose, and lipids; altered starch granule size	Bhattacharyya et al. (1990)
Solanum tuberosum	*amf*/*wx*	*gss1* (251)/GBSS I (2.2.4.1.21:1)	Negligible levels of amylose	Hovenkamp-Hermelink et al. (1987)
Sorghum bicolor	*wx*	*gss1* (251)/GBSS I (2.2.4.1.21:1)	Negligible levels of amylose	Hseih (1988)

Table 1. (Continued)

Plant	Mutant[a]	Gene (ISPMB No.)/ gene product (E.C. No.)	Phenotype	Reference
Triticum aestivum	*wx*	*gss1* (251)/GBSS I (2.2.4.1.21:1)	Negligible levels of amylose	Nakamura et al. (1983)
Zea mays	*sh-2*	*agp2* (247)/ADP-G-PP (2.2.7.7.27:2)	High levels of sucrose and low levels of starch	Bhave et al. (1990); Hannah and Nelson (1976); Tsai and Nelson (1966)
	bt-2	*agp1* (246)/ADP-G-PP (2.2.7.7.27:1)	High levels of sucrose and low levels of starch	Preiss et al. (1990); Creech (1965); Cameron and Teas (1954)
	wx	*gss1* (251)/GBSS I (2.2.4.1.21:1)	Negligible levels of amylose	MacDonald and Preiss (1985); Tsai (1974); Sprague et al. (1943); Weatherwax (1922)
	du	Regulatory gene for SSS II and SBE II	Low levels of starch and high levels of amylose	Boyer and Preiss (1981); Yeh (1981); Boyer and Preiss (1978a)
	su	*sde* (261)/DBE	High levels of sugars and low levels of starch; highly branched amylopectin	Pan and Nelson (1984)
	su-2	Not characterised	Amylopectin with increased short chains	Takeda and Preiss (1993)
	ae	*sbe2*/SBE II a and b (2.4.1.18:2a, 2.4.1.18:2b)	High levels of amylose, abnormal amylopectin	Singh and Preiss (1985); Hedman and Boyer (1982); Boyer and Preiss (1978a)

[a]Abbreviations: *wx, waxy; shx/sh, shrunken; ae, amylose extender, du, dull; flo, floury; r, rugosus; amf, amylose free; bt, brittle; su, sugary; gss 1*/GBSSI, granule-bound starch synthase; *agp*/ADP-G-PP, ADP-glucose pyrophosphorylase; *sss*/SSS, soluble starch synthase; *sbe*/SBE, starch branching enzyme; *sde*/DBE, starch debranching enzyme.

sequence similarity. This suggests that there might be two slightly dissimilar subunits for plant ADP-G-PP and that this mutation affects the largest of the two units. It has also been speculated, based on identity at the DNA level, that the two different coding genes may have a common origin via gene duplication and divergence.

b) Brittle-2

Brittle-2 (*bt-2*) is another mutation that affects the activity of ADP-G-PP in *Zea mays* endosperm. Both the activity of ADP-G-PP and the starch content are reduced in this mutant. Northern analysis of the mutant and wild-type mRNA established that an mRNA transcript was missing in the mutant (Barton et al. 1986). Immunological analysis using the antibodies raised against the spinach (*Spinacia oleracea* L.) ADP-G-PP small subunit and spinach ADP-G-PP large subunit showed that the mutation at the *bt-2* allele resulted in the loss of the ADP-G-PP small subunit (Preiss et al. 1990).

c) rb

Mutation at the *rb* locus of *Pisum sativum* results in pea seeds which are wrinkled when mature. These seeds have a decreased starch content and increased levels of sucrose. Analysis of the enzymes involved in starch synthesis from the wild-type and *rb* mutant line showed that the mutation dramatically reduced the activity of ADP-G-PP. The mutation appears to eliminate one of the ADP-G-PP subunits (Smith et al. 1989; Hylton and Smith 1992). Superficially, this mutation is analogous to the *sh-2* and *bt-2* mutations of maize endosperm.

d) adg-1: adg-2

These are mutations that affect the ADP-G-PP activity in *Arabidopsis thaliana* leaves. The activity of ADP-G-PP is less than 2% and 5% in *adg-1* and *adg-2* respectively. Mutant line *adg-1* was devoid of both subunits of ADP-G-PP, while plants of the mutant line *adg-2* were devoid of the large subunit. Plants of the *adg-1* line and *adg-2* line had 2% and 40% starch levels respectively.

e) Waxy

Waxy (*wx*) are perhaps the best studied among the starch mutants. There has been a long-term interest in the *wx* locus due to its importance in agriculture and genetics. Studies using the *wx* mutants of maize (*Zea mays* L.) revealed that endosperm having the *wx* alleles contains amylopectin exclusively. Very little starch synthase activity was measured, especially in older kernels, even though the amount of starch in both the wild-type and homozygous *wx* genotypes remained the same (Nelson and Rines 1962). Starch synthase activity increased with the increase in the *wx* gene dosage (Tsai 1974), and in most of the *waxy* maize mutants analyzed, a 60-kDa protein was found to be missing from the starch granules of endosperm, pollen and embryo sac (Echt and Schwarz 1981). The granule-bound starch synthase activity in *wx* mutants correlated with the abundance of the 60-kDa protein (MacDonald and Preiss 1985). This correlative evidence suggests that the *waxy* protein might be granule-bound starch synthase.

Mutants of the waxy phenotype have also been identified in *Amaranthus hypochondriacus* (expressed in the perisperm), *Chlamydomonas reinhardtii* (expressed during both autotrophic and heterotropic growth), *Hordeum vulgare, Oryza sativa, Sorghum bicolor* (all expressed in the endosperm) and *Solanum tubersum* (expressed in tubers). The *wx* mutants in *Chlamydomonas* were obtained using UV mutagenesis, and in *Solanum* after screening minitubers derived from X-ray mutagenized tubers. The mutant in *Solanum,* which is expressed in the dicot storage tissue, is referred to as *amf* (amylose-free) mutant. Polyclonal antibodies prepared against the 60-kDa *wx* protein from potato cross-reacted with the *wx* proteins of maize and *Amaranthus* (Vos - Scheperkeuter et al. 1986). Although the antiserum neutralized about 40% of the *Amaranthus* granule-bound starch synthase activity, it was unable to inhibit the potato granule-bound starch synthase activity.

Purification of the 60-kDa protein from the maize *wx* line enabled the isolation of the cDNA clone (Shure et al. 1983). The physical map was deduced at the molecular level and it correlated well with the genetic map (Wessler and Varagona 1985). Some of the maize *waxy* mutants are phenotypically unstable, and are attributed to the insertion of controlling elements such as *activator* (*Ac*) (Federoff et al. 1983; Behrens et al. 1984; Müller-Neumann et al. 1984; Pohlman et al. 1984), *dissociation* (*Ds*) (Federoff et al. 1983), *enhancer* (*En*) (Pereira et al. 1985), and *suppressor-mutator* (*Spm*) (Schwarz-Sommer et al. 1984). Excision of the *Ds* element from the *wx* mutant resulted in the isolation of two new alleles, S5 and S9, whose product resembled *wx* proteins and exhibited altered starch synthase activities (Wessler et al. 1986). A DNA sequence of the genomic DNA including the *wx* locus of maize mapped all of the transposon-induced mutations at the *wx* locus that were analyzed at the molecular level (Klösgen et al. 1986). Structural analyses of the *wx* locus from *Hordeum* (Rohde et al. 1988), and *Oryza* (Wang et al. 1990) have been made. All *wx* loci have from 11 to 13 introns as well as a putative transit peptide. Using the rice genomic clone for the *wx* locus, rice cDNA libraries were screened. Interestingly, a second and larger transcript was noted in two lines tested. However, this second transcript seemed to be the result of improper splicing of the first intron in the granule-bound starch synthase locus (Okagaki 1992). The *wx* locus in *Solanum* has been sequenced, and the *amf* mutation was shown to be a deletion of a single AT base pair in the region coding for the transit peptide. This frameshift mutation causes premature termination of translation and, consequently, the synthesis of a truncated peptide (Leij et al. 1991). The mature proteins from all species so far analyzed share identities between 70 to 87% and similarities between 84 to 95% at the amino acid level. Recently, the gene encoding the *wx* protein in wheat has also been cloned, and it was shown to differ from all others in that it contained an 11 amino acid insertion towards the N-terminus (Ainsworth et al. 1993).

The validity of the identity of the major granule-bound starch synthase as the *wx* protein was contested when another distinct isoform of granule-bound starch synthase, with measurable activity *in vitro*, was noted in *Pisum sativum* and *Solanum* (Smith 1990). This second form of granule-bound starch synthase contained a novel 203-amino-acid domain at the N-terminus, and was expressed during early pea development or potato tuber development (Dry et al. 1992). However, studies of *Chlamydomonas* showed that this protein is the gene product of the *wx* locus that controls the biosynthesis of amylose (Delrue et al. 1992). A more recent study, using pea embryos, has shown that the second form of granule-bound starch synthase might actually be an isoform of soluble starch synthase (Denyer et al. 1993). This study also demonstarated that the granule-bound starch synthase I activity can be recovered when the protein is solubilized in a denatured form and then renatured. The data from the *Chlamydomonas* studies also indicate that the *wx* locus, at least in *Chlamydomonas* besides controlling the level of amylose production, also controls the synthesis of an altered form of amylopectin (Delrue et al. 1992). The best evidence that the *wx* protein is granule-bound starch synthase I comes from the ability to stop granule-bound starch synthase I activity by the antisense granule-bound starch synthase I RNA from potato (see Sect. 5.c; Visser et al. 1991) and *Manihot esculenta* (see Sect 5.d; Salehuzzaman et al. 1993) in transgenic potatoes.

f) Shrunken

This mutation was noted in *Hordeum vulgare* and is characterized by a shrunken endosperm. The *Shrunken* (*shx*) mutation is transmitted as a single Mendelian recessive allele displaying xenia (Ahokas 1991). Although the activities of several of the starch-synthesizing enzymes were reduced in this mutant, it is believed that the primary effect of the mutation is on soluble starch synthase I. The polypeptide composition of the endosperm of the *shx* mutant resembled that of the younger wild-type, and was similar to the morphological slower maturation of *shx* grains (Tyynelä and Schulman 1993). The relationship between these enzymological changes and the changes in starch granule size distribution observed for *shx* (Schulman et al. 1994), however, needs to be established.

g) st-3

The *st-3* mutation in *Chlamydomonas reinhardtii* leads to an apparent increase in amylose with a considerable modification of the residual amylopectin. These mutants were developed by X-ray mutagenesis. They accumulate low amounts of starch (20 to 40% of wild-type levels), and are olive green in colour. The amylopectin fraction displayed very short (2–7 glucose units) chains and a concomitant decrease in intermediate (8–10 units) glucan chains. There were no defects in branching enzyme activities. The *st-3* mutants were specifically defective for a subset of soluble starch synthase activities that can be compared to higher plant-soluble starch synthase II. The activity of the soluble starch synthase II is proportional to the number of wild-type alleles in vegetative diploids, suggesting that *st-3* is the structural gene for SSSII.

h) Dull

Dull (*du*) was presumed to be a mutant line for soluble starch synthase in *Zea mays. du* displays a 30% reduction in starch accumulation and a net increase in amylose content (Yeh et al. 1981). This mutant is found to be defective not only for soluble starch synthase II, but also for branching enzyme II (Boyer and Preiss 1978a; Boyer and Preiss 1981). As a result, it is assumed that *du* is a regulatory gene affecting both enzymes, rather than a structural gene.

In rice, *du* is transmitted as a monogenic recessive allele (Kaushik and Khush 1991). *du* mutants in rice were found to be non-allelic based on the segregation pattern for the endosperm appearance and amylose content of individual F_2 grains. In rice, *du* has been designated as a series beginning from 1 to 5 (Satoh 1985; Satoh and Amano 1987). These alleles lie on different chromosomes (Yano et al. 1988; Kaushik and Khush 1991).

i) Sugary

The *sugary* locus, *su* of *Zea mays* is expressed in the endosperm and seedling. This locus is responsible for the activity of starch debranching enzyme, as the activity is proportional to the copy number of the *su* gene. The endosperm of mutants at this locus is characterized by high sugar levels and low starch levels. Also, the starch is more highly branched.

j) Sugary-2

Sugary-2 (*su-2*) in *Zea mays* is a mutation that increases both the amylose content and the λ_{max} of the amylopectin iodine complex. Amylopectin fractions of the starch displayed an increase in the short chains (2–7 units) and the relative decrease in intermediately sized glucans. However, the enzymological defect or the location of the gene remains unknown.

k) Rugosus

Rugosus (*r*) is probably the first mutant in the starch metabolic pathway that was ever studied, when Mendel used the shape of the mature *Pisum sativum* as a marker in his studies on inheritance (Mendel 1865). Plants that are homozygous for *r* locus have lowered starch levels, enhanced amylose, sucrose and lipid levels and the starch grains are small and compound. The *r* lesion is associated with the complete absence of one of the isoforms for SBE from early development of the pea embryo (Smith 1988). A major polymorphism in the SBEI gene between near-isogenic *RR* and *rr* lines shows 100% cosegregation with the *r* locus, establishing that the *R* locus is the structural gene for SBEI. An aberrant transcript for SBEI is produced in *rr* embryos, due to a 0.8-kb DNA insertion that is similar to the *Ac*/*Ds* family of transposable elements from maize (Bhattacharyya et al. 1990).

l) Amylose extender

Amylose extender (*ae*) mutants are characterized by seeds containing high levels of amylose, and abnormal amylopectin. The amylopectin has longer than normal inner and outer branches. Also, an intermediate material with a lower molecular weight than amylose is noted (Baba and Arai 1984). The *ae* mutants in *Zea mays* are deficient in starch branching enzyme-IIb. A near-linear relationship between the dosage of the recessive *ae* allele and the SBEII activity was observed (Hedman and Boyer 1982). Later, it was shown that SBEIIb is identical to SBEIIa in maize (Singh and Preiss 1985). It is clear whether the SBEIIb form is solely responsible for the abnormally branched starch molecules found in *ae* mutants.

SBE III activity was lost in two of the five *ae* mutants of rice (Mizuno et al. 1993). The *ae* allele is transmitted as a recessive trait in rice (Kaushik and Khush 1991). Even though proteins corresponding to the 87-kDa SBE III enzyme were present in the other three *ae* mutants, their activity was greatly reduced, suggesting that they might be inactive forms of the enzyme (Mizuno et al. 1993). Identification of the substitutions in the mutations might lead to the elucidation of the *cis* element(s) required for the tissue-specific gene expression.

m) Floury

Chemical mutagens were employed to isolate the *floury* (*flo*) mutants in rice. Five of them were found to have a high amylose content (Yano et al. 1985). The *floury* mutants are characterized by their soft white endosperm, which crumbles easily into powder. Based on segregation patterns, it was determined that there are two genes, *flo-1* and *flo-2* (Kaushik and Khush 1991). The *opaque* gene *o* (Rutger et al. 1985) is found to be allelic to *flo-1* (Kaushik and Khush 1991). Based on the physical and ultrastructural analysis of the starch grains, it was concluded that the starch from *flo* mutants is similar to the starch from the *ae* mutants in maize (Yano et al. 1985).

n) White core

The *white core* mutants of rice were produced using chemical mutagens and are characterized by a central white portion in the rice endosperm. The *white core* mutants are found to be non-allelic to each other (Kaushik and Khush 1991).

o) Risø

Mutant barley lines were produced at the Risø Research Centre, Risø, Denmark, from Bomi and Carlsberg II backgrounds using chemical and physical mutagens. These mutants were shown to have impaired starch synthesis, resulting in a lowered starch content and in altered granule sizes.

5. Transgenic Studies

a) Expression of *E. coli* ADP-Glucose Pyrophosphorylase in Potato

ADP-G-PP is sensitive to allosteric effectors and has been hypothesized to play an important role in both starch and glycogen biosynthesis. Stark et al.(1992), using transgenic potato plants carrying the *glgC* gene encoding the homotetrameric ADP-G-PP enzyme, demonstrated that ADP-G-PP is indeed the key enzyme in starch metabolism. Using ADP-G-PP constructs that are sensitive to the activator fructose 1,6-bisphosphate at different levels, they showed that ADP-G-PP activity is rate-limiting in starch biosynthesis and that the regulatory properties of ADP-G-PP, not the amount of enzyme protein, make it rate-limiting.

b) Inhibition of ADP-Glucose Pyrophosphorylase in Potato Using an Antisense ADP-G-PP Gene Coding for the Small Subunit

Müller-Röber et al. (1992) introduced a chimeric gene containing the coding region of the small subunit of ADP-G-PP in the antisense orientation with respect to the CaMV 35S promoter into potato plants. ADP-G-PP activity was partially inhibited in the transgenic plant leaves. In the transgenic tubers, the activity of ADP-G-PP was completely inactivated and there was a low accumulation of starch (2–5% of the control). Transgenic potato plants produced a large number of tubers, most of which were small in size resulting in increased fresh weight. However, there was a decrease in total dry weight. These small tubers were found to contain high levels of sucrose (30% of dry weight) and glucose (8% of dry weight). The inhibition of ADP-G-PP expression did not seem to influence the expression of other genes involved in starch biosynthesis, but was found to have a profound effect on the expression of the sucrose phosphate synthase gene and genes coding for the patatin family of proteins.

c) Expression of an Antisense Gene for Potato Granule-Bound Starch Synthase in Potato

Full-length GBSS cDNA in reverse orientation was fused between the CaMV 35S promoter and the nopaline synthase terminator (Visser et al. 1991). Inhibition of the GBSS activity in potato was brought by *Agrobacterium*-mediated transformation. The expression of the antisense construct of GBSS varied in stomatal guard cells and columella cells of the root tips, but was stable in the tubers. The activity and/or amount of GBSS protein in the tubers was affected, and the starch content varied between 70 and 100%. Some of the transformed lines of potato expressing the antisense GBSS had a phenotype closely resembling amylose-free mutants.

d) Expression of an Antisense Gene for Cassava Granule-Bound Starch Synthase in Potato

Attempts were made to inhibit the single copy gene expression for GBSS in potato using the 5′ region of the maize genomic GBSS gene sequence (Visser et al. 1989). The antisense effects were only partial and not accompanied by any phenotypic changes in the tuber starch granules. Later, an attempt was made to inhibit the same gene in potato using the *Manihot esculenta* Crantz. gene for GBSS, as it had a high degree of identity with the potato gene for GBSS (74% identity) (Salehuzzaman et al. 1993). This heterologous gene was fused in an antisense orientation with the potato GBSS promoter and nopaline synthase terminator. *Agrobacterium*-mediated transformation of potato plants resulted in

10 out of 42 plants showing an antisense effect. Three of the transformed lines had null activity for GBSS, and resulted in an almost complete amylose-free potato starch.

6. Prospects

Starch is an important raw material for industrial applications, both for food and non-food purposes, and the market is increasing. Of particular interest is the use of starch as a non-petroleum chemical feedstock for the manufacture of biodegradable polymers such as "plastics", and as a non-cellulose feedstock in the paper industry. These markets have enormous growth potentials, especially given the fact that crude oil and cellulose resources are shrinking, and also taking into account increasing environmental awareness. Since starch is not only totally biodegradable but also inexpensive, renewable and readily available, the replacement of certain petroleum products with starch-based products has stimulated intense research activity with respect to increased usage of starch as an industrial raw material.

Starch-based "plastics" currently on the market range from starch-polyethylene blends to more complex "plastics" such as Novon™ and Fertec™ with up to 80% starch. Low-starch polymers are mainly used as components in films, whereas high-starch polymers have a large number of material applications. Annual EC starch production is approaching 10 million tons and grows 4–5% annually, with 80% isolated from cereals and 20% from potatoes. Of this, 45% is used as either native or modified starch in the food, polymer, paper, chemical, building, and pharmaceutical industries, the rest being used as starch hydrolysates.

It has long been recognized that the development of new uses and markets for starch will benefit from the advent and analyses of starch mutants. However, the number of described and characterized mutants is still relatively low. Two classes of mutations that have been exploited in the starch-consuming industry are the *wx* (Sect. 4.e) and *ae* (Sect. 4.1). Waxy starch from maize and potato is used extensively in the food industry, and the high-amylose starch from the maize *ae* mutant is preferred in the production of films and "plastics". Another example of the exploitation of starch mutants is sweet corns, where various genetic lesions have led to increased sucrose concentrations at the expense of starch (Hannah 1990).

Concomitant with the increased interest in starch as an industrial raw material is a growing demand for the production of modified starch in transgenic plants (Shewmaker and Stalker 1992; Kishore and Somerville 1993; Visser and Jacobsen 1993). This approach offers the possibility to replace much of the post-harvest separations and chemical modifications, which are sometimes environmentally hazardous, expensive, and time-consuming. It also makes possible the production, *in planta*, of novel carbohydrates. A desirable product in transgenic plants would be all- or high-amylose starch for the polymer industry. This could be achieved by inactivation of all *sbe* genes (encoding SBE). Such a genotype would be difficult to obtain by conventional breeding or by random mutagenesis due to the presence of several isoforms of SBE. Selective inactivation of *sbe* genes would be expected to result in starches with altered branching patterns. Inactivation of the genes for GBSS should result in all-amylopectin starch for the food industry. Examples of post-harvest modifications that could be replaced by a transgenic approach are production of cross-linked starch for increased stability, production of starch with well-defined granule sizes, and substitutions such as hydroxypropylation, methylation, carboxylation and phosphorylation. Production of novel carbohydrates in transgenic plants has been demonstrated by the synthesis of cyclic dextrines from starch (Shewmaker and Stalker 1992).

In conclusion, carbohydrate metabolic engineering (Stitt 1994), including the production of modified starch in transgenic plants, has great future potentials. However, an absolute requirement for transgenic applications is a deeper understanding of carbohydrate metabolism and partitioning, and of the structure, function and regulation of the enzymes involved in starch metabolism.

Acknowledgement. This work was supported by the Swedish Agriculture and Forestry Research Council (SJFR) and the Swedish Council for Industrial and Technological Development (TFR).

References

Ahokas H (1991) Hereditas 114:281–284. **Ainsworth C**, Clark J, Balsdon J (1993) Plant Mol Biol 22:67–82.

Baba T, Arai Y (1984) Agric Biol Chem 48:1763–1775. **Barton C**, Yang L, Galvin M, Sengupta-Gopalan C, Borelli T (1986) In: Shannon JC, Knievel DP, Boyer CD (eds) Regulation of carbon and nitrogen reduction and utilization in maize. American Society of Plant Physiologists, pp 363–365. **Behrens U**, Fedoroff N, Laird A, Müller-Neumann M, Starlinger P, Yoder J (1984) Mol Gen Genet 194:346–347. **Bhattacharyya MK**, Smith AM, Ellis THN, Hedley C, Martin C (1990) Cell 60:115–122. **Bhave MR**, Lawrence S, Barton C, Hannah LC (1990) Plant Cell 2:581–588. **Boyer CD**, Preiss J (1978a) Carbohydr Res 61:321–334. **Boyer CD**, Preiss J (1978b) Biochem Biophys Res Commun 80:169–175. **Boyer CD**, Preiss J (1981) Plant Physiol 67:1141–1145.

Cameron JW, Teas HJ (1954) Am J Bot 41:50–55. **Creech RG** (1965) Genetics 52: 1175–1186.

Delrue B, Fontaine T, Routier F, Decq A, Wieruszeski JM, Vandenkoornhuyse N, Maddelein ML, Fournet B, Ball S (1992) J Bacteriol 174:3612–3620. **Denyer K**, Sidebottom C, Hylton CM, Smith AM (1993) Plant J 4:191–198. **Dry I**, Smith A, Edwards A, Bhattacharyya M, Dunn P, Martin C (1992) Plant J 2:193–202.

Echt CS, Schwarz D (1981) Genetics 99:275–284.

Federoff N, Wessler S, Shure M (1983) Cell 35:235–242. **Fontaine T**, Dhulst C, Maddelein ML, Routier F, Pepin TM, Decq A, Wieruszeski JM, Delrue B, Vandenkoornhuyse N, Bossu JP, Fournet B, Ball S (1993) J Biol Chem 268:16223–16230.

Hannah LC (1990) Corn Utilization Conference III, St. Louis, MO, USA, National Corn Growers Association. **Hannah LC**, Nelson OE (1976) Biochem Genet 14:547–560. **Hannah LC**, Tuschall DM, Mans RJ (1980) Genetics 95:961–970. **Hedman KD**, Boyer CD (1982) Biochem Genet 20:483–492. **Hovenkamp-Hermelink JHM**, Jacobsen E, Ponstein AS, Visser RGF, Vos-Scheperkeuter GH, Bijmolt EW, de Vries JN, Witholt B, Feenstra WJ (1987) Theor Appl Genet 75:217–221. **Hseih J** (1988) Bot Bull Acad Sini 29:293–299. **Hylton C**, Smith AM (1992) Plant Physiol 99:1626–1634.

Kaushik RP, Khush GS (1991) Theor Appl Genet 83:146–152. **Keeling PL**, Wood JR, Tyson RH, Bridges IG (1988) Plant Physiol 87:311–319. **Kishore GM**, Somerville CR (1993) Curr Opin Biotechnol 4:152–158. **Klösgen RB**, Gierl A, Schwarz-Sommer Z, Saedler H (1986) Mol Gen Genet 203:237–244. **Konishi Y**, Nojima H, Okuno K, Asaoka M, Fuwa H (1985) Agric Biol Chem 49:1965–1971. **Kooistra E** (1962) Euphytica 11: 357–373. **Kreis M** (1986) Philos Trans R Soc Lond B 314:355–365.

Leij FRvd, Visser RGF, Ponstein AS, Jacobsen E, Feensta WJ (1991) Mol Gen Genet 228:240–248. **Lim PY**, Perata P, Pozueta-Romero J, Akazawa T, Yamaguchi J (1992) Biosci Biotechnol Biochem 56:695–696. **Lin T-P**, Caspar T, Sommerville C, Preiss J (1988a) Plant Physiol 86:1131–1135. **Lin T-P**, Caspar T, Somerville C, Preiss J (1988b) Plant Physiol 88:1175–1181.

MacDonald FD, Preiss J (1985) Plant Physiol 78:849–852. **Mendel G** (1865) Verh Naturforsch Ver Brünn 4:3–47. **Mizuno K**, Kawasaki T, Shimada H, Satoh H, Kobayashi E, Okumura S, Arai Y, Baba T (1993) J Biol Chem 268:19084–19091. **Morrison WR**, Karkalas J (1990) In: Methods in plant biochemistry, vol 2. Academic Press, San Diego, California, USA, pp 323–352. **Müller-Neumann M**, Yoder JI, Starlinger P (1984) Mol Gen Genet 198:19–24. **Müller-Röber B**, Sonnewald U, Willmitzer L (1992) EMBO J 11:1229–1238.

Nakamura T, Yamamori M, Hirano H, Hidaka S (1993) Biochem Genet 31:75–86. **Nelson OE**, Rines HW (1962) Biochem Biophys Res Commun 9:297–300.

Okagaki RJ (1992) Plant Mol Biol 19:513–516. **Okuno K**, Sakaguchi S (1982) J Hered 73:467.

Pan D, Nelson O (1984) Plant Physiol 74:324–328. **Pereira A**, Schwarz-Sommer Z, Gierl A, Bertram I, Peterson PA, Saedler H (1985) EMBO J 4:17–23. **Pohlman RF**, Federoff NV, Messing J (1984) Cell 37:635–643. **Pozueta-Romero J**, Frehner M, Viale AM, Akazawa T (1991a) Proc Natl Acad Sci USA 88:5769–5773. **Pozueta-Romero J**, Ardila F, Akazawa T (1991b) Plant Physiol 97:1565–1572. **Preiss J** (1991) Oxf Surv Plant Mol Cell Biol 7:59–114. **Preiss J**, Danner S, Summers PS, Morell M, Barton CR, Yang L, Nieder M (1990) Plant Physiol 92:881–885.

Rohde W, Becker D, Salamini F (1988) Nucleic Acids Res 16:7185–7186. **Rutger JN**, Azzini LE, Brookhouzen PJ (1985) In: Rice genetics. International Rice Research Institute, Los Banos, Philippines, pp 261–271.

Salehuzzaman SINM, Jacobsen E, Visser RGF (1993) Plant Mol Biol 23:947–962. **Sano Y** (1984) Theor Appl Genet 68:4567–4573. **Satoh H** (1985) Gamma Field Symposia, Institute of Radiation Breeding NIAR MAAF Ohmiya-Machi, Nakagu, lbaraki-ken, Japan, pp 319–22. 17–35. **Satoh H**, Amano E (1987) Rice Genet Newslett 4:38–40. **Satoh H**, Omura T (1981) Jpn J Breed 31:316–326. **Schulman AH**, Ahokas H (1990) Physiol Plant 78:583–589. **Schulman AH**, Tester RF, Ahokas H, Morrison WR (1994) J Cereal Sci 19:49–55. **Schwarz-Sommer Z**, Gierl A, Klösgen RB, Wienand U, Peterson PA, Saedler H (1984) EMBO J 3:1021–1028. **Shewmaker CK**, Stalker DM (1992) Plant Physiol 100:1083–1086. **Shure M**, Wessler S, Federoff N (1983) Cell 35:225–233. **Singh BK**, Preiss J (1985) Plant Physiol 79:34–40. **Smith AM**, Denyer K (1992) New Pytol 122: 21–33. **Smith AM**, (1988) Planta 175:270–279. **Smith AM** (1990) Planta 182:599–604. **Smith AM**, Bettey M, Bedford ID (1989) Plant Physiol 89:1279–1284. **Sprague GF**, Brimhall B, Nixon RM (1943) J Am Soc Agron 35:817–822. **Stark DM**, Timmerman KP, Barry GF, Preiss J, Kishore GM (1992) Science 258:287–292. **Stitt M** (1994) Curr Opin Biotechnol 5:137–143.

Taiz L, Zeiger E (1991) Plant physiology. Benjamin Cummings, Redwood City, pp 158 –168. **Takeda Y**, Preiss J (1993) Carbohydr Res 240:265–275. **Tester RF**, Morrison WR, Schulman AH (1993) J Cereal Sci 17:1–9. **Tsai CY** (1974) Biochem Genet 11:83–96. **Tsai CY**, Nelson OE (1966) Science 151:341–343. **Tyson RH**, AP Rees T (1988) Planta 175:33–38. **Tyynelä J**, Schulman AH (1993) Physiol Plant 89:835–841.

Visser RGF, Jacobsen E (1993) Trends Biotechnol 11:63–68. **Visser RGF**, Hergersberg M, Van DLFR, Jacobsen E, Witholt B, Feenstra WJ (1989) Plant Sci 64:185–192. **Visser RGF**, Somhorst I, Kuipers GJ, Ruys NJ, Feenstra WJ, Jacobsen E (1991) Mol Gen Genet 225:289–296. **Vos-Scheperkeuter GH**, Boer Wd, Visser RGF, Feenstra WJ, Witholt B (1986) Plant Physiol 82:411–416.

Wang TL, Hedley CL (1991) Seed Sci Res 1:3–14. **Wang Z-y**, Wu Z-L, Xing Y-Y, Zheng F-g, Guo X-I, Zhang W-g, Hong M-m (1990) Nucleic Acids Res 18:5898. **Weatherwax P** (1922) Genetics 7:568–572. **Wessler SR**, Baran G, Varagona M, Dellaporta SL (1986) EMBO J 5:2424–2432. **Wessler SR**, Varagona MJ (1985) Proc Natl Acad Sci USA 82:4177–4181.

Yano M, Okuno K, Kawakami J, Satoh H, Omura T (1985) Theor Appl Genet 69: 253–257. **Yano M**, Okuno K, Satoh H, Omura T (1988) Theor Appl Genet 76:183–189. **Yeh J**, Garwood DL, Shannon JC (1981) Stärke 33:222–230.

P. Sathish
Chauanxin Sun
Anders Lönneborg
Christer Jansson
University of Stockholm
Arrhenius Laboratory, Dept of Biochemistry
Svante Arrhenius Väg
106 91 Stockholm, Sweden

IV. Function: Histones in Higher Plants

By Eva Zyprian

1. Introduction: The Fundamental Role of Histones

Every eukaryotic cell contains its major genetic information in long DNA molecules constituting the chromosomes located in its nucleus. In higher plant cells, this nuclear DNA can easily be 1 m or more in total length and its concentration can reach 0.1 mg/ml or more (Grierson and Covey 1988). In order to fit into the nucleus and to function as genetic information specifying growth and development, this material has to be packaged in an organized way. DNA packaging and condensation also is a prerequisite for ordered distribution in cell divisions involving mitosis or meiosis.

The principal cellular components organizing this packaging of DNA in chromatin are the histones. Histones are small basic proteins found associated with DNA in the nucleus of all eukaryotic cells. There are five different histones: H1, H2a, H2b, H3, and H4. H1 is very rich in lysine, H2a and H2b contain less lysine, and H3 and H4 are rich in arginine. Two copies each of the four different histone proteins, H2a, H2b, H3, and H4 aggregate to an octameric cylinder, the nucleosome (Fig. 1). The DNA is wrapped around the nucleosome in 1.8 to 2 complete left-handed superhelical turns (Fig. 1). Histone H1 does not participate in the nucleosome but aggregates with the DNA in between such nucleosomal "beads" and seals them together (Grierson and Covey 1988; Kaudewitz 1992).

The nucleosome has dimensions of 11 × 11 × 5.5 nm. It is built from three parts and assembled in three steps: at first, a tetramer, containing two copies each of H3 and H4, will bind to roughly 120 bp of DNA. Once the DNA is wrapped around this tetramer, the second tetramer will associate, extending the range of DNA bound to about 160 bp. This second tetramer contains two copies of H2a and two copies of H2b. The DNA is on the surface of such a nucleosomal core. Once core formation is complete, H1 can bind to the 70-bp "linker" region between the cores and lead to further condensation. In this way the 11-nm DNA fiber will be coiled into a 30-nm-thick "solenoid" structure (Fig. 2) (Kaudewitz 1992; Wolffe 1994).

Most of our knowledge on histones, their genes, and their effects on other genes comes from studies on animals or yeast. A compilation of characterized histones dated 1989 (Wells and McBride) lists only few plant histones

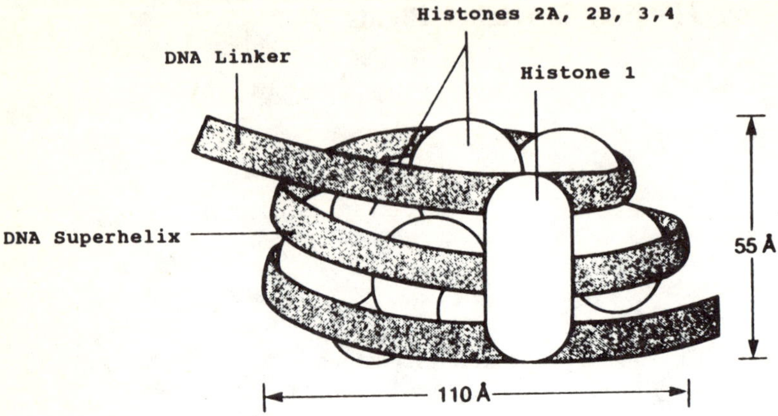

Fig. 1. Schematic picture of the nucleosome. (Kaudewitz 1992)

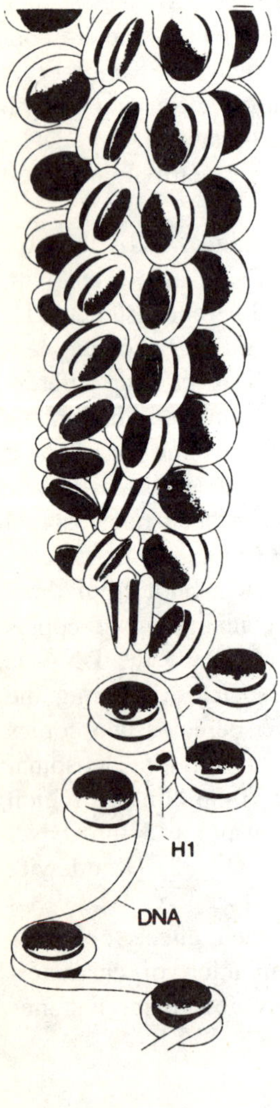

Fig. 2. Solenoid structure of the 11 nm nucleosomal fiber. (Kaudewitz 1992)

(cf. Table 1). This chapter will deal with recent data on the production of histones in higher plants, their genetic organization, and the fine modulation of genetic activities by histones resulting in phenotypic modulations.

For general literature see: Old and Woodland 1984; Weintraub 1985; Chaubet et al. 1987; Wolffe 1994.

Table 1. Histone genes studied in higher plants

Arabidopsis thaliana	
H1	Gantt and Lenvik (1991)
H3/H4	as cited in Wells and McBride (1989); Chaubet et al. (1987); Chaubet et al. (1992a)
Hordeum vulgare	
H3/H4	as cited in Wells and McBride (1989); as cited in Chaubet et al. (1989)
Lycopersicon esculentum	
H2a	Koning et al. (1991)
Oryza sativa	
H2a	Thomas and Padayatty (1983)
H2b	Thomas and Padayatty (1983)
H3/H4	Thomas and Padayatty (1983); as cited in Wells and McBride (1989); Raghavan et al. (1992)
Picea abies	
H2a	Sundas et al. (1993)
Pisum sativum	
H1	Berednikov et al. (1993)
H2a	Koning et al. (1991); Tanimoto et al. (1993)
H3/H4	as cited in Wells and McBride (1989)
Triticum aestivum	
H1	as cited in Wells and McBride (1989); Yang et al. (1991); Nasuda et al. (1993)
H2a	as cited in Wells and McBride (1989); Lindsey et al. (1991)
H2b	Yang et al. (1991); Nasuda et al. (1993)
H3/H4	as cited in Wells and McBride (1989); Tabata et al. (1984)
Vicia faba	
H1	Rogers and Bendich (1992)
Vicia unijuga	
H1	Berednikov et al. (1992)
Zea mays	
H1	Razafimahatratra et al. (1991)
H2b	Joanin et al. (1992)
H3/H4	as cited in Wells and McBride (1989); Chaubet et al. (1987); Chaubet et al. (1992b); Lepetit et al. (1993)

2. Structure and Organization of Histone Genes

a) The Inner Core Histones: H3 and H4

The tetramer $(H3–H4)_2$ is the first to complex with DNA during the assembly of nucleosomes. This explains the strong selective pressure acting on these proteins and the high conservation observed for these histones throughout the animal and plant kingdom. Their genes have been studied best of all histone genes in plants.

The first histone gene sequenced from higher plants was H3 from wheat (*Triticum aestivum*). It was isolated together with an H4 gene. Both genes occupy adjacent positions and are transcribed in the same direction (Tabata et al. 1984).

These genes are equipped with typical RNA polymerase II promoter elements such as TATA- and CCAAT boxes. In addition, they carry a GATCC box typical for histone genes from animals, just upstream of the TATA element.

Hexaploid wheat carries 80 to 100 copies of the H3 gene and 100 to 125 copies for H4. Not all H3 genes were found linked to the H4 genes. Therefore the histone genes in plants do not necessarily form repeating units (Tabata et al. 1984). The wheat H3 gene is specifically expressed during the S-phase of the cell cycle (type 3a, see below) (Nakayama et al. 1992).

How this cell-cycle-dependent expression is achieved is far from clear. The wheat H3 promoter contains several sequence motifs which are candidates for *cis*-acting control elements: the hexamer ACGTCA binding wheat transcription factors HBP1a and HBP1b, the nonamer CATCCAACG binding transcription factor HBP2 and the octamer CGCGGATC found conserved in all known histone genes. The hexamer and the octamer function independently and both bind transcription factors, but without any additive effects (Nakayama et al. 1992).

The wheat transcription factors HBP1a and HBP1b belong to the leucine zipper family of transcription factors and form homodimers (Tabata et al. 1991). Their involvement in cell-cycle-dependent expression is not understood, as they also recognize the hexamer motif present on other plant promoters, including genes on the T-DNA of *Agrobacterium tumefaciens* or the 35S promoter of CaMV. (Tabata et al. 1991). In the case of HBP1b, phosphorylation of this factor may play a role (Takase et al. 1991).

The H4 gene of rice (*Oryza sativa*) was found together with H2a and H2b genes in a cluster. This arrangement is unique for higher plants (Thomas and Padayatty 1983). However, H4 is transcribed from one DNA strand, and H2a and H2b in the rice cluster are transcribed from the opposite DNA strand (Thomas and Padayatty 1983). The H3 gene of rice was not found to be active mainly in dividing cells (or in cells that undergo endoduplication). Its mRNAs accumulate transiently in nondividing cells of the rice grain. In *Hyoscyamus niger*, however, transcripts corresponding to the rice H3 gene follow a tissue-specific temporal expression pattern (Raghavan et al. 1992).

In maize (*Zea mays*), various H3 and H4 genes have been characterized (Chaubet et al. 1987). Three different H3 genes (H3C2, H3C3, and H3C14) show the same primary structure of about 150 amino acid residues (amino acid no. 90 varies when compared to wheat, barley, and rice H3 polypeptides). All H4 genes encode the same polypeptide.

The H3 protein is synthesized with a special codon usage particular to maize (but also seen in wheat, rice, and barley H3 and H4 genes). The promoter regions of H3- and H4 genes in maize contain the TATA element (about 100 nt upstream the initiation codon). The CCAAT box is difficult to identify. The octamer reported for wheat histone genes CGCGGATC (see above) is present also in maize.

As wheat (and barley), maize contains multiple H3 and H4 genes with 30 to 40 H3 gene copies and 50 to 60 H4 gene copies. Although coding for the same product, the various genes differ by their special environment (specific 5′-flanking sequences). On the basis of similarities between these genetic surroundings, they can be classified into families and subfamilies (Chaubet et al. 1989, 1992b). Plants related to maize such as teosinte (*Zea diploperennis*), coïx, sorgho (*Sorghum*) and sugarcane (*Saccharum*), as well as millet (*Pennisetum*), contain similar families of H3 and H4 genes (Chaubet et al. 1992). Three of the maize H3 and H4 subfamilies have been localized on chromosomes 2, 1 and 1 or 5, but additional H3 and H4 gene families are present on most other, if not even all maize chromosomes (Chaubet et al. 1992b).

The H3 and H4 genes of maize are active during the early steps of germination and silent in dry seeds. In adult plants, the highest activities are found in plant organs containing meristematic tissues (Chaubet et al. 1987, 1991). One H4 variant, H4C7, shows replication-independent activity.

A nonameric sequence in its promoter region might be responsible as *cis*-acting element for the basal expression of this H4 variant from maize (Lepetit et al. 1993).

In *Arabidopsis thaliana*, two H3 and H4 genes analyzed show the same protein structure as the maize variants. On nucleotide sequence level, however, the homology is only 76%, due to the specific codon usage in maize not seen in *Arabidopsis*. There are only five to seven copies of these genes in the small genome, and at least some of them were found clustered (H4A748 and H3A713 are localized within only 11 kb distance to each other). As in maize, the *Arabidopsis* genes can be classified into families based on their "genetic environment".

The conserved sequence motifs in the promoter regions are essentially the same as identified in maize H3 and H4 gene variants (Chaubet et al. 1987).

The major H3 genes of *Arabidopsis* are typically active in organs containing a high proportion of meristematic tissues.

Two replication-independent "minor" H3 gene variants were identified in *Arabidopsis,* H3.3I and H3.3II (Chaubet et al. 1992a). Both code for the same histone protein.

In contrast to the major H3 genes H3.1 and H3.2 of *Arabidopsis,* these minor variant genes are separated by only 1 kb of noncoding sequence. Both contain introns. The polypeptides specified differ from the major H3 proteins in only four amino acid residues. On nucleotide sequence level, both H3.3 genes are 88.8% homologous to each other and 78% homologue to the intronless major genes. The TATA boxes are located just upstream of the first intron. The histone gene specific octamer CGCGGATC is found 260 or 310 nt upstream of the 5′-intron on the reverse strand. Regarding promoter organization, both minor genes thus resemble the major H3 genes (Chaubet et al. *1992a*).

In *Vicia faba*, the gene copies for H3 and H4 were found to vary considerably in copy number between individuals (average copy number is 19 per haploid genome; this number varies over a 27-fold range). However, no large variation is seen between the tissues of an individual plant. Phenotypic effects of these variations in gene dosage did not become evident (Rogers and Bendich 1992).

The same variation is seen for rDNA in *Vicia faba.* However, both multigene families vary independently from each other. They might be caused by unequal recombination occurring in tandemly repeated satellite DNA separating the H3 and H4 gene copies. (Rogers and Bendich 1992).

b) The Outer Core Histones: H2a and H2b

Histones H2a and H2b are generally less conserved than H3 and H4. In plants, only little information on H2a and H2b is available. In 1983, Thomas and Padayatty reported the isolation of H2a and H2b genes from rice (*Oryza sativa*), where they were found clustered together with an H4 gene.

The genes for H2a and H2b are coded on one strand, while the H4 gene is located on the opposite strand and separated from H2a/H2b by 3.1 kb. H4 is transcribed in the other direction than H2a and H2b. Such clustering and bidirectional transcription is unique within the plant histone genes studied (Thomas and Padayatty 1983).

Plant H2a and H2b proteins differ from the animal counterparts in their C- and N-terminal regions. This was deduced from information gained from a tomato and a pea cDNA encoding H2a and isolated by Koning et al. (1991). In tomato, H2a is encoded by a small multigene family. In both tomato (Koning et al. 1991) and pea (Tanimoto et al. 1993), these genes are most active in cells actively dividing and hence passing through the cell cycle, or in cells showing endopolyploidization. Both in pea and in tomato the mRNAs for H2a are polyadenylated.

An H2a gene isolated from Norway spruce (*Picea abies*) shows the typical activity in propagating tissues and thus represents the replication-dependent type (Sundas et al. 1993). It is similar to other eukaryotic H2a genes known (nucleotide sequence homology 60 to 65%).

It also contains a nine amino acid "H2a box" conserved in other H2a proteins. Its transcript is polyadenylated. However, this gene seems to be an exception to the rule, as it contains an intron of 891 bp in the coding region, atypical of the replication-dependent histone gene class (Sundas et al. 1993; see Sect. 3a).

In wheat, the H2a protein is longer as compared to the animal counterparts, but still is similar to them. This made possible an experiment by Lindsey et al. (1991), constructing hybrid nucleosomal octamers from chicken histones replacing the chicken H2a by two different H2a variants from wheat. One of these, H2a(3), carries extensions at both the C- and the N-terminal protein sides, while wheat H2a(1) shows only the C-terminal extension. The hybrid nucleosomes reconstructed both bound more DNA than the "pure" chicken octamer core nucleosome. This and additional evidence presented by Lindsey et al. (1991) suggest that the C-terminal extension of 19 amino acids of the wheat H2a proteins is involved in binding the linker DNA.

Two H2b variants were isolated from maize (Joanin et al. 1992). They are built from 150 (H2b-1) and 149 (H2b-2) amino acids and exhibit the classical organization of H2b proteins:

A hydrophobic C-terminal region (resembling animal H2b histones) of 90 amino acid residues, joined to the N-terminal domain containing a specific pentapaptide repetition through nine basic amino acids. As in wheat H2b, also maize H2b is longer in the N-terminal part as compared to the animal H2b proteins. The H2b protein in maize is encoded by a multigene family. The same is found for wheat with an extreme dispersal of H2b gene copies on almost all chromosomes (Nasuda et al. 1993). A 152 amino acid residue H2b was recently isolated from wheat and represents one of the major class H2b variants (Yang et al. 1991).

c) The Linker Histone H1

α) General Structure

Although highly polymorphic, the lysine-rich histone H1 has a general composition from three parts: a central globular domain of about 80 residues in length is flanked by disordered N- and C-terminal parts. The central globular domain interacts with the nucleosome, where the DNA enters and exits from the core particle. The N- and C-terminal domains are presumed to neutralize charged phosphates along the DNA backbone. This permits the chromatin to assume a more compact structure (Gantt and Lenvik 1991; Razafimahatratra et al. 1991).

The amino acid sequence of the central globular domain is rather well-conserved, while the highly-charged, random coiled N- and C-terminal domains are highly variable. As already mentioned, higher eukaryotes have several H1 subtypes, differing in their primary structure, and possibly involved in developmental regulation.

β) H1 Gene Organization

Arabidopsis thaliana most likely contains only three H1 genes, and two of them have been characterized (Gantt and Lenvik 1991). H1-1 encodes a protein of 274 amino acid residues, H1-2 encodes a histone 1 composed of 273 amino acids.

Both are very rich in lysine, alanine, and proline, and show a high lysine/arginine ratio. The central globular domains are more similar to each other (nucleotide sequence homology 92%) as the N- and C-terminal domains. A chicken H1-type histone, H5, has been shown to fold its central globular domain into helices. The two *Arabidopsis* H1 variants fit well in that model.

H1-1 has an intron and a poly-A signal located 30 nt upstream the polyadenylation site. Its promoter shows classical TATA (at –31) and CCAAT (at –78) boxes of pol II promoters. At position –101 there is the octamer GATCCGCG common to other plant histone genes (see above). Overall sequence homology between H1-1 and H1-2 is 56% (for more details see Gantt and Lenvik 1991).

The first monocot H1 gene reported in its complete encoded amino acid sequence was from maize (Razafimahatratra et al. 1991). The structure is basically the same as described for *Arabidopsis* H1-1 and H1-2, except a special codon usage in maize already mentioned for maize H3- and H4 genes (see above). The protein is composed of 245 amino acids (25.5 kDa). It is encoded by a small multigene family. Its cDNA cross-hybridized with teosinte, coix, sorgho, millet, barley, wheat, and oat, but not with *Arabidopsis*. Its expression pattern follows type a described under Section 3. A similar wheat H1 gene was isolated from a seedling library (Yang et al. 1991). In *Triticum aestivum* cv. Chinese spring, H1 genes were localized on different sets of homologous chromosomes 5a, 5b, and 5d (Nasuda et al. 1993).

γ) Phenotypic Modulations by Histone H1

The first evidence for phenotypic effects caused by highly variable H1 histones as it should be postulated from the effects on gene expression described below, comes from research on legumes:

In legumes, H1 is polymorphic and encoded by several alleles. In the garden pea (*Pisum sativum*), even all subtypes of H1 are polymorphic. Three loci were mapped: a cluster of five tightly-linked genes on chromosome 1, another single gene on chromosome 1, and one on chromosome 5. The various alleles can slightly modify the phenotypes: allele 1 of subtype 5, for example, is correlated with the sum of the aerial temperature through the vegetation period. One allelic variant of subtype 1 is absent to the north of the 44th parallel, and allele 1 of subtype 6 is present all over the world except in Asia and China (Berednikov et al. 1993). A natural population of the wild species *Vicia unijuga* was studied for the possibility that certain alleles may provide selective advantage. Screening of four H1 subtype f variants (due to allelic variation) in the surroundings of the town Akademgorodok showed equal proportions of the four allelic variants in the central parts of the city, but a strong prevalence for one variant (f3) in the periphery. It is assumed that a factor connected to urbanization and human influence caused this alteration in allelic frequencies. Insecticide treatment shifting the natural cross-pollination through bumblebees to self-pollination is hypothesized as a cause (Berednikov et al. 1992).

3. Production of Histones

Due to their basic structural function in DNA packaging and their influence on gene activity, actively dividing plant cells (or cells undergoing endopolyploidization) have an enormous demand for histones. At the same time, they need to modulate nucleosomal composition to allow expression of other essential genes. In general, there is hence a multiplicity of histone variants coded for by multiple genes. The histone genes follow three different activity patterns and can be subdivided (Old and Woodland 1984; Chaubet et al. 1992a):

a) Replication-Dependent Histone Genes

Expression of this type of histone genes is strictly coupled to the S-phase of cells. They are the most abundant class and the corresponding histone proteins predominate in actively dividing tissues. They are required in large amounts to package newly-synthesized DNA. In general, the genes encoding such histones do not contain any introns and the message does not contain a poly-A tail.

b) Replacement Histone Genes

A minor class of replication-independent histone genes producing "replacement" histones, these are synthesized throughout the cell cycle at a basal constitutive level. Significant amounts of the corresponding mRNAs are found only in nondividing (mature) tissues. Their genes differ from the replication-dependent ones by the absence of introns and a few amino acid residues in the encoded polypeptides. Their mRNAs are polyadenylated.

c) Tissue-Specific Histone Genes

A third class of tissue-specific histone genes yields histone variants specific for a certain tissue or developmental stage, their synthesis is not cell-cycle-dependent. The corresponding mRNAs are not spliced, but will generally be polyadenylated (Chaubet et al. 1992a).

4. Function of Histones

a) General Inhibition of Gene Activity

DNA association with histones not only leads to condensation and stabilization of the genetic material, but also has profound effects on its function. Gene expression is generally hindered by DNA packaging, as signals controlling gene activity (like promoters or regulatory sequences) are not accessible to the recognizing enzymes such as RNA polymerase or to positively acting protein factors which need to bind their target DNA ("closed" chromatin structure). Disruption of the nucleosome/DNA complexes is often required to gain gene activity ("open" chromatin structure) (Weintraub 1985; Wolffe 1994).

For example, one of the factors isolated from wheat germ extracts and known to inhibit yeast RNA polymerase III was identified as histone H1. It excluded the yeast transcription machinery from its template DNA (Furter and Hall 1991).

An investigation on the genes encoding the ribosomal constituents (rDNA) in tomato leaf cells and rapidly growing suspension culture cells showed that most of this rDNA is in a "closed" chromatin formation (80%). Only 20% were found in the "open" nucleosome-free (but not necessarily histone-free) configuration. The nucleosomal repeat length was 184 +/−5 bp and did not vary in a comparison of three different tomato organs (Conconi et al. 1992).

b) Activation of Gene Activity

DNA condensation through nucleosomes not only hinders genetic activity, but may also activate genetic loci by sterically altering DNA structure and hence exposing positive signals. In cases where several signals must be recognized to activate genes, they might be brought into the correct distance to each other by the DNA bending caused through the nucleosome formation (reviewed by Wolffe 1994).

c) Fine Modulation of Genetic Activity Through Histone Modifications

Histone proteins are subject to considerable modifications such as acetylation, phosphorylation, ADP–ribosylation, or the binding of ubiquitin. They can interact with HMG (high mobility group) regulatory proteins which are also associated in the chromatin (Weintraub 1985); these reversible changes can mediate a fine regulation of gene expression. They cause subtle conformational changes which may influence the interaction between the nucleosomal subunits and/or their contacts with DNA.

Lopez-Rodas et al. (1993) proposed a model stating that transient deacetylation (which neutralizes a positive charge) of H2a and H2b in their short, flexible N-terminal parts enables early transcriptional activation (permitting regulatory proteins to access to nucleosomal structures). The deacetylation is accomplished by HD1A, a histone deacetylase from maize embryos. This enzyme itself is modified by phosphorylation. Upon this modification it changes substrate specificity from deacetylating H3 and H4 to acting preferentially on H2a and H2b in the phosphorylated state. HD1A thus could be involved in the cellular signal transduction pathway (Lopez-Rodas et al. 1993).

d) Regulation of Gene Activity Through Histone H1

Histone H1 is a prominent candidate for playing a regulatory role in gene expression. As it is rather loosely attached to the nucleosomal core, it might reversibly associate or dissociate from the DNA. H1 is bound to inactive regions in such a way that it cross-links adjacent nucleosomes ("closed chromatin"). In active chromatin, however, it fails to hold the nucleosomes together (maybe because H1 becomes modified or replaced by a different H1 variant). This "open chromatin" configuration leads to DNaseI-hypersensitive sites in active genes and changes the torsional strain on DNA (Weintraub 1985). Most organisms possess several types of H1, and the individual types are extremely polymorphic. This observation of many H1 variants still adds to the complexity of gene regulation by nucleosomes.

A cDNA encoding a DNA binding protein was isolated from soybean embryos. The protein encoded has domains resembling the regulatory HMG (high mobility group) nuclear proteins, as well as a domain resembling H1. Such "bifunctional" proteins may easily interact with the chromatin structure (Laux et al. 1991).

5. Conclusions

Only few plant species have been studied for their histones (Table 1); but even from these few, a complex picture of gene arrangement and regulated expression emerges:

1. A multiplicity of genes required to satisfy the huge demand of cells for histones need to be expressed in a regulated way. However, there is no general scheme of organization for histone genes. Actively dividing cells must dispose of large histone quantities to package the newly synthesized DNA. In mature tissue, histones need to be replaced after damage or to allow certain genes to gain activity (enabling transcription). In differentiated tissues, specific histone variants are involved in maintaining a specific gene activity. The characteristic features of the corresponding histones are just beginning to be understood.

2. The coupling of the major histone synthesis to the S-phase of the cell cycle is not yet clear. First hints on the involvement of transcription factors being themselves regulated by modification might show the track.
3. Histones have the potential to influence gene activities in a general way and evidence for phenotypic changes through variable frequencies of certain histones became available. Histones thus should be a factor considered when studying gene expression.

References

Berednikov VA, Bogdanova VS, Rozov SM, Gorel SM (1992) Can J Bot 70:1591–1595.
Berednikov VA, Bogdanova VS, Rozov SM, Kosteriin OE (1993) Heredity 72:199–209.

Chaubet N, Chaboute M-E, Philipps G, Gigot C (1987) Dev Genet 8:461–473. **Chaubet N**, Philipps G, Gigot C (1989) Mol Gen Genet 219:404–412. **Chaubet N**, Clement B, Philipps G, Gigot C (1991) Plant Mol Biol 17:935–940. **Chaubet N**, Clement B, Gigot C (1992a) J Mol Biol 225:569–574. **Chaubet N**, Philipps G, Gigot C, Guitton C, Bouvet N, Freyssinet G, Schneerman H, Weber DF (1992b) Theor Appl Genet 84:553–559. **Conconi A**, Sogo JM, Ryan CA (1992) Proc Natl Acad Sci USA 89:5256–5260.

Furter R, Hall BD (1991) Plant Mol Biol 17:773–786.

Gantt JS, Lenvik TR (1991) Eur J Biochem 202:1029–1039. **Grierson D**, Covey SN (1988) Plant molecular biology, 2nd edn. Blackie, Glasgow.

Joanin P, Gigot D, Philipps G (1992) Plant Mol Biol 20:581–588.

Kaudewitz F (1992) Genetik 2 Aufl. Ulmer, Stuttgart. **Koning AJ**, Tanimoto EY, Kiehme K, Rost T, Comai L (1991) Plant Cell 3:657–666.

Laux T, Seurinck J, Goldberg RB (1991) Nucleic Acids Res 19:4768. **Lepetit M**, Ehling M, Atanassova R, Chaubet N, Gigot C (1993) Plant Sci 89:177–184. **Lindsey GG**, Orgeig S, Thompson P, Davies N, Maeder DL (1991) J Mol Biol 218:805–813. **Lopez-Rodas G**, Brosch G, Georgieva EI, Sendra R, Franco L, Loidl P (1993) FEBS Lett 317:174–180.

Nakayama T, Sakamoto A, Yang P, Minami M, Fujimoto V, Ito T, Iwabuchi M (1992) FEBS Lett 300:167–170. **Nasuda S**, Liu YG, Sakamoto A, Nakayama T, Iwabuchi M, Tsunewaki K (1993) Plant Mol Biol 22:603–614.

Old RW, Woodland HR (1984) Cell 38:624–626.

Raghavan V, Jiang C, Bimal R (1992) Am J Bot 79:778–783. **Razafimahatratra P**, Chaubet N, Philipps G, Gigot C (1991) Nucleic Acids Res 19:1491–1496. **Rogers SO**, Bendich AJ (1992) Theor Appl Genet 84:617–623.

Sundas A, Tandre K, Kvarnheden A, Engstrom P (1993) Plant Mol Biol 21:595–605.

Tabata T, Fukasawa M, Iwabuchi M (1984) Mol Gen Genet 196:397–400. **Tabata T**, Nakayama T, Mikami K, Iwabuchi M (1991) EMBO J 10:1459–1468. **Takase H**, Tabata T, Mikami K, Iwabuchi M (1991) Plant Cell Physiol 32:1195–1204. **Tanimoto EY**, Rost TL, Comai L (1993) As cited in DIMDI abstracts. **Thomas G**, Padayatty JD (1983) Nature 306:82–84.

Weintraub H (1985) Cell 42:705–711. **Wells D**, McBride C (1989) Nucleic Acids Res 17:r311 (seq suppl) **Wolffe AP** (1994) Cell 77:13–16.

Yang P, Katsura M, Nakayama T, Mikami K, Iwabuchi M (1991) Nucleic Acids Res 19:5077.

Dr. Eva Zyprian
Bundesanstalt für Züchtungsforschung
an Kulturpflanzen
Institut für Rebenzüchtung Geilweilerhof
76833 Siebeldingen, Germany

V. Extrachromosomal Inheritance: Yeast Linear Killer Plasmids as a Tool in Genetic Engineering

By Jürgen Schründer and Friedhelm Meinhardt

1. Introduction

In principle, every stably replicated DNA species can serve as a vector for cloning foreign DNA in yeast or any other organism, in which such DNA is found. In strains of *Saccharomyces cerevisiae* a high copy number, autonomously replicating, extrachromosomal DNA element, the 2 μm DNA, was found to exist (Sinclair et al. 1967). This covalently closed circular DNA element resembles in many respects bacterial plasmids. Since it is present in approximately 50 copies per cell it is relatively stably maintained. It was used to construct vectors for cloning and expression of foreign genes in *Saccharomyces cerevisiae* (for review see Hollenberg 1982). The occurrence of circular plasmids is not restricted to baker's yeast, because similar elements were also detected in *Zygosaccharomyces* (Tohe and Utatsu 1985) and in *Kluyveromyces drosophilarum* (Chen et al. 1986). The latter plasmid, pKD1, is 1.6 μm in size and was also transferred and stably maintained in other yeast species of the genus, i.e., *Kluyveromyces lactis* (Bianchi et al. 1987).

Other circular vectors currently in use for yeast transformation are hybrid molecules between *E. coli* plasmids and yeast DNA. Two types of vectors can be distinguished, i.e., integrating vectors and freely replicating vectors. In the former instance, foreign DNA is integrated into a yeast chromosome either at homologous or heterologous sites, in the latter, vectors carry so-called ARS sequences which confer the ability for autonomous replication (Hinnen and Meyhack 1982).

Linear extrachromosomal DNA elements are ubiquitous among eukaryotes, they are especially common in yeasts but also in filamentous fungi. For a compilation of linear plasmids detected so far see a previous review in this series (Meinhardt and Rohe 1992). In contrast to the circular episomal elements from pro- and eukaryotes linear plasmids have characteristic structural features, such as terminal inverted repeats (TIRs) and 5′ covalently attached proteins (tPs). With respect to the structure, linear plasmids are similar in their genetic organization to the adenoviruses (Challberg and Kelly 1982; Challberg et al. 1982) and also to linear bacteriophages (Jung et al. 1987b). Besides these overall structural similarities, computer-aided sequence comparisons of the amino acid and nucleotide sequences of DNA- and RNA-polymerases encoded by these elements exhibited

Progress in Botany, Vol. 56

phylogenetic relations between linear plasmids and viral genomes, indicating that all linear elements evolved from a common ancestor (Kempken et al. 1992; Rohe et al. 1992; Meinhardt and Rohe 1992). Most of the linear plasmids detected so far are cryptic, i.e., no phenotype could be attributed to them. However, despite the lack of selective pressure, these elements replicate efficiently and are stably maintained in their hosts. Phylogenetic calculations suggest that these genetic traits represent a group of degenerated viruses, which have lost all viral functions during evolution, existing as intracellularly parasiting selfish DNA.

Linear plasmids can also frequently be observed in representatives of the prokaryotic Actinomycetales. In this taxon they often confer to their hosts the ability to produce antibiotics or resistance against them; or provide other advantagous genetic information such as H_2 autotrophy (for a review see Schründer et al. 1994). Very recently, a conjugative linear plasmid was isolated from *Rhodococcus erythropolis*, which encodes isopropylbenzene and trichloroethene catabolism as well as the arsenite resistance and the mercury resistance of this strain (Dabrock et al. 1994).

There are only very few eukaryotic linear plasmids to which a phenotype can be attributed, i.e., the senescence causing elements in certain strains of *Neurospora* (Bertrand et al. 1985; Court et al. 1991) and the plasmid-encoded killer systems of *Pichia inositovora* (Hayman and Bolen 1991) and *Kluyveromyces lactis* (Gunge et al. 1981; Stark et al. 1990). At present, the most thoroughly investigated system is represented by the killer plasmids of the yeast *Kluyveromyces lactis*, serving as a model for basic as well as applied research.

2. Structure and Function

a) Detection and Localization

Encouraged by the detection of extrachromosomal genetic elements in the yeast *Saccharomyces cerevisiae* (Sinclair et al. 1967), in the ascomycetous mold *Podospora anserina* (Stahl et al. 1978), and in the higher plant *Zea mays* (Pring et al. 1977), a great number of yeasts and other eukaryotes were analyzed for the presence of similar genetic traits. Among 70 different yeast strains from 17 genera, additional extrachromosomal DNA was found to exist only in some representatives of *Kluyveromyces lactis* (Gunge et al. 1981). Further characterization by restriction enzyme analysis and electron microscopy revealed that these new elements were linear molecules. Since all prokaryotic and eukaryotic plasmids were thought to be circular at that time (with the exception of the elements from corn), this finding was surprising. In the strains under investigation two linear plasmids differing in size were detected. They were designated pGKL1 and pGKL2, which are 8.9 and 13.4 kb in size, respectively.

The plasmids originally detected and designated by N. Gunge and coworkers, were subsequently also described by Wesolowski et al. (1982) and named k1 and k2, respectively.

K. lactis cells, which contain the linear plasmids differ from all other cells in a way, that they have the ability to kill other sensitive yeasts belonging to the same or different species (Gunge et al. 1982). Since the loss of the plasmids is accompanied by the loss of the killing ability, these plasmids were suggested to be the physical basis of the killer phenotype (for details, see Sect. 3). Besides, the hitherto seldom observed linearity also the localization of these elements represents a peculiarity. In contrast to all other plasmids, they are located in the cytoplasm, as could be shown by fluorescence staining techniques and density centrifugation of organelles (Gunge et al. 1982; Stam et al. 1986).

b) Genomic Organization

Both linear plasmids, pGKL1 and pGKL2, are totally sequenced. These data, as well as transcript analyses, revealed four open reading frames (ORFs) of the smaller pGKL1 plasmid, which is 8874 bp in size. The larger plasmid pGKL2 has ten ORFs and is 13457 bp long (Hishinuma et al. 1984; Stark et al. 1984; Sor and Fukuhara 1985; Romanos and Boyd 1988; Tommasino et al. 1988; Tommasino 1991). Both plasmids are evidently tightly packed with information, since the coding capacities are extremely high, i.e., 94% for pGKL1 and even 97% for pGKL2 (see also Fig. 1).

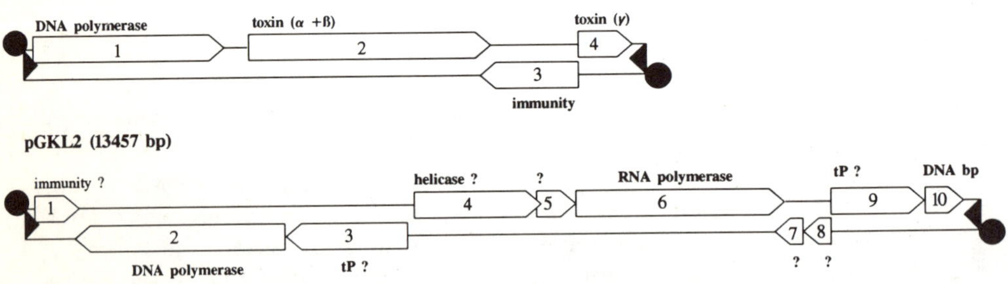

Fig. 1. Genetic organization of the linear killer plasmids of *Kluyveromyces lactis*. Open reading frames (ORFs) are given as *arrows*, the *direction of the arrows* gives the direction of transcription. Where functions can be assigned to genes they are indicated. *Black circles* represent the 5′ terminal proteins (*tP*); *black triangles* indicate terminal inverted repeats. The figure is based on sequence data and transcript analysis derived from Hishinuma et al. (1984); Stark et al. (1984); Sor and Fukuhara (1985); Romanos and Boyd (1988); Tommasino et al. (1988), and Tommasino (1991). Numbering of ORFs is according to Tommasino et al. (1988)

The function of the four ORFs of pGKL1 is mainly understood. ORF1 codes for a DNA polymerase (Jung et al. 1987a), which is necessary for replication of pGKL1; ORFs 2 and 4 encode subunits of the killer toxin (Stark and Boyd 1986); ORF3 is essential for immunity to the toxin (Tokunaga et al. 1987a). Our knowledge of gene function of pGKL2 ORFs is rather poor. However, pGKL1 is strictly dependent on pGKL2 or, in other words, pGKL1 cannot exist without pGKL2. From computer analysis of the pGKL2 sequence, it is suggested that ORF2 encodes a pGKL2 specific viral B-type DNA polymerase (Tommasino et al. 1988) and ORF6 codes for a RNA polymerase (Wilson and Meacock 1988). The ORF4 gene product is similar to the DNA helicase of the vaccinia virus (Stark et al. 1990). Since there are similarities between the gene product of ORF1 from pGKL2 and that of ORF3 from pGKL1, this ORF is suggested to be involved in the immunity expression (Stark et al. 1990). Only for ORF10 a biochemical analysis was performed. The encoded protein is a 14-kDa histon-like polypeptide, which binds to both plasmids pGKL1 and pGKL2, presumably acting as a structural stabilizer rather than a regulatory protein (Tommasino 1991).

As characteristic structural components both plasmids carry covalently attached terminal proteins, which have a molecular weight of 28 and 36 kDa for pGKL1 and pGKL2, respectively (black circles in Fig. 1). Both plasmids have also characteristic terminal inverted repeats (TIRs), 202 bp in size for pGKL1 and 184 bp for pGKL2 (Gunge 1986; Kikuchi et al. 1984; Sor and Fukuhara 1985; Stam et al. 1986). Our knowledge of the terminal proteins and the attachment of these to the DNA is rather scarce; all we know about the terminal proteins is by analogy to the adenoviruses (Challberg et al. 1980, 1982; Sussenbach 1984). Since on the plasmids there is no ORF detected which was identified to encode the terminal proteins, it was suggested that pGKL2 encodes both proteins (Gunge 1988; Stark et al. 1990; Kämper et al. 1991). Alternatively, the proteins may be derived from the DNA polymerases which have large extensions at their NH_2-termini compared to adenoviruses and linear bacteriophages (Chan et al. 1991; Meinhardt and Rohe 1992). According to the latter hypothesis, both proteins (DNA polymerase and terminal protein) are derived from the same ORF by either posttranscriptional or posttranslational processing.

c) Replication

Killer strains of *K. lactis* always contain both linear plasmids. pGKL1 is not able to exist without pGKL2. However, pGKL2 is replicated and can be maintained solely. Strains carrying pGKL2 alone have no killing ability and cannot be differentiated phenotypically from plasmid free strains (Gunge et al. 1982; Wesolowski et al. 1982). This strict dependence is due to the pGKL2 encoded RNA polymerase, which is necessary for transcription of the ORFs of both plasmids (Stark et al. 1990). Although experimental proof is lacking, it is generally accepted that the replication of killer plasmids follows the adenoviral mechanism, as it is given in Fig. 2. Replication is initiated by the attachment of a terminal protein to the end of a plasmid at the TIR region. The terminal protein has a covalently linked nucleotide serving as the primer for the DNA polymerase. From this protein-polymerase complex DNA, synthesis proceeds unidirectionally in 5′–3′ direction, thereby replacing one of the parental strands. As a consequence, a totally replicated plasmid and the displaced parental strand are generated. Base pairing of the terminal inverted repeats (TIRs) gives rise to a hairpin structure from which, after attachment of an additional terminal protein, DNA synthesis is again initiated in the same way as described for the other strand. Since the attachment of the terminal protein-DNA polymerase complex is sequence specific, this might explain the existence of the two DNA polymerases, one encoded by each plasmid (Tommasino et al. 1988). Electron microscope investigations support the strand displacement replication mechanism, intermediates could be detected which fit into this model (Fujimura et al. 1988). Both plasmids have a copy number of

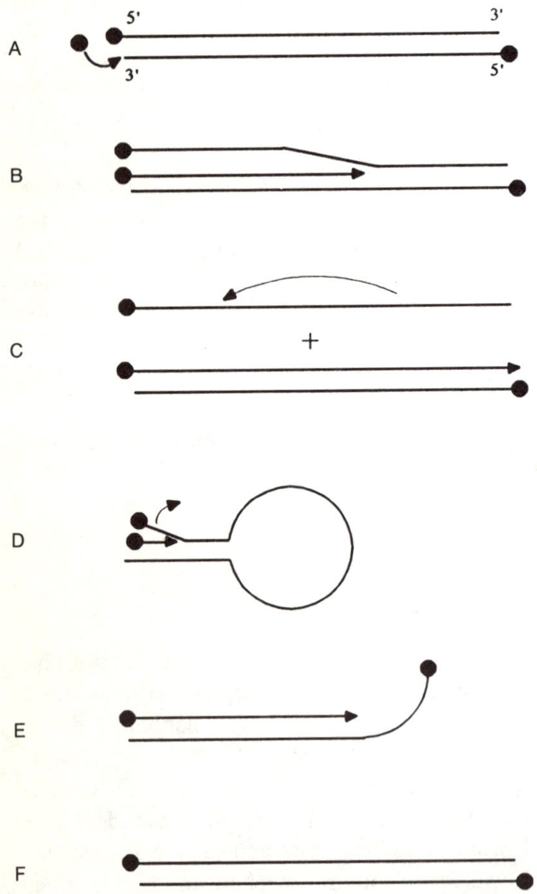

Fig. 2A–F. Schematic representation of adenoviral like replication of linear genomes. (After Challberg and Kelly 1982). Replication is initiated by the terminal protein and the DNA polymerase (**A**). DNA replication occurs in 5′–3′ direction while the parental strand of the same orientation is displaced (**B, C**). The displaced parental strand creates a partial double strand by homologous base pairing in the region of the terminal inverted repeats where the second strand displacement starts (**D**). Replication continues until the second strand is also completely replicated (**E, F**)

about 50–100 per cell. The killer plasmids are stably maintained in *K. lactis* and even without selective pressure they are not lost (Gunge et al. 1981). Nothing is known, however, of the way in which the copy number is regulated.

d) Expression of ORFs

The fact that the linear killer plasmids and other linear plasmids from yeasts (Shepherd et al. 1987; Bolen et al. 1992; Gunge et al. 1993) are localized in the cytoplasm implicate the existence of an independent expression system that is not governed by the nucleus or by compartimental systems (i.e., mitochondria and chloroplasts). A central role in the expression of the cytoplasmically localized information plays the RNA polymerase encoded by ORF6 of pGKL2. This enzyme is a rather peculiar polypeptide because it carries β and β' residues in a single peptide chain (Wilson and Meacock 1988; see also Stark et al. 1990). Upstream from the ORFs in the 5′ noncoding regions, all ORFs of the killer plasmids have in common certain upstream conserved sequences (UCS) (Sor and Fukuhara 1985; Romanos and Boyd 1988; see Fig. 3). These UCS sequences represent specific cytoplasmic promoters. The eukaryotic RNA polymerases I–III do not recognize and use these promoter sequences. Thus, genes governed by these UCS cannot be expressed when localized in the nucleus and vice versa (Stam et al. 1986; Romanos and Boyd 1988; Wilson and Meacock 1988; Kämper et al. 1989a, b, 1991). Transcript analysis suggest UCSs of ORFs 5 and 6 to be the most active transcription signals (Tommasino et al. 1988; Tommasino 1991). The involvement of other gene products in transcription initiation of pGKL2 ORFs is not known. However, the putative gene products of ORF4 of pGKL2, a helicase, could be instrumental in unwinding the DNA double helix (Stark et al. 1990).

Up to now, essentially nothing is known about translation of transcripts derived from both of the linear plasmids. It has not yet been proven whether transcripts possess caps at their 5′ ends, which are normally necessary for binding to the ribosomes (Kozak 1983). If there are indeed no caps, other components are to be postulated for the initiation of translation. All ORFs of pGKL1 have in common the sequence AAA adjacent to the start codon; A at the position −3 is necessary for efficient transcription (Cigan and Donahue 1987). However, ORFs 1 and 3 of pGKL2 differ from this rule, because there is a T at this position.

UCS pGKL1 5' ··· ACT^{A}/$_{T}$AATATATGA ····· 24 - 38 bp ····· ATG

UCS pGKL2 5' ··· TAA(N)TNTGA ········· 28 - 109 bp ····· ATG

Fig. 3. Consensus sequences of upstream conserved sequences (*UCS*) of pGKL1 and pGKL2. In addition to the UCSs, also the distance between these putative promoters and the ATG start codon is given

3. Phenotypic Manifestation

a) The Killer Phenotype

Killer phenomena exist in nearly all microorganisms, e.g., in bacteria where several strains have the ability to kill other sensitive bacteria (see Chap. 10 in Birge 1988). Certain strains of the brewer's yeast *Saccharomyces cerevisiae* carry virus-like particles (VLPs) in which dsRNA is encapsulated. Those strains having dsRNA containing particles with a L-dsRNA and M-dsRNA kill other sensitive yeast cells (For details concerning the dsRNA viruses of *S. cerevisiae* see Wickner 1992). The physical basis of the killer phenotype of *Kluyveromyces lactis*, however, are the linear dsDNA plasmids. They encode a heterotrimeric exotoxin, which inhibits growth of several yeasts, including *Candida intermedia, Candida utilis, Kluyveromyces thermotolerans, Kluyveromyces vanudenii, Saccharomyces cerevisiae, Saccharomyces italicus, Saccharomyces rouxii, Torulopsis glabrata and Zygosaccharomyces rouxii* (Gunge and Sakaguchi 1981; Gunge et al. 1981).

b) Structure of the Toxin and Mode of Action

The *K. lactis* killer toxin is composed of three subunits: α (99 kDa), β (30 kDa), and γ (27.5 kDa). DNA sequence analysis of pGKL1 and the purification of the killer toxin from the culture medium of *K. lactis* have led to the conclusion that the genes encoding the killer toxin subunits are organized as shown in Fig. 1 (Stark and Boyd 1986; Tokunaga et al. 1987b). ORF2 encodes a 128-kDa polypeptide which is the precursor of the α- and β-subunits, whereas ORF4 encodes the precursor of the γ-subunit (see also Table 1). It has been assumed that the 128-kDa precursor protein is cleaved by the *K. lactis* chromosomal *KEX2* gene product (Stark and Boyd 1986).

Wesolowski et al. (1982) isolated a class of mutants which had lost the killer phenotype while maintaining intact plasmids. The mutation is localized in a single locus of the host chromosome, which was called *KEX1*, in analogy to the *kex* mutations described for the dsRNA killer system of *Saccharomyces cerevisiae* (Wickner and Leibowitz 1976). Further analysis revealed that the *KEX1* gene of *K. lactis* appears to be functionally related to the *KEX2* gene of *S. cerevisiae* and that it is involved in the production of the mature aminotermini of the α- and β-killer toxin subunits.

In addition to the toxin, pGKL1 also carries an immunity function (Tokunaga et al. 1987a) which functions as a self-protecting agent in toxin-producing cells.

The toxin causes an irreversible arrest of sensitive cells in the unbudded (G1) phase of the cell cycle (Sugisaki et al. 1983; see Fig. 4 and Table 2). Despite the complex structure, the γ-subunit appears to be the only component required

Table 1. Listings of the subunits of the killer toxin and mode of action

Toxin subunits	pGKL1 coding regions	M_r [kDa]	Glyco-sylation	Mode of action
α	ORF2	99	+	Chitinase
β	ORF2	30	–	Interaction with the cell membrane entry of the presumed, promoting toxin into the cell
γ	ORF4	27.5	–	Toxin, causes G1 arrest

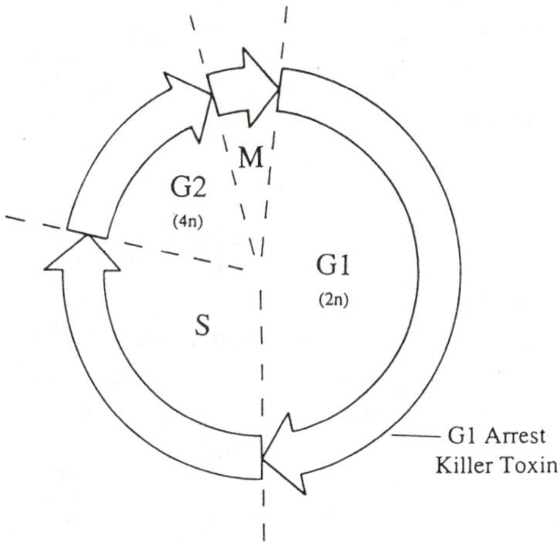

Fig. 4. Schematic representation of the cell cycle of yeasts (for further explanation, see Table 2)

Table 2. Cell cycle of yeasts

Phase	Definition	Duration (h)
G1 (gap)	Time between cell division and DNA synthesis	Up to 12
S (synthesis)	DNA synthesis	6–9
G2 (gap)	Time between DNA synthesis and mitosis	3–4
M	Mitosis	> 1

to arrest proliferation because intracellular expression of γ-polypeptide alone in a sensitive yeast strain mimicks the effect of the exogenously applied native toxin (Tokunaga et al. 1990; Butler et al. 1991b). The α-subunit shows sequence similarities to some chitinases; allosamidin, a specific inhibitor of chitinases, abolishes the biological activity of the toxin (Butler et al. 1991a). The function of the β-subunit is currently unknown, but it is remarkably hydrophobic, and that is why the α- and β-subunits might play a rôle in binding and uptake of the γ-subunit to sensitive cells.

Production of the toxin is strictly coupled to growth; maximal levels occur in the late exponential phase (Sugisaki et al. 1984). The toxin has its optimum activity at pH 7.0 and is still fully active after exposure to temperatures up to

50 °C (Stark et al. 1990). In early studies on the toxin's mode of action, it was shown to inhibit cyclic AMP production in isolated yeast membranes (Sugisaki et al. 1983). However, recent work shows that the mode of action is far from being understood, and strongly argues against the involvement of adenylate cyclase in toxin action (White et al. 1989). The arrested cells increase in volume by some 50% after several hours of toxin action, suggesting that cell growth continues following treatment (Stark et al. 1990).

The mode of action is very different to that of the other yeast killer toxin, i.e., the dsRNA encoded killer protein of *Saccharomyces cerevisiae*, which acts as a ionophore and causes membrane permeability changes and cell shrinkage.

c) Transfer into Other Yeasts

As outlined above, in killer strains of *Kluyveromyces lactis*, two different linear DNA molecules, pGKL1 and pGKL2, can be found. Both plasmids were transferred by protoplast fusion, cytoduction, or by transformation to different yeast species, such as *Saccharomyces cerevisiae*, *Kluyveromyces fragilis* and *Candida pseudotropicalis* (Gunge and Sakaguchi 1981; Gunge et al. 1982; Sugisaki et al. 1985; Tokunaga et al. 1990). In all these yeast species the expression of the killer phenotype is to be observed. However, in *K. lactis* the plasmids are most stable. In *Saccharomyces cerevisae* they are incompatible with the mitochondrial DNA, they can only replicate in petit strains, in which no mtDNA is present (ρ^0 strains), or most of the DNA is deleted (ρ^- strains) (Gunge and Yamane 1984; Gunge 1988). In addition, the replication of linear plasmids is greatly hampered in diploid *S. cerevisiae* strains, because the gene product of the mating type loci (MATa/MATα) encodes a protein (a1-α1) which functions as a repressor by binding to sequences of pGKL2. Since pGKL1 strictly depends on pGKL2, replication of both plasmids is disturbed in diploid *S. cerevisiae* (Gunge et al. 1992).

4. Vector Construction

a) Circular Vectors

Since the commonly used prokaryotic *Escherichia coli* is unable to perform the general eukaryotic posttranslational modification of expressed polypeptides, yeast cells became preferred hosts for heterologous gene expression of biologically active proteins from eukaryotic source. Yeasts, being nonpathogenic microorganisms with established applications in food industry, can be grown rapidly on simple media and to a high cell density. For the best characterized species, *Saccharomyces cerevisiae*, several types of plasmids were developed to introduce exogenous DNA into the cells (see also Table 3; for a review see Gellissen et al. 1992):

Vectors of the YIp type (yeast integrating plasmids) contain yeast chromosomal DNA to target integration by homologous recombination, as well as a selectable marker and a bacterial replicon. The first transformed yeast resulted from such a homologous recombination at the *LEU2* mutant locus of a leucin-auxotrophic *Saccharomyces cerevisiae* strain (Hinnen et al. 1978). The integrants are quite stable, the typical rate of vector loss being $<1\%$ per generation in the absence of selective pressure. However, integrative transformation itself is a rare event (1–10 transformants/μg DNA). A system based on homologous recombination of expression vectors into the iterative ribosomal DNA units of *Saccharomyces cerevisiae* has been developed by Lopes et al. (1989). This so-called MIRY (**M**ultiple **I**ntegration into the **R**ibosomal DNA of **Y**east) enabled expression-vector copy numbers of 100–200. The copy number of the plasmids was stable during nonselective growth for 70 generations. Recently, an analogous vector system for high copy number integration into the ribosomal DNA of *K. lactis* has also been described (Bergkamp et al. 1992; Rossolini et al. 1992) and was called MIRK (**M**ultiple **I**ntegration into the **R**ibosomal DNA of ***K***. *lactis*). After transformation with vectors containing a portion of *K. lactis* rDNA, these plasmids were found to be present in about 60 copies per cell; they were stably maintained during growth under nonselective conditions.

Higher transformation frequencies were achieved by using plasmids containing yeast autonomous replication sequences (ARS), which function as origin of replication in *S. cerevisiae*. After transformation, these YRp vectors (yeast-replicating plasmid) remain in an extrachromosomal state and are present in 3–20 copies per cell (Struhl et al. 1979). When using an ARS sequence to introduce foreign DNA into *Kluyveromyces lactis* strains, only a heterologous integration into the host's genome was observed (Das and Hollenberg 1982). When a genuine sequence for autonomous replication (KARS: *Kluyveromyces lactis* autonomous replication sequence) was present, vectors remained in a plasmid state. Gunge (1988) first looked for autonomously replicating sequences in the *Kluyveromyces lactis* pGKL plasmids, which may function as the origin of DNA replication in *Saccharomyces cerevisiae*. As a screening vector the integrative plasmid YIp32 was used. YIp32 transforms *Saccharomyces cerevisiae* (*leu2*) to Leu$^+$ phenotype at a very low frequency as a result of chromosomal integration, but if combined with ARS element(s) the transformation efficiency increases 100 to 1000 times. Based on this principle, several restriction fragments of the killer plasmids were found to contain ARS elements and sequence analysis of this fragments revealed the existence of the ARS consensus sequence (A/T) TTTAT (T/G) TT (A/T) reported for *Saccharomyces cerevisiae* (Broach et al. 1982).

The ARS and KARS vectors are mitotically highly instable: plasmid free cells accumulate at a rate of up to 20% per generation without selective pressure, due to inefficient transmission to daughter cells during cell division (Murray

Table 3. Compilation of integrative and episomal circular vectors for yeast. Copy number and stability are also indicated

	Type	Description	Copy number	Stability[a]
Integrative	YIp	**Y**east **I**ntegrating Plasmids: homologous integration into chromosomal DNA	1	> 99%
	MIRY	**M**ultiple **I**ntegration into the **R**ibosomal DNA of **Y**easts	100–200	Stable
	MIRK	**M**ultiple **I**ntegration into the **R**ibosomal DNA of ***K**luyveromyces lactis*	60	Stable
Episomal	YRp	**Y**easts **R**eplicating Plasmids, including ARS or KARS sequences	3–20	80%
	YCp	**Y**east **C**entromeric Plasmids: inclusion of a yeast centromeric sequence (CEN)	1–2	
	YACs	**Y**east **A**rtificial **C**hromosomes; addition of telomeric structures		
	YEp	**Y**east **E**pisomal Plasmids (native yeast plasmids such as: *S. cerevisiae* 2-μm DNA)	10–40	> 99.9%

[a] Percentage of cell which maintains the recombinant DNA after cell division

and Szostak 1983). Plasmid stability can be improved by inclusion of a yeast centromer sequence (CEN) but the copy number of the so-called YCp vectors (yeast centromeric plasmids) is reduced to 1 or 2 per cell (Clarke and Carbon 1980). The addition of telomeric structures in so-called YACs (**Y**east **A**rtificial **C**hromosomes) lead to linear vectors which can harbor long DNA fragments of up to 200 kb in size (for a review see Burke et al. 1987).

The most commonly used expression vectors in yeast genetic engineering are *Escherichia coli* shuttle vectors based on the *Saccharomyces cereviseae* 2-μm plasmid, termed YEp vectors (**Y**east **E**pisomal plasmid) (Beggs 1978). They can transform yeast at a frequency of 5000–20000 recombinants per μg DNA, have a copy number of 10 to 40 copies per cell, and are stably inherited: plasmid free cells arise at the rate of 1 in 10^4 per generation (Struhl et al. 1979; Hollenberg 1982; Futcher and Cox 1983).

A circular plasmid with a functional organization similar to the 2-μm DNA has recently been isolated from *Kluyveromyces drosophilarum* and transferred into *Kluyveromyces lactis* (Chen et al. 1986; Bianchi et al. 1987). Expression vectors derived from this 1.6-μm circular plasmid pKD1 have been developed using the *Saccharomyces cerevisiae URA3* gene or a Tn903-derived kanamycin resistance sequence as selectable markers (Bianchi et al. 1987).

Since yeasts secrete only few proteins into the growth medium, purification and processing of heterologous proteins can be greatly facilitated by efficient secretion of the protein of interest into the culture fluid. Using a 2-μm or a pKD1-derived vector system, the heterologous protein can be secreted if fused to a suitable signal sequence. This can be either a foreign signal, often derived from the protein being secreted, or a yeast signal. In both cases, a heterologous signal was found to function very poorly and that is why the efficiency of secretion signals derived from linear plasmid pGKL1 was tested (Stark et al. 1990). In fact, both the $\alpha\beta$ and the γ secretion signals of the toxin encoding ORFs 2 and 4 respectively, can be used successfully to direct secretion of several proteins (see Table 4) in both *Saccharomyces cerevisiae* and *Kluyveromyces lactis.*

b) Linear Vectors

The two cytoplasmically localized linear killer plasmids of *Kluyveromyces lactis* have all the parameters sought for yeast vectors: a broad host range, a high copy number of about 50–100 per cell, and a high stability without selective pressure. Vectors based on the pGKL plasmids are therefore potentially advantageous to conventional circular yeast vectors.

First attempts at transformation of a uracil-requiring *Kluyveromyces lactis* strain with in vitro recombined pGKL1-DNA, using the *Saccharomyces cerevisiae URA3* gene as a selectable marker resulted in the formation of circular hybrid plasmids only (DeLovencourt et al. 1983). Using the linear DNA plasmid pGKL2 Fujimura et al. (1987) in vitro constructed a hybrid plasmid carrying

Table 4. Compilation of secreted exoproteins using signal sequences from the killer toxin ORFs. Secretion can be achieved in *Saccharomyces cerevisiae* and *Kluyveromyces lactis*

Yeast	Signal sequence	Protein	Yield (mg/l)	Reference
S. cerevisiae	ORF2 (first 16 codons)	Human interleukin 1β	1–2	Baldani et al. (1987)
	ORF2 (first 20 codons)	Mouse α-amylase		Tokunaga et al. (1987)
	ORF2 (first 35 codons)	Bact. β-lactamase		Mileham et al. (in prep.)
	ORF4	Mouse α-amylase	1.2	Tokunaga et al. (1988)
K. lactis	ORF2 (first 16 codons)	Human interleukin 1β	80	Fleer et al. (1991a)
	ORF2	Human serumalbumine	300	Fleer et al. (1991b)
	ORF2	Diphtheria toxin A-hormone fusion protein		Perenties et al. (1988)
	ORF2	Tissue plasminogen activator (tPA)		Yeh et al. (1990)
	ORF2	Tissue-specific inhibitor of metalloproteinase (TIMP)		Yeh et al. (1990)

a *LEU2* gene of *S. cerevisiae* as a selectable marker. After transformation of a leucin-auxotrophic *S. cerevisiae* strain, hybrid plasmids were detected which were able to propagate autonomously only as circular molecules. Attemps to integrate the LEU2 coding region into the smaller plasmid pGKL1 by in vitro techniques also resulted exclusively in circularized hybrid plasmids (Kämper et al. 1989a).

An alternative way to obtain the desired linear plasmids is by in vivo recombination. For this purpose the yeast *LEU2* gene was flanked by sequences from pGKL1 (Kämper et al. 1989a). As the recipient, a Leu$^-$ *Saccharomyces cerevisiae* strain was used, that harbors both native killer plasmids, pGKL1 and pGKL2, thus providing the potential recombination sites. Circular plasmids were frequently detected in Leu$^+$ transformants; however, four clones carried new linear plasmids in addition to pGKL1 and pGKL2. Molecular analysis revealed that the insertion of the *LEU2* gene into pGKL1 occurred at the predicted position. Surprisingly, however, an extension of several hundred base pairs could be observed at the ends of the hybrid molecules. Eventually, it turned out from sequence analysis, that the observed extensions in a hybrid plasmid (pLS1) were due to the attachment of telomeric TG motifs.

These telomere sequences, together with the **A**utonomously **R**eplicating **S**equences (ARS) from pGKL1, suggested that the recombinant plasmids might replicate in the nucleus, and indeed they were found to exist in the nucleus. The telomere plasmids are probably formed because *LEU2* is a nuclear gene which cannot be expressed in the cytoplasm, its promoter and terminator being recognized solely in the nucleus. The attachment of telomeres facilitates the localization of the plasmids in the nucleus, and hence the expression of the gene. As a corollary, one would expect that under the control of a pGKL promoter, *LEU2*, should be expressed in the cytoplasm. In front of ORF2 there is an **U**pstream **C**onserved **S**equence (UCS) which has also been found in the upstream regions of several other pGKL genes. They are probably promoter elements (Stark et al. 1984; Sor and Fukuhara 1985; Romanos and Boyd 1988; Wilson and Meacock 1988). Applying molecular genetic techniques the *LEU2* coding region was flanked with promoter and terminator sequences. To distinguish this recombinant *LEU2* gene from the native version, it was designated *LEU2**. Transformants obtained with this fragment contained in addition to the killer plasmids a new hybrid molecule, called pJKL1 (see Fig. 5), which like pGKL plasmids is located in the cytoplasm, has terminal inverted repeats, and possesses covalently linked proteins at the 5′ ends (Kämper et al. 1989a, b, 1991).

Similar approaches to introduce foreign DNA into the linear plasmids subsequently led to several analogous systems, in which the bacterial *aph* gene (Tanguy-Rougeau et al. 1990) and the *Saccharomyces cerevisiae HIS3* gene (Gallo and Galeotti 1990) served as selectable markers.

In order to find out whether the *LEU2** gene cartridge is of general applicability, and to simplify the transfer system, we tried to find out in which locus on the large plasmid pGKL2 foreign DNA can be inserted without affecting stability and expression of the killer plasmid system. In the above outlined experiments (Kämper et al. 1989a, b, 1991) the killer toxin ORF2 of pGKL1 was the tar-

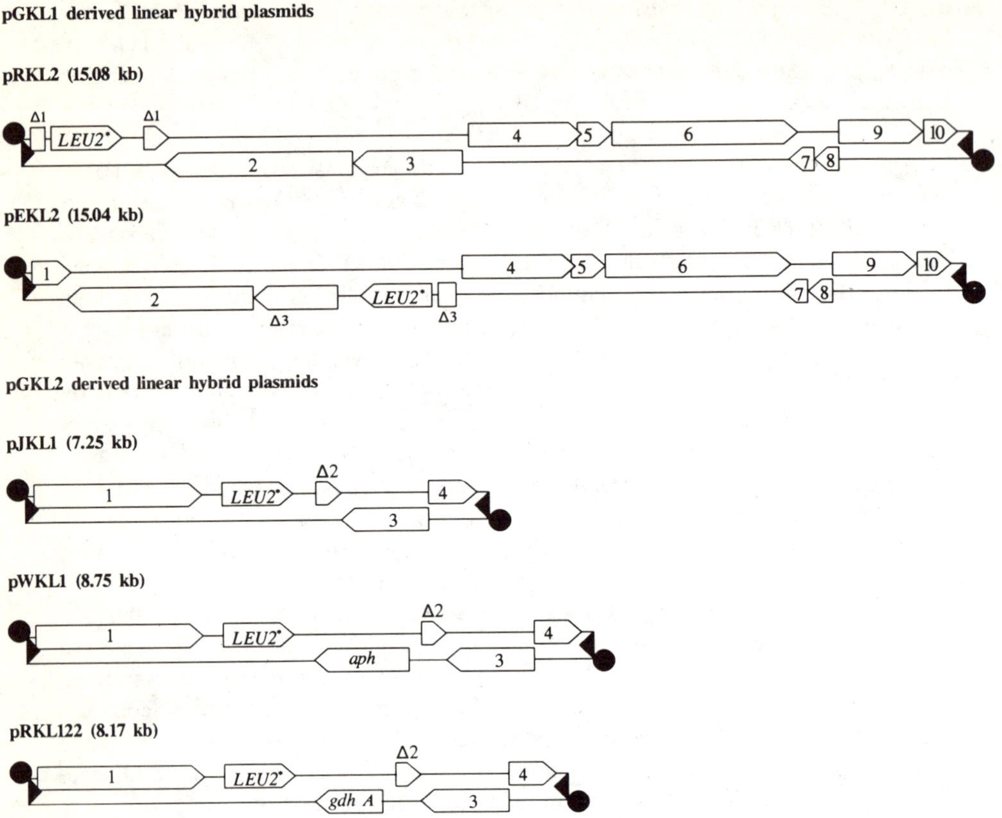

Fig. 5. Selection of linear hybrid molecules based on linear killer plasmids of *Kluyveromyces lactis*. For explanation of symbols, see legend to Fig. 1. Δ designates parts of the respective ORFs. *LEU2** refers to the cytoplasmically expressible *LEU2* gene. *aph* Gene encoding aminoglycoside phosphotransferase from Tn903; *gdh A* gene encoding the glucose dehydrogenase A from *Bacillus megaterium* M1286

get for integration. In further experiments, ORF1 of pGKL2 was chosen for the integrative experiments due to its putative involvement in immunity expression, a property which can easily be checked in a bioassay. In order to disrupt the ORF1 of pGKL2, the cytoplasmically expressible *LEU2** gene cartridge was isolated and flanked with appropriate sequences from ORF1. After transformation and in vivo recombination, a linear hybrid molecule was obtained (Schaffrath et al. 1992). Restriction enzyme and hybridization analyses made evident that the native linear plasmid pGKL2 and the newly derived hybrid molecule pRKL2 (see Fig. 5) segregated, resulting in the total loss of resident pGKL2 in subclones. At the same time, these results prove the deletion of the central ORF1 region of pGKL2 via homologous recombination. Since it has been suggested that ORF1 of pGKL2 might be involved in the expression of the immunity phenotype

(Stark et al. 1990), we checked a strain harboring recombinant pRKL2 in a bioassay (Butler et al. 1991b). No detectable differences were noted between strains carrying native or linear hybrid plasmids. Stability of the hybrid plasmids is also not affected, as it is in the same range as the wild-type killer plasmids (Schaffrath et al. 1992).

In vivo recombination is therefore of general applicability for disruption of different plasmid-encoded ORFs. As outlined above, there is no detectable influence on stability and maintenance of the plasmids when pGKL2 ORF1 is disrupted. Thus, ORF1 is the locus of choice to integrate foreign DNA into pGKL2. The integration of foreign DNA into a dispensable locus of pGKL2 may also be of great importance in gene technology for another reason, because it is no longer necessary to use a system with two plasmids. In previous experiments, integration of foreign DNA was confined to pGKL1. However, replication and maintenance of pGKL1 is strictly dependent on pGKL2.

5. Application

The above-described one-step gene transplacement procedure offers the possibility of addressing basic problems concerning the biology of the linear killer plasmids as well as the development of an expression system for heterologous genes.

a) Functional Analysis of Cytoplasmic ORFs

Our knowledge about the functional significance of most of the ORFs of pGKL2 is rather poor. Gene knockout procedures are generally well suited to find out whether a specific ORF is essential for maintenance of the system or not. Since we do not know the genetic basis of the terminal proteins involved in plasmid replication, we were especially interested to disrupt ORF3 and ORF9, because these ORFs have the capacity to encode the terminal proteins having molecular weights of 28 and 36 kDa, respectively. For this purpose, the cytoplasmically expressible *LEU2** gene cartridge was flanked with appropriate ORF3 or ORF9 derived sequences, in order to facilitate homologous recombination eventually resulting in replacement of the native ORFs by the disrupted copy. However, the desired linear hybrid plasmids could only be found after transformation experiments using the ORF3 directed recombination vector.

Attemps to disrupt ORF9 were all not successful, only nonspecific integrations of the marker gene into the chromosomal DNA were obtained. The reasons for this remain totally obscure at present.

The stability of the hybrid plasmid pEKL2 (see Fig. 5) carrying a disrupted ORF3 was measured after sequential subculturing of the transformed yeast strain for 145 h (about 60 generations) with and without selective pressure. Subcultivation of the recombinant yeast strain without selective pressure resulted in the total loss of the hybrid plasmid, whereas DNA isolates of the strain growing under selective pressure showed the linear hybrid plasmid to exist in addition to the parental plasmids in gel electrophoretic analysis. At the same time, these results prove that segregation has not occurred, and that the hybrid plasmid cannot functionally replace the native pGKL2 plasmid, but is needed to express the *LEU2** gene when growing in selective medium. Thus, ORF3 of pGKL2 is essential for maintenance of pGKL2.

Up to now, only speculations about the coding region and the structure of terminal proteins of linear plasmids exist, because no amino acid sequence of a terminal protein is available. A hypothesis that the large, cryptic N-terminal domain of the DNA polymerase, which is found to exist in all linear plasmids sequenced so far, might function as the terminal protein has been put forward by Chan et al. (1991). We have tried to disrupt the pGKL1 DNA polymerase ORF via homologous recombination, introducing the *LEU2** marker gene between the coding region and the enigmatic cryptic N-terminal domain. After transformation of *Kluyveromyces lactis* with vectors containing an in vitro recombined DNA polymerase, the replacement of the native ORF1 by the disrupted copy has not occurred until now. Further investigations indicated that the obtained leucine prototrophic transformants carry the selectable marker on circular molecules. Bulk DNA preparations from these yeast strains transformed *E. coli* to ampicilline resistance; plasmid DNA isolated from the bacterial transformants showed the same restriction pattern as the originally used in vivo recombination vector.

Thus, at present the origin of the terminal proteins of linear plasmids still remains obscure.

b) Analysis of Yeast Telomeric Sequences

The observed attachment of telomeres after transformation of a *LEU2*$^{-}$ *Saccharomyces cerevisiae* strain using the homologous *LEU2* gene under control of its own promoter as a selectable marker (Kämper et al. 1989a) offers the possibility of analyzing the formation of telomeres in different yeasts.

The DNA at the ends of linear eukaryotic chromosomes usually consists of tandemly repeated simple sequences. This so-called telomeric DNA is essential for chromosome stability. A given eukaryotic species has a characteristic telomeric DNA sequence, although the same sequence may occur in more than one species. Telomeric repeats can be regular, consisting of perfect tandem repeats, or irregular, consisting of length or sequence variations of a basic repeat unit, such as the characteristic TG_{1-3} motif of *Saccharomyces cerevisiae* telomeres (for a recent review on telomeres and telomerases see Blackburn 1992).

The very last 5 bp of the TIRs of pGKL1, i.e., TGTGT, show strong homology to yeast telomeres TG_{1-3}; the attachment of telomeric DNA might be brought about by the action of a specific telomere terminal transferase (telomerase) that recognizes this motif and adds nucleotides one by one. Since it is possible to transfer this linear telomeric hybrid plasmid via protoplast fusion into other yeast species (Meinhardt, unpubl.), it would be appropriate to test whether the *Saccharomyces cerevisiae* telomeres exercise their function in different hosts, or if there are changes of the telomeric DNA sequence motif in the different yeast strains.

c) Heterologous Gene Expression

As outlined in Section 4b, "linear vectors" for cytoplasmic expression of a heterologous gene, an in-phase fusion of the start codon from the cytoplasmic ORF to the coding region of the nuclear *LEU2* gene was necessary. In order to investigate whether an in-phase fusion is an absolute prerequisite for cytoplasmic expression of a heterologous gene, we followed a totally different approach to achieve expression of a bacterial antibiotic resistance gene, i.e., the *aph* gene from Tn903 encoding the aminoglycoside phosphotransferase. A hybrid plasmid was constructed, that consisted of pGKL1 into which, in addition to the previously mentioned cytoplasmically expressible *LEU2** selection marker, the phosphotransferase encoding bacterial antibiotic resistance gene was integrated. In the hybrid plasmid (pWKL1; see also Fig. 5) this gene was preceded by a putative transcriptional signal from pGKL2, i.e., the upstream conserved sequence of ORF5 (UCS5). All foreign genes so far integrated and expressed in linear plasmids required in addition to the cytoplasmic promoter (UCS) an in-phase fusion to the corresponding open reading frame. In pWKL1 the *aph* coding region was intentionally not combined in phase, in order to find out whether expression can occur under these conditions. *Kluyveromyces lactis* transformants carrying pWKL1 were highly resistant to the antibiotic geneticine (G418) and exhibited elevated levels of APH activity. Expression of the gene was strictly dependent on the presence of the UCS5 containing fragment, which spanned 110 bp (Meinhardt et al. 1994, in press).

Since this was the first example that, concomitant with a selectable marker gene, foreign DNA can be integrated and expressed in a linear plasmid without an in-phase fusion, experiments were performed aiming at a generalization of this finding. For this purpose, the *gdhA* gene from *Bacillus megaterium*, encoding glucose dehydrogenase (Heilmann et al. 1988; Meinhardt et al. 1989), was arbitrarily placed downstream the cytoplasmic promoter (UCS5). The *LEU2** gene cartridge served again as a selectable marker. Linear hybrid plasmids were frequently formed after transformation of *K. lactis*, but expression of the *gdhA* gene might at least be designated poor. However, when a linear plasmid (pRKL122, see Fig. 5) was constructed that carried an in-phase fusion of the *gdhA* gene to plasmids ORF5, high levels of expression could be observed.

Thus, for efficient expression of genes using linear cytoplasmic vectors, an in-phase fusion is generally required. However, as outlined above (*aph* expression), exceptions to this rule might occur.

Hybrid plasmids were also transferred to a haploid *Saccharomyces cerevisiae* strain, in which nearly the same level of expression was to be observed (unpubl. data).

d) Conclusions

Linear cytoplasmically localized killer plasmids from *Kluyveromyces lactis* are generally very well suited for the construction of hybrid molecules from which efficient expression of heterologous genes can be achieved. Like the native plasmids, the linear hybrid vectors can be transferred and stably maintained in other yeast species. The copy number of 50–100 molecules per cell, on the one hand, causes the extreme stability (nearly 100%) and concomitantly a high dose of the gene to be expressed is ensured.

The cytoplasmic transcriptional mechanism of linear elements is far from being understood. The functionality of cytoplasmic promoters (UCS) is mainly postulated from sequence analysis/comparisons. The above-mentioned hybrid plasmids offer the possibility of investigating functionology of UCS sequences and measuring promoter activity and strength, eventually also leading to the improvement of heterologous gene expression.

Acknowledgement. J.S. was supported by a grant of the Max Buchner-Forschungsstiftung, Frankfurt am Main, Germany.

References

Baldari C, Murray JAH, Ghiara P, Cesareni G, Galeotic CL (1987) EMBO J 6: 229–234. **Beggs JD** (1978) Nature 275:104–108. **Bergkamp RJM**, Kool IM, Geerse RH, Planta RJ (1992) Curr Genet 21:365–370. **Bertrand H**, Chan BSS, Griffith AJF (1985) Cell 41:877–884. **Bianchi MM**, Falcone C, Jie CX, Wesolowski-Louvel M, Frontali L (1987) Curr Genet 12:185–192. **Birge EA** (1988) Bacterial and bacteriophage genetics, an introduction. 2nd edn, Springer, Berlin Heidelberg New York. **Blackburn EH** (1992) Annu Rev Biochem 61:113–129. **Bolen PL**, Kurtzman CP, Ligon JM, Mannarelli BM, Bothast RJ (1992) Antonie Leeuwenhoek Int J Gen Mol Microbiol 61:195–205. **Broach JR**, Li YY, Felfman J, Jayaram M, Abraham J, Nasmyth KA, Hicks JB (1982) Cold Spring Harbor Lab Symp Quant Biol 47:165–1173. **Burke ET**, Carle GS, Olson MV (1987) Science 236:806–812. **Butler AR**, O'Donnell RW, Martin VJ, Gooday GW, Stark MJR (1991a) Eur J Biochem 199:483–488. **Butler AR**, White JH, Stark MJR (1991b) J Gen Microbiol 137:1749–1757.

Challberg MD, Kelly TJ (1982) Annu Rev Biochem 51:901–934. **Challberg MD**, Desiderio SV, Kelly TJ (1980) Proc Natl Acad Sci USA 77:5105–5109. **Challberg MD**, Ostrove JM, Kelly TJ (1982) J Virol 41:265–270. **Chan B**, Court DA, Vierula JP, Bertrand H (1991) Curr Gen 20:225–237. **Chen XJ**, Saliola M, Falcone C, Bianchi MM, Fukuhara H (1986) Nucleic Acids Res 14:4471–4481. **Cigan AM**, Donahue TF (1987) Gene 59: 1–18. **Clarke L**, Carbon J (1980) Nature 257:504–509. **Court DA**, Griffith AJF, Kraus SR, Russell PJ, Bertrand H (1991) Curr Genet 19:129–137.

Dabrock B, Keβeler M, Averhoff B, Gottschalk G (1994) Appl Environ Microbiol 60: 853–860. **Das S**, Hollenberg CP (1982) Curr Genet 6:123–128. **DeLouvencourt L**, Fukuhara H, Heslot H, Wesolowski M (1983) J Bacteriol 154:737–742.

Escarmis C, Salas M (1982) Nucleic Acids Res 10:5785–5798.

Fleer R, Yeh Y, Amellal N, Yeh Y, Fournier A, Guinet F, Gault N, Faucher D, Folliard F, Fukuhara H, Mayaux JF (1991a) Gene 107:285–295. **Fleer R**, Chen XJ, Amellal N, Maury I, Fournier A, Bacchetta F, Baduel P, Jung G, L'Hote H, Becquart J, Fukuhara H, Mayaux JF (1991b) Bio/Technology 9:968–975. **Fujimura H**, Hishinuma F, Gunge N (1987) Curr Gene 12:99–104. **Fujimura H**, Yamada T, Hishinuma F, Gunge N (1988) FEMS Microbiol Lett 49:441–444. **Futcher AB**, Cox BS (1983) J Bacteriol 154:612–622.

Gallo E, Galeotti CL (1990) Transplacement of a nuclear gene into the cytoplasmatic killer plasmids of *Kluyveromyces lactis*. Abstr 15th Int Conf on Yeast genetics and molecular biology, 547pp. **Gellissen G**, Melber K, Janowicz ZA, Dahlems UM, Weydemann U, Piontek M, Strasser AWM, Hollenberg CP (1992) Antonie Leeuwenhoek Int J Gen Mol Microbiol 62:78–93. **Gunge N** (1986) Yeast 2:153–162. **Gunge N** (1988) *Kluyveromyces* linear DNA plasmids. In: Koltin Y, Leibowitz MJ (eds) Viruses of fungi and simple eukaryotes. M Dekker, New York, pp 265–282. **Gunge N**, Kitada K (1988) Eur J Epidemiol 4:409–414. **Gunge N**, Sakaguchi K (1981) J Bacteriol 147:155–160. **Gunge N**, Yamane C (1984) J Bacteriol 159:533–539. **Gunge N**, Tamaru A, Ozawa F, Sakaguchi K (1981) J Bacteriol 145:382–390. **Gunge N**, Murata K, Sakaguchi K (1982) J Bacteriol 151: 462–464. **Gunge N**, Fukuda K, Morikawa S, Murakami K, Takeda M, Miwa A (1993) Curr Genet 23:443–449.

Hayman GT, Bolen PL (1991) Curr Genet 19:389–393. **Heilmann HJ**, Maegert HJ, Gassen HG (1988) Eur J Biochem 174:485–490. **Hinnen A**, Meyhack B (1982) Vectors for cloning in yeast. In: Hofschneider PH, Goebel W (eds) Gene cloning in organisms other than *E. coli*. Current topics in microbiology and immunology 96, Springer, Berlin Heidelberg New York, pp 101–117. **Hinnen A**, Hicks JB, Fing GR (1978) Proc Natl Acad Sci USA 75:1929–1933. **Hishinuma F**, Nakamura K, Hirai K, Nishizawa R, Gunge N, Maeda T (1984) Nucleic Acids Res 12:7581–7597. **Hollenberg CP** (1982) Cloning with 2 μm DNA vectors and expression of foreign genes in *Saccharomyces cerevisiae*. In: Hofschneider PH, Goebel W (eds) Gene cloning in organisms other than *E. coli*. Current topics in microbiology and immunology 96. Springer, Berlin Heidelberg New York, pp 119–144.

Jung G, Leavitt MC, Ito J (1987a) Nucleic Acids Res 15:9088. **Jung G**, Leavitt MC, Hsieh JC, Ito J (1987b) Proc Natl Acad Sci USA 84:8287–8291.

Kämper J, Meinhardt F, Gunge N, Esser K (1989a) Mol Cell Biol 9:3931–3937. **Kämper J**, Meinhardt F, Gunge N, Esser K (1989b) Nucleic Acids Res 17:1781. **Kämper J**, Esser K, Gunge N, Meinhardt F (1991) Curr Genet 19:109–118. **Kempken F**, Herrmanns

J, Osiewacz HD (1992) J Mol Evol 35:503–513. **Kikuchi Y**, Hirai K, Hishinuma F (1984) Nucleic Acids Res 12:5685–5692. **Kozak M** (1983) Microbiol Rev 47:1–45.

Lopes TS, Klootwijk J, Veenstra AE, van der Aar PC, van Heerikhuizen H, Raué HA, Planta RJ (1989) Gene 79:199–206.

Meinhardt F, Rohe M (1992) Prog Bot 54:318–333. **Meinhardt F**, Stahl U, Ebeling W (1989) Appl Microbiol Biotechnol 30:343–350. **Meinhardt F**, Kempken J, Kämper J, Esser K (1990) Curr Genet 17:89–95. **Meinhardt F**, Wodara C, Larsen M, Schickel J (1994) A novel approach to express a heterologous gene on *Kluyveromyces lactis* linear killer plasmids: Expression of the bacterial *aph* gene from a cytoplasmic promoter fragment without *in phase* fusion to the plasmid open reading frame. Plasmid (in press). **Milcham AJ**, Wales R, Boyd A, Secretion of bacterial *β*-lactamase from *Saccaromyces cerevisiae*, (in prep.). **Murray AW**, Szostak JW (1983) Cell 34:911–970.

Perenties JP, Genbauffe FS, Veldman SA, Galeotti CL, Livingston DM, Bodley JW, Murphy JR (1988) Proc Natl Acad Sci USA 85:8386–8390. **Pring DR**, Levings III CS, Hu WWL, Timothy DH (1977) Proc Natl Acad Sci USA 74:2904–2908.

Rohe M, Schründer J, Tudzinsky P, Meinhardt F (1992) Curr Genet 21:173–176. **Romanos MA**, Boyd A (1988) Nucleic Acids Res 16:7333–7350. **Rossolini GM**, Riccio ML, Gallo E, Galeotti CL (1992) Gene 119:75–81.

Salas M (1988) Curr Top Microbiol Immunol 136:71–88. **Schaffrath R**, Stark MJR, Gunge N, Meinhardt F (1992) Curr Genet 21:357–363. **Schründer J**, Meinhardt F, Rohe M, Kämper J (1994) BioEngineering 2:29–39. **Shepherd HS**, Ligon JM, Bolen PL, Kurtzman CP (1987) Curr Genet 12:297–304. **Sinclair JH**, Stevens BJ, Sanghavi P, Rabinowitz M (1967) Science 156:1234–1237. **Sor F**, Fukuhara H (1985) Curr Genet 9:147–156. **Stahl U**, Lemke PA, Tudzynski P, Kück U, Esser K (1978) Mol Gen Genet 162:341–344. **Stam JC**, Kwakman J, Meijer M, Stuitje AR (1986) Nucleic Acids Res 14:6871–6884. **Stark MJR**, Boyd A (1986) EMBO J 5:1995–2002. **Stark MJR**, Mileham AJ, Romanos MA, Boyd A (1984) Nucleic Acids Res 12:6011–6030. **Stark MJR**, Mileham AJ, Romanos MA, Boyd A (1990) Yeast 6:1–29. **Struhl K**, Stinchkomb DT, Scherer S, Davis RW (1979) Proc Natl Acad Sci USA 76:1035–1039. **Sugisaki Y**, Gunge N, Sakaguchi K, Yamasaki M, Tamura G (1983) Nature 304:464–466. **Sugisaki Y**, Gunge N, Sakaguchi K, Yamasaki M, Tamura G (1984) Eur J Biochem 141:241–245. **Sugisaki Y**, Gunge N, Sakaguchi K, Yamasaki M, Tamura G (1985) J Bacteriol 164:1373–1375. **Sussenbach JS** (1984) The structure of the genome. In: Gingsberg HS (ed) The adenoviruses. Plenum Press, New York, pp 35–124.

Tanguy-Rougeau C, Chen XJ, Wesolowski-Louvel M, Fukuhara H (1990) Gene 91: 43–50. **Tohe A**, Utatsu I (1985) Nucleic Acids Res 13:4267–4282. **Tokunaga M**, Wada N, Hishinuma F (1987a) Nucleic Acids Res 15:1031–1046. **Tokunaga M**, Wada N, Hishinuma F (1987b) Nucleic Acids Res 16:8097–8112. **Tokunaga M**, Wada N, Hishinuma F (1987c) Biochem Biophys Res Commun 144:613–619. **Tokunaga M**, Wada N, Hishinuma F (1988) Nucleic Acids Res 16:7499–7511. **Tokunaga M**, Kawamura A, Kitada K, Hishinuma F (1990) J Biol Chem 265:17274–17280. **Tommasino M** (1991) Yeast 7:245–252. **Tommasino M**, Ricci S, Galeotti CL (1988) Nucleic Acids Res 16:5863–5878.

Wesolowski M, Algeri A, Goffrini P, Fukuhara H (1982) Curr Genet 5:191–197. **White JH**, Butler AR, Stark MJR (1989) Nature 341:666–668. **Wickner RB** (1992) Annu Rev

Microbiol 46:347–375. **Wickner RB**, Leibowitz MJ (1976) Genetics 82:429–442. **Wilson DW**, Meacock PA (1988) Nucleic Acids Res 16:8097–8110.

Yeh P, Fleer R, Maury I, Mayaux JF (1990) Abstr D14, 6th Int Symp Genetics of Industrial Organisms, Strasbourg.

Prof. Dr. Friedhelm Meinhardt
Dipl. Biol. Jürgen Schründer
Institut für Mikrobiologie
Westfälische Wilhelms-Universität Münster
Correnstraße 3
48149 Münster, Germany

D. Taxonomy

I. Systematics of the Bryophytes

By Jan-Peter Frahm

1. General Aspects

An impression of current research topics in bryology can be obtained from the abstracts of the contributions presented at International Congresses. This concerns the bryology sessions during the International Botanical Congress in Tokyo in 1993, a symposium on East Asiatic Bryology in 1992 in Helsinki (published in Bryobrothera, Vol. 1) and a symposium on bryology in Latin America in 1992 in Mexico City (published in Tropical Bryology 6).

In 1992, the eminent Japanese hepaticologist Sinske Hattori died at the age of 76. Volume 74 of the Journal of the Hattori Botanical Laboratory was devoted to commemorating him. The 25 contributions show a profile of the current research in bryology.

2. Morphology, Anatomy

Peristome morphology has been used for moss taxonomy since the time of Philibert at the end of the last century. Philibert was the first who grouped the broad variation of peristome teeth into a few basic types (nematodontous with teeth consisting of whole cells as in Tetraphidaceae and Polytrichaceae, and arthrodontous teeth consisting of remnants of cell walls as in all other systematic groups, divided into haplolepideous and diplolepideous peristomes). At the beginning of this century, Fleischer used this morphological information for the familial classification of mosses. In a series of papers, Shaw et al. (1989) dealt with the peristome development of mosses in various systematic groups. They could show that the earliest developmental stages of all types are identical and uniform in all mosses. Comparative studies of the development of peristomes of the haplolepideous and diplolepideous type show that the teeth of the haplolepideous peristome are homologous to the cilia of the diplolepideous peristome and not to the endostome segments, as previously thought.

Progress in Botany, Vol. 56

Algae were reported to grow endophytically in living tissues of mosses. The identity of the alga is uncertain, but it is commonly named *Chlorochytrium.* A species of *Chlorochytrium* is known to live endophytically in plants of *Lemna.* Recently, this alga was found in two more species of mosses collected in the southern United States (Reese 1992). So far, *Chlorochytrium* has been recorded from six different moss species in four families.

Margins of costal lamellae on the leaves of the species of *Polytrichadelphus* have been examined by SEM (Proctor 1992). Lamellae margin characters allow separation of species of *Polytrichadelphus* from those of *Polytrichum,* which before was possible only by sporophytic characters. The lamellae of *Polytrichadelphus* are covered by flakes of epicuticular wax, which resemble those of the species of *Dawsonia,* but differ from all other Polytrichaceae. Based on the lamellae character, *Polytrichadelphus lyallii,* so far the only representative of the genus in North America, is excluded from the genus. *Polytrichadelphus* has thus a southern hemisphere distribution, ranging from Sulawesi to New Guinea and New Zealand, and southern South America through the Andes to Costa Rica. This type of distribution is interpreted as being derived from a continuous range in the late Cretaceous with radiation towards the north in the late Tertiary.

3. Chemotaxonomy

Monoclea is a primitive large thalloid hepatic of dubious systematic position. Two species were recognized within the genus, mainly based on thallus size and geographical distribution, the robust *M. forsteri* in New Zealand and southern South America, and *M. gottschei* in northern tropical America. Because of the obscure differentiation, there was a tendency to synonymize the two. In a study of morphology and flavonoid and terpenoid contents, again two species, although of different circumscription, could be separated (Gradstein et al. 1992); however, with a different circumscription. *Monoclea forsteri* is poor in secondary metabolites and is confined to New Zealand. *Monoclea gottschei* has the largest amount of bisbibenzyls ever encountered in a liverwort, a great variety of flavonoids, and is confined to the New World. Within *M. gottschei,* two subspecies are distinguished, one in Chile and one in tropical America. Both species can also be differentiated by sporophyte characters. Morphological and chemical data support a placement of *Monoclea* in the Marchantiidae.

The rate of evolution of bryophytes is usually regarded as very slow. This view depends on evidence from morphology, fossil records, and interpretation of ranges, especially of disjunct taxa. Several electrophoretic studies show, however, that the level of gene diversity is much higher than expected and comparable to that of angiosperms. This is supported by a study on the isozyme variation in the moss *Meesia triquetra* from Canada (Montagnes et al. 1993). Population of *Meesia triquetra* from boreal, subarctic, and high arctic regions of Canada have different genetic composition and reflect different phytogeographic histories. High arctic populations have highest genetic identities, which are interpreted as the result of recent immigrations from subarctic populations. Gene diversity in subarctic populations is high and probably due to isolation during glaciation. The boreal populations are descendants from relatively old, interbreeding populations and thus show a rather low gene diversity.

4. Systematics and Evolution

a) General

Due to the old age of bryophytes, their origin and their relation to green algae and tracheophytes remains a matter of speculation. Morphological, ultrastructural, and chemical studies in the past led to the knowledge that the major groups of bryophytes (hornworts, liverworts, and mosses) are not of monophyletic origin, and that mosses are closest to tracheophytes. A molecular approach was now undertaken using RNA sequences. The value of this method was discussed by Albert and Mishler (1992) and Albert et al. (1992). In a practical approach, the chloroplast RNA of two genes with 170–200 nucleotides was compared from 11 species of bryophytes with those of green algae, maize, and tobacco (Mishler et al. 1992). A cladistic analysis of the data revealed that hepatics and mosses are separate, but mosses are more closely related to tracheophytes.

In contrast to flowering plants, bryophytes are known to have broad geographic ranges, low levels of endemism, and frequent disjunctions without apparent morphological differentiation. These observations have led to the hypothesis that bryophytes undergo low morphological differentiation. However, studies of biochemical variability show that many bryophytes are genetically polymorphic. These microevolutionary aspects of bryophytes are summarized by Shaw (1991a, 1992). The same author documented the low morphological variability in widely separated sterile populations of the copper moss *Scopelophila cataractae* (Shaw 1993a) and the phenotypic (morphology of gametophyte and sporophyte) and genotypic (enzyme variation, heavy metal tolerance) variation of populations of *Funaria hygrometrica* (Shaw 1991b).

A sensation in bryophyte taxonomy during the past years was the discovery of sporophytes of *Takakia* (Smith and Davison 1993). The genus was first described in 1958 based on a collection made in the Japanese Alps in 1951. The species *T. lepidozioides* was known only from sterile material and looked totally unlike any other plant, resembling hypothetic early landplants with subterraneous stolons and erect stems without rhizoids, two to four pronged phyllidia, presence of slime hairs, and a chromosome number of n = 4. First the discovery of archegonia, although naked and stalked, confirmed that *Takakia* is a bryophyte. In 1963, a second species, *T. ceratophylla,* was described. The ranges comprise the

Himalaya, Borneo, Japan, the Aleutian Islands, SE-Alaska, and British Columbia. *Takakia* was regarded as hepatic and usually placed into the Calobryales. The discovery that one of the species has an apical meristem instead of an apical cell supported the theory that *Takakia* is apparently a living fossil and a relic of an early land plant, and led to the separation of *Takakia* in a division of its own, the Takakiophyta. In 1988, antheridia were found for the first time. An ultrastructutal study revealed that the spermatozoids are those of mosses and not of hepatics. This find initiated the search of female plants to breed both sexes and to produce sporophytes. At least, mature sporophytes were found in 1990 in the Aleutian islands. They confirmed that *Takakia* is a moss in contrast to its placement in the hepatics for nearly 40 years. The capsules open, however, with a longitudinal slit, which is typical for the class Andreaeopsida, whereas the sporophytes of most mosses open with an operculum (or as a reduction not at all). Thus the authors place *Takakia* in a separate order, Takakiales in the Andreaeopsida.

The first part of the results of the third German Bryotrop Expedition to Rwanda and Zaire was published in the journal Tropical Bryology, Vol. 8. It includes 26 contributions dealing with a description of the collecting sites, the history of the bryological exploration of Central Africa and treatments of numerous systematic groups of mosses and hepatics as well as checklists for the hepatics and mosses of Central Africa. Although Central Africa is one of the best-explored areas in tropical Africa and all areas studied had been visited previously by bryologists, numerous species could be recorded as new, not only to Zaire and Rwanda, but also to Africa, and numerous species (mostly pleurocarpous mosses) were described as new to science.

Numerous genera of mosses and hepatics were monographed or revised in 1992–93, which shows the slow but steady progress towards knowledge of the inventory of the world. The number of species of bryophytes in the world is still dubious. Mosses: *Cladopodanthus* (Magill 1993), *Rigodium* (Zomlefer 1993, the genus comprised 26 species before, which were reduced to 6), *Philophyllum* (Buck 1992), *Papillaria* in the Pacific (Streimann 1992), *Neckeropsis* in Africa (Enroth 1993), *Isopterygium* in the Neotropics (Ireland 1992), *Herzogiella* in North America (Ireland 1991), *Holodontium* (Ochyra 1993), Calymperaceae and Leucophanaceae for the Flora Neotropica (Reese and Salazar Allen 1993), *Potamium* (Florschütz-de Waard 1992), *Leucobryum* in Asia (Yamaguchi 1993), *Orthotrichum* in SE Asia (Lewinsky 1992). Hepatics: Geocalycaceae (Engel 1992a,b), *Pleurozia* (Thiers 1993), *Thysananthus, Dendrolejeunea, Fulfordianthus* gen. nov. (Gradstein 1992a), *Riccia* in Latin America (Jovet-Ast 1993), the European and African species of *Marchantia* (Bischler-Causse 1993).

A conspicuous increase in the number of species described as new to science could be observed during the past years, especially in mosses. During 1992–93, 44 species and 4 genera of mosses were described as new, and 20 species and 3 genera of hepatics. Most of the species of mosses were described from the tropics (e.g., 14 species of *Sphagnum* and three other mosses from Brazil), but also 8 species from Europe (*Orthotrichum ibericum, Seligeria irrigata, Racomitrium lusitanicum, Thamnobryum cataractarum, Bryum demaretianum, Sphagnum rubiginosum, S. isoviitae, S. olafii*).

Part of the species recently described as new would not have been described as new on a species level some decades before, since taxonomical concepts tend to be wider or smaller, a trend which goes up and down periodically. Basically, most of the bryophyte species in Europe were recognized towards the end of the last century. Then more and more microspecies were described. This is especially demonstrated by species of *Sphagnum* during the Warnstorf era. Mönkemeyer included 183 species in the genus *Bryum* in 1927 in the supplement volume to *Rabenhorst's Kryptogamenflora* for Europe. In contrast, Gams listed 52 (sometimes still dubious) species in 1973 in the fifth edition of the bryophyte volume in the *Kleine Kryprogamenflora.* A checklist of the European mosses by Corley et al. in 1981 again includes 68 species. The increase is due to the recognition of characters such as rhizoidal gemmae ("*Bryum erythrocarpum* complex") or axillary leaf bulbils ("*Bryum bicolor* complex") and the consequent recognition of new species.

Another recent trend in bryophyte taxonomy is that infraspecific categories seem to be no more in use. Every taxon, regardless of the taxonomic value of the character states, is described as a new species but not as a subspecies or variety. In this way, the authors of such new species miss the chance to express phylogenetical relationships. For example, another species of the "*Bryum erythrocarpum* complex", *Bryum demaretianum*, was described as new from a single locality in Belgium (Arts 1992). This complex of very closely related species differs merely by the color, shape, or size of their rhizoidal gemmae. This concerns also the *Racomitrium-canescens* and the *Racomitrium heterostichum* complex, which were split several years ago into smaller species.

The genera of the so-called *Warnstorfia-Calliergon* group were revised (Hedenäs 1993). The genera *Loeskhypnum* (for *Drepanocladus badius*) and *Warnstorfia* (for some species of *Drepanocladus*) are re-established, *Calliergon stramineum* is separated in the new genus *Straminergon* and *Calliergon sarmentosum* is placed in the genus *Warnstorfia*. *Straminergon* is based on sparse axillary hairs, whereas they are longer in *Calliergon*. The genera *Loeskhypnum* and *Warnstorfia* had not been accepted since the time they were created 50–80 years ago, and it is doubtful whether that will be different now. The separation of a monotypic genus *Straminergon* based on characters such as axillary hairs seems to be doubtful.

On the generic level, there is a trend to enlarge the number of genera. This effect is mainly based on the raising of previously infrageneric categories to generic rank or establishing new genera which would have been regarded as subgenera before.

b) Hepatics

Many bryophyte species are not typified. This concerns especially species described during the last century, for which no type specimen was designated. The species concept is based merely on the description (often only consisting of a short Latin phrase) or the interpretations of it, which are published in larger floras. For example, the status of the hepatic species *Marsupella aquatica* was controversely discussed by various authors as species, microspecies, subspecies, variety, or modification of *M. emarginata*. The partly lengthy discussions of this problem were quite philosophical, since a type was not designated by Lindenberg in 1829. Therefore a lectotypification was undertaken (Frahm 1993a). The only material in the Lindenberg herbarium does not however, match, the previous circumscription. Mixed tufts with *Marsupella emarginata* were found in the field, which show that both are genotypically distinct. These populations of mixed tufts are separable by statistical methods, the variation of characters of both taxa is, however, overlapping. Since both taxa are sympatric and show no ecological differentiation (as shown by the mixed tufts), *Marsupella aquatica* seems to be merely a "morphospecies".

In a long paper of 200 pages, Schuster (1992a) discusses the oil bodies in hepatics and describes the oil bodies in Lejeuneaceae.

c) Mosses

Over the past two centuries, several attempts have been made to compile lists of all known mosses, starting with *Species Muscorum* by Hedwig in 1801, followed by *Synopsis Muscorum* by C. Müller in 1848, Jaeger and Sauerbeck with their *Adumbratio florae muscorum* 1870–78, and the *Index Muscorum* by Paris 1898–1904. The last census of mosses was the *Index Muscorum*, including all names and references for mosses, published between 1959 and 1969 in five volumes. It is still a standard work, but includes data only up to 1962. Six supplements to the *Index Muscorum* were published biannually in the journal *Taxon*. It started with additions published in the years 1974–75, but the period 1963–73 was never included. Now the authors of the supplements have published these as well as the additions for 1963–73 in one book, the *Index of Mosses 1963–1989* (Crosby et al. 1992). The *Index Muscorum* proposed taxonomic opinions, and therefore accepted names were differentiated from synonyms, often reflecting the personal view of the authors. In this way it was attempted to set a taxonomic standard, which was, however, not successful since taxonomic opinions are not stable even over short periods of time. The authors of the *Index of Mosses* did not follow this concept, but only indicated other uses of names. The number of mosses worldwide is estimated to be 10 000 species, the total number of specific names described is 57 000. Exact numbers will be available when a world list of mosses is completed, which is under preparation at the Missouri Botanical Garden. It will include all accepted names and synonyms.

Searching for new characters for bryophyte taxonomy, the use of branch development pattern has been tested with Bartramiaceae (Akiyama and Nishimura 1993). Akiyama has found that dormant branch primordia of the *Bryum* type are characteristic for acrocarpous mosses and dormant branch buds called *Climacium* type are characteristic of pleurocarpous mosses. Species of the acrocarpous Bartramiaceae (with the exception of the monotypic *Catoscopium*) as well as Timmiaceae have, however, a *Climacium*-type branch development and thus should be excluded from the Bryineae.

d) Cytology

Electrophoretic data made evident that the North American *Plagiomnium medium* ssp. *curvatulum* is an allopolyploid and therefore treated as a species (Wyatt et al. 1993a). *Plagiomnium ellipticum* is considered as one of the progenitors. The other progenitor is either extinct or the European endemic *P. elatum*, which matches the required genetic profile best. An origin derived through hybridization had been postulated previously, based on morphological characters for *P. medium* and supported by isozyme analysis (Wyatt et al. 1992). By the same techniques, it was discovered that *Rhizomnium pseudopunctatum* is an allopolyploid of *R. gracile* and *R. magnifolium* (Wyatt et al. 1993b).

e) Fossil Bryophytes

During the past years, amber has proved to be a rich source for fossil bryophytes from the Tertiary, especially amber from the Dominican Republic, which has an age of 20–40 million years. Except for a few fossils from Mexican amber, Dominican amber has so far provided the only information about the bryoflora of the Tertiary in tropical regions. New fossil hepatics from Dominican amber were described by Gradstein (1993) and Grolle (1993). Of the ten species recognized by Gradstein, six were described as new but extinct species. For the first time, however, three fossil hepatics could be recognized as extant species. These are the oldest fossil records of living hepatics, and demonstrate that these species are at least 20 million years old.

For the first time also mosses preserved in Dominican amber were studied; These are the first fossil mosses known from the tropics (Frahm 1993c). Nine species could be recognized in fifteen specimens, four of them could be attributed tentatively to extant species, five to extant genera. This shows a remarkable consistency in the moss flora of the Caribbean over the past 20–40 million years. Surprisingly, all fossil specimens of mosses are 50% smaller than the present-day specimens, which lacks any explanation.

Ignatov (1992) described a remarkable fossile moss genus from Mongolia. The genus called *Bryohutuliinia* includes two species from the Jurassic and Cretaceous. Both species are characterized by the lack of costa, limbate leaves, elongated laminal cells, and distichous leaves, a combination of characters which is not found in extant genera.

5. Floristics

a) General

A census of the mosses of western Australia (Stoneburner et al. 1993) comprises 192 species, 56 of them Pottiaceae, reflecting the arid environment of this area. Twenty-six percent are endemic to Australia, 40% endemic to Australia and New Zealand, 10% are bipolar in distribution and probably the result of introduction from the northern hemisphere. Sixty percent of the species include South America or South Africa in their distribution, demonstrating the strong Gondwanan element and origin of the Western Australian bryoflora.

The table mountains ("tepuis") of the Guayana highland are known for their extremely high rate (75%) of endemic vascular plants. Although the vascular flora of the tepuis is well documented, there is less information on the bryophyte flora. Based on personal field data, Buck (1991) and Buck and Crum (1993) indicate a poor and relatively depauperate bryophyte flora.

The moss flora of Eastern North America by Crum and Anderson (1981), as well as the ancient moss flora of North America by Grout (1936–39), do not include keys to the genera and families, and this made these comprehensive works difficult to use. This gap was closed by a key to the moss genera of North America north of Mexico (Vitt and Buck 1992). The key includes 313 genera of mosses which are keyed out in 444 alternatives.

In a status report on the progress on the bryophyte flora of Western Melanesia, Koponen et al. (1992) summarize the results of this enormous project. Based on 17 000 collected specimens, 156 genera of mosses and 85 genera of hepatics were treated. Two genera and 60 species were described as new, 300 names were reduced to synonymy. Thirty-eight genera and 168 species were added to the flora.

b) Floras

A new mossflora for New Zealand was published by Beever et al. (1992) describing and illustrating 75 common species and the distinguishing characters for another 380 species, which are about 80% of the species of the country. Eighty-nine species are included in a study of the pleurocarpous mosses of Madeira by Hedenäs (1992). The mosses of Iceland are treated in a series of nine publications (Johannsson 1989–93). Although the descriptions are in Icelandic, all species are fully illustrated and their distribution within Iceland is shown on dot maps. It is the first modern survey of the mossflora of Iceland since Hesselbo's flora published in 1912–18.

The third volume of the *Illustrated mossflora of the Nordic* (= Scandinavia in a wider sense) was published covering Bryaceae and Mniaceae (Nyholm 1993).

For North Carolina, a liverwort flora was published comprising 195 species (Hicks 1992). The hepatic flora of eastern North America was completed with Volumes 5 and 6 (Schuster 1992b). The complete work consists of more than 5000 pages and includes more than 1000 plates, which is quite exceptional for a regional flora.

c) Atlases

The distribution of all 490 species of mosses from Ontario was mapped and published by Ireland and Ley (1992). The third and last volume of the atlas of the Bryophytes of Britain and Ireland was published (Hill et al. 1993). It includes dot maps for 366 species of mosses. The atlas comprises the mapping activities of the British Bryological Society. Britain is the first country in Europe for which a bryophyte atlas has been completed. Complete distribution maps are also part of a mossflora of the Netherlands. Uncompleted atlases including only selections of species exist for the mosses of the Iberian Peninsula and Poland. For Poland, Vol. 8 of the atlas was published (Ochyra et al. 1992b), for the Iberian Peninsula Vol. 3 (Casas et al. 1992).

d) Checklists

From Peru, 889 species of mosses are included in a checklist (Menzel 1992). Regrettably, such simple tasks are still mostly undertaken by scientists from non-tropical countries. The author calls his checklist a preliminary one, although four checklists were previously published, but it is still preliminary in the sense that it is far from being complete. The increase in the knowledge of the tropical bryoflora can be demonstrated by the example of Peru. In a first compilation of the mosses of Peru in 1951, 345 species were listed. In 1975, the number increased to 568, to 834 in 1985, and 903 in 1987, and now went down to 889. The decrease is the result of recent revisions and new synonyms.

A checklist of the mosses and hepatics of Malawi lists 197 species of mosses and 106 species of hepatics (O'Shea 1993).

Ten years after publication of the first checklist of the bryophytes of Macaronesia, a checklist of the bryophytes of the Canary Islands was published by Dirkse et al. (1993). It is based, however, on the old data but adds 24 new records for the Canary Islands, 15 of which are new records for Macaronesia, as well as the floristic data of the authors. Sixty species were excluded from the flora of the Canary Islands, which were doubtful or revised records. At present, 140 species of hepatics and 309 species of mosses are known from the Canary Islands. According to the first checklist of the bryophytes of Luxemburg, 518 species were found on this relatively small area (Werner 1993).

e) Conservation

Except for commercial use of peat mosses or the collection of mosses for gardening or decoration, bryophytes seemed not to be endangered in the past. However, conservation of bryophytes has become topical recently. For the first time, a congress was devoted to the conservation of bryophytes. The trend to indicate endangered species leads to compilations such as a selection of 108 bryophyte species from Madeira for which the degree was evaluated to which they are threatened (Sergio et al. 1992a).

As shown in Germany, activities of conservationist can get out of control. In Germany, all species of the genera *Sphagnum, Dicranum, Rhytidiadelphus, Leucobryum,* and *Hylocomium* are totally protected by law. Not only collecting, but also possession of specimens is forbidden, as well as the use of species of these genera (e.g., *Sphagnum*!) in university practicals.

A symposium held in Uppsala (Sweden) in 1992 focused on endangered bryophytes. The proceedings were published as a special issue of the journal Biological Conservation 59(2–3), 1992. A European committee for the Conservation of Bryophytes met in Funchal, Madeira, in 1992 to prepare a Red List of European Bryophytes (Urmi 1992). Twenty percent of the European species of mosses and hepatics were included in the Red List, 40% were excluded. The status of 40% of the species remained unclear.

The ministers of the European Union approved the habitat directive as an addition to the Bern convention in 1992 (Raeymaekers 1993). Annex 1 lists 167 different endangered habitats. Annex 2 includes species that need protection, 392 species of seed plants, 16 species of ferns, and also 21 species of bryophytes. *Leucobryum glaucum* and all *Sphagnum* species are listed in Annex 3 as endangered by exploitation. The list of species in Annex 2 is quite heterogenous. It includes species confined to few localities on Madeira as well as *Bruchia vogesiaca* growing in pastures or *Dicranum viride* growing in man-made forests, species which are rare but do not grow in endangered habitats.

6. Geography

North America and Europe have a common geological origin as part of the Laurasian continent, which split 50 Million years ago. Thus the mossflora of both continents has a common origin. In order to estimate the differences between the mossfloras of both continents, a comparison was undertaken (Frahm and Vitt 1993). From North America, 1170 species of mosses of 272 genera in 58 families are known, and 1084 species in 232 genera and 51 families from Europe. Although North America has direct connections to the Neotropics, has phytogeographical relations to eastern Asia, extends much higher into the Arctic, and is much larger, it has only 86 species, 40 genera, and 7 families more than

Europe. Sixty-eight genera are represented in North America and Europe with the same species, representing the old stock of Laurasian taxa. Sixty-three genera (mainly neotropical and Pacific in distribution) are found only in North America, compared with 20 genera confined to Europe. Forty-three percent of the species from North America are not found in Europe and 30% of the European species not in North America. The higher percentage in North America is interpreted as a result of lower rates of extinction during the Pleistocene, and better availability of refugia.

In a survey of the origin, evolution, and distribution of the SE Asian species of the genus *Campylopus,* Frahm (1992b) showed that Sri Lanka and SE India is the center of speciation of this region. It is supposed that the species rafted on the Indian plate from the Gondwana continent northwards. Speciation of SE Asian taxa took place by eastward migration of the species on the Indian plate influenced by climatic changes and sea level changes during the Pleistocene and orogenesis during the Tertiary. There is less evidence for a evolutionary pathway to SE Asia via Australia. Therefore this bryophyte genus in SE Asia seems to have a history similar to pitcher plants (*Nepenthes*) or chameleons.

In his famous textbook on the geography of bryophytes, Herzog in 1926 noticed that disjunctions between tropical Africa and America are present only on the genus level, similarly to flowering plants, but (with a few exceptions of pantropical bryophytes) not on the species level. Monographs and revisions during the past 20 years revealed, however, more and more disjunctions between tropical Africa and the Neotropics on the species level. Now Delgadillo (1993) has summarized 334 species and varieties which the neotropical mossflora shares with Africa. These species represent 8% of the neotropical mossflora. Since the list is still incomplete and only a small part of the genera has so far been revised, the number of species disjunct in tropical Africa and America will increase. As explanations, both the results of ancient land connections and long distance dispersal are discussed and considered. At least part of the 334 species, such as lowland species without or with limited spore production (species with large spores, cleistocarpous capsules, sporophytes without seta often covered by leaves), can be assumed to have persisted unchanged for more than 180 million years, which allows us to measure the low evolutionary rate in at least part of the mosses. New disjunctions between tropical Africa and America are continuously being discovered. For example, the moss *Adelothecium bogotense*, common in the neotropics between Mexico and Brazil, was now found in the Uluguru Mountains, Tanzania (Ochyra et al. 1992). Twelve species of *Fissidens* (of about 1000 described) are common in the Neotropics, Asia, and Africa (Pursell et al. 1992).

Computer databases facilitate phytogeographic comparisons. In a comparison of the moss species reported for Mexico and Colombia using the LATMOSS database, Delgadillo (1992a) found that 371 of the 946 species reported for Mexico and 937 reported for Colombia are common in both countries. The number is interpreted not just as a floristic relationship, but in relation to the phytogeographical role of the Central American bridge. Since most of these species common in both countries are montane in distribution, most species have migrated at higher elevations. Of the 16 genera represented in Mexico and Colombia with five and more species there are only four tropical genera. Therefore the land bridge functioned mainly for N-S migrations.

Seven floristic corridors through Mexico were distinguished for the floristic interchange between N and S America. There are, however, 15 species amongst the mosses occurring in these belts which are cleistocarpous (Delgadillo 1992b). The capsules of these species do not open, the spores are released by rotting of the capsule wall, and thus do not allow long-distance dispersal by wind. Their method of dispersal remains doubtful.

In 1976, Pócs published a first list of bryophyte species being disjunct between tropical Africa and Asia. At that time, he listed 35 hepatics and 73 mosses showing this type of disjunction. Field work in both parts as well as revisions of genera, revealed new data, which were now summarized (Pócs 1992). According to this new list, there are 70 species of hepatics and 108 species of mosses common in tropical Africa and tropical Asia. Compared with Afro-American hepatic disjunctions (52 species), the floristic affinity between tropical Africa and Asia is stronger (70 species) and seems to support the concept of "Paleotropics" for hepatic distribution.

Another representative of the Asian bryoflora, *Bryochenea sachalinensis,* was found in Alaska (O'Brien 1993). This is the first example of a species found in previously glaciated areas. As explanation, long-distance dispersal is regarded as unlikely because of the dioeceous condition of the species and the oceanic climate with high precipitation, which would not allow spore dispersal over more than 2000 km. Therefore, the Alaska locality is considered remnant of a formerly wider range in warmer periods of the Holocene.

The knowledge of species ranges is increasing with new floristic records and new synonymies as a result of revisions. With increasing knowledge, the known ranges of bryophytes usually become larger. *Calycularia crispula* is a primitive thalloid hepatic known from the Himalayas, Thailand, Taiwan, Korea, and Japan, and also from tropical Africa and a single record in Mexico. This species has now been found in British Columbia, SE Alaska, and the Aleutian Islands (Davison and Smith 1992). The new North American records are interpreted as relics of a formerly continuous amphipacific range. The isolated Mexican record is explained by dispersal from the source areas in northwestern North America, perhaps from areas in the Rocky Mountains, which are today too arid for the species. A similar gap was closed also in the distributional data of the moss *Campylopus japonicus* (Frahm 1993b). This species is also widespread in E Asia and was also known before from Mexico. Now it has been found in the Queen Charlotte Islands in British Columbia.

The phytogeographical affinities of the hepatic floras of Australia and New Zealand, Western Melanesia, Borneo, the Phlippines, Bhutan, China, Taiwan, Japan, and Korea were studied using statistical methods, based on recent checklists including 2906 species (Piippo 1992). Australia and New Zealand have a strikingly different hepatic flora with many endemic genera, which is explained by its Gondwanalandic origin and its long isolation. Western Melanesia has

Gondwanan as well as Laurasian elements without any endemic genera. Eastern Asia has endemic but also Gondwanan genera which reached the northern hemisphere via the Indian and Cathaysian plate.

Bipolar disjuncts amongst bryophytes are a phytogeographical problem in bryology. It is generally assumed that they are of holarctic origin and have reached the southern hemisphere (except by introduction) by mountain chains (in America) or mountain hopping (in Africa and the Malesian islands). About 24 species are known for this distribution type, a number which has been reduced to 12 by Ochyra (1992). As a new example, the holarctic moss *Amblyodon dealbatus* has been recorded from Tierra del Fuego.

Recent results of studies on the bryoflora of SE Asia were presented at a Congress of East Asiatic Bryology in Helsinki, 1990, of which the proceedings have been published (Koponen and Hyvönen 1992).

Grimmia mesopotamica, a desert bryophyte so far known only from the Near East, was found new to Europe in southern Spain (Guerra et al. 1993). It is another example for the extension of Irano-Turanian or Saharo-Arabian ranges to southern Spain as relics of the old Tethyan floristic element of xeromorphic species. Another presumably Tethyan element was reported as new to Europe, again from Spain (Cano et al. 1992). *Crossidium seriatum* was only known before from two localities in Arizona and California, and was now found in SW Spain. Such disjunctions are explained as relics of Pangaean ranges and are of preangiosperm origin.

Orthotrichum flowersii, a species so far known only from Utah and Idaho in North America, was found in the Durance valley in France (Boudier and Pierrot 1992). In the same region, Boudier (1992) collected *Tortula rigescens* for the first time in Europe, a species so far only known from three localities in the Near East. The Durance valley seems thus to be a refugium for xerothermic elements of Tethyan origin, which was not known before.

7. Ecology

a) General

There are many statements in textbooks on the ecology of bryophytes which are based on local studies, but these results are easily generalized. Such statements concern the host specifity of epiphytic bryophytes in Europe, the species richness of different host trees, and the pH preferences of epiphytic bryophytes. In a study of the epiphytic bryophytes of the Vosges Mountains, France, based on more than 1000 floristic records, Frahm (1992a) showed that there is generally no host specifity. In contrast to observations from other parts of Europe, beech trees have the richest epiphytic flora. Also, the number of epiphytic species is not correlated with the pH of the bark of the host trees and not higher on trees with subneutral bark, but higher on oak trees with low pH than on ash or elder with a higher pH. Near waterfalls, even basiphilous bryophyte species are found on the bark of oak with a pH of about 5. All these effects are supposedly due to the higher humidity in the Vosges Mountains, and it can be concluded that host specifity is decreasing with humidity (and goes to zero in aquatic habitats, where bryophytes are growing on beer tins or water turtles). It is shown that most epiphytic species are found in forests with rocks or cliffs. Nearly all epiphytic species are facultative epiphytes which can live on rocks and can easily recolonize trees even after a complete clearing. Forests without rocks harbor only a few species of epiphytic bryophytes which can propagate by spores and are dispersed over longer distances. Thus, species richness of epiphytic bryophytes is a matter of habitat structure and not always of human influence or air pollution.

Spore dispersal is much stressed for long-distance dispersal to explain long range disjunctions, e.g., those between tropical America and Africa. However, experimental studies on spore dispersal are scarce. In a study of the spore dispersal of the moss *Atrichum angustatum*, Stoneburner et al. (1992) could show that 94% of all spores released fell within a radius of 2 m from the colony. Only 1% (but 2.4–4.5 million) of spores were recovered in 15 m distance. No spores were found on traps at 1.5 m height directly above the colony! Although million of spores do not disperse beyond a few meters, it can be estimated that thousands of spores are still dispersed over longer distances, and one spore is sufficient to establish a new colony. Similar experiments were performed by Miles and Longton (1992) with colonies of *Atrichum undulatum* and *Bryum argenteum*. Spore deposition showed a steep gradient from 14 000 spores in the center of the colony to less than 10 in 2 m distance. It could be shown that fewer spores of *Bryum argenteum* with a spore size of 8–14 μm fall within a radius of 2 m than of *Atrichum undulatum* with a spore size of 16–20 μm (and all spores of *Archidium alternifolium* with a spore size of 200 μm remain within the parent perichaetium).

The ecology of tropical rainforest bryophytes was treated by several authors. Gradstein (1992b) summarized dispersal, ranges, and conservation of rainforest bryophytes. The phytomass of epiphytic bryophytes and species richness was studied by Wolf (1993) on 59 canopy trees along a transect through the Colombian Cordillera Central between 1000 and 4130 m altitude. The general increase of phytomass coincides with the rise in humidity. Hofstede et al. (1993) determined the epiphytic load of a *Weinmannia* tree in Riseralda, Colombia. An extrapolation from this specimen results in 44 tons per hectare of epiphytic biomass with suspended soil. A determination of the nutrients in the green phytomass of epiphytes revealed that phosphorus is concentrated at higher levels in the soil suspended in the epiphytes than in the terrestrial soil. This demonstrates the role of epiphytes in the nutrient economy of the forest ecosystem.

A basic question in reproductive biology is the reason for rarity of bryophytes. For the moss *Scopelophila cataractae* it could be shown that all isolated populations in North America (although male and female plants are present in North America) are unisexual and therefore sterile (Shaw 1993b). It can be assumed that all populations are derived from single unisexual diaspores and that unisexuality of populations can maintain the population over long periods of time. In culture experiments, the growth was enhanced by highly metal-contaminated soil.

Along a transect from Campbell Island to Auckland Islands, Lord Howe Island, and Espiritu Santo, Vitt (1991) studied the changes in the mossflora with regard to latitude and elevation. In general, the species richness is constant across latitudes. There is, however, an increase of the number of species with elevation on all islands. Acrocarpous mosses are more common in nonforested areas, pleurocarpous in low elevation forest habitats. The number of pleurocarpous species increases with towards the tropics, and they show the highest frequency of derived characters.

b) Pollution

Twenty years ago, the Rhine was one of the most polluted rivers in Central Europe. Subsequent installation of purification plants along the Rhine and its tributaries has substantially improved the water quality. Like other organisms, also bryophytes reacted to this effect. In 1972, aquatic bryophyte distribution was mapped along the Lower Rhine, which showed large areas without any, only one, or only a few bryophyte species. This survey was repeated 20 years later in 1992 (Frahm and Abts 1993). Compared with 1972, eight species were recorded as new along this part of the Rhine and the quantity of ten species has considerably, partly explosively, increased.

Because of its cation exchange capability, *Sphagnum* plants can easily accumulate large quantities of heavy metals. This effect is being used in the "moss bag technique" developed by Mäkinen in Finland. *Sphagnum* stuffed in permeable bags are exposed in polluted areas for a certain time, and the heavy metal contents can be determined with an atomic absorption spectrometer. This method has usually been used for determination of the degree of air pollution in and around cities or steel plants. Now the moss bag technique was also used in aquatic transplantation experiments in Portugal (Sergio et al. (1992b). In Portugal, epiphytic bryophyte specimens from unpolluted areas were planted in and around Lisbon, and the uptake of heavy metals was determined after 6 months (Maguas et al. 1991). The uptake was higher in the hepatic *Frullania dilatata* as compared with the moss *Leucodon sciuroides*, showing a different suitability of bryophyte species for such tests. The different uptake of heavy metals in different species may also serve as an explanation for the different resistance of these species.

Air pollution or drought are stressed as explanations for the distribution of epiphytic bryophytes. Both factors have been controversially discussed in the past. In Portugal, the epiphytic bryophyte vegetation on *Olea europaea* at 76 stands was studied and was compared in a multivariate analysis with bioclimatic data (Sergio et al. 1990). Species diversity and composition are especially related to the Gaussen Index of Dry Days, and to a lesser extend to the temperature of the Emberger Index.

The use of aquatic bryophytes as indicators for the acidification of rivers was studied by Tremp and Kohler (1991) and Tremp (1992) in the Black Forest, Germany. Different pH ranges of the species allow a determination of the acidity of rivers in the field. Experimental studies showed that short period changes of the water pH had no effect on the bryophytes. Therefore, they can be used as indicators for the mean acidity of streams. The accumulation of aluminum and iron in the shoots of bryophytes was used for a staining method which allowed a qualitative determination in the field.

The changes in the Dutch bryoflora related to air pollution were studied by Greven (1992) based on transplantation and fumigation experiments, and comparison of floristic data. The frequency of certain epiphytic species has declined, but other species (predominantly, but not only, acidophytes) have increased. Transplantations of species regarded as sensitive to air pollution showed that at least some of these can survive in polluted areas. The lack of such species in polluted areas is therefore due to damages of the protonemal stage or dispersal units (spores, brood bodies).

References

Akiyama H, Nishimura N (1993) Bryologist 96:185–191. **Albert VA**, Mishler BD (1992) Cladistics 8:73–83. **Albert VA**, Mishler BD, Chase MW (1992) In: Soltis PD et al. (eds) Molecular systematics of plants, New York, pp 369–403. **Arts T** (1992) J Bryol 17: 263–268.

Beever J, Allsion KW, Child J (1992) The mosses of New Zealand. The University of Otago Press Dunedin. **Bischler-Causse H** (1993) Bryoph Bibl 45:1–129. **Boudier P** (1992) Cryptogam Bryol Lichénol 13:1–6. **Boudier P**, Pierrot RB (1992) Cryptogam Bryol Lichénol 13:165–170. **Buck WR** (1991) In: Herben T, McQueen CB (eds) Proc 6. Meeting Central and East Europ Bryol Working Group, Průhonice, pp 46–51. **Buck WR** (1992) Bryologist 95:334–337. **Buck WR**, Crum H (1993) Briitonia 45:17–20.

Cano MJ, Guerra J, Ros RM (1992) Bryologist 95:280–283. **Casas C**, Brugues M, Cros RM, Sergio C (1992) Cartografia de Briofits fasc. III. Barcelona (Institut d'Estudis Catalans). **Crosby MR**, Magill RE, Bauer CR (1992) Index of mosses. Missouri Botanical Garden St Louis. **Crum HA**, Anderson LE (1981) Mosses of eastern North America. Columbia University Press, New York.

Davison PG, Smith DK (1992) Bryologist 95:266–269. **Delgadillo MC** (1992a) Bryologist 95:261–265. **Delgadillo MC** (1992b) J Bryol 17:313–316. **Delgadillo MC** (1993) Bryologist 96:604–615. **Dirkse GM**, Bouman AC, Losada-Lima A (1993) Cryptogam Bryol Lichénol 14:1–47.

Engel JJ (1992a) J Hattori Bot Lab 72:105–115. **Engel JJ** (1992b) Contrib Univ Mich Herb 18:107–111. **Enroth J** (1993) J Hattori Bot Lab 73:159–173.

Florschütz-de Waard J (1992) Trop Bryol 5:109–121. **Frahm JP** (1992a) Herzogia 9:213–228. **Frahm JP** (1992b) Bryobrothera 1:221–230. **Frahm JP** (1993a) Fragm Florist Geobot 38:489–501. **Frahm JP** (1993b) Bryologist 96:142–144. **Frahm JP** (1993c) J Hattori Bot Lab 74:249–259. **Frahm JP**, Abts UW (1993) Limnologica 23:123–130. **Frahm JP**, Vitt DH (1993) Nova Hedwigia 56:307–333.

Gradstein SR (1992a) Bryologist 95:42–51. **Gradstein SR** (1992b) In: Bates FW, Farmer AM (eds) Bryophytes and lichens in a changing environment. Clarendon Press Oxford, pp 234–258. **Gradstein SR** (1993) Nova Hedwigia 57:353–374. **Gradstein SR**, Klein R, Kraut L, Mues R, Spörle J, Becker H (1992) Plant Syst Evol 180:115–135. **Greven HC**, (1992) Diss Bot 194:1–237. **Grolle R** (1993) J Hattori Bot Lab 74:71–76. **Grout AJ** (1936–1939) Moss flora of North America North of Mexico. 3 vols. Privately publ. by the author, Newfane, Vermont. **Guerra J**, Ros RM, Martinez-Sanchez JJ, Frey W (1993) Bryologist 96:245–247.

Hedenäs L (1992) Flora of Madeiran pleurocarpous mosses. Bryoph Bibl 44. Schweizerbart, Stuttgart. **Hedenäs L** (1993) J Bryol 17:447–480. **Hicks ML** (1992) Guide to the liverworts of North Carolina. Duke University Press, Durham. **Hill MO**, Preston CD, Smith AJE (1993) Atlas of the bryophytes of Britain and Ireland, vol 3. Martins Harley Books, Essex. **Hofstede RGM**, Wolf JHD, Benzing DH (1993) Selbyana 14:37–45.

Ignatov MS (1992) J Hattori Bot Lab 71:377–388. **Ireland RR** (1991) Lindbergia 17: 111–115. **Ireland RR** (1992) Trop Bryol 6:111–132. **Ireland RR**, Ley LM (1992) Syllogeus 70:1–138.

Johannsson B (1989–93) Islenskir mosar. Fjölrit Náttúrufraedistofnunar 12,13,15,16,19, 20,21,22,24. Reykjavik. **Jovet-Ast S** (1993) Cryptogam Bryol Lichénol 14:219–302.

Koponen T, Hyvönen J (1992) Proc Congress of East Asiatic Bryology, Helsinki 1990. Bryobrothera 1:1–332. **Koponen T**, Norris DH, Piippo S (1992) Bryobrothera 1:157–160.

Lewinsky J (1992) J Hattori Bot Lab 72:1–88.

Magill RE (1993) Bryologist 96:233–241. **Maguas C**, Sergio C, Catarino F, Branquinho C, Kratz W (1991) In: Framer JG (ed) Heavy metals in the environment. Edinburgh, pp 235–239. **Menzel M** (1992) J Hattori Bot Lab 71:175–254. **Miles CJ**, Longton RE (1992) J Bryol 17:355–368. **Mishler BD**, Thrall PH, Hopple JS, De Luna E, Vilgalys R (1992) Bryologist 95:172–180. **Montagnes RJS**, Bayer RJ, Vitt DH (1993) J Hattori Bot Lab 74:55–170.

Nyholm E (1993) Illustrated flora of Nordic mosses Fasc. 3. Nordic Bryol Soc, Lund.

O'Brien TJ (1993) Bryologist 96:205–209. **Ochyra R** (1992) Fragm Flor Geobot 37:251–259. **Ochyra R** (1993) Fragm Flor Geobot 38:75–98. **Ochyra R**, Bednarek-Ochyra H, Pócs T, Crosby MR (1992a) Bryologist 95:287–295. **Ochyra R**, Szmajda P, Bednarek-Ochyra H (1992b) Atlas of the geographical distribution of mosses in Poland part 8. Poznan-Kraków. **O'Shea BJ** (1993) J Bryol 17:645–670.

Piippo S (1992) J Hattori Bot Lab 71:1–35. **Pócs T** (1992) Bryobrothera 1:35–47. **Proctor MCF** (1992) J Bryol 17:317–334. **Pursell RA**, Bruggeman-Nannega MA, Iwatsuki Z (1992) Bryobrothera 1:49–55.

Raeymaekers G (1993) Bryol Times 72:1. **Reese WD** (1992) J Phycol 28:707. **Reese WD**, Salazar Allen N (1993) Flora Neotropica Monographs, NY Botanical Garden New York, pp 1–110.

Schuster RM (1992a) J Hattori Bot Lab 72:151–360. **Schuster RM** (1992b) The Hepaticae and Anthocerotae of North America, East of the hundredth meridian, vols 5 & 6. Field Museum of Natural History, Chicago. **Sergio C**, Sim-Sim M, Santos-Silva C (1990) An Jard Bot Madr 46:457–467. **Sergio C**, Schumacker R, Fontinha, S, Sim-Sim M (1992a) Biol Conserv 59:223–231. **Sergio C**, Seneca A, Maguas C, Branquinho C (1992b) Cryptogam Bryol Lichénol 13:155–163. **Shaw J** (1991a) Adv Bryol 4:29–74. **Shaw J** (1991b) Evolution 45:1260–1274. **Shaw J** (1992) In: Bates JW, Farmer AM (eds) Bryophytes and lichens in a changing environment. Clarendon Press, Oxford, pp 362–380. **Shaw J** (1993a) Syst Bot 18:525–537. **Shaw J** (1993b) Am J Bot 80:1034–1041. **Shaw J**, Mishler BD, Anderson LE (1989) Bryologist 92:314–325. **Smith DK**, Davison PG (1993) J Hattori Bot Lab 73:263–271. **Stoneburner A**, Lane DM, Anderson LE (1992) Bryologist 95: 324–328. **Stoneburner A**, Wyatt R, Catcheside DG, Stone IG (1993) Bryologist 96: 86–101. **Streimann H** (1992) J Hattori Bot Lab 71:83–111.

Thiers BM (1993) Bryologist 96:517–554. **Tremp H** (1992) Hohenheimer Umwelttagung 24:143–153. **Tremp H**, Kohler A (1991) VDI Berichte 901:1123–1136.

Urmi E (1992) Bryol Times 70:6–7.

Vitt DH (1991) In: Nimis PL, Crovello TJ (eds) Quantitative approaches to phytogeography. Kluwer Amsterdam, pp 205–231. **Vitt DH**, Buck WR (1992) Contrib Univ Mich Herb 18:43–71.

Werner J (1993) J Bryol 17:489–500. **Wolf JHD** (1993) Ann Mo Bot Gard 80: 928–960. **Wyatt R**, Odrzykoski IJ, Stoneburner A (1992) Syst Bot 17:532–550. **Wyatt R**, Odrzykoski IJ, Stoneburner A (1993a) Lindbergia 18:49–58. **Wyatt R**, Odrzykoski IJ, Stoneburner A (1993b) Mem Soc Torrey Bot Club 25:20–34.

Yamaguchi T (1993) J Hattori Bot Lab 73:1–124.

Zomlefer WB (1993) Bryologist 96:1–72.

Prof. Dr. Jan-Peter Frahm
Botanisches Institut der Universität
Meckenheimer Allee 170
53115 Bonn, Germany

II. Paleobotany of Land Plants

By Paul Kenrick and Else Marie Friis

1. Introduction

The impact of paleobotanical data on land plant systematics is greatest for ancient cladogenic events, resulting in isolated extant groups of divergent morphology. Paleobotanical data are therefore ideally suited to addressing problems in land plant macrosystematics, an area of botanical research that has received considerable attention over the last 10 years. Macrosystematics deals with the recognition and definition of patterns of relationship among major groups of organisms. Current interest stems from important advances in systematic theory that began during the 1960s and have developed into the widely used "cladistic approach" (Nelson and Platnick 1981). Macrosystematic studies make use of data from a number of disciplines ranging from general comparative morphology to plant ultrastructure and biochemistry. Recently, the application of molecular biological techniques to systematic problems has generated a large and important new set of data. Because none of these sources of data produce unambiguous results, it is important to retain a critical multidisciplinary approach to macrosystematic problems. Furthermore, the prevalence of extinction in the history of land plants means that paleobotanical data contribute the only source of information on a huge range of forms, and many aspects of relationship, homology, and the history of terrestrial life can only be understood through the window provided by the fossil record.

This review summarizes important recent paleobotanical contributions to land plant systematics, focusing on the years 1992 to 1993. All major groups except angiosperms have been reviewed. Angiosperm phylogeny and floral evolution is an area of rapid expansion, and there is a considerable amount of new information on the structure and organization of early angiosperms and related anthophytes (Endress and Friis, 1994). These advances, together with new developments in the systematics and developmental morphology of extant groups, have resulted in an extensive revision of angiosperm classification. In order to give a more comprehensive account of the scientific progress in this area, anthophyte systematics and phylogeny will be treated in a separate chapter in the 1995 volume.

Several textbooks in paleobotany have been published during the past decade, and two new editions of earlier treatments appeared in 1993. *Paleobotany and the Evolution of Plants* by Stewar and Rothwell (1993) is an expanded sec-

Progress in Botany, Vol. 56

ond edition of Stewart's textbook from 1983. *The Biology and Evolution of Fossil Plants* by Taylor and Taylor (1993) is based on Taylor's textbook from 1981 but is considerably expanded, including additional chapters on the structure and organization of vascular plants and on plant-animal interactions. Both books provide an excellent introduction to the fossil record. That of Stewart and Rothwell (1993) is aimed at undergraduates and has a particularly consistent form of presentation; Taylor and Taylor (1993) is aimed at upper-level undergraduate and graduate courses and is more encyclopedic. Both books are also invaluable reference sources for botanists and paleobotanists.

Information on stratigraphic occurrences of plants is essential to dating clades and charting temporal changes in floral diversity. A useful handbook treating the fossil record of plants and animals by Benton (1993) provides information on stratigraphic ranges of groups at the family level. The book includes four chapters devoted to the fossil record of plants: Bryophyta (Edwards 1993), Pteridophyta (Cleal 1993a), Gymnospermophyta (Cleal 1993b), and Magnoliophyta (Collinson et al. 1993). Short comments indicate whether a record is reliable or in need of further documentation.

2. Systematics

Recent developments in systematic theory have led to an important period of reevaluation, where the aim of many studies has been the reinterpretation of current knowledge from a new theoretical perspective. In botany, much recent work has focused on relationships among major groups of land plants. These studies have clarified many important aspects of relationship and homology and have brought the major systematic problems into sharper focus. Cladistic studies using paleobotanical data have addressed problems of relationship among major groups of seed plants. Classic early work adopted a synthetic approach that integrated paleobotanical and neobotanical data on comparative morphology (Crane 1985; Doyle and Donoghue 1986). Subsequent investigations have dealt with one or more seed plant groups in greater detail or reexamined the more general questions of relationship using updated morphological data or new molecular data sets. Recently, the role of paleobotanical data in seed plant macrosystematics has been examined in greater detail. Loconte and Stevenson (1990) analyzed relationships in extant seed plants based on comparative morphology and concluded that paleobotanical data are unnecessary to resolve relationships among *extant* groups. In response to this study, Doyle and Donoghue (1992) designed an explicit test of the effects of paleobotanical data and found that additional information from the fossil record did alter relationships and also had important implications for hypotheses of homology. Certain characters that appear to be homologous when viewed from a neobotanical perspective are best interpreted as convergences when fossils are considered. Although seed plant systematics can be addressed at the neobotanical level, excluding the extensive, well-documented fossil record may lead to erroneous conclusions.

At the more general level of vascular plants and embryophytes, cladistic studies have focused mainly on neobotanical data (Mishler and Churchill 1984, 1985; Graham et al. 1991; Garbary et al. 1993). Integration of the large database on Paleozoic and Mesozoic fossils has begun, but no detailed cladistic synthesis has yet been published. Preliminary studies that include data from the early fossil record support vascular plant monophyly and an early divergence of lycopods from other tracheophytes (Kenrick and Crane 1991; Gensel 1992). The early fossil record is consistent with previous cladistic studies based on neobotanical evidence that support bryophyte paraphyly with a basal position for liverworts in embryophytes and a sister-group relationship between either mosses or hornworts and tracheophytes (Mishler and Churchill 1984, 1985; Fig. 1). The paraphyletic status of bryophytes has been challenged recently in a study based on male ga-

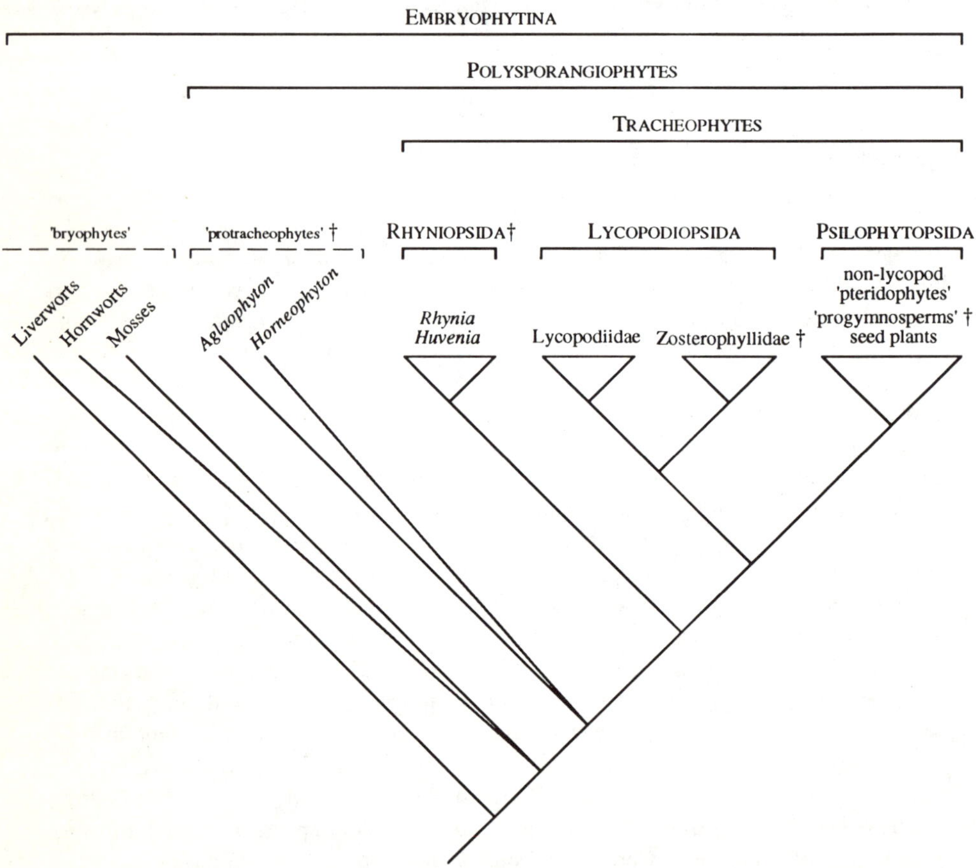

Fig. 1. Cladogram summarizing relationships among major basal land plant groups including some early fossils. † = extinct group. (Kenrick and Crane 1992)

mete ultrastructure. Garbary et al. (1993) support monophyly of bryophytes, but their data also suggest controversially that bryophytes are nested within vascular plants and that lycopods are paraphyletic to all other embryophytes. These results are inconsistent with other aspects of morphology: lycopod monophyly is well supported by general neobotanical and paleobotanical morphological data (DiMichele and Skog 1992).

a) Embryophytes

The early fossil record of embryophytes (land plants) is dominated by a diverse range of microfossils including dispersed spores, cuticles, and other putative plant fragments (see reviews by Gensel et al. 1990; Gray and Shear 1992). Studies over the past 20 years indicate a probable mid-Ordovician origin for land plants documented by distinctive spores with an obligate tetrad morphology. Significant changes in spore morphology occurred rapidly around the mid-Silurian with the diversification of single trilete spores. These microfossil data indicate a marked alteration in composition of the terrestrial flora at this time (Gray 1991), but the meaning of this change in terms of major land plant groups is still unclear. The early microfossil record probably reflects a terrestrial flora at the bryophyte level of organization, but the systematic affinities of the spore producers are problematic because of the generalized spore morphology and the absence of associated macrofossils.

The earliest and most abundant macrofossil evidence of land plants comes from the Late Silurian and Early Devonian of northern Europe and North America (Edwards and Berry 1991). Information from these regions is particularly rich because of the historical development of paleobotany in research institutions in these areas. New early fossils from paleogeographically remote parts of China promise to redress this bias and are beginning to provide important comparative data. Cai et al. (1993) described an interesting Upper Silurian (Pridoli) compression fossil assemblage from northwest China. The assemblage includes probable basal tracheophytes or protracheophytes of slightly greater morphological complexity than those in contemporaneous northern European assemblages.

Recent discoveries of minute fossil sporophytes from Lower Devonian sediments of the Welsh Borderland, Britain, document a diversity of early land plants at the tracheophyte and "protracheophyte" (stem-group tracheophytes, Fig. 1) levels of organization. Many specimens are extremely well preserved, and remarkable cellular details of sporangia, epidermal structures, and internal anatomy have been observed (Fanning et al. 1992; Edwards et al. 1992). The fragmentary nature of this material makes more specific systematic assignments problematic.

In contrast to vascular plants, the early macrofossil record of "bryophytes" is poor (see recent review by Taylor and Taylor 1993). In many groups of liverworts, the poor fossil record has been attributed to the delicate nature of the thallus (Krassilov and Schuster 1984), but taphonomic factors are unlikely to be important in other bryophyte groups, where collector bias and failure to recognize stem-group taxa probably also contribute to the poor early record (Kenrick 1994). Discovery of further Devonian taxa, in particular early relatives of mosses and hornworts, may prove decisive in clarifying relationships among basal embryophytes.

Well-documented observations on the gametophyte generation are extremely rare in the fossil record of vascular plants, but recent paleobotanical discoveries are beginning to clarify early stages in the evolution of this aspect of the plant life cycle. Exceptionally well-preserved Lower Devonian plants with intact gametangia have been described from the Rhynie Chert, Scotland, and these provide the first direct evidence of gametophytes and life cycles in early land plants (Remy et al. 1993). Life cycle reconstruction has been possible because the Rhynie Chert fossils are preserved in growth position, and linkage with associated sporophytes is achieved through anatomical comparisons. Observations on the Rhynie Chert have been extended to compression fossils at other Devonian localities. General features of the gametophyte generation in these early taxa include a simple, leafless, branched axial (as opposed to thalloid) morphology, terminal gametangia borne on an expanded disc-shaped or cup-shaped gametangiophore, and a level of anatomical differentiation comparable to associated tracheophyte or protracheophyte sporophytes (i.e., cuticle, stomates, and vascular tissues) (Remy et al. 1993). Cladistic analysis places taxa with isomorphic life cycles in basal tracheophytes and the tracheophyte stem-group, implying that the anisomorphic alternation of generations that characterizes extant vascular plants evolved from an isomorphic ancestral state. There is some corroborating evidence for gametophyte reduction in vascular plants in the complex gametophyte morphologies of extant Lycopodiaceae (Kenrick 1994).

b) Tracheophytes

Recent cladistic treatments of vascular plants recognize an early dichotomy between lycopods (Lycopodiopsida) and all other extant groups including *Equisetum*, "ferns", and seed plants (Psilophytopsida) (Kenrick and Crane 1991; Raubeson and Jansen 1992; Fig. 1). Relationships among basal members of the Psilophytopsida clade (nonlycopod vascular plants) are poorly understood, but this is an area in which the fossil record is currently making a significant contribution.

Recent paleobotanical studies on basal members of Psilophytopsida have emphasized the importance of the well-documented record of land plant soft tissues, which is unparalleled in any other group of organisms (Stein 1993). A number of important anatomical characters were identified and analyzed by Beck and Stein (1993) and Stein (1993) including xylem shape (deeply lobed), xylem development, protoxylem anatomy, appendage anatomy, presence of secondary xylem, and root ontogeny. Many taxa in basal Psilophytopsida are characterized by a deeply lobed primary xylem, and within this group there are two distinctive types of stelar architecture. One group, "radiate protoxylem", is characterized by protoxylem distributions that radiate outwards from a central point along the arms of the primary xylem strand (e.g., *Pertica*, *Gothanophyton*, Stenokoleales, Aneurophytales, and basal seed plants) (Fig. 2). A second group, "permanent protoxylem", is characterized by mesarch protoxylem restricted to the tips of the lobes of the primary xylem. In this group, protoxylem also often contains parenchymatous inclusions (e.g., Iridopteridales, Cladoxylopsida: early groups related to Sphenopsids and "ferns") (Fig. 2). The Beck and Stein (1993) anatomical data support an *Equisetum*-fern clade that is sister group to lignophytes (seed plants plus extinct "progymnosperms"). Although primarily addressing relationships at a general level in Psilophytopsida, these observations are also relevant to basal seed plants. Beck and Stein (1993) confirm that most data are consistent with lignophyte monophyly, but they emphasize that there are few unequivocal characters supporting this clade. In particular, the bifacial cambium and other wood characteristics form a character complex with elements that may be more general in Psilophytopsida. Furthermore, the question of whether the protostelic condition observed in some early seed plants is plesiomorphic or autapomorphic has not yet been fully resolved. Further evaluation of lignophyte monophyly requires critical examination of the anatomy of closely related Paleozoic fossils. One such potentially relevant taxon is *Stenokoleos holmesii* from the Upper Devonian of New York. *S. holmesii* is a member of the Stenokoleaceae, a rather poorly understood group with significant anatomical similarities to early seed plants. Reproductive structures of Stenokoleaceae are unknown, but stem anatomy places the group within lignophytes and possibly early seed plants (Matten 1992).

Most macrofossil data on early vascular plants comes from Euramerican assemblages, but recent descriptions of Devonian fossils from China document

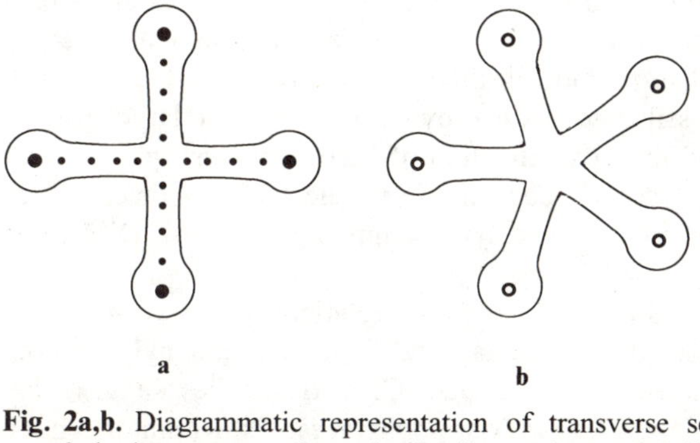

a b

Fig. 2a,b. Diagrammatic representation of transverse section of xylem showing lobed morphologies common in early fossil members of Psilophytopsida (nonlycopod vascular plants). *Black circles* represent protoxylem positions; *black circles with white centers* represent protoxylem with parenchyma inclusions. **a** "Radiate protoxylem": protoxylem distributed along lobed arms of xylem strand. **b** "Permanent protoxylem": protoxylem confined to ends of xylem lobes an interesting diversity of forms that provide important new comparative data. *Eophyllophyton bellum* (Hao and Beck 1993) documents the earliest evidence of planation in leaf-like appendages and provides the first evidence for G-type tracheids in the Psilophytopsida clade. The G-type tracheid is a distinctive wood cell that is widespread in early members of the Lycopodiopsida (Kenrick and Crane 1991). The documentation of this cell type in basal Psilophytopsida suggests that the G-type tracheid is probably ancestral for all crown group vascular plants.

Many early vascular plant fossils represent the remains of the distal portions of larger plants, and except under conditions of in situ preservation (e.g., coal measures) rooting structures are rarely observed attached. *Lorophyton goense* (Fairon-Demaret and Li 1993) is significant because it represents a rather complete small tree with roots and crown of branches. It is related to the Cladoxylopsida, an interesting early fossil group related to ferns.

c) Lycopods

Lycopods comprise a tiny fraction of extant land plant diversity but the group has a large and important fossil record. Lycopodiopsida (Fig. 1) first appear unequivocally in the Lower Devonian (Hueber 1992), where small herbaceous leafless taxa (Zosterophyllidae) dominate early plant assemblages and co-occur with other leafy forms (Lycopodiidae) that strongly resemble extant *Huperzia*. Lycopods flourished through the Devonian and, by the Lower Carboniferous, almost half of all known fossil plants are assignable to the group (Niklas et al. 1985). A diverse range of heterosporous arborescent forms dominated lowland peat-forming communities during the Carboniferous, and it is their remains that are largely responsible for the economically important coals of this age. Megafossil data suggest that lycopod diversity declined through the Upper Carboniferous and Mesozoic, but recent quantitative analysis of fossil megaspores suggests that diversity was greater in the Mesozoic than previously thought (Kovach and Batten 1993). The decline in lycopod megafossils is explained by extinction of the large swamp-dwelling arborescent forms and the diversification of herbaceous groups. Because of the important fossil record, lycopods continue to be the focus of much paleobotanical research, as reflected in a recent symposium organized by DiMichele and Skog (1992).

Recent cladistic treatments confirm a sister-group relationship between the extinct Zosterophyllidae and extant Lycopodiidae and support monophyly of both groups (Gensel 1992; Kenrick and Crane 1992). Cladistic studies suggest that the formerly enigmatic early fossil group Barinophytaceae is within the main Zosterophyllidae clade (Kenrick and Crane 1992). This conclusion indicates that some zosterophylls developed heterospory convergently with lycopods and increases knowledge of zosterophyll diversity in the Upper Devonian and Lower Carboniferous.

Although zosterophylls are abundant and widespread in many Early Devonian localities, the group has only recently been recognized in the most famous terrestrial Devonian lagerstätte, the Rhynie Chert (Lyon and Edwards 1991). Exceptionally well-preserved but highly fragmentary remains attributed to the group show certain cellular details of sporangial morphology and stem anatomy that have not been observed in other more common forms of petrifaction. *Trichopherophyton teuchansii* has the circinate vernation typical of the zosterophyll clade, but other features of stem anatomy suggest that it is a plesiomorphic member of the group.

Recent systematic studies of Lycopodiidae confirm monophyly of the heterosporous, ligulate group and a close relationship between Isoetaceae and arborescent Carboniferous forms (Bateman 1992; Bateman et al. 1992; Pigg 1992). A detailed systematic study focusing on extinct arborescent forms found that taxa with bisporangiate cones (i.e., *Paurodendron*, *Oxroadia*, *Anabathra*, *Chaloneria*) form a poorly resolved basal paraphyletic group subtending a clade characterized by monosporagiate cones (Fig. 3). The monosporangiate clade contains three well-defined families: Sigillariaceae, Lepidodendraceae, and the recently recognized Diaphorodendraceae (DiMichele and Bateman 1992). Progressive reduction of megaspore number per sporangium within this clade results in seed-like reproduction in Diaphorodendraceae and Lepidodendraceae. Despite their substantial size, the morphology of arborescent lycopods is relatively simple, and the bauplan can be reduced to a relatively small number of modular meristematic elements some of which appear to be essential (i.e., rhizomorphic root system, stem, cones) and others optional (i.e., lateral branches and crown branches). Based on systematic and morphological analyses, Bateman et al. (1992) suggest that arborescent lycopods are predisposed to nonadaptive saltational evolution through small changes in growth regulation early in ontogeny.

One of the most complete reconstructions of an early rhizomorphic lycopod is the pseudoherbaceous *Oxroadia*. Reconstructions of two species are based on a rich assortment of well-preserved material from lowermost Carboniferous (Upper Tournasian) limestone horizons in Scotland (Bateman 1992). *Oxroadia* is clearly a rhizomorphic lycopod because of its shoot-like rooting system with secondary growth. It is a plesiomorphic member of this clade (Fig. 3) and has an unusual growth form resulting from early dichotomy of the main stem followed by repeated branching of a prostrate cone-bearing crown. The woody anatomy, growth form and systematic position of *Oxroadia* suggest that this unusual pseudoherbaceous habit may have evolved through saltation from a tree-like ancestor involving a highly precocious dichotomy of the apical meristem giving an abbreviated stem and a much smaller plant (Bateman 1992).

d) Ferns

Relationships among Marattiales, Ophioglossales, leptosporangiate ferns, and other major Paleozoic clades of Psilophytopsida including sphenopsids and seed

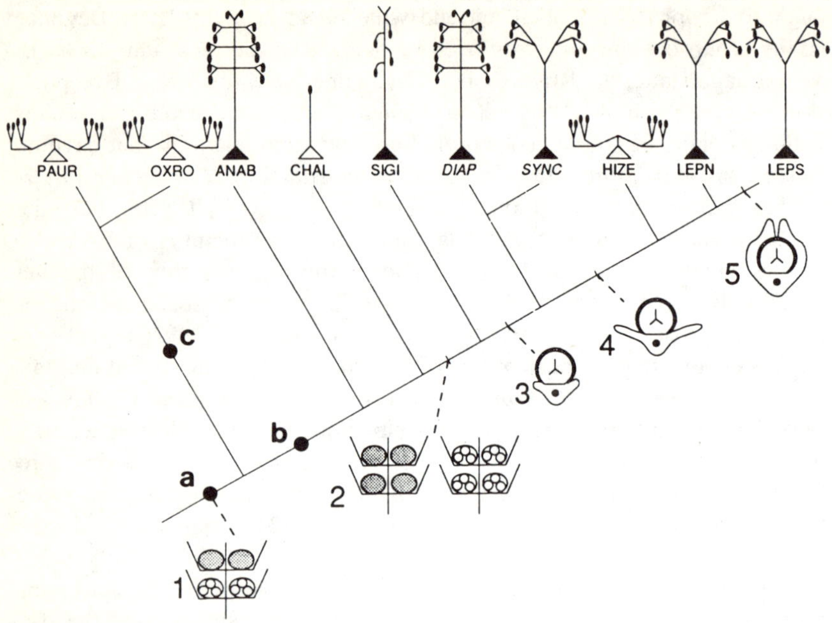

Fig. 3. Cladogram summarizing relationships among extinct arboreous rhizomorphic Lycopodiopsida. Reproductive innovations: *1* bisexual cone; *2* segregation of megasporangia and microsporangia into separate cones; *3* reduction to single functional megaspore; *4* lateral expansion of megasporophyll; *5* enclosure of megasporangium by megasporophyll. Growth habit (not to scale) illustrated on *upper part* of cladogram: *Paurodendron* (*PAUR*); *Oxroadia* (*OXRO*); *Anabathra* (*ANAB*); *Chaloneria* (*CHAL*); *Sigillaria* (*SIGI*); *Diaphorodendron* (*DIAP*); *Synchysidendron* (*SYNC*); *Hizemodendron* (*HIZE*); *Lepidodendron* (*LEPN*); *Lepidophloios* (*LEPS*). (Bateman 1992)

plants are still poorly understood. Cladistic treatments at the macrosystematic level have focused on extant taxa and are preliminary (e.g., Bremer 1985; Bremer et al. 1987). No cladistic study based on extant groups has yet exhausted the general data of comparative morphology. Furthermore, problems of homology recognition among extant taxa are exacerbated because many of the major cladogenic events occurred in the Devonian, and there has been significant subsequent morphological divergence. Over the last 30 years there has been a growing realization that many Carboniferous ferns assigned to the extinct Coenopteridales actually represent ancient members of the Filicales (Rothwell 1991). Further clarification of relationships and homologies at this systematice level requires a synthetic approach modeled on recent macrosystematic studies of seed plants.

Marattiales and leptosporangiate ferns have an extensive Late Paleozoic and Mesozoic fossil record. The history of marattialean ferns extends back to the Lower Carboniferous, where pinnules bearing distinctive abaxial exannulate synangia are the commonest form of fossil evidence. Late Carboniferous and Permian marattialean ferns were diverse and form the focus of many studies, including description of new taxa (Wan and Basinger 1992; Gao and Thomas 1993) and reevaluations of previously described material. *Grandeuryella* is a permineralized marattialean fossils from the Late Carboniferous and Early Permian of France. The presence of synangiate organs rather than clusters of free sporangia in specimens originally described by Renault in 1883 demonstrates a greater similarity to modern *Angiopteris* than had previously been supposed (Lesnikowska and Galtier 1991, 1992). Cladistic analysis suggests that most Late Palaeozoic fossils are stem-group taxa and that the marattialean crown group first appears in the Mesozoic (Hill and Camus 1986). Most late Paleozoic stem-group fossils are extinct by the Early Mesozoic, but *Scolecopteris antarctica*, recently described from the early Middle Triassic of Antarctica, represents one of the longest surviving examples (Delevoryas et al. 1992).

Taxa assignable to the Osmundales first appear unequivocally in the Permian and are commonly preserved as permineralized stems. Recent descriptions include late Permian taxa from China (Li 1993), Mesozoic taxa from Australasia (Tidwell 1991; Tidwell et al. 1991; Tidwell and Pigg 1993), and Cenozoic taxa from Spitsbergen (Kvaček and Manum 1993).

Leptosporangiate ferns underwent their main diversification during the Mesozoic, and the history of the group is documented by numerous dispersed spores and abundant macrofossils. Van Konijnenburg-van Cittert discussed the evolutionary history of Matoniaceae (van Konijnenburg-van Cittert 1993) and Schizaeaceae (van Konijnenburg-van Cittert 1991, 1992a) based on a study of in situ spores from both extant and extinct taxa. The Matoniaceae, now restricted to the Malaysian Archipelago, were widespread during the Jurassic and Early Cretaceous in both Southern and Northern Hemispheres. Spores of Matoniaceae may be separated into three groups, with those of modern genus *Phanerosorus* being the most derived (van Konijnenburg-van Cittert 1993). The Schizaeaceae are now mainly a Southern Hemisphere family, but the group was a common element in Northern Hemisphere floras during the late Mesozoic. van Konijnenburg-van Cittert (1992a) suggests that forms with heavily ornamented, trilete spores developed earlier than other forms. The oldest unequivocal schizaeaceous remains are *Lygodium*-like fossils from the Jurassic and Early Cretaceous. *Anemia*-like and *Mohria*-like plants are recorded from the Early Cretaceous. A more comprehensive analysis of relationships among extant and extinct species of *Anemia* and species of the extinct genera *Ruffordia* and *Pelletixia* showed that *Anemia* section *Coptophyllum* is paraphyletic (Skog 1992). The inclusion of fossil taxa further provided information on the timing of divergence.

The macrofossil record of Marsileales is poor, but the group is thought to extend back into the Early Cretaceous based on megaspores of *Molaspora* (Collinson 1991). This hypothesis has recently been confirmed by the first description of unequivocal megafossils of *Marsilea* from the Mesozoic (mid-Cretaceous) (Skog and Dilcher 1992, 1994). These fossil data document a nearly complete plant with roots, rhizome, leaflets, and possible sporocarps, showing morphology and habit that is almost identical to extant *Marsilea*. Megaspores characteristic of extant Salviniales occur in the earliest Tertiary, and other similar forms can also be traced to the early Cretaceous (Collinson 1991). The Tertiary megafossil record is better than for Marsiliales, but the earliest megafossils resembling extant *Salvinia* and *Azolla* have recently been described from the late Cretaceous Deccan Intertrappean beds, India (Nambudiri and Chitaley 1991).

e) Sphenopsids

Sphenopsid diversity is greatest in Lower Carboniferous coal-swamp floras, where the group comprises both arborescent and herbaceous taxa. Species numbers declined through the Upper Carboniferous to relatively low levels that have persisted to the present day, where the group is represented by the small but widespread genus *Equisetum*. Relationships among sphenopsids including Archaeocalamitaceae, Calamitaceae and Equisetaceae, and putative Devonian presphenopsids such as *Arachnoxylon*, *Calamophyton*, *Cladoxylon*, *Hyenia*, *Ibyka*, and *Pseudosporochnus* have been widely discussed (Stein et al. 1984). Archaeocalamitaceae are the earliest unequivocal sphenopsids and form an important link between Calamitaceae, Equisetaceae, and putative Devonian ancestors. Recent descriptions of permineralized Archaeocalamitaceae strobili and vegetative axes from the Lower Carboniferous of Scotland have clarified certain aspects of sporangium morphology and demonstrate that at least some members of this group were heterosporous (Bateman 1991). Preliminary cladistic analysis of relationships among the three families supports the plesiomorphic position of Archaeocalamitaceae with sphenopsids and suggests that both heterospory and secondary xylem may have been lost in *Equisetum*. A considerable extension of the stratigraphic range of Archaeocalamitaceae from Upper Carboniferous (Namurian) to mid-Lower Permian (Artinskian) has recently been reported (Mamay and Bateman 1991). The appearance of *Archaeocalamites lazarii* some 55 Ma after the last recorded occurrence of the group in the Carboniferous is probably the greatest known paleobotanical stratigraphic hiatus.

f) Early Seed Plants

The early fossil record of seed plants has attracted considerable attention over the last few years with the discovery and characterization of several important taxa from Upper Devonian sediments (Rothwell and Scott 1992; Rothwell and Serbet 1992; Stewart and Rothwell 1993; Taylor and Taylor 1993; Erwin et al. 1994). Most early seed plants are characterized by fern-like leaves bearing ovules and pollen-producing structures on relatively slender stems with small amounts of primary vasculature and manoxylic secondary wood. The earliest taxa have hydrasperman reproduction, which has been interpreted as the plesiomorphic condition for all seed plants (Rothwell 1986). Hydrasperman reproduction refers to the special structures and processes involved in pollen capture, fertilization, and subsequent dispersal of fertilized ovules in early seed plants. One of the most distinctive features of plants with hydrasperman reproduction is the elaborate structure that develops from the distal portion of the ovule nucellus and which functions in pollen capture and fertilization (Stewart and Rothwell 1993). Evidence from the early fossil record currently supports seed plant monophyly (Rothwell and Scott 1992; Rothwell and Serbet 1992; Serbet and Rothwell 1992), but lack of resolution in the relationships of basal seed plants (Doyle and Donoghue 1992), character conflicts involving vegetative morphology and anatomy (Galtier 1992), and uncertainties concerning important aspects of morphology in other potentially closely related Upper Paleozoic groups (Galtier 1992; Beck and Stein 1993) point to continued uncertainty over many aspects of relationship and homology.

Most information on Upper Devonian seed plants comes from isolated ovules, permineralized plant fragments or larger but nonpermineralized compression fossils. While these data have considerable interest, it is widely recognized that further clarification of the early stages of seed plant phylogeny requires more complete reconstructions involving anatomical and morphological characterization of vegetative and fertile parts. A series of important recent studies have focused on reconstructing one of the earliest seed plants, *Elkinsia polymorpha*, in some detail (Rothwell et al. 1989; Serbet and Rothwell 1992; Rothwell and Serbet 1992). *E. polymorpha* has helically arranged dimorphic fronds. Fertile fronds branch dichotomously and bear terminal ovulate cupules (hydrasperman type) or pollen organs comprising six to eight partially fused, radially arranged sporangia. Vegetative fronds are planar, have up to four orders of branching, and bear pinnate laterals that terminate in pinnules of sphenopterid type. Stem anatomy comprises a three-ribbed protostele, secondary xylem in larger axes, and a prominent sclrenchymatous cortex. Fronds have two C-shaped xylem strands. The reconstruction of *E. polymorpha* represents an important step in understanding early seed plant morphology, and this taxon will figure prominently in ongoing phylogenetic studies on the origin of seed plants.

Current understanding of early seed plant morphology is based on megafossils from North America and Europe. Ovulate and associated microsporangiate structures have been described recently from the Late Carboniferous of Peru (Erwin et al. 1994). *Oclloa cesariana* has hydrasperman reproduction and is interesting because the ovules are terminal but not borne in a cupule, as with most other early seed plants. Because many aspects of morphology and anatomy are still unknown, the systematic position of *O. cesariana* is equivocal. It is not yet possible to determine whether the absence of a cupule is a retained plesiomorphic feature or a loss.

Late Paleozoic seed ferns including taxa such as medullosans, lyginopterids, and *Callistophyton* have played an important role in understanding relationships in seed plants and have figured prominently in several recent cladistic treatments (see literature cited in Doyle and Donoghue 1992). Cladistic studies have brought the systematic problems in basal seed plants into sharper focus, but relationships among these groups and extant cycads, *Ginkgo*, conifers, and anthophytes are still the subject of much debate. Co-occurring with these taxa is a diverse range of less well known but important fossil material that is likely to contribute information critical to further clarification of relationships and homologies. Recent research has focused on Calamopityaceae, an abundant but rather poorly understood group of Lower Carboniferous plants that share generalized features of several Paleozoic seed ferns. Calamopityaceae are known mainly from fragments of stem anatomy, and one of the major challenges is reconstructing stem, branch, and leaf architecture. A major revision of the group is currently in progress (Beck et al. 1992; Galtier 1992; Galtier and Beck 1992; Galtier et al. 1993; Stein and Beck 1992).

g) Cycads

Studies based on extant plants indicate that cycads are sister group to all other seed plants (Loconte and Stevenson 1990; Doyle and Donoghue 1992). The systematic position of the group is less clear in more comprehensive analyses that include well-known fossils, in particular "pteridosperms" and "progymnosperms". Phylogenetically informative cycad reproductive material is rare in the fossil record, and while the abundant leaf record provides important information on quantitative diversification and geographic distributions of the group, it has had little impact on cycad systematics. An interesting rediscovery of the cycadalean ovule *Cycadeocarpus columbianus* was recently reported from the Middle Jurassic of Queen Charlotte Island, British Columbia, Canada (Chaloner and Hemsley 1992). The ovules are permineralized and excellently preserved with pollen grains inside the micropyle. Pollen is of general *Cycadopites* type and monocolpate. Associated with the ovules are leaf petioles and leaflets with cycad anatomy. The large size of the ovules combined with the radiosymmetrical organization of the ovule wall clearly indicate a cycadalean affinity and distinguish them from the platyspermic ovules of ginkgos.

Extant cycads have a relict distribution, but the group was a ubiquitous and cosmopolitan element of Mesozoic vegetation. During the late Mesozoic and Tertiary, the importance of cycads declined through loss of diversity and contraction of geographic range. A recent study of leaves assigned to *Eostangeria ruzinciniana* from Romania indicates that cycads survived in Europe at least to the end of the Miocene, probably in vegetation adapted to subxeric conditions (Palamarev and Usunova 1992). *Eostangeria ruzinciniana* was originally assigned to the fern genus *Angiopteris*, but was transferred to cycads based on epidermal characters of the leaflets. Extinct cycads assigned to the genus *Pterostoma* from high latitude floras in the Tertiary of Australia and New Zealand provide further evidence for a greater diversity and wider distributions in the Southern Hemisphere (Pole 1992; Rowett 1992; Hill and Pole 1994). *Pterostoma* was widespread particularly in Early Tertiary floras and became extinct by the Late Tertiary.

h) Pteridosperms

Pteridosperms are a heterogenous group of fossils comprising a grade of organization characterized by plants with ovules and pollen producing organs borne either on leaves, as in many Paleozoic taxa, or on simple or branched axes, as in many Mesozoic taxa. A study of medullosan pteridosperms from China has revealed significant new details on leaf morphology and reproductive organization in two widespread Paleozoic genera, *Paripteris* and *Linopteris.* Most significant is probably the discovery that the microsporangiate organ (*Potoniea*) associated with *Paripteris* is borne on planate, dicotomously branched axes in a dense spiral arrangement forming a large compound structure (Laveine et al. 1993). This arrangement is very different from that in other medullosans and may be of importance in future phylogenetic studies.

Permineralized fossils discovered during the past few years in silicified Permian and Triassic peats from Antarctica constitute one of the most important sources of information on structure and organization in Gondwanan pteridosperms. Anatomically preserved leaves of *Glossopteris* from the Permian and of *Dicroidium* from the Triassic were reported in Progress in Botany, Volume 53, and several new publications on this silicified material have appeared recently. One step towards a more comprehensive understanding of the *Glossopteris* plant is the discovery of silicified axes with leaves attached from the Transantarctic Mountains (Pigg and Taylor 1993). This is the first direct link between anatomically preserved leaves and axes for *Glossopteris.* In compression material leaf arrangement sometimes appears to be whorled, but the permineralized specimens clearly show a spiral arrangement of densely crowded leaves.

The recent discovery of permineralized ovulate structures is one of the most significant findings in the Antarctic *Glossopteris* flora (Taylor and Taylor 1993). Although reproductive structures of *Glossopteris* are extremely common in Late Paleozoic Gondwanan floras, most are impression fossils and their organization is poorly understood. Lack of critical information on the ovulate structure has caused problems in scoring reproductive characters in glossopterids, resulting in weakly supported hypotheses of relationship. Taylor and Taylor (1993) were able to document that in these Antarctic glossopterids ovules were borne adaxially on laminate structures that completely matched *Glossopteris* leaves previously described from the same locality. Taylor and Taylor (1993) figured one ovulate structure with several ovules, but they also noted that uni-ovulate structures were recovered from the same horizon. A further addition to our knowledge of the *Glossopteris* plant comes from a silicified Permian peat in East Antarctica. This peat has yielded well-preserved glossopterid roots (*Vertebraria*) with evidence for stages in root ontogeny (Neish et al. 1993).

Several new descriptions of glossopterid plants from other Gondwanan regions are based on impression and compression fossils (e.g., Chandra and Singh 1992; Maheshwari and Tewari 1992; Srivastava 1991; Mcloughlin 1994a,b). Although these fossils generally do not provide any new information of phylogenetic importance, they document the abundance of glossopterids in Late Paleozoic Gondwanan floras. They also demonstrate problems in specific delimitation for the huge number of species described from the various areas.

Permineralized Triassic floras from Antarctica are providing important new information on corystosperms. It has long been debated whether the microsporangiate organs were compound microsporophylls or branch systems. Discovery of anatomically preserved material from Antarctica clearly demonstrates that at least some corystosperms had helical arrangements of microsporangiate units (Yao et al. 1992). This material is beautifully preserved and also provides details on in situ pollen morphology (eusaccate) and ultrastructure (Osborn and Taylor 1993). Microsporangiate units are associated with abundant leaves of *Dicroidium* described in detail earlier (see Progress in Botany, Vol. 53). New findings of small stems with leaf bases attached show anatomical similarities to these *Dicroidium* leaves and provide the first direct evidence for a link between *Dicroidium* leaves and *Rhexoxylon* stems (Meyer-Berthaud et al. 1992, 1993).

Another significant discovery in the Mesozoic floras of Antarctica is the microsporangiate organ of *Caytonanthus* (*Caytonia* plant). This is the first report of *Caytonia* from a Gondwanan flora (Rees 1993). The material comes from the classic Hope Bay and Botany Bay floras, northern Graham Land, Antarctica. It is interesting that *Caytonanthus* in Antarctica is associated with the *Caytonia* leaf, *Sagenopteris*, as it is in many Northern Hemisphere floras. The occurrence of the *Caytonia* plant in Gondwana is consistent with the close relationship between *Caytonia* and the glossopterids inferred from several phylogenetic analyses (e.g., Nixon et al. 1994).

A new microsporangiate structure, *Brenneria potomacensis*, containing pollen grains of the *Decussosporites* type has recently been discovered from Early Cretaceous strata of eastern North America (Pedersen et al. 1993). *Brenneria* is believed to be related to other Mesozoic pteridosperms but also shares some similarities with conifers. *B. potomacensis* consists of a central axis bearing microsporangiate units in a helical arrangement similar to several other pteridosperms (*Antevsia*, *Harrisiothecium*: Friis and Pedersen unpubl.; *Pteruchus*: Yao et al. 1992; Kirchner and Müller 1992). Microsporangiate units bear two laterally fused sporangia and are in this respect simpler than other pteridosperms. Pollen grains are striate and disaccate with infilling of sacci resembling *Caytonanthus*, *Harrisiothecium* and some glossopterids. Associated dispersed seeds have been linked to *Brenneria* on the basis of in situ *Decussosporites* pollen. At present, too many critical features of the *Brenneria* plant are missing to allow a more detailed analysis of relationship (Pedersen et al. 1993).

i) Ginkgos

The phylogenetic position of Ginkgoales is one of several unresolved questions in seed plant systematics. Meyen (1987) grouped the Ginkgoales in its own class,

Ginkgoopsida, together with a number of extinct plants traditionally classified as pteridosperms. Alternatively, several cladistic analyses indicate that Ginkgoales may be more closely related to conifers and cordaits (e.g., Doyle and Donoghue 1992; Nixon et al. 1994). However, in these analyses only the extant species *Ginkgo biloba* was included. It has long been clear from the study of fossil leaves that variation within the group was considerable, particularly in the Mesozoic. New studies, mainly on Chinese material, now provide evidence for considerable variation in the organization of reproducture structures, and this material may eventually help in clarifying the phylogenetic position of the group. Two ovulate structures described by Zhou were reported earlier (Progress in Botany, Vol.50 1989) and one new ovulate structure, *Yimaia recurva*, was described recently (Zhou and Zhang 1992). This fossil is particularly interesting because it has been possible to link ovulate branches with leafy branches and pollen. The female branch of *Yimaia* comprises eight to nine densely spaced, sessile and orthotropous ovules. In contrast to *Ginkgo*, the ovules are not borne in a cupule. *Yimaia* is associated with leafy branches of *Baiera hallei*. Pollen grains have been observed adhering to both ovules and leaves. Zhou (1993) has added further ultrastructural details of the megaspore membrane in *Yimaia* and the fossil *Ginkgo yimaensis*. This study also includes a detailed analysis of developmental stages in the megaspore membrane of extant *Ginkgo biloba*.

Further details on epidermal characters have also emerged from cuticular studies of fossil *Ginkgo* leaves from the Late Jurassic and Early Cretaceous of Mongolia (Sun 1993; Zhao et al. 1993). A preliminary phylogenetic analysis of Ginkgoales based on living and fossil material by Zhou (1991) used Cordaites and early conifers as outgroups. This analysis does not address questions of relationships to other nonginkgoalean seed plants but may be a useful basis for further macrosystematic studies.

The Mesozoic *Czekanowskia* group is sometimes classified with Ginkgoales, but the systematic status of the group is equivocal because reproductive organs are poorly understood. *Czekanowskia* is a very conspicuous element in many Mesozoic floras of Eurasia. In a recent monograph, Samylina and Kirtchkova (1991) provide a detailed survey of the genus. This study includes the description of 62 species of *Czekanowskia* from the former USSR, and most of these are new. The paper is written in Russian, but descriptions are documented by numerous line drawings and good quality SEM micrographs as well as transmitted light micrographs of epidermal details. A key is provided for all species, and there are a number of tables and maps to illustrate geographic and stratigraphic distributional patterns. Several of these maps and tables were published later in a shorter paper in English that summarizes the characters and history of the genus (Samylina and Kirtchkova 1993).

A variety of large *Gingko*-like, flabelliform leaves from the Late Paleozoic are sometimes collectively referred to the highly heterogenous Palaeophyllales. Some leaves may be classified with the Ginkgoales, while the position of other taxa is uncertain. A new taxon, *Velizia inconstans*, related to this morphological complex has been described from the Late Paleozoic of Argentina (Césari and Hünicken 1992). This material is significant because it includes leaves and reproductive structures attached to shoots. Although preservation does not permit detailed reconstruction of the plant, it is clearly distinct from the Ginkgoales in organization of vegetative and reproductive organs. The leaves of *Velizia* show morphological resemblance to leaves of *Rhipidopsis*. Interestingly, Maheshwari and Bajpai (1992) reported new specimens of *Rhipidopsis* from the Permian Gondwana flora of the Rajmahal Basin, India, which also showed attachment of the shoot, documenting a spiral phyllotaxis and the apparent absence of short shoots as observed in *Velizia*.

j) Cordaitopsids and Coniferopsids

During the past few years considerable progress has been made towards a better understanding of Palaeozoic coniferopsids. The discovery of new material together with reinvestigation of critical key fossils has identified a much wider diversity of early coniferopsids than previously known, and reinterpretations of ovulate structures in members of the extinct order Voltziales indicate that classification and evolutionary models may have to be modified. Reinvestigation of the fossil *Walchiostrobus* (Kerp and Clement-Westerhof 1991) has shown that the ovulate dwarf shoot had three ovules rather than one, as originally described by Florin (1940). Furthermore, ovules are adaxially attached as in other Walchiaceae. This has some phylogenetic implications since a derivation of ovulate cones such as *Pseudovoltzia* with abaxially attached ovules from *Walchiostrobus* is not as straightforward as anticipated in classical phylogenetic hypotheses (Kerp and Clement-Westerhof 1991).

New findings of cheirolepidiaceous conifers provide further knowledge of structure and diversity in this enigmatic family (Clement-Westerhof and van Konijnenburg-van Cittert 1991; Raubeson and Gensel 1991; Srinivasan 1992; Thévenard 1993). Well-preserved ovuliferous material of *Hirmeriella* from the Early Jurassic of Germany was described by Clement-Westerhof and van Konijnenburg-van Cittert (1991) and forms the basis for a detailed discussion of ovuliferous structures in the Cheirolepidiaceae and a reevaluation of previous interpretations of cone structure in the family. Comparisons among the various taxa are visualized with instructive line drawing, and characters are documented in tabular form. The authors conclude the *Hirmeriella* possesses the most primitive characters within the family, while *Frenelopsis* may be regarded as a derived form. They also suggest that the Cheirolepidiaceae are derived from the Paleozoic volzialean family Majonicaceae (Clement-Westerhof and van Konijnenburg-van Cittert 1991).

A rich assemblage of fossil conifers has been described from calcium carbonate nodules in the Late Cretaceous of Hokkaido, Japan (Nishida 1991). Preservation is excellent and reveals a wealth of systematically important details. Recent descriptions include a sciadopityaceous seed cone assigned to a new form genus

Sciadopityostrobus (Saiki 1992), a taxodiaceous seed cone also assigned to a new form genus *Mikasastrobus* (Saiki and Kimura 1993), an arauracarian seed cone (Stockey et al. 1992), and leafy branches of *Brachiophyllum* (Ohana and Kimura 1993). The araucarian seed cone from Hokkaido has well-preserved seeds with embryos, and the fossil can be assigned to modern *Araucaria* based on well-documented features of seed anatomy. This is the first record of *Araucaria* cones in Asia, and this discovery extends the Northern Hemisphere stratigraphic record of *Araucaria* into the Late Cretaceous. New descriptions of Early Cretaceous arauracarian fossils from other parts of the world include leaves and pollen cones from Spain (Barale 1992) and a rich assemblage of leaves from Australia (Cantrill 1992). The Spanish leaves and pollen cones were assigned to the same species based on epidermal features. The material shares features of *Agathis* as well as *Araucaria* and was placed in the form genus *Dammarites* (Barale 1992). The Australian material includes six new species of leaf fossils referable to *Araucaria*. Based on epidermal features, Cantrill (1992) was able to separate these into three groups, one group comparable to modern members of section *Eutacta*, one group comparable to section *Columbea*, and one group (with one species) showing characters of both *Eutacta* and *Columbea*.

A new species of *Pinus* (subsection *Oocarpae*), closely related to modern Californian closed pines, has been described from well-preserved permineralized material in a Miocene chert, Washington State, USA (Miller 1992). The material comprises permineralized pinaceous leaves, dwarf shoots, pollen, and seed cones. Reconstruction of the whole plant based on organ associations was possible because only one type of each organ was present in the locality and the material had not been transported. Closed cone pines have also been reported from younger deposits in California (Axelrod and Cota 1993). Another new pinaceous cone was described from the Miocene of Idaho, USA, by Mente and Brack-Hanes (1992). This material is preserved as compression fossils and information is restricted mainly to morphological features and epidermal details. A comparison between the fossil material and modern Pinaceae is presented in tabular form and illustrates that the fossil has a mosaic of features found in several extant taxa. For this reason, Mente and Brack-Hanes (1992) were unable to place the material in any of the modern genera and referred it to the form genus *Pityostrobus*. The fossil history of Pinaceae was briefly summarized by Mente and Brack-Hanes (1992), and a more detail account of phylogenetic and spacial patterns in the radiation of the genus *Pinus* was given by Millar (1993) with special emphasis on events during the Eocene.

A recent study by Ferguson (1992) provides important new information on the fossil history of the taxaceous genus *Amentotaxus*. The work includes a detailed survey of the extant species focusing on characters that may be useful in identifying fossil material. Ferguson (1992) concluded that *Amentotaxus* is basal in the Taxaceae, which is consistent with the long fossil record dating from the Palaeocene. The genus shows great morphological stability for the first 55 million years, but considerable evolutionary change took place during the Quaternary. Recent morphological evolution was attributed to geological changes occurring in eastern Asia during the last few million years (Ferguson 1992).

k) Anthophytes

Under this heading we include angiosperms and closely related gymnospermous seed plants such as the Gnetales and the extinct order Bennettitales, in addition to several unassigned forms recognized as possible anthophytes during the past few years (see Progress in Botany, Vol. 54). A recent study by van Konijnenburg-van Cittert (1992b) has added a significant new member to the anthophytes group by documenting pollen of *Ephedripites* in situ in the Liassic microsporangiate organ *Piroconites*. *Piroconites* consists of numerous densely packed and three-loculed synangia borne on one surface of a larger bract-like structure (*Bernettia*). The individual synangium shows some resemblance to the synangia of extant *Welwitschia*, but the dense clustering is unlike anything known in members of the Gnetales. A comparable clustering of synangia is reported for the microsporangiate structure *Erdtmanitheca*, which has *Eucommiidites* pollen in situ. *Erdtmanitheca* is also considered to be related to anthophytes and especially close to the Gnetales (Pedersen et al. 1989). *Piroconites*, however, differs from *Erdtmanitheca* in its planate organization.

A monograph of Bennettitales from the English Wealden documents the vast diversity of this group in the Early Cretaceous (Watson and Sincock 1992). The work includes well-documented and well-illustrated descriptions of about 50 taxa of leaves, flowers, bracts, and stems. More than half of the species described are new.

Structure, homology and relationship within anthophytes are currently widely debated and many studies have been published over the last few years. These studies include classical morphological and anatomical investigations of fossil and living material, molecular investigations of extant taxa and theoretical computer-based analyses. A more comprehensive account of progress in this area will be given in Progress in Botany, Volume 57.

3. Discussion

Paleobotanical data play an important role in elucidating relationships among major groups of land plants. Much land plant diversity is extinct, and many extant groups represent the last vestiges of once diverse and important Paleozoic and Mesozoic clades. Recent advances in systematic theory have brought the role of paleobotanical data into sharper focus by highlighting the importance of the fossil record for understanding relationships and homology, particularly at higher systematic levels. Much recent work has focused on reinterpreting current knowledge from the new theoretical perspective offered by cladistic analysis. Early studies focused on seed plants and were characterized by their synthetic approach that integrated paleobotanical and neobotanical comparative data. This approach is currently being extended to lycopods and other "pteridophyte" groups with an extensive and well-documented fossil record. Cladistic analysis has helped to clarify important macrosystematic problems and to identify those areas where an empirical approach is likely to have greatest impact. The most interesting new data are likely to come through detailed reconstructions of plants in the stem groups and basal crown groups of major clades such as angiosperms, seed plants, leptosporangiate ferns, tracheophytes and, in particular, mosses, hornworts, and liverworts.

Systematic studies provide the essential framework for interpreting other paleobotanical and neobotanical patterns of land plant diversity. In this chapter, we have focused on the systematic phylogenetic aspects of paleobotanical research, but information from fossil plants has much wider applications in areas such as stratigraphy, paleoecology, paleoclimatology, paleogeography, biogeography, and historical aspects of species diversity. The synthesis of data from these diverse disciplines provides an overall picture of life history over time. A recent synthesis has been published in a multiauthor volume edited by Behrensmeyer et al (1992) which discusses the evolution of Phanerozoic terrestrial ecosystems focusing on the taphonomy and paleoecology of land-based faunas and floras. *Terrestrial Ecosystems Through Time* reconstructs the general components of past communities and deals with general patterns of change over long time periods from a broad historical perspective. Geographical, ecological, climatological, and stratigraphic aspects of the Paleozoic fossil record have recently been analyzed in a book edited by Cleal (1991).

The origins of modern diversity have attracted much attention recently, and the problem has been approached from a number of different perspectives. Recent trends in books such as Ricklefs and Schluter (1993) have been towards an integration of data from the fossil record with modern biogeographic and ecological analysis. This approach acknowledges the importance of historical processes, understood through systematics, biogeography, and paleontology, and ecological processes, such as interactions among extant populations, in the origin and maintenance of species diversity. The origin of modern distributions and community structure were treated in a book by Tallis (1991), focusing mainly on the late Cenozoic. Much recent analysis and synthesis deals with general trends and large-scale pattern, but other studies have focused on narrower, more specific issues. Details of community structure such as the historical development of plant-animal interactions have also attracted attention, and recent approaches have been summarized in a Royal Society Symposium (Chaloner et al. 1991). More specific biological questions recently addressed using paleobotanical data include the origins of polyploidy in angiosperms using variations in stomatal guard cell size as a measure of total DNA content (Masterson 1994). Measures of carbon isotope abundances in fossil material are also providing data on the origins of major photosynthetic pathways (Cerling et al. 1993; Bocherens et al. 1994; Morgan et al. 1994).

Reconstructing climatic change through the Cenozoic has attracted much attention recently and is of great potential importance to understanding modern climates and predicting the effects of future climatic changes. Data on the temporal and geographic distributions of temperature-sensitive groups of plants and animals are critical to testing computer-based climate models for the Cenozoic (Spicer and Corfield 1992; Taylor et al. 1992; Wing and Greenwood 1993). Analyses of isotope abundances may also prove to be useful in tracing changes in major atmospheric components through time. Changes in stomatal densities are currently being tested as potential indicators of historical changes in atmospheric carbon dioxide concentrations (Beerling and Chaloner 1992; Beerling et al. 1992; Van der Burgh et al. 1993).

References

Axelrod DI, Cota J (1993) Am J Bot 80:743–751.

Bateman RM (1991) Palaeontogr B223:1–59. **Bateman RM** (1992) Palaeontogr B228:29–103. **Bateman RM**, DiMichele WA, Willard DA (1992) Ann Mo Bot Gard 79:500–559. **Barale G** (1992) Rev Palaeobot Palynol 75:53–64. **Beck CB**, Stein WE (1993) Palaeontogr B229:115–134. **Beck CB**, Galtier J, Stein WE (1992) Rev Palaeobot Palynol 75:1–32. **Beerling DJ**, Chaloner WG (1992) Holocene 2:71–78. **Beerling DJ**, Chaloner WG, Huntley B, Pearson JA, Tooley MJ, Woodward FI (1992) Philos Trans R Soc Lond 336:215–224. **Behrensmeyer AK**, Damuth JD, DiMichele WA, Potts R, Sues H-D, Wing SL (eds) (1992) Terrestrial ecosystems through time. University of Chicago Press, Chicago. **Benton MJ** (ed) (1993) The fossil record 2. Chapman & Hall, London. **Bocherens H**, Friis EM, Mariotti A, Pedersen KR (1994) Lethaia 26:347–358. **Bremer K** (1985) Cladistics 1: 369–385. **Bremer K**, Humphries CJ, Mishler BD, Churchill SP (1987) Taxon 36:339–349.

Cai Chong-Yan, Dou Ya-Wei, Edwards D (1993) Geol Mag 130:155–170. **Cantrill DJ** (1992) Int J Plant Sci 153:622–645. **Cerling TE**, Wang Y, Quade J (1993) Nature 361:344–346. **Césari SN**, Hünicken M (1992) Palaeonotgr B224:121–129. **Chaloner WG**, Hemsley AR (1992) Cour Forschungsinst Senckenb 147:233–239. **Chaloner WG**, Harper JL, Lawton JH (eds) (1991) The evolutionary interaction of animals and plants. Cambridge University Press, London. **Chandra S**, Singh KJ (1992) Rev Palaebot Palynol 75:183–218. **Cleal CJ** (ed) (1991) Plant fossils in geological investigation: the Palaeozoic. Ellis Horwood, Chichester. **Cleal CJ** (1993a) In: Benton MJ (ed) The fossil record 2. Chapman & Hall, London, pp 779–794. **Cleal CJ** (1993b) In: Benton MJ (ed) The fossil record 2. Chapman & Hall, London, pp 795–808. **Clement-Westerhof JA**, Van Konijnenburg-Van Cittert JHA (1991) Rev Palaeobot Palynol 68:147–179. **Collinson ME** (1991) In: Blackmore S, Barnes SH (eds) Pollen and Spores, Systematics Association Spec Vol 44. Clarendon Press, Oxford, pp 119–150. **Collinson ME**, Boulter MC, Holmes PL (1993) In: Benton MJ (ed) The fossil record 2. Chapman & Hall, London, pp 809–841. **Crane PR** (1985) Ann Mo Bot Gard 72:716–793.

Delevoryas T, Taylor TN, Taylor EL (1992) Rev Palaeobot Palynol 74:101–107. **DiMichele WA**, Bateman RM (1992) Am J Bot 79:605–617. **DiMichele WA**, Skog JE (1992) Ann Mo Bot Gard 79:447–449. **Doyle JA**, Donoghue MJ (1986) Bot Rev 52: 321–431. **Doyle JA**, Donoghue MJ (1992) Brittonia 44:89–106.

Edwards D (1993) In: Benton MJ (ed) The fossil record 2. Chapman & Hall, London, pp 775–778. **Edwards D**, Berry C (1991) In: Cleal CJ (ed) Plant fossils in geological investigation. Ellis Horwood, Chichester, pp 117–153. **Edwards D**, Davies KL, Axe L (1992) Nature 357:683–685. **Endress PK**, Friis EM (eds) (1994) Plant Syst Evol Suppl 8. **Erwin DM**, Pfefferkorn HW, Alleman V (1994) Rev Palaeobot Palynol 80:19–38.

Fairon-Demaret M, Li Cheng-Sen (1993) Rev Palaeobot Palynol 77:1–22. **Fanning U**, Richardson JB, Edwards D (1991) In: Blackmore S, Barnes SH (eds) Pollen and Spores, Systematics Association Spec Vol 44. Clarendon Press, Oxford, pp 25–47. **Fanning U**, Edwards D, Richardson JB (1992) Bot J Linn Soc 109:161–188. **Ferguson DK** (1992) Cour Forschungsinst Senckenb 147:255–285. **Florin R** (1940) Palaeontogr 85B:176–243.

Galtier J (1992) Cour Forschungsinst Senckenb 147:119–125. **Galtier J**, Beck CB (1992) Palaeonogr B224:1–16. **Galtier J**, Meyer-Berthaud B, Beck CB (1993) Palaeontogr B230:59–79. **Gao Zhifeng**, Thomas BA (1993) Palaeonotology 36:81–89. **Garbary DJ**, Renzaglia KS, Duckett JG (1993) Plant Syst Evol 188:237–269. **Gensel PG** (1992) Ann Mo Bot Gard 79:450–473. **Gensel PG**, Johnson NG, Strother PK (1990) Palaios 5:520–547. **Graham LE**, Delwiche CF, Mishler BD (1991) Adv Bryol 4:213–244. **Gray J** (1991) In: Blackmore S, Barnes SH (eds) Pollen and Spores, Systematics Association Spec Vol 44. Clarendon Press, Oxford pp 49–87. **Gray J**, Shear W (1992) Am Sci 80:444–456.

Hao Shou-Gang, Beck CB (1993) Palaeontogr B230:27–41. **Hill CR**, Camus JM (1986) Bull Br Mus Nat Hist (Bot) 14:219–300. **Hill RS**, Pole MS (1994) Rev Palaeobot Palynol 80:123–130. **Hueber FM** (1992) Ann Mo Bot Gard 79:474–499.

Kenrick P (1994) Biol Rev 69:293–330. **Kenrick P**, Crane PR (1991) Bot Gaz 152:335–356. **Kenrick P**, Crane PR (1992), Congr Proc 4th IOP, Paris. **Kerp H**, Clement-Westerhof JA (1991) N Jb Geol Paläont Abh 183:257–268. **Kirchner M**, Müller A (1992) Palaeontogr B224:63–73. **Kovach WL**, Batten DJ (1993) Paleobiology 19:28–42. **Krassilov VA**, Schuster RM (1984) In: Schuster RM (ed) New manual of bryology 2. Hattori Botanical Laboratory, Miyazaki, pp 1172–1193. **Kvaček Z**, Manum SB (1993) Palaeontogr 230B:169–181.

Laveine J-P, Lemoigne Y, Zhang Shanzhen (1993) Palaeontogr B230:81–139. **Lesnikowska A**, Galtier J (1991) Rev Palaeobot Palynol 67:141–152. **Lesnikowska A**, Galtier J (1992) Rev Palaeobot Palynol 72:299–315. **Li Zhong Ming** (1993) Rev Palaeobot Palynol 77:51–63. **Loconte H**, Stevenson DW (1990) Brittonia 42:197–211. **Lyon AG**, Edwards D (1991) Trans R Soc Edinb Earth Sci 82:323–332.

Maheshwari HK, Bajpai U (1992) Palaeontogr B224:121–149. **Maheshwari HK**, Tewari R (1992) Palaeobotanist 39(1990):338–380. **Mamay SH**, Bateman RM (1991) Am J Bot 78:489–496. **Masterson J** (1994) Science 264:421–424. **Matten LC** (1992) Cour Forschungsinst Senckenb 147:75–85. **McLoughlin S** (1994a) Palaeontogr 231B:1–29. **McLoughlin S** (1994b) Palaeontogr 231B:31–62. **Mente RF**, Brack-Hanes SD (1992) Bot J Linn Soc 108:333–344. **Meyen SV** (1987) Fundamentals of Palaeobotany. Chapman and Hall, London. **Meyer-Berthaud B**, Taylor EL, Taylor TN (1992) Geobios 25:341–344. **Meyer-Berthaud B**, Taylor TN, Taylor EL (1993) Palaeontology 36:337–356. **Millar CI** (1993) Ann Mo Bot Gard 80:471–498. **Miller CN** (1992) Am J Bot 79:754–760. **Mishler BD**, Churchill SP (1984) Brittonia 36:406–424. **Mishler BD**, Churchill SP (1985) Cladistics 1:305–328. **Morgan ME**, Kingston JD, Marino BD (1994) Nature 367:162–164.

Nambudiri EMV, Chitaley S (1991) Rev Palaeobot Palynol 69:325–336. **Neish PG**, Drinnan AN, Cantrill DJ (1993) Rev Palaeobot Palynol 79:221–243. **Nelson G**, Platnick N (1981) Systematics and biogeography. Columbia University Press, New York. **Niklas KJ**, Tiffney BH, Knoll AH (1985) In: Valentine JW (ed) Phanerozoic diversity patterns: profiles in macroevolution, Princeton University Press, Princeton, pp 97–128. **Nishida H** (1991) Bot Mag Tokyo 104:253–273. **Nixon KC**, Crepet WL, Stevenson D, Friis EM (1994) Ann Mo Bot Garden 81:484–533.

Ohana T, Kimura T (1993) Bull Natl Sci Mus (Tokyo) C 19:41–64. **Osborn JM**, Taylor TN (1993) Rev Palaeobot Palynol 79:205–219.

Palamarev E, Usunova K (1992) Cour Forschungsinst Senckenb 147:287–293. **Pole MS** (1992) J R Soc NZ 22:303–306. **Pedersen KR**, Crane PR, Friis EM (1989) Grana 28: 279–294. **Pedersen KR**, Friis EM, Crane PR (1993) Grana 32:273–289. **Pigg KB** (1992) Ann Mo Bot Gard 79:589–612. **Pigg K**, Taylor TN (1993) Am J Bot 80:500–516.

Raubeson LA, Gensel PG (1991) Bot Gaz 152:380–391. **Raubeson LA**, Jansen RK (1992) Science 255:1697–1699. **Rees P McA** (1993) Geobios 26:33–42. **Remy W**, Gensel PG, Hass H (1993) Int J Plant Sci 154:35–58. **Ricklefs RE**, Schluter D (eds) (1993) Species diversity in ecological communities. University of Chicago Press, Chicago. **Rothwell GW** (1986) In: Spicer RA, Thomas BA (eds) Systematic and taxonomic approaches in palaeobotany, Systematics Association Spec Vol 31. Oxford University Press, Oxford, pp 137–162. **Rothwell GW** (1991) Am J Bot 78:782–788. **Rothwell GW**, Scott AC (1992) Rev Palaeobot Palynol 72:273–289. **Rothwell GW**, Serbet R (1992) Cour Forschungsinst Senckenb 147:225–231. **Rothwell GW**, Scheckler SE, Gillespie WH (1989) Bot Gaz 150:170–189. **Rowett AI** (1992) Trans R Soc S Aust 116:95–107.

Saiki K (1992) Am J Bot 79:989–995. **Saiki K**, Kimura T (1993) Rev Palaeobot Palynol 76:83–96. **Samylina VA**, Kirtchkova AI (1991) The genus *Czekanowskia* (systematics, history, distribution and stratigrapic significance). Nauka, Leningrad, 139 pp (in Russian). **Samylina VA**, Kirtchkova AI (1993) Rev Palaeobot Palynol 79:271–284. **Serbet R**, Rothwell GW (1992) Int J Plant Sci 153:602–621. **Skog JE** (1992) Rev Palaeobot Palynol 70:279–295. **Skog JE**, Dilcher DL (1992) Am J Bot 79:982–988. **Skog JE**, Dilcher DL (1994) Rev Palaeobot Palynol 80:1–18. **Spicer RA**, Corfield RM (1992) Geol Mag 129:169–180. **Srinivasan V** (1992) Rev Palaeobot Palynol 72:245–255. **Srivastava AK** (1990) Palaeobotanist 39:281–302. **Srivastava AK** (1991) Geobios 24:383–386. **Stein WE** (1993) Int J Plant Sci 154:229–263. **Stein WE**, Beck CB (1992) Rev Palaeobot Palynol 72:73–102. **Stein WE**, Wight DC, Beck CB (1984) Syst Bot 9:102–118. **Stewart WN**, Rothwell GW (1993) Paleobotany and the evolution of plants. Cambridge University Press, Cambridge. **Stockey RA**, Nishida H, Nishida M (1992) Rev Palaeobot Palynol 72:27–40. **Sun Ge** (1993) Palaeontogr B230:159–168.

Tallis JH (1991) Plant community history. Chapman and Hall, London. **Taylor EL**, Taylor TN, Cúneo NR (1992) Science 257:1675–1677. **Taylor TN**, Taylor EL (1993) The biology and evolution of fossil plants. Prentice Hall, Engelwoods Cliffs. **Thévenard F** (1993) Rev Palaeobot Palynol 78:145–166. **Tidwell WD** (1991) Palaeontogr B223:81–90. **Tidwell WD**, Munzing GE, Banks MR (1991) Palaeontogr B223:91–105. **Tidwell WD**, Pigg KB (1993) Palaeontogr B230:141–158.

Van der Burgh J, Visscher H, Dilcher DL, Kürschner WM (1993) Science 260:1788–1790. **van Konijnenburg-van Cittert JHA** (1991) In: Blackmore S, Barnes SH (eds) Pollen and spores – patterns of diversification. Clarendon Press, London, pp 103–118. **van Konijnenburg-van Cittert JHA** (1992a) Cour Forschungsinst Senckenberg 147:

109–117. **van Konijnenburg-van Cittert JHA** (1992b) Rev Palaeobot Palynol 71:239–254. **van Konijnenburg-van Cittert JHA** (1993) Rev Palaeobot Palynol 78:235–267.

Wan Zhihui, Basinger JF (1992) Rev Palaeobot Palynol 75:219–238. **Watson J**, Sincock CA (1992) Bennettitales of the English Wealden. The Palaeontographical Society, London. **Wing SL**, Greenwood DR (1993) Philos Trans R Soc Lond 341:243–252.

Yao X, Taylor TN, Taylor EL (1992) Org Int Paléobot IV Conf OFP Inf Spéc 16-B:178.

Zhao Li-Ming, Ohana T, Kimura T (1993) Trans Proc Palaeontol Soc Jpn NS 169:73–96. **Zhou Zhiyan** (1991) Rev Palaeobot Palynol 68:203–216. **Zhou Zhiyan** (1993) Rev Palaeobot Palynol 78:167–182. **Zhou Zhiyan**, Zhang Bole (1992) Palaeontogr B224: 151–169.

Dr. Paul Kenrick
Prof. Else Marie Friis
Department of Palaeobotany
Swedish Museum of Natural History
Box 50007
S-104 05 Stockholm
Sweden

E. Geobotany

I. Plant Geography

By Eckehart J. Jäger

The *trends of investigation* in plant geography have not changed much since the time of the last report. One of the most important problems remains what will happen with the plants when climate changes. Therefore the distribution limits caused by climatic factors are of special interest.

Another question is to what degree the local floras will become impoverished and intermingle following man's activities. The rate of introduction and spread of invaders are difficult to predict, the same is true for the evolutionary response to the impoverishment of floras.

A sound use of computers reveals an objective phytogeographic subdivision of the earth and classifications of floristic elements.

The interest in floristic exploration seems to be increasing, but the publication of many good floras cannot obscure the fact that the contributing forces would not be sufficient to describe the rich tropical floras even by the end of the next century.

1. Taxonomic and Floristic Foundation of Plant Geography

a) New Floras

Encouraging progress in floristic exploration is the publication of the first volumes of six large new floras in 1993–94:

Flora of North America is scheduled to comprise 14 volumes, the first two appeared in 1993. Volume 1 contains the introduction (physical geography, vegetation, floristic regions after Takhtajan, evolution of the flora), volume 2 the ferns and gymnosperms. For every species a small but instructive distribution map is provided.

Flora Mesoamericana covers the five southernmost states of Mexico and the Central American countries down to the Panama/Colombia border. With about 18000 species to be treated in seven volumes, it will be among the largest floras of the world. (For comparison, Europe 12 000 species, Peru 18 000, North America north of Mexico 20 000, Australia 18–25 000, the former USSR 21 000, South Africa 22 000, Malesia 28 000, China 30 000, Fl. Neotropica 100 000.) Volume 6 (1994), which will be published first, contains 1891 species, including the Poaceae and Cyperaceae.

Progress in Botany, Vol. 56

Flora of Arabian Peninsula and Socotra (Miller and Cope 1994) treats an area with many blank spaces on the botanical map. Volume 1 (of six scheduled) contains distribution maps of 695 species.

Flora of Somalia (Thulin 1993) geographically adjoins the last mentioned flora. In the first volume (of four), 1092 species of ferns, gymnosperms, and dicots (from Annonaceae to Fabaceae) are treated.

The new *Flora of Japan* (in English, Iwatsuki and Yamazaki 1993) begins with volume 3a (Sympetalae). Seven volumes are scheduled, all illustrated with distribution maps.

Flora of Victoria will be complete in four volumes. The first (Foreman and Walsh 1993) provides a comprehensive treatment of physical geography, paleobotany, vegetation, with chapters on the exotic flora and on the threatened plants, distribution maps, and proposals for specific means of conservation.

The following list of floras that have been continued during the past 3 years may be impressive, but it should not obscure the fact that there are many gaps and many stagnating projects. (For the characterization and full citation of these floras see the last reports, especially Prog. Bot. 54:430–433 and 51:314–316.)

Europe: Fl. von Württemberg (vols. 3+4, 1992: Droseraceae to Apiaceae); Fl. of Czechia (Bohemia and Moravia, vol. 3, 1993: Brassicaceae to Malaceae); Fl. Slovenska (vol. 4/3, 1992: Rosales); Fl. Polski (vol. 3, 1992: Fagales-Salicales-Urticales-Polygonales); Fl. Iberica (vol. 3, 1993: Plumbaginaceae to Capparaceae, vol. 4, 1993: Cruciferae to Monotropaceae);

Asia: Fl. Sibiri (vol. 5, 1992: Salicaceae to Amaranthaceae) vol. 6, 1993: Portulacaceae to Ranunculaceae); Fl. of Russian Far East (Sosud. rast. vol. 5, 1991: ferns, Boraginaceae, Solanaceae, Scrophulariaceae and others, vol. 6, 1992: Asteraceae); Fl. of Middle Asia (Opred. rast. sredn. Azii vol. 10, 1992: Asteraceae); Fl. of Central Asia (Rastenija Centr. Azii vol. 10, 1993: Araliaceae, Umbelliferae, Cornaceae); Fl. Iranica (fasc. 169–171, 1992: Violaceae, Liliaceae, Ranunculaceae); Fl. of Pakistan (fasc. 193, 1991: Ranunculaceae); Fl. Reipublicae Popularis Sinicae (vol. 76/2, 1991: Anthemideae 2, vol. 51, 1991: Violaceae, vol. 13/1, 1991: Palmae, vol. 34/2, 1992: Saxifragaceae 1, vol. 61, 1992: Oleaceae, Loganiaceae, vol. 57/3, 1992: Ericaceae, vol. 55/3, 1992: Umbelliferae); Fl. Sichuanica (vol. 5/2, 1991: Gramineae) Fl. Yunnanica (vol. 5, 1991: Eupteleaceae, Betulaceae, Caprifoliaceae, Campanulaceae, and others); Fl. du Cambodge, du Laos et du Vietnam (vol. 26, 1992: Rhoipteleaceae, Primulaceae, Styracaceae), Fl. of Thailand (vol. 5/3, 1991, vol. 5/4, 1993: Amaranthaceae to Caryophyllaceae, Umbelliferae);

Malesia-Australia: Fl. Malesiana (vol. 11/1, 1992: Mimosoideae, vol. 11/2, 1993: Rosaceae, Amaryllidaceae, Alliaceae); Fl. of Australia (vol. 54, 1992: Lichens, vol. 35, 1992: Brunoniaceae to Goodeniaceae, vol. 50, 1993: oceanic islands); Fl. of New South Wales (vol. 3, 1992, vol. 4, 1993: complete), Fl. de la Nouvelle Caledonie (vol. 17, 1991: Euphorbiaceae, vol. 18, 1992: Myrtaceae Leptospermoideae, vol. 19, 1993: Ebenaceae); Fl. of Micronesia (vol. 5, 1993: Bignoniaceae to Rubiaceae);

Africa: Fl. of Ethiopia (vol. 2, 1994: Cycadaceae to Euphorbiaceae); Fl. du Gabon (vol. 33, 1992: Connaraceae); Fl. of Tropical East Africa (issues published 1991: Rubiaceae complete, Burseraceae, Meliaceae, Boraginaceae, 1992: Verbenaceae, 1993: Balanophoraceae, Pandanaceae, Musaceae, Proteaceae, Vitaceae); Fl. Zambesiaca (vol. 9/6, 1991: Ulmaceae, Ceratophyllaceae and others, vol. 6/1, 1992: Asteraceae, vol. 12/4, 1993: Index); Fl. of South Africa (vol. 5/3, 1992: Dracaenaceae to Smilacaceae, vol. 19/3, 1993: Anacardiaceae: *Rhus)*; Fl. de Mocambique (fasc. 114, 1990: Cuscutaceae, 120, 1990:

Bignoniaceae), Fl. de Madagascar et des Comores (1990: Gentianaceae, 1991: Iridaceae, Proteaceae, Montiniaceae and others);

Central and South America: Fl. Novo Galiciana (vol. 17, 1992: ferns and gymnosperms, vol. 13, 1993: Limnocharitaceae to Typhaceae); Fl. of Chiapas (fasc. 3, 1990: Malvaceae); F1. of Dominica (vol. 2, 1991: Dicotyledoneae, complete); Fl. of Costa Rica (fasc. 97–103, 1991: Krameriaceae to Zygophyllaceae); Fl. de Colombia (fasc. 13, 1990: Asteraceae-Heliantheae 1, see also Diaz-Piedrahita and Velez-Nauer 1993: Asteraceae-Barnadesieae and Mutisieae); Fl. of Ecuador (fasc. 45, 1992: Clethraceae, Cyrillaceae, 46, 1993: Phytolaccaceae to Achatocarpaceae, 47, 1993: Rubiaceae-Anthospermae, 48, 1993: Cecropiaceae); Fl. of the Guianas (Ser. B fasc. 3–6, 1991–1993: ferns, ser. A fasc. 11, 1992: Urticales, Casuarinaceae, 12, 1992: Lecythidaceae, 13, 1993: Melastomataceae, 14, 1993: Olacaceae, Opiliaceae); Fl. of Paraguay (fasc. 17, 1992: Apocynaceae).

Some other important examples of new floras ought to be mentioned: In the second edition of volume 1 of *Flora Europaea* (1993), 350 taxa (about 10%!) are added, the habitat and distribution data are revised. The need for a new edition of the other volumes is still being discussed. Long awaited was the publication of an excursion flora of Austria, that follows the last one of Fritsch after 72 years (Adler et al. 1994). A new flora of the British Isles (Stace 1991) includes many introduced plants that have hitherto been neglected. The new excursion flora of Scandinavia is nicely illustrated in colour and contains small, but exact maps, that incorporate the floristic progress after Hultén's atlas (Mossberg et al. 1992). Also for the Canary Islands we now have a beautiful excursion flora (Hohenester and Welss 1993). A new checklist of the flora of Tenerife (Schönfelder et al. 1993) includes 71 species new to the island, dot maps of the endangered plants and an exhaustive bibliography.

A new annotated checklist for the Caucasus region is being published family by family in Botaničeski žurnal (Micheev 1992). A voluminous illustrated book on the Flora of Korea (in Korean) gives short summaries of the distribution within the country (Lee 1989). In a space of only 3 years, three fully illustrated volumes (in six voluminous books) on the flora of Vietnam appeared (in Vietnamese with short English summaries: Ho 1991, 1992, 1993), a complete treatment of a flora of 10500 species.

The actual checklist of all tropical African seed plants (Lebrun and Stork 1991, 1992) is with two volumes and 11547 species listed halfway to completion. Many species have not yet been included in any flora. A flora of Mauritania (Barry and Celles 1991, complete in two volumes) covers a large area at the boundary between the Holarctic and the Paleotropical kingdom. The Asteraceae, with about 800 species the largest family of Zaire, Rwanda, and Burundi, have been treated by Lisowski (1991, 1992).

The most important novelty from the USA next to Flora of North America is the new, illustrated edition of the Jepson Manual of California (Hickmann 1993) that comprises 5867 species, including 1169 endemics. The state of floristic exploration in the United States is quite uneven. Kentucky has only an annotated checklist, at least with raw distribution data (Browne and Athey 1992), Tennessee only a checklist (Wofford and Kral 1993, 2267 native plus 518 naturalized species of higher plants). The new ambitious flora of Cuba begins with a small family (Loranthaceae, Leiva 1992), but with an extraordinarily good layout: excellent illustrations, maps for every species on a 10×10 km grid.

Despite all these efforts, the rate of investigation is not sufficient. Toledo and Sosa (1993) evaluate the numbers of collections and botanists in Latin America and conclude that in the area with highest biodiversity the existing forces cannot win the race against extinction. In the 30 years since the appearance of the first volume, Flora Neotropica has covered 5334 species of flowering plants in 55 volumes. The work will take 300 years to be completed (Mori 1992).

b) Plant Distribution Mapping

The current intensive mapping work is stimulating investigations on the historical development of this method in phytogeography. Stork and Lebrun (1988) evaluate the oldest distribution maps, beginning in 1780 with two maps drawn by Soulavie that show the distribution of *Castanea, Vitis, Olea* and *Abies* in southern France (43 years before the first maps in the atlas of Schouw!), and proceeding in the second half of the last century by activities of the publishing house Justus Perthes (*Petermanns Mitteilungen*, *Berghaus*' *Atlas*). Schnedler (1992) pays tribute to the work of H. Hoffmann, the author of the first dot- and grid maps.

The increasing number of plant distribution maps is listed in different bibliographies: for the whole world in the *Index Holmiensis* (vol. 7, letters K-M, in press), for Africa by Stork and Lebrun (1988), for Malesia by van Balgooy (1993), for Spain by Fernandez Casas et al. (1992, 25000 maps cited), for Cuba by Manitz (1992), for the former Czechoslovakia by Chrtek (1991), and for Central Germany by Hilbig (1992).

The possibility of printing the maps with the computer encourages local amateur botanists, flora writers, and institutions to publish more and more flora atlases of different scales, now also for mosses (Germany: Düll and Meinunger 1989, Ontario: Ireland and Ley 1992), mushrooms (Krieglsteiner 1991: macroscopic mycota of Germany), and lichens (progress report for Europe and 108 maps: Wirth and Oberhollenzer 1990). An atlas of the flora of Finland is the first that is available on compact disc only (Botanical Museum, University of Helsinki 1993). Most authors of such atlases miss the opportunity to interpret the actually quite interesting distributional patterns. In the above-mentioned atlas of the 490 mosses of Ontario, the authors confine themselves to assigning the patterns to the categories southern, northern, eastern, western and widespread. Also the 1197 beautiful large-scale maps of plants of New Foundland (Rouleau and Lamoureux 1992) are presented without interpretation. On the other hand, the maps of the Swiss atlas have been used by Wohlgemuth (1993) to deduce the dependence of species number on factors like rock type and altitude.

The number of dots per species is one of the criteria of the quality of the maps. It is very high (in some species over 1000) in the atlas of the higher plants and mosses of Thuringia, a one-man work by Meinunger (1992, 1672 maps on a 1.2 × 1.2-km grid), whereas it reaches only 70 in Poldini's (1991) atlas of Friuli, that is nevertheless an important contribution to the knowledge of a species-rich part of Italy. Another, not less important criterion is the consideration of critical taxa, an advantage of the plant distribution atlas of Carinthia (Hartl et al. 1992,

2457 grid maps, e.g., 83 *Hieracium*, 24 *Rubus*, 19 *Taraxacum*, hybrids, and ephemerophytes listed). Useful is also the possibility of checking the sources, but these are given only rarely, e.g., by Molina and Rubio-Sanchez (1992, 15 maps for Spain on a 10 × 10 km grid). During the mapping, one always finds species new to the area or may correct old data. In Carinthia, about 100 species had been given erroneously or their occurrence is dubious, whereas many others have been found for the first time for the country (Hartl et al. 1992). Boratynski et al. (1992) found on an excursion of only 10 days on Zakinthos island (Greece) 72 species of woody plants, including 15 new to the island. A similarly astonishing increase resulted from an excursion through Euboea (30 woody plants new to the island, 154 dot maps, Boratynski et al. 1990). On only one excursion in April, 42 plant species not known earlier on Karpathos island were detected. These observations are included in an atlas of higher plants of the Cretan area, a result of 10 years' field work (Turland et al. 1993, 1624 maps on an 8 × 8-km grid).

Many good maps are published in taxonomic monographs. Typical of tropical orchids are the very narrow ranges, as is shown by the dot maps of Pleurothallidinae (Luer 1993). Some of these orchids have an interesting humid-montane southeast Brasil + Andean disjunction.

Special favourites of plant mapping seem to be not only orchids and trees (e.g., Ulmaceae, Boratynska 1991), but obviously also ferns. They have been mapped with differentiation of time of records for Wales (Hutchinson and Thomas 1992, 84 maps), with differentiation of abundance for France (Prelli and Boudrie 1992, 120 maps of provinces), for Rwanda (Kornaś et al. 1993, 86 dot maps) and for South Africa (Burrows 1990, 343 maps on a 50 × 50-km grid). Fern distribution maps are also included in the new flora of Japan (Nakaike 1992), as are generally plant distribution maps in many modern floras (see above).

c) Chorology and Critical Taxa

Investigations on taxonomically critical groups are often aided by chorological arguments. To Gottschlich and Raabe (1991), a sound range pattern is a test of a good delimitation of *Hieracium* taxa. The taxa of the *Silene pusilla* aggregate investigated by Melzheimer and Polatschek (1992) fit well the general pattern of ranges in the Alpes. The taxonomic structure of *Poa* sect. *Stenopoa* in Siberia is understood by a chorological study of Olonova (1993). *Poa nemoralis* and *P. palustris* as the oldest species have built up hybrids such as *P. krylovii*. The development continued in the spread of the *P. botryoides* group that is distributed in central Asiatic stony steppes. Small and Fawzy (1992) found no evident discontinuity within the east-west cline of the *Medicago monantha* complex, the delimitation of the traditionally recognized infraspecific taxa therefore seems to be unjustified.

To draw maps of a genus like *Hieracium* is a specialist's task (dot maps: Beaman 1990 for Mexico, Gottschlich and Raabe for Westfalen). This relates also to the distribution of the *Dryopteris dilatata* complex, the three species of which differ distinctly in their vertical and horizontal ranges as well as in ecological requirements (Piekos-Mirkowa 1991); or to *Rubus* subgen. *Rubus*, that is mapped on a very detailed grid (3 × 3 km) in central Europe (Weber 1992). The European specialist for *Oenothera* finds in Portugal the South American species of the genus to be the most common ones (Rostanski 1991). Also in mosses, chorology helps to clarify taxonomic classification. From 78 Eurasiatic taxa of *Leucobryum* given in the literature, only 19 remained revision, only *L. glaucum* (boreal zone) and *L. juniperoideum* (south Europe-south Asia) have wide ranges, all others are concentrated in tropical southeast Asia and Malesia (Yamaguchi 1993).

2. Differentiation of Taxa in Time and Space

a) Reconstruction of Migration on the Basis of Recent Distribution

It was always tempting to throw light on the dark past of plant migration with arguments derived from recent plant distribution; but it has also often been stressed that reconstructions of place of origin and of migration routes are of low value when not considering the fossils.

Thus, the progress in investigations like those of Liu and Ho (1992) on *Lomatogonium* and of Wang (1992) on migration routes in East Asia is to be seen mainly in the good description of recent distribution and not in the statement that, e.g., *Lomatogonium* arose in the Quinlin-Hengduan mountain region or that many angiosperm genera migrated from southwest China through Manchuria to Siberia. The oldest fossils of temperate flora in China are not from the southwest (Tao 1992). Hwang (1993) is more realistic in locating only the present center of diversity of the genus *Deutzia* in the same Quinling-Hengduan range. China harbours 52 of the 65 species of *Deutzia*. In *Amyema* (Loranthaceae), major centers of species richness are in the Philippines, New Guinea, and Australia. Barlow (1992) mapped the genus on a 1° grid and regards it as being of Gondwanan derivation and as having diversified in the northern part of the Australian plate.

Ecological features may help to judge the direction of geographic progression. The species pair *Carlina vulgaris-C. biebersteinii* is ecogeographically differentiated: the first species is a submediterranean-central European plant, it is drought-resistant and frost-sensitive; the second is confined to the north and the mountains in central Europe, but stretches to central Siberia. It tolerates frost and is content with a short, comparatively moist vegetation period. Meusel (1993) supports a general use of such differences in spatial behaviour as taxonomical characters of plants.

The genus *Wahlenbergia* comprises four groups in Australia that differ in growth form, habitat preference, and distribution pattern (Smith 1992). In *Nectandra* (Lauraceae), character gradients, compared against geographic distribution, habitat preferences, and reproductive phenology, suggest an origin from a Northern Hemisphere matrix and a dramatic allopatric diversification aided by ornithochory after immigration into South America in the Pliocene (Rohwer and Kubitzki 1993).

Since its penetration into phytogeography some 15 years ago, vicariance biogeography has not yet yielded the promised innovation of that science. Areagrams of recent species and of their ancestors like that of *Lepidopetalum* (van Welzen et al. 1992) are still regarded with reserve (e.g., Henderson 1991). A cladogram based on chloroplast DNA variation suggests that *Picea* originated in North America and that the colonization of Eurasia occurred through separate intercontinental migrations (Sigurgeirsson and Szmidt 1993). The historical biogeography of pinyon pines coincides with Axelrod's hypothesis of plant differentiation during the late Tertiary aridization in the western USA (Malusa 1992).

Caryological progressions support the reconstruction of range development of *Lonicera* subsect. *Caeruleae* (diploids in southern mountains, polyploids in an wide boreal range, but see also Skvortsov, Prog. Bot. 51:322) and the differentiation of *Parnassia palustris* in Britain, where the distribution is loosely correlated with the limits of Devensian glaciation (Gornall and Wendworth 1993). Very complicated is the connection of chromosome numbers and distribution in *Galium* sect. *Leptogalium* (Krendl 1993). Most of the 15 diploid species have small relictary ranges; *G. pumilum* (8x) is broadly distributed, in *G. anisophyllum* the level of polyploidy increases from the Carpathians to the West Alps, but in the Apennine mountains di-, tetra-, hexa- and dekaploids grow together.

A multivariate comparison between three northwest European populations of *Artemisia norvegica* supports the explanation that an ancestor survived south of the glacier. Then the plant followed the retreat of the glacier northwards and differentiated into distinct populations during postglacial time (Ovstedal and Mjaavatten 1992). Polyploid arctic + alpine *Draba* species are supposed to have migrated repeatedly into the nordic area. Thereby allopolyploidy served as an escape from genetic depauperation caused by inbreeding at the diploid level (Brochmann et al. 1992). Evidence from isoelectric focusing of Rubisco on the history of *Lepidium* in Australia is presented by Mummenhoff et al. (1992). The data fit best with an immigration by long-distance dispersal by birds from Asia and/or from South America during late Tertiary or Quaternary.

b) Evidence from Fossils

The development of African–South American connections in biogeography is summarized in context with plate tectonics and fossil data by George and Lavocat (1993) and by Goldblatt (1993); that of early dispersal of the boreal flora by Budantsev (1992).

The late Quaternary vegetational history of northwestern Greece saw dry open oak woodland or steppe in the lowland, but (in contrast to previous assumptions) at a height of 1800 m in small, isolated pockets a dense mixed woodland of *Fagus, Ostrya, Carpinus*, and *Abies* with at least 16 tree species (Willis 1992). In New Zealand, *Agathis australis* could spread during the Holocene with a speed of 197 m/year (Ogden et al. 1992, maps of distribution since 40000 B.P.). The migration of *Picea* in Norway in the valleys upwards to the climatically conditioned altitudinal boundary was not completed until the Viking period (Hafsten 1991, 8 maps from the birth of Christ to the present). The spread of halophytes and steppe plants to Central Germany during interglacial times indicates open vegetation at the beginning of these periods (Mai 1992).

Pollen of *Ambrosia artemisiifolia* has been found in paleolithic settlementts in the Russian Far East, but to judge the occurrence of the species in that time may be premature (Verchovskaja and Esipenko 1993). Palynology supports the survival of *Juglans regia* in the western Mediterranean area during last glaciation (Carrion and Sanchez Gomez 1992, the walnut has been regarded as being introduced there), whereas human action is responsible for the establishment of *Quercus ilex* in the climax vegetation of Corsica and for the introduction of *Pinus halepensis* on that island (Reille 1992).

3. Factors of Distribution

a) Climatic Factors

The interest in plant geography is increasing with the dread of climatic change. The predicted warming will not simply be followed by an upheaval of zones, because the plant species respond individualistically (Rozema et al. 1993). Since 8000 B.P., the tree limit of *Pinus sylvestris* in the Scandinavian mountains has retracted by 165 m, whereas birch and alder expanded their ranges (Kullmann 1993a).

Moreover, the general warming will be geographically differentiated. In southern France, southern floristic elements have extended their ranges northward during the last decades (Maudin 1993), whereas in central Sweden the tree limit of *Betula pubescens* subsp. *tortuosa* has rather decreased after the warm 1950s. Unless drastic warming occurs, stem dieback is predicted for the near future (Kullmann 1993b). A July mean of 11.2 °C and a period of 128 days with a mean over 0 °C are best correlated with the northern tree limit (Malyshev 1993), but not evenly for all tree species occurring there.

The influence of low temperature on the distribution limits is quite complex. The dwarf birch *Betula glandulosa* produces at its northern limit in America about 80% less pollen, 50% less female flowers, and the viability of the pollen is lowered to 68%. These and other effects result in $< 5\%$ viable fruits and a nearly complete absence of sexual reproduction (Weis and Hermanutz 1993). The summer isotherms in England are paralleled by many distribution limits, e.g., that of *Sorbus torminalis*, that at the northwestern limit bears no ripe fruits (Roper 1993, density maps). With the aim of using plant taxa as indicators of bioclimates, Grandjouan et al. (1993) compared the mean pollen number per m^2 and year of 80 species at 9 sites on a transect from Abisco to Oran. Some of the investigated taxa were indicative along the whole range.

When estimating the phytogeographical consequences of warming, one has to consider also the direct influence of the increasing CO_2 content of air (Rozema et al. 1993) or the possibility of deeper thawing of permafrost soil, that may result in edaphic drought (Callaghan et al. 1992 in an investigation on response of plants to warming at high latitudes). On the other hand, a higher evaporation of warmed ocean will be followed by a higher amount of precipitation. This will also alter altitudinal limits, e.g., those of five montane pines in Arizona, because the drought resistance of their seedlings is correlated with the elevational position (Barton and Teeri 1993).

The same relates to many distribution limits on Tenerife, as is shown by Fernandes Palacios (1992), who establishes that the distribution of 44 vascular plant species is correlated to the variation of temperature and precipitation. Humidity is also responsible for the concentration of the ferns of Rwanda to the west of the country, where the yearly precipitation amounts to over 1100 mm. The sequence of disappearance of the species with increasing drought is the same there as along a north-south gradient in Zambia (Kornaś et al. 1993). The disjunct distribution of *Polystichum alfarii* in Mesoamerica is related to the ferns' unique preference for seasonally dry forest at altitudes between 1300 and 1700 m (Barrington 1992). Drastic temporary changes in diversity of the flora were observed by Halwagy and El Saadawi (1992) after successive years of drought in a Kuwait Wadi: the species number of higher plants decreased from 71 to 26. Richel (1990) uses many different hygric and thermic parameters to describe the climatic range of plants in Zaire along ecological profiles, thereby taking into account also the actual state of botanical exploration. So he tries to distinguish between ecologically and historically conditioned distribution limits.

Two observations on the effect of drought on evergreen and deciduous trees seem to reveal contradictory results. Pigott and Pigott (1993) confirm the generally accepted assumption that evergreens are less damaged by summer drought than summergreen trees. *Quercus ilex* shows no injury when water potential of the shoots falls to −3.5 MPa, whereas the summergreen *Tilia* wilts, sheds leaves, and has reduced growth when water potential falls to −2.5 MPa. The authors conclude that both the small- and large-scale pattern of distribution near the southern limit of *Tilia cordata* is controlled by water supply. However Geldenhuys (1993) observed during a period of severe drought that when after 40 months of 236 trees 47 died, 45 of these had been evergreen ones. Thus his study seems to support the hypothesis that deciduous species in general are better adapted than evergreens to survive severe droughts. It may be that the South African trees are better adapted to survive episodic long-term drought than trees in Europe, where precipitation is more reliable.

b) Dispersal and Distribution

The facts of distribution, especially that of plants on young oceanic islands, confirm the importance of atypical single dispersal events, that are difficult to quantify, and discourage experimental investigations on the effectivity of dispersal.

In arid central Australia, a comparison of range width with dispersal means revealed that only bird-dispersed plants tend to have wider ranges (Oakwood et al. 1993). Thereby the coevolution rate is not high, especially in temperate and cold regions. In New Jersey, 33% of the bird-dispersed species are not native. Especially the late-fruiting plants are dispersed by native and introduced birds with which they have not coevolved (White and Stiles 1992). In Japan, the season of most frequent avian dispersal is from January to March. Most seeds were carried to the vicinity, only a few over 1000 m. Also, in savanna grasses, most of the seeds fall down round the mother plant. The dispersal types of these grasses are investigated by Ernst et al. (1992). For the palm genus *Leopoldinia* Kubitzki (1991) describes an interesting evolution from species with heavy fleshy fruits to species that bear small buoyant fruits, which occasionally also are dispersed by fish. Judging from its large geographical distribution, the small-fruited *L. pulchra* is the most successful species in this respect.

Long-distance wind dispersal to the Canary Islands and subsequent evolutionary pappus reduction are described by Shmida and Werger (1992). Ward and Brookfield (1992) test the probability of coconuts drift to the west coast of Panama with a stochastic model of the winds, currents, and distances from Pacific islands. They suggest that dispersal by humans is more likely, because no nut germinated after marine flotation longer than 110 days. In the *Solanum* species of Madagascar, only *S. richardii*, that germinates after marine flotation, perhaps crossed the Mozambique channel by sea. Half of the representatives of Solanaceae were introduced there (D'Arcy 1992).

4. Man's Impact on Plant Distribution

Until the end of the last century, man's impact on plant distribution was regarded to be only a disturbance of the facts that were to be investigated. Recently, special attention is being paid to it, with respect to extinction, anthropogenic evolution, and changes that will follow climatic change.

a) Spread of Invaders

Under new, changing conditions, young taxa seem to be especially successful (Collins and Wein 1993 for species of *Polygonum* in Canada). One of the new conditions, the general fertilization of soils, supports high, fast-growing forbs like the Caucasian *Heracleum mantegazzianum*. As an alien, it expanded rapidly through central and north Europe and reached Tromsö (Alm and Jensen 1993) and the Shetland islands, in North America, Ontario and Michigan (Case and Beaman 1992). Its history in the former Czechoslovakia is described by Pyšek (1991). Another high forb that has expanded recently in Scandinavia is *Ligusticum scoticum* (Cedergren-Gelting 1993), whereas *Petasites hybridus* in Norway could not escape from its medieval monastery sites, because only androdynamic plants were introduced (Faegri 1992). The spread of *Veronica filiformis*, a Pontic-Caucasian-Armenian endemic, was accelerated by the expansion of frequently cut lawns in parks and gardens in Central Europe. This creeping species is best adapted to these sites (Müller and Sukopp 1993).

It is not certain that the mild winters during the past 10 years are already due to anthropogenic climatic change; be it as it may, the plants that are recently expanding their range are mainly thermophilous species, e.g., in Switzerland *Geranium purpureum* and *Sedum hispanicum* (Huber 1992). Also, the spread of *Geranium pyrenaicum* depends on warm winters (Krausch 1991, in an analysis of the history of *Rudbeckia laciniata, Oenothera syrticola*, and others in Brandenburg). The range expansion of *Reynoutria japonica* and *Impatiens glandulifera* in Britain is stopped by low summer temperatures in the northwest (Beerling 1993). Gautier (1992) draws the potential range of the tropical weed *Chromolaena odorata* on the basis of climate and recent distribution.

An important precondition for the spread of most invaders is that competition of native species is reduced, e.g., by application of herbicides in cornfields, supporting the weed *Panicum dichotomiflorum* (Huber 1992). On the other hand, some invaders may alter germination and growth of native species. The introduced *Myrica faya* inhibits with its thick litter the germination of the dominating Hawaiian tree *Metrosideros polymorpha* (Walker and Vitousek 1991).

To predict the rate of spread of an invader is almost impossible. There may be a lag phase between introduction and spread, and the potential range is usually not fulfilled by a steady moving of a front line, but by single dispersal events, that depend on dispersal means, dispersal vectors, and special site conditions. The time between introduction and spread varies from 1 to over 100 years (Kowarik 1992a). In semiarid northern Australia, the rate of spread of *Mimosa pigra* was in close correlation with the rainfall in the previous wet season, perhaps because of greater dispersal distances by flotation. Lonsdale (1993) tests a model for range expansion against these observations and considers it to be an inadequate description of spread for this invading species.

b) Urban Floras

The dimension of future extinction of species and mingling of floras may be judged from the study of urban floras. Since the 1970s, the floras of many cities in Central and West Europe, Russia, North and South America, and New

Zealand have been investigated and compared with older literature and with each other, e.g., those of Berlin, Braunschweig, Osnabrück, Münster, Saarbrücken, Halle, Leipzig, Frankfurt, Munich, Dortmund, Kassel, Cologne, Würzburg, Prag, Brno, Pilzen, Kasan, Kiev, Lublin, Wroclaw, Opole, Warsaw, London, Auckland, Tokyo, Yokohama, Durham/NC, and Lima (Wittig 1991 in a textbook on urban floras). New examples are that of Ufa, the capital of Bashkiria (Šibirdina and Šibirdin 1993), where 72 from 834 species are extinct and 77 are new, compared with a list of 1910, and the floras of Hongkong and Singapore, the differences of which are explained mainly by the colder winters of Hong Kong (Corlett 1992). The synanthropic flora of Gorlic (SE Poland, 28500 inhabitants) is typical of such small towns, with 67.5% apophytes and 32.5% aliens. In big cities this proportion is about 50:50 (Zajac and Makowiec 1992).

On the basis of an exhaustive historical documentation of the flora of Warsaw (Sudnik-Wojcikowska 1987, 1416 grid maps of all plant species, 34% of them being native and hemerophobic, 33% apophytes, 21% ephemerophytes) and other Polish cities, Moraczewski and Sudnik-Wojcikowska (1994) and Sudnik-Wojcikowska and Moraczewski (1993) develop a method to judge the anthropogenic pressure and to compare the different urban floras with minimum spanning trees (dendrites). The proposed index of synanthropization is based on the proportion of native species that are lacking on anthropogenic habitats, of apo-, archeo, keno- (=neo-), epoeco-, ephemero-, and ergasiophygophytes.

c) The Proportion of Status Groups in Different Floras

The assignment of the plant species to these status groups is certainly not easy. Particularly in countries with old agriculture and trade, one is in danger of underrating the number of archeophytes and agriophytes, whereas in Australia the late starting time of introduction makes it possible to know all invaders and their times of arrival. This may partly explain the very high proportion of aliens in Victoria (Australia) compared with that of the flora of Israel. The aliens contribute 28% to the flora of Victoria (1221 species of a total of 3144, the proportion increasing rapidly during the last decades, Foreman and Walsh 1993). In contrast, 88% of the 2780 higher plant species of the flora of Israel are confined to mainly natural habitats (as far as such habitats exist there!); of the remaining, only 5% are obligate synanthropics (Danin 1991). This percent is different in single regions of Israel: Jerusalem has 76% anthropophytes, on nests of harvesting ants grow 35%, on the Judean foothills 10% and in the Negev desert only 5% of the respective flora.

Also, the semidesert hill Great Bogdo east of the lower Volga river is not rich in aliens (Golub and Sinjakina 1992). This hill was visited already by Pallas in 1774. Only 20 introduced plant species have been recorded since then in a total of 238 species. Rather poor in newcomers are also the higher mountain belts. On the Jesenik mountain (northern Czech Republic) most of the 330 aliens remain below 980 m (Bures et al.

1992). The increase of the number of aliens is there correlated to intensity and frequency of disturbation: 1960: 122, 1970: 252, 1980: 281, 1990: 330 species. Profit and loss of the flora of Bavaria since the beginning of floristic tradition 200 years ago is summarized by Schönfelder (1993). Of 2200 taxa, 800 are extinct (70) or endangered. Since 1974 the percent of extinct plants has increased by 100%. The anthropogenic change of the Ukrainian flora has been treated by Protopopova (1991) and Burda (1991).

d) Distribution, Extinction, and Conservation

The romantic idea of prehistoric man as being in harmony with nature is certainly far from true. There is no doubt that man has been contributing to the extinction of plants for millennia, although this is not so well documented by fossils as is, for example, the extinction of all great mammals on the islands of the Mediterranean. The dimension of anthropogenic extinction has, however, increased dramatically since about 150 years and is still constantly increasing. The rate of extinction is extremely high in the epiphytes of the tropics. It may be judged from a projection from average range extension, that leads Koopowitz (1992) to the assumption that of 1000 species of orchids in Madagascar, 500–750 are already extinct.

When judging the threat of extinction, one has to consider the general distribution. Some plants such as, e.g., *Viola epipsila*, are vanishing or are extinct over the whole European area. Hybridization may be an additional threat to the survival of small populations (Kuta 1991). Other species that survived in Central Europe in small relict populations from the time of late Quaternary or early Holocene are common in their main area, e.g., *Tephroseris integrifolia* in Siberia, which possibly can be conserved in Bavaria by horticultural methods only (Krach and Krach 1991). The application of such methods may be a matter of discussion, but for some man-made species, e.g., for the weeds of flax fields, they must be accepted as being the only way of conservation.

Another question that has been much discussed in the past years is the need to conserve also "invaders", especially neophytes, plants escaped from cultivation (ergasiophygophytes), and native plants that grow on anthropogenic sites (apophytes). Kowarik (1992b) supports a uniform, objective attitude in conserving all plants growing constantly in the wild, not only the native ones. He argues against an assignment of aliens to "second-class plants" and claims to consider all established plant taxa in preparing the red lists of endangered plants. As criterion of plants being naturalized, he proposes a minimum survival time of 25 years, that will include the climatic extremes to be expected, and as a second criterion the development of different age stages and the regular regeneration by genets or ramets.

Man has always contributed to the rise of new taxa that are now constant members of the "native" flora, e.g., in *Rosa*, *Crataegus*, and *Rubus*. Hybridization of native species with aliens may be undesirable from the viewpoint of species conservation, but it has long been a reality. Some seemingly native plants have passed through a stage of cultivation, e.g., *Acer platanoides* in large parts of central Germany; others have changed over to man-made sites, such as *Taxus baccata*, that has only a few relict native populations there. In consideration of the importance of sites disturbed by man or by natural events, conservation of dynamic processes in habitats is required. Also the natural turnover of floras has hitherto not been sufficiently considered (Reichholf 1993).

The study of invasions is important also in predicting the ecological risks of release of transgenic cultigens, particularly the probability of their becoming feral, and moreover the probability of undesired ecological long-term effects in man-made or natural ecosystems. As models for such risks Sukopp and Sukopp (1993) list many examples of conventional cultigens becoming feral in central Europe, and case histories of invasions of aliens and their unwanted ecological consequences.

5. Analyses of Floras and Classification of Distribution Patterns

A quantitative analysis of a flora is often undertaken with the aim of revealing the origin, geographic position, affinities, and peculiarities of the flora, and to confirm phytogeographic boundaries. Malysev (1991) develops empirical formulas for the representativity and originality of floras, and discusses the relevance of these concepts for nature conservation.

Interesting results of a computer classification of European ranges of all 242 plant species in Italian beech forests have been obtained by Nimis and Bolognini (1993). The joint distribution of different chorotypes revealed from a matrix of 531 geographic units show good relationships to ecology. Some of these types are concentrated in the main distributional centers for the beech forest flora in southern Europe, 20% of the species were not able to expand their ranges from there in postglacial times. A similar computer classification for the ranges of southern Yukon flora resulted in seven phytogeographic groups (Lausi and Nimis 1991). Most of the species groups obtained by numerical classification of phytogeographic data in Fuegia correlated to the main types of habitat, e.g., steppe, forest, and bog communities (Tretiach et al. 1989). Plants with wide bipolar ranges are mainly those of aquatic habitats. That different habitats may harbor different chorological elements is confirmed also by Morefield (1992): wetlands in the White Mts. are unusually rich in widespread taxa.

The distribution of 333 montane species of the Ukrainian Carpathians is classified and interpreted by Malinovskij (1990, 120 maps). The close connection between the flora of the Carpathians and eastern European lowlands is explained by similar ecological conditions (short, moist summer) by Parfenov (1991). The majority of the montane element prefer the plains of Predpolessye. In a floristic analysis of the islands around Kasos (Greece), Raus (1989) confirms the enigmatic fact that almost every small islet harbors one to six islet specialists, i.e., species which exclusively occur on such islets and have never been found on larger islands (e.g., *Allium commutatum, Asparagus stipularis*).

6. Phytogeographic Subdivision of the Continents and Endemism

Phytogeographic subdivision is achieved on two ways: the classical search for hierarchically arranged, quasihomogenous kingdoms, regions, and provinces, and the search for centers of endemism separated by transition zones.

Centers of endemism have been the base of a map of Africa by White that was the chorological framework of the AETFAT vegetation map of this continent. White (1993) sees the rationale underlying this chorological classification as confirmed. He discusses the fundamental doubts and difficulties: the difficulty of interpreting historical elements, the fact that previous chorological proposals for Europe differ radically, the hierarchical nature of schemes that are proposed, the problem of weighting species that are of different importance in the vegetation, the difficulty of delimitation of the chorological boundaries, and the problem of dealing with the flora of the mountains. White does not deny the value of perception, but intuitive discernment has to be followed by rigorous analysis.

The method of centers of endemism is adopted by Prance (1989) for a subdivision of South America into 18 phytochoria, 11 of which are regional centers of endemism, and the same method will be applied in constructing a new plant geographical map of southwest Asia on the base of a data bank in UTM-degrees with a geographical information system (Danin et al. 1992). The value of generic endemism in floristic subdivision is discussed with the example of the Caucasus region also by Chochrjakov (1991).

The phytochorological spectra of the three Iranian deserts Dasht e Kavir, Dasht e Lut, and Jaz Muien allow the assignation of the first two to the Irano-Turanian center of endemism, the latter to the Saharo-Sindian center (Léonard 1993). A map of Arabia displays concentration of endemics in the east, southeast, and J. Dibbagh (Miller and Nyberg 1991). The Sultanate of Oman is subdivided chorologically by Ghazanfar (1992); 7.7% of the 1204 higher plant species are endemics; most of them are concentrated in the south. Also, the 15 provinces of New Zealand recognized by Wardle (1991) are not arranged hierarchically. The number of endemics is especially high there in the northern part of the southern island.

In an exhaustive treatment of the endemic genera of seed plants of China, three centers of endemism emerge, all in the southwest: Sichuan-Western Hubeh, Yunnan-Western Guangxi, and Sichuan-Northwestern Yunnan (Ying et al. 1993). The distribution patterns of the dot-mapped endemic genera are compared with a map of floristic regions after Wu, who discerns in China 7 subkingdoms and 22 regions, thereby certainly assessing the hierarchical rank too high. The number and percent of generic endemism of some countries are compared:

China	243	endemic genera	7.8% of the flora
New Zealand	93	" "	9.9% " " "
Australia	538	" "	32.0% " " "
Japan	17	" "	1.5% " " "
India	134	" "	7.7% " " "

In Chile 67 of 1008 genera (6.6%) are endemic (Marticorena 1990), in Mexico 10% of the genera of higher plants (Rzedowski 1991).

The concept of centers of endemism is of special use in concentrating conservation activities on the most important areas. Obviously, this method would give no result in subdividing most of the temperate and boreal lowlands.

The hierarchical arrangement of floristic areas is defended by McLaughlin (1992). Both the division based on ranges of species and that based on taxonomic rank in his investigation of the flora of Western United States showed a high degree of hierarchical structure. It appears, therefore, that a hierarchical system of phytochoria is a useful and natural way of depicting floristic relationships. In the western USA, the Cordillerean and Madrean regions are subdivided into 5 provinces (Californian, Cordillerean, Intermountain, Madrean, Sonoran), 11 subprovinces, and 20 districts.

The comparison of hundreds of distribution maps leads to a new hierarchical subdivision of the Namib region (Jürgens 1991). The floristic boundaries in this region are closely connected to the rhythm of precipitation. The limit of influence of winter rain parallels most of the distribution limits, it extends farther north than hitherto assumed. Also in Minnesota, the migration of most species has reached an equilibrium with ecological factors. A major floristic boundary in the southern and western portions of the state is defined by the present climate (Wheeler et al. 1992). On the other hand, one of the most discussed boundaries in Australasia, the Wallace line, is connected with the extent of deep sea and the historically changing sea level. This line is an important limit also for the genus *Euchresta* (Leguminosae, Chen et al. 1992). The position of Langtang Himal at the western border of the east Himalayan region is interpreted climatically by Miehe (1990, many dot maps). Recently, the migration of montane and especially alpine plants along the Himalayan chain has been hampered by the deep valleys.

Mosses with wide distribution are a tool for comparing altitudinal zonation of tropical rain forests (Frahm and Gradstein 1991, five zones discerned).

References

Adler W, Oswald K, Fischer R (1994) Exkursionsflora von Österreich. Ulmer, Stuttgart, 736 pp. **Alm T**, Jensen C (1993) Blyttia 51:61–69.

van Balgooy, MMJ (1993) Pacific plant areas. Vol 5. Rijksherbarium, Leiden, 251 pp. **Barlow BA** (1992) Blumea 36:239–381. **Barrington DS** (1992) Rhodora 94:327–339. **Barry JP**, Celles JC (1991) Flore de Mauritanie. University Press CRDP, Nice, 550 pp. **Barton AM**, Teeri JA (1993) Am J Bot 80:15–25. **Beaman JH** (1990) Revision of Hieracium (Asteraceae) in Mexico and Central America. Syst Bot Monogr 29:1–77. **Beerling DJ** (1993) J Biogeogr 20:45–53. **Boratynska K** ("1989" 1991) Arbor Kornickie 34: 3–29. **Boratynski A**, Browicz K, Zielinski J ("1988" 1990) Arbor Kornickie 33:13–74. **Boratynski A**, Browicz K, Tomlik A, Zielinski J ("1990" 1992) Arbor Kornickie 35:47–76. **Brochmann C**, Soltis DE, Soltis PS (1992) Plant Syst Evol 182:35–70. **Browne ET**,

Athey R (1992) Vascular plants of Kentucky: an annotated checklist. University Press, Lexington, 180 pp. **Budantsev LY** (1992) Bot Rev 58:1–48. **Burda RI** (1991) Antropogennaja transformacija flory (Anthropogenic transformation of floras). Naukova dumka, Kiev, 168 pp. **Bures L**, Klimes L, Kralik J (1992) Preslia 64:63–77. **Burrows JE** (1990) Southern African ferns and fern allies. Frandsen, Sandton, 359 pp.

Callaghan TV, Sonesson M, Somme L (1992) Philos Transact R Soc Lond B 338:279–288. **Carrion JS**, Sanchez-Gomez P (1992) J Biogeogr 19:623–630. **Case MA**, Beaman JH (1992) Mich Bot 31:152–154. **Cedergren-Gelting K** (1993) Sven Bot Tidskr 87:255–261. **Chen Cheih**, Sun Hang, Mizuo Mizuno (1992) Acta Phytotax Sin 30:43–56. **Chochrjakov AP** (1991) Bjull Mosk Obšč Ispyt Prir Biol 96/6:90–109. **Chrtek J** (1991) Severočesk Přirod Litomerice 25:1–76. **Collins B**, Wein GR (1993) Can J Bot 71:939–945. **Corlett RT** (1992) J Biogeogr 19:421–430.

Danin A (1991) In: Engel T, Frey W, Kürschner H (eds) Contributiones selectae ad floram et vegetationem orientis. Cramer, Berlin, pp 95–103. **Danin A**, Hedge IC, Léonard J, Weinert E, White F (1992) J Veg Sci 3:566. **D'Arcy WG** (1992) Ann Mo Bot Gard 79:29–45. **Diaz-Piedrahita S**, Velez-Nauer C (1993) Monogr Jard Bot Jose Celestino Mutis 1:1–162. **Düll R**, Meinunger L (1989) Deutschlands Moose. IDH, Bad Münstereifel, 368 pp.

Ernst WHO, Veenendaal EM, Kebakile MM (1992) Vegetatio 102:1–11.

Faegri K (1992) Blyttia 50:115–119. **Fernandez Casas J**, Gamarra Gamarra R, Morales Abad J (1992) Treballs Inst Bot Barc 15:1–422. **Fernandez-Palacios JM** (1992) J Veget Sci 3:595–602. Flora Europaea, vol 1, Psilotaceae to Platanaceae (1993) Cambridge University, Cambridge, XLVI + 581 pp. Flora Mesoamericana, vol 6 (1994) Mo Bot Gard, St Louis, 543 pp. Flora Neotropica, vol 60 (1993) 333 pp, vol 61 (1993) 46 pp, New York Bot Gard, New York. Flora of North America north of Mexico (1993) vol 1, XXI + 372 pp, vol 2, XVI + 475 pp, Oxford University, New York. **Foreman DB**, Walsh NG (1993) Flora of Victoria, Inkata, Melbourne, Sydney, 320 pp. **Frahm JP**, Gradstein SR (1991) J Biogeogr 18:669–678.

Gautier L (1992) Candollea 47:645–662. **Geldenhuys CJ** (1993) South Afr J Bot 59: 522–534. **George W**, Lavocat R (1993) The Africa-South America connection. Clarendon, Oxford, 184 pp. **Ghazanfar SA** (1992) Global Ecol Biogeogr Lett 2:189–195. **Goldblatt P** (1993) Biological relationships between Africa and South America. Yale University Press, New Haven, X + 630 pp. **Golub VB**, Sinjakina VV (1992) Bot Žurn 77/12:98–102. **Gornall RJ**, Wentworth JE (1993) New Phytol 123:383–388. **Gottschlich G**, Raabe U (1991) Abh Westfäl Mus Naturkd 53/4:3–140. **Grandjouan G**, Cour P, Gros R (1993) Vegetatio 109:107–124.

Hafsten U (1991) Blyttia 49:171–181. **Halwagy M**, El-Saadawi W (1992) Acta Bot Neerl 41:183–195. **Hartl H**, Kniely G, Leute GH, Niklfeld H, Perko M (1992) Verbreitungsatlas der Farn- und Blütenpflanzen Kärntens. Naturwiss Ver, Klagenfurt, 451 pp. **Henderson IM** (1991) Austral Syst Bot 4:59–71. **Hickman JC** (1993) The Jepson manual: Higher plants of California. University of California, Berkeley, XVII + 1400 pp. **Hilbig W** (1992) Wiss Z Univ Halle Math-Nat 41/1:91–141. **Ho Pham-hoang** (1991, 1992, 1993) Cayco Vietnam, an illustrated flora of Vietnam, vol 1–3. Library of Congress, Santa Ana, 3570 pp. **Hohenester A**, Welss W (1993) Exkursionsflora für die Kanarischen Inseln mit Ausblicken auf ganz Makaronesien. Ulmer, Stuttgart, 374 pp. **Huber W** (1992) Bot Helv 102: 93–108. **Hutchinson G**, Thomas BA (1992) Watsonia 19:1–19. **Hwang Shu-mei** (1993) Acta Phytotax Sin 31:105–126.

Ireland RR, Ley LM (1992) Atlas of Ontario mosses. Can Mus Nature, Ottawa, 138 pp. **Iwatsuki K**, Yamazaki T (1993) Flora of Japan, vol 3a Kodansha, Tokyo, XII + 482 pp.

Jürgens N (1991) Vegetatio 97:21–38.

Koopowitz H (1992) Selbyana 13:115–122. **Kornas J**, Nowak KA, Matyjaszkiewicz M (1993) Zesz Nauk Uniw Jagiellonsk Kraków 1087, Prace Bot 25:1–137. **Kowarik I** (1992a) Verh Bot Ver Berl Brandenburg Beih 3:1–188. **Kowarik I** (1992b) Schriftenr Vegetationskd 23:175–190. **Krach B**, Krach JE (1991) Ber Bayer Bot Ges 62:181–182. **Krausch HD** (1991) Gleditschia 19:297–308. **Krendl T** (1993) Chromosomenzahl und geographische Verbreitung in der Gattung *Galium* (sect. *Leptogalium*, Rubiaceae). Österr Akad Wiss, Wien, Biosyst Ecol Ser 4:51–112. **Krieglsteiner GJ** (1991) Verbreitungsatlas der Großpilze Deutschlands (West), vol 1a. Ulmer, Stuttgart, 416 pp. **Kubitzki K** (1991) Nord J Bot 11:429–432. **Kullmann L** (1993a) Global Ecol Biogeogr Lett 2:181–188. **Kullmann L** (1993b) J Veget Sci 4:765–772. **Kuta E** (1991) Veröff Geobot Inst Rübel Zürich 106:257–265.

Lausi D, Nimis S (1991) In: Nimis PL, Crovello TJ (eds) Quantitative approaches to phytogeography. Kluwer, Dordrecht, pp 35–122. **Lebrun JP**, Stork AL (1991, 1992) Enumeration des plantes a fleurs d'Afrique tropicale, vol 1, 249 pp, vol 2, 257 pp. Conserv J Bot, Genève. **Lee Tchang Bok** (1989) Illustrated flora of Korea. ISBN 89–7187–047–8–96480, 990 pp. **Leiva A** (1992) Fontqueria 34:1–16. **Léonard J** (1993) Bull Jard Bot Natl Belg 62:389–396. **Lisowski S** (1991) Fragm Flor Geobot 36 Suppl 1/1+2:1–627. **Lisowski S** (1992) Fragm Flor Geobot 37:275–369. **Liu Shang-Wu**, Ho Ting-Nung (1992) Acta Phytotax Sin 30:289–319. **Lonsdale WM** (1993) J Ecol 81: 513–521. **Luer CA** (1993) Icones Pleurothallidinarum, X. Missouri Botanical Garden, St Louis, 244 pp.

Mai DH (1992) Gleditschia 20:57–85. **Malinovskij AK** (1990) Montannyj element flory Ukrainskich Karpat (The montane element of the Ukrainian Carpathians). Nauk dumka, Kiev, 240 pp. **Malusa J** (1992) Syst Bot 17:42–66. **Malyshev L** (1991) In: Nimis PL, Crovello TJ (eds) Quantitative approaches to phytogeography. Kluwer, Dordrecht, pp 15–33. **Malyshev L** (1993) Vegetatio 10:175–186. **Manitz H** (1992) Beitr Phytotax Jena 15:127–165. **Marticorena C** (1990) Gayana Bot 47:85–113. **Maudin JP** (1993) Acta Bot Gallica 140:81–90. **McLaughlin SP** (1992) J Biogeogr 19:21–32. **Meinunger L** (1992) Haussknechtia Jena Beih 3/1+2:423+840 pp. **Melzheimer V**, Polatschek A (1992) Phyton 31:281–306. **Meusel H** (1993) Flora 188:73–84. **Micheev AD** (1992) Bot Žurn 77/10: 68–74. **Miehe G** (1990) Langtang Himal. Cramer, Stuttgart, 529 pp. **Miller AG**, Cope TA (1994) Flora of the Arabian Peninsula and Socotra, vol 1 (Koeltz Koenigstein), 600 pp. **Miller AG**, Nyberg JA (1991) In: Engel T, Frey W, Kürschner H (eds) Contributiones selectae ad floram et vegetationem orientis. Cramer, Berlin, pp 263–279. **Molina A**, Rubio-Sanchez A (1992) Bot Complutensis 17:143–201. **Moraczewski IR**, Sudnik-Wojcikowska B (1994) Flora 189/3: **Morefield JD** (1992) J Biogeogr 19:33–50. **Mori SA** (1992) Brittonia 44:372–375. **Mossberg B**, Stenberg L, Ericsson S (1992) Den nordiska floran. Wahlström & Widstrand, Stockholm, 696 pp. **Müller N**, Sukopp H (1993) Tuexenia 13:399–413. **Mummenhoff K**, Hurka H, Bandelt HJ (1992) Plant Syst Evol 183:99–112.

Nakaike T (1992) New flora of Japan: Pteridophyta. Shibundo, Tokyo, 868 pp. **Nakanishi H** (1991) Hikobia 11:73–83. **Nimis PL**, Bolognini G (1993) Vegetatio 109:125–143.

Oakwood M, Jurado E, Leishman H, Westoby H (1993) J Biogeogr 20:563–572. **Ogden J**, Wilson A, Hendy C, Newnham RM (1992) J Biogeogr 19:611–622. **Olonova MB** (1993) Bot Žurn 78/9:50–59. **Ovstedal DO**, Mjaavatten O (1992) Plant Syst Evol 181:21–32.

Parfenov V (1991) Veröff Geobot Inst Rübel Zürich 106:80–95. **Piekos-Mirkowa H** (1991) Veröff Geobot Inst Rübel Zürich 106:282–287. **Pigott CD**, Pigott S (1993) J Ecol 81:557–566. **Plechanova MN**, Solov'eva LV, Močalova OV (1992) Bot Žurn 77/9:1–11. **Poldini L** (1991) Atlante corologico delle plante vascolari nel Friuli-Venezia Giulia. Direzione Reg Foreste Parchi, Udine, 899 pp. **Prance GT** (1989) In: Lieth H, Werger MJA (eds) Tropical rainforest ecosystems: biogeographical and ecological studies. Elsevier, Amsterdam, pp 99–132. **Prelli R**, Boudrie M (1992) Atlas écologique des fougéres et plantes alliées. Lechevalier, Paris, 272 pp. **Protopopova VV** (1991) Sinantropnaja flora Ukrainy i puti ee razvitija (Synanthropic flora of Ukraine and the ways of its development). Nauk dumka, Kiev, 202 pp. **Pyšek P** (1991) Folia Geobot Phytotax Praha 26:439–454.

Raus T (1989) Bot Chron 9:19–39. **Reichholf JH** (1993) Rundgespräche Kommission Ökol Bayer Akad Wiss 6:75–84. **Reille M** (1992) New Phytol 122:359–378. **Richel T** (1990) Mitt Inst Allg Bot Hamb 23b:767–776. **Rohwer JG**, Kubitzki K (1993) Bot Acta 106:88–99. **Roper P** (1993) Watsonia 19:209–229. **Rostanski K** (1991) Bol Soc Brot Sér 2 64:5–33. **Rouleau E**, Lamoureux G (1992) Atlas des plantes vasculaires de l'ile de Terre Neuve et des iles de Saint-Pierre-et-Mique-lon. Fleurbec, Saint-Henri-de-Lévis, 777 pp. **Rozema J**, Lambers H, Geijn SC, Cambridge ML (1993) Vegetatio 104/105:1 ff. **Rzedowski J** (1991) Acta Bot Mexicana 15:47–64.

Schaminée JHJ, van Duuren L, de Bakker AJ (1992) Gorteria 18:57–96. **Schnedler W** (1992) Oberhess Naturwiss Z 54:9–40. **Schönfelder P** (1993) Rundgespräche Kommission Ökol Bayer Akad Wiss 6:39–48. **Schönfelder P**, Leon Arencibia MC, Wildpret W (1993) Itinera Geobot 7:375–404. **Shmida A**, Werger MJA (1992) Vegetatio 102:183–199. **Šibirdina LM**, Šibirdin AP (1993) Bot Žurn 78/3:1–10. **Sigurgeirsson A**, Szmidt A (1993) Nord J Bot 13:233–346. **Small E**, Fawzy M (1992) Can J Bot 70:1292–1301. **Smith PJ** (1992) Telopea 5/1:91–175. **Stace C** (1991) New flora of the British Isles. Cambridge University, Cambridge, 1226 pp. **Stork AL**, Lebrun JP (1988) Inst Élevage Méd Vét Pays Tropic Étude Bot 13:1–128. **Sudnik-Wojcikowska B** (1987) Flora miasta Warszawy i jej przemiany w ciagu XIX i XX wieku. Wyd Uniw Warszaw, Warszawa, 242+435 pp. **Sudnik-Wojcikowska B**, Moraczewski IR (1993) Feddes Rep 104:81–92. **Sukopp U**, Sukopp H (1993) Gaia 2:267–288.

Tao Jun-Rong (1992) Acta Phytotax Sin 30:25–43. **Thulin M** (1993) Flora of Somalia, vol 1, Roy Bot Gard, Kew, VIII+493 pp. **Toledo VM**, Sosa V (1993) Taxon 42:355–364. **Tretiach M**, Lausi D, Poelt J (1989) Stud Geobot 9:3–72. **Turland NJ**, Chilton L, Press JR (1993) Flora of the Cretan area. Annotated checklist & atlas. HMSO, London, 439 pp.

Verchovskaja NB, Esipenko LP (1993) Bot Žurn 78/2:94–101.

Walker LR, Vitousek PM (1991) Ecology 72:1449–1455. **Wang Wen-Tsai** (1992) Acta Phytotax Sin 30:97–117. **Ward RG**, Brookfield M (1992) J Biogeogr 19:467–480. **Wardle P** (1991) Vegetation of New Zealand. Cambridge University, Cambridge, 672 pp. **Weber HE** (1992) Flor Rundbriefe 26/2:116–124. **Weis IM**, Hermanutz LA (1993) Am J Bot 80:1021–1027. **van Welzen PC**, Piskaut P, Windadri FI (1992) Blumea 36:439–465. **Wheeler GA**, Cushing EJ, Gorham E, Morley T, Ownbey GB (1992) Can J Bot 70:319–333. **White DW**, Stiles EW (1992) Can J Bot 70:1689–1696. **White F** (1993) Bull Jard Bot Natl Belg 62:225–281. **Willis KJ** (1992) New Phytol 121:101–155. **Wirth V**, Oberhollenzer H (1990) Lichen mapping in Europe. Staatl Mus Naturkunde, Stuttgart, 199 pp. **Wittig R** (1991) Ökologie der Großstadtflora. Fischer, Stuttgart, 261 pp. **Wofford BE**, Kral R (1993) Sida Bot Misc 10:1–66. **Wohlgemuth T** (1993) Bot Helv 103:55–71.

Yamaguchi T (1993) J Hattori Bot Lab 73:1–123. **Ying Tsun-Shen**, Zhang Yu-Long, Boufford DE (1993) The endemic genera of seed plants of China. Science Press, Peking 813 pp.

Zajac EU, Makowiec M (1992) Zeszyty Nauk Univ Jagiell Prace Bot 24:71–89.

Prof. Dr. Eckehart J. Jäger
Institut für Geobotanik und Botanischer Garten
Martin-Luther-Universität Halle
Neuwerk 21
06108 Halle, Germany

II. Salt Resistance in Herbaceous Halophytes and Mangroves

By Marianne Popp

1. Introduction

One of the now widely used definitions of halophytes is that of Jennings (1976), describing them as "the native flora of saline habitats". This is very close to the description by Warming (1909 English Edition, 1895 *Plantesamfund* Danish Edition), who coined this and a larger number of still commonly used terms like hydro-, meso-, and xerophytes. According to Warming (1909), "a certain amount of soluble salts must be present before halophytic vegetation is called into existence; but the nature of the salts seems to be a matter of indifference". Following Munns et al. (1983), such a habitat is one "containing saline water with an osmotic pressure (π) of more than 3.3 bar" (equivalent to 70 mM monovalent salts).

Referring to this definition of halophytes, there should be no doubt that mangroves, which are trees and shrubs of the tropical and subtropical tidal zone, belong to this group of plants. However, reviews and books focus always much more on herbaceous halophytes (e.g., Flowers et al. 1986; Ungar 1991). To compensate for this lack of recognition, it is the aim of this chapter to review the recent literature on salt resistance, emphasizing the comparison between herbaceous and woody halophytes.

2. Terminology

a) Halophytes, Nonhalophytes

Dealing with halophytes immediately poses the question of how to call those plants which are not able to survive in a saline habitat. Warming (1909) called them "halophobous", a term rarely used nowadays, but nevertheless much more logical than the widely used term "glycophyte". The contrast of a saline soil is not a sweet one, but a nonsaline one. Thus the very clear and consequent terminology introduced by Warming (1909) to give the main characteristic of

Progress in Botany, Vol. 56

the edaphic conditions on which a plant lives should not be obscured by the use of "sweet plants" for plants growing on soils of moderate ion content. It is also a misuse of the warming terminology to call the opposite of a halophyte a mesophyte (Jennings 1976; Cheeseman 1988), since this is the definition for a plant that flourishes in an environment with moderate amounts of moisture. Although it is neither very original nor very satisfactory in respect to semantics, there is no better way than to differentiate into halophytes and non-halophytes (Osmond 1980).

b) Salt Tolerance, Salt Resistance

A further semantic problem is the use of the term "tolerance" and "resistance" in combination with salt or other stresses. In Levitt's (1980) concept "resistance" is the generic term, which is divided into stress tolerance and stress avoidance, where the latter means that the plant prevents the stress by physical or metabolic means. Stress tolerance describes the state when the plant comes to thermodynamic equilibrium with the stress without an injury or if an injury occurs it is repaired. To quote, Larcher (1983); "salt resistance is the ability of a plant to withstand the presence of excess salts (especially Na^+, Cl^-, and SO_4^{2-}) without serious impairment of vital function", while "salt tolerance is a property of the protoplasm that enables it to tolerate more or less well, depending on species, tissue type, and level of vitality, the changed ionic ratios associated with salt stress and the toxic and osmotic effects associated with increased ion concentration". Again for Larcher (1983), "salt resistance" is the overriding term which in his scheme comprises mechanisms of salt regulation on the whole plant level and traits of salt tolerance on the cellular level. Thus, one should be cautious when calling a halophyte "salt-resistant". However, the agronomists have developed a different point of view, saying that "in an agricultural or horticultural context, a more salt-tolerant cultivar is one that has a higher yield of marketable product in salinity conditions than does a less tolerant one" [use of drought tolerance by Jones (1983) has been replaced by salt tolerance].

This use of the term "tolerance" is rather unfortunate, since the real meaning of the word is to endure a situation as it is without further reaction. The best example for a real salt-tolerant organism are the halophytic bacteria, which thrive in 1 to 3 M NaCl solutions. The situation of halophytic plants is much better described by "resistance", which according to the Dictionary of Science and Technology (1992) is "the fact of opposing or acting against".

This chapter will follow the concept of Larcher (1983) and treat the different components of salt resistance at the whole plant (salt regulation) and cellular level (salt tolerance).

3. Salt Regulation

a) Salt Exclusion

α) Salt Excluders, Salt Includers

To differentiate between those halophytes which have a low or high salt content in their shoots they are described as either salt-excluders or salt includers and salt-accumulators respectively. These are quite useful terms, which make clear that the latter group of plants achieves its osmotic adjustment to the low water potential of saline soils mainly by the storage of Na^+ and Cl^-, whereas the former group stores other osmolytes like K^+ and sugars to a greater degree (Albert 1982).

Although osmotic adjustment by storage of organic solutes is regarded as being more costly in terms of energy than storage of Na^+ and Cl^- (Yeo 1983), in salt-excluding species the more salt-resistant ecotypes are those with a greater restriction of Na^+-accumulation in the shoots. Comparing two populations of *Hordeum jubatum* from a saline and one of a nonsaline habitat, Wang et al. (1992) found lower Na^+ concentrations in the shoots of the plants originating from the saline sites, when they were treated with 148 mol m^{-3} Na^+ in hydroponic solutions. This confirms earlier findings of Rozema (1976) on ecotypes of *Juncus* and by Ahmad and Wainwright (1977) on *Agrostis stolonifera.*

This trait is also very important for salt resistance in nonhalophytes. Screening 415 accessions of *Triticum tauschii* on a nutrient solution of 150 mol m^{-3} NaCl, Schachtmann et al. (1991a) came to the conclusion that "sodium exclusion from leaves may be the mechanism of salt tolerance in cereals".

Even for a salt-secreting species like *Spartina alterniflora*, detailed investigations revealed that ion exclusion at the root level accounted for 91 to 97% of the theoretical maximum ion uptake as a result of transpiration and growth (Bradley and Morris 1991).

The case studies listed so far should not give the impression that salt exclusion at the root level is restricted to those halophytes which are called salt-excluders because of their low salt content in the shoots.

β) Salt Exclusion from the Xylem

The still most striking example for salt-exclusion from the xylem sap are the mangroves, which have in their leaves NaCl concentrations as high as or even exceeding those of seawater. Using energy-dispersive X-ray microprobe analyses, Werner and Stelzer (1990) showed radial ion profiles of vacuolar Na^+, Cl^- and K^+ of transversely fractured root cells of *Rhizophora mangle*, indicating a steep gradient from the outside growth medium to the xylem vessels. By the same method, they calculated for plants rooted in 200 mol m^{-3} NaCl xylem sap concentrations for Na^+ of 26.7 mol m^{-3} and for Cl^- of 34 mol m^{-3} for the roots and 20.1 mol m^{-3} Na^+ and 37.6 mol m^{-3} Cl^- for the shoots. Even lower values of Na^+ and Cl^- were reported by Popp et al. (1993) for the mangal associate *Conocarpus erectus* which contained at nearly full seawater strength (465 mol m^{-3} Cl, 362 mol m^{-3} Na^+) only 17.6 mol m^{-3} Cl and 7.5 mol m^{-3} Na^+ in the xylem sap of twigs.

Time-course measurements of root ion content indicated the possibility of a sodium-excluding mechanism in roots of *Kosteletzkya virginica*, a Malvaceae of brackish tidal marshes (Blits and Gallagher 1990). Even after 30 days of exposure to 255 mol m^{-3} NaCl, the Na^+/K^+ ratio was 0.82 in roots. This low Na^+/K^+ ratio was brought about not only by exclusion of Na^+, but also by a strong K^+ affinity, which the authors ascribed to a change of the sterol/phospholipid ratio in salinized roots.

All grades of K^+/Na^+ selectivity and sodium exclusion were demonstrated for halophytic and nonhalophytic members of the Chenopodiaceae (Reimann 1992; Reimann and Breckle 1993), the two halophytes *Suaeda maritima* and *Atriplex prostrata* showing the most pronounced Na^+ accumulation in the shoots. However, high levels of Na^+ accumulation are not necessarily a prerequisite of high salt resistance in halophytes of the accumulation type.

Glenn et al. (1992) investigated low-sodium and high-sodium subspecies of *Atriplex canescens* over a range of salinities up to 720 mol m^{-3} NaCl. Clear differences in Na^+ and K^+ accumulation and Na^+/K^+ ratios were not attended by a difference in salt resistance.

γ) Salt Exclusion from the Leaves

Besides salt exclusion in the roots, low NaCl levels in leaves can be also achieved by NaCl retention in lower plant parts (cf. Albert 1982). The brown beetle grass [*Diplechna fusca* (L.) Beauv.] sequestered high levels Na^+ and Cl^- in the sheath, thus keeping the concentration of these ions low in the leaf blade and maintaining a high selectivity of K^+ over Na^+ in the photosynthetic active tissue (Warwick and Halloran 1991).

For mangroves, it would be obvious to store Na^+ and Cl^- away in roots or lower stem parts, but neither investigations of 3-year-old glasshouse-grown plants of *Aegiceras corniculatum* (Popp et al. 1993) nor samples from adult trees from natural sites (Popp and Polania 1989) gave any indication for such a retention. Na^+ and Cl^- values calculated on tissue water basis of *A. corniculatum* showed a clear increase from the roots via lower and upper stem to the leaves.

b) Elimination

α) Salt Glands and Bladder Hairs

One of the most obvious signs of salt regulation are the salt crusts on leaves and shoots of those halophytes which possess either salt glands or salt hairs. Nevertheless, there is some confusion in the literature as to which mangrove species possess salt glands and which do not. Liphschitz and Waisel (1982) listed *Sonneratia* and *Rhizophora* within mangroves as possessing salt glands, wrongly interpreting Walter and Steiner (1936), who described glandular structures for these two species, but clearly pointed out that no salt is extruded by all these cells. Similarly, Biebl and Kinzel (1965) only suspected that the structures they described for leaves of *Langucularia racemosa* might have the function of salt glands. Recent investigations have shown that *L. racemosa* does not secrete salt, but mannitol, by these glands (Popp, unpubl.) and thus they could be called extrafloral nectaries. Salt glands in mangroves are restricted to the genera *Aegialitis, Aegiceras, Avicennia*, and *Acanthus*.

In the discussion of the functioning of salt glands, again a semantic problem arises: are they excreting (e.g., Rozema et al. 1991), recreting (e.g., Waisel 1991; Freitas and Breckle 1992), or secreting (e.g., Flowers et al. 1990; Marcum and Murdoch 1992)? Although these terms are clearly defined (not like the case with resistance and tolerance), they are variously used by different authors.

Aware of this existing confusion, Waisel (1991) cites Frey-Wissling, who described "recretion" as the removal of those ions that pass through the plant without being metabolized or changed. (Excretion means the elimination of metabolic waste, while secretion describes the release of substances which will fulfil a further task for the organism, e.g., nectar).

Irrespective of the prefix used, all the papers cited above and others point to a marked quantitative importance of the amounts of salt recreted by glands or hairs. Use of different methods of measuring and calculation using all kinds of bases make comparison difficult. Table 1 tries to compile some data from which it was possible to estimate recretion rates on a leaf area basis and to relate the recretion to the internal salt content, which is the most decisive factor for the efficacy of the salt glands. This is rather high in all cases listed in Table 1, as well as in other studies, for instance for the halophytic wild rice, *Porteresia coarctata* Tateoko, it was also reported that salt recretion is generally in excess of 50% of the salt accumulated by the leaves (Flowers et al. 1990). In this latter paper, and also in one on *Sporobolus virginicus* (Marcum and Murdoch 1992), it is pointed out that the high portion of Na^+ in the recretion fluid helps to keep the K^+/Na^+ ratio inside the leaves in favor of K^+. Comparing *Zoysia matrella*, a more salt-resistant species, and *Zoysia japonica*, a more salt-sensitive member of the Chloridoideae (Poaceae), Marcum and Murdoch (1990) ascertained that the *Z. matrella* had a higher density of larger glands and secreted more Na^+ per unit leaf mass.

Table 1. Recretion rates (given in nmol m^{-2} leaf area s^{-1}) and recretion in relation to internal salt content of various halophytes

Plant	NaCl mol m^{-3} root medium	Recretion nmol m^{-2} leaf area s^{-1} Na^+	Cl^-	NaCl	Recretion in relation to internal salt content	Reference
Diplachne fusca 19 °C	250	12	9		Excretion 5 times the rate of accumulation in 24 h	Gorham (1987)
" 39 °C	250	258	283			"
Spartina alterniflora	690			41	Approximately half of the ions taken up	Bradley and Morris (1991)
Spartina anglica	500	12			About every 6–18 days the amount of Na^+ present in the shoot is removed	Rozema et al. (1991)
Avicennia marina	natural site			119		Ish-Shalom-Gordon and Dubinsky (1993)
Aegialitis annulata						
Young leaves	50	180	200		26% of the internal	Popp et al. (1993)
Old leaves	50	80	120		2% salt content of	
Young leaves	500	440	420		38% leaves is removed	
Old leaves	500	280	250		29% in 12 hours	

Also the bladder hairs of different Chenopodiaceae contribute essentially to the salt balance in leaves. Using a new method for sampling bladders (Breckle et al. 1990), Freitas and Breckle (1992) investigated *Atriplex prostrata, Atriplex halimus*, and *Halimione portulacoides* from a Portuguese salt marsh. The two perennial plants, *A. halimus* and *H. portulacoides*, accumulated more than 80% of ions in relation to the leaves in the bladders. In the annual *Atriplex prostrata*, the contribution of the bladder hairs to total NaCl content was somewhat lower.

A special case was put forward by Waisel (1991), who argued that the salt glands of *Tamarix aphylla* not only serve for the removal of excess salt, but also trap CO_2 in the alkaline fluid recreted by the salt glands as $CaCO_3$, which increases the CO_2 concentration around the twigs in the early morning hours and improves photosynthesis. By labeling experiments with ^{14}C, the author was able to show that $^{14}CO_2$ released from the outside chalk crust was assimilated by the twigs; however, the quantitative importance of this process awaits further investigation.

Much further investigation will also be needed to improve our understanding of the manner in which salt glands function. There is little to be added to the excellent review on salt glands given by Thomson et al. (1988). Adenosine triphosphatase localization in the stalk and recreting cells of *Avicennia marina* confirmed the involvement of active transport processes (Drennan et al. 1992), whereas studies on the *Limonium* salt gland brought further evidence that ion movement to the glands is a combination of symplastic and apoplastic routes (Shachar-Hill 1990).

Considering the huge differences in recretion rates per leaf area (Table 1), it is tempting to ask if they have something to do with the structure of the salt glands, e.g., that more complicated gland complexes lead to higher recretion rates. This is probably not the case. *Diplachne fusca*, with its two-celled gland type of grasses, reaches at high temperatures recretion rates similar to *Aegialitis annulata*, which has a multicellular gland with 40 secretory cells (Thompson et al. 1988; Table 1). However, as mentioned above, proper comparison is very difficult in face of the multitude of factors influencing salt recretion (leaf age, light, temperature, root medium, etc.).

Irrespective of the recretion rate per leaf area, in all the examples listed here the amount of recreted salt was a very high proportion of the internal salt content. This raises the question of alternative mechanisms in those halophytes which do not have salt glands, which is the case in the majority of mangrove species.

β) Retranslocation

Retranslocation of Na^+ and Cl^- via the phloem from the leaves to the root, and extrusion into the medium were demonstrated for *Trifolium alexandrium* (Winter 1982) using radioactive label Na^+ and Cl^- applied to leaves. In a split-root system where $^{22}Na^+$ was applied to one half of the root system of *Pappophorum pappiferum* and *Pappophorum philippianum*, again efflux of $^{22}Na^+$ into the non-labeled solution could be demonstrated (Taleisnik 1989). After 48 h, 35% of the Na^+ accumulated in the shoots was retranslocated and released to the root medium in *P. pappiferum* treated with 50 mM NaCl. In *P. philippianum* the percentage of recirculated Na^+ was much lower, with only 2.8% in the same salt treatment (Taleisnik 1989).

For nonsalt-recreting mangroves, retranslocation was postulated as a means for salt regulation by Clough et al. (1982). The long life spans (up to 18 months) and the rather constant ion concentrations of mangroves leaves imply the presence of a further salt-controlling component. To test the possibility of retranslocation of ions via the phloem, twigs of *Rhizophora mangle* were bark-ringed, and the ion content of leaves was compared to that of nontreated twigs of the same insertion after 17 days (Popp et al. 1993). The outcome of this experiment was equivocal, with one plant showing marked enhanced concentrations of Na^+, K^+, and Cl^- in the leaves of bark-ringed twigs, one plant exhibiting a slight increase, and one plant showing no effect compared to the controls. Since it is not known to what extent phloem-to-xylem exchange might take place in mangroves, the quantitative importance of salt retranslocation is difficult to assess.

c) Succulence

Dilution of the accumulated salt by increasing water content per leaf area or shoot volume is a well-known phenomenon in herbaceous halophytes, as well as in mangroves (cf. Albert 1982). The essential role of increasing succulence for the salt balance of nonsecreting mangroves species became evident from an investigation on *Rhizophora mangle, Laguncularia racemosa*, and *Conocarpus erectus* in comparison to *Avicennia germinans*, (Popp et al. 1993). Cl^- content per leaf area was highly correlated to water content per leaf area in the three nonsecreting species, while no correlation at all could be stated for *A. germinans*. In contrast to this clear evidence from field-grown trees, no relationship could be observed between different salt treatments (full seawater, 1/10 of seawater) and degree of succulence in leaves of *L. racemosa* of plants grown in the glasshouse (Polania 1990). However, also *Aster tripolium* grown under glasshouse conditions did not show increased succulence at high salinity, which was ascribed to the low light intensity used (Shennan et al. 1987). While there is an ongoing discussion as to which of the ions (Na^+ or Cl^-) is more effective in inducing succulence, it is clear from the above-mentioned results that it is not the ion concentration alone, but also additional factors leading to the increase in water content. Again, a comparison of two ecotypes showed the importance of this trait in salt resistance of certain species. Studying two growth forms of *Suaeda calceoliformis*, Youngman and Heckathorn (1992) observed that the prostrate form occurring in the more saline areas of the marsh achieved a higher tissue water content than the erect form from less saline sites. Lower tissue elasticity, along with greater hydraulic conductance, enabled the prostrate ecotype to adapt to increasing salinity by higher water storage.

This agrees well with a study on *Rhizophora mangle* and two mangal associates. *Conocarpus erectus* and *Coccoloba uvifera*, where it was also demonstrated that the most salt-resistant species, *Rh. mangle*, had much more rigid cell walls than the other two species (Rada et al. 1989).

Evidently, more studies are needed to prove the connections between turgor relations and succulence.

4. Components of Salt Tolerance

a) Transport and Compartmentation of Ions

The phenomena described in Sections 3.a and 3.b on the whole plant level have their explanation in ion transport processes at the cellular level. A comprehensive review on plant cell membranes and salinity was given by Lüttge (1993) only recently. Thus, only a few main points will be covered here.

α) Membrane Composition

Although there are contradicting results on the lipid composition of various halophytes under saline and nonsaline conditions (cf. Lüttge 1993), it goes without questioning that the degree of saturation and the presence of planar stereols influence the passive permeability of a membrane. Such changes in favor of a decreased permeability of the membrane were observed (as mentioned above) in *Kosteletzkya virginica* at different salinity levels (Blits and Gallagher 1990). Leach et al. (1990) determined membrane fluidity of tonoplasts and plasmalemma of *Suaeda maritima* by electron spin resonance. Fluidity was lower in the tonoplast, which was consistent with the high degree of saturation of the fatty acid chains of this membrane. The resulting low salt leakage from the vacuole can be taken as a principal adaptation of salt-accumulating halophytes.

β) Ion Channels

For salt accumulators like *Suaeda maritima*, it is generally accepted that Na^+ and Cl^- are compartmentalized in the vacuole (500–600 mM) and cytosolic concentrations of NaCl do not exceed 150 mM (Flowers et al. 1986). In face of recent reports on slow vacuolar ion channels, which conduct both Na^+ and Cl^- in non-halophytes (Hedrich and Schröder 1989), the question arises as to how halophytes keep such NaCl concentration gradients across the tonoplast.

Using the patch clamp technique, Schachtman et al. (1991b) tested root tonoplast from *Triticum aestivum* (salt-resistant) and *Triticum turgidum* (salt-sensitive), and could detect no difference in K^+/Na^+ selectivity of the cation channel. Investigating the tonoplast ion channel of *Sueada maritima*, again no special adaptation was found (Maathuis et al. 1992). It was concluded by these authors that NaCl leakage from the vacuole is prevented by a very low opening frequency of these channels. This agrees with findings by Maathuis and Prins (1990), who compared root tonoplast channels from *Plantago media* and *Plantago maritima*. Although these two species differ in ion uptake and distribution, ion selectivity single-channel conductance and gating characteristics of the tonoplast channels was quite similar. However, opening probability of the channels, in both species, was strongly reduced when the plants were grown on NaCl (25–100 mM).

γ) Na^+/H^+ Antiporters

Na^+/H^+-exchange carriers or antiporters are a prerequisite to keep the Na^+-level in the cytoplasm low. Being located either at the plasmalemma and/or at the tonoplast they extrude Na^+ out from cytoplasma to the outside or to the vacuole driven by the electrochemical gradient established by H^+-ATPases.

Looking for Na^+/H^+ antiport activity, the group working with the two different *Plantago* species observed a marked difference (Staal et al. 1991): Na^+/H^+ exchange carrier was only present in the tonoplast of *P. maritima*, which was grown with addition of 50 mM NaCl. There was no Na^+/H^+ antiport activity in the salt-free controls of *P. maritima* and in none of the *P. media* plants. While these clearcut results were obtained with tonoplast vesicles, other investigations, using for instance intact tissue and following the net fluxes of $^{22}Na^+$ at varying pH gradients, arrive at quite different findings. Using the latter method, Mennen et al. (1990) concluded that no Na^+/H^+ antiport was present in *Gossypium hirsutum*, whereas Hassidim et al. (1990), working with root membrane vesicles, obtained evidence for the presence of the antiporters in plasmalemma and tonoplast.

Another line of evidence for the activity of the Na^+/H^+ antiporter comes from in vivo nuclear magnetic resonance studies (Spickett et al. 1993; Martinez and Läuchli 1993).

Exposing maize roots to salt concentrations greater than 200 mM resulted in a cytoplasmic alkalinization of 0.1 to 0.2 pH units and an increase of vacuolar pH of 0.6 units. In root tips of *Spartina anglica*, which accumulated sodium more slowly than did maize, no change in cytoplasmic pH was observed, and vacuolar pH increased by only 0.3 pH units (Spickett et al. 1993).

Such an alkalinization of vacuolar pH also occurred in barley roots after salt treatment, but was reduced in the presence of high Ca^{2+} (6 mM; Martinez and Läuchli 1993). This fits very well to former findings on the ameliorating effect of Ca^{2+} during salinity stress (cf. Rengel 1992).

δ) H^+-ATPases

Neither for the H^+-ATPase of the plasmalemma nor for the H^+-ATPase of the tonoplast have pronounced differences so far been reported between halophytes and nonhalophytes (cf. Lüttge 1993). However, quantitative and qualitative changes of the tonoplast H^+-ATPase were observed in leaf cells of *Mesembryanthemum crystallinum* during adaptation to salinity.

Since NaCl stress induces in *M. crystallinum* also a shift from C_3-photosynthesis to CAM, detailed studies were needed to demonstrate that the increased amounts of tonoplast H^+-ATPase occurred in connection with salt sequestration to the vacuole, whereas a new subunit was related to CAM activity (Lüttge 1993).

A study by Koyro et al. (1993) points also in the direction that quantitative changes take place in ATPase activities after salt treatment. Investigations of membrane fine structures as well as biochemical tests demonstrated enhanced ATPase activities in rhizodermal membranes of *Sorghum bicolor* and *Spartina townsendii* after exposure to 40 mol m^{-3} NaCl.

b) Compatible Solutes

Considering the difference in Na^+ and Cl^- concentrations between cytoplasm and vacuole, as given above for *Suaeda maritima*, raises the question of how intracellular osmotic adjustment is achieved between these compartments. It is now generally accepted that organic, so-called compatible solutes serve this purpose (cf. Stewart and Ahmad 1983), comprising several groups of substances like amino acids (proline), quaternary ammonium compounds (glycine betaine), acyclic polyols (sorbitol), and cyclic polyols (pinitol).

Although the beneficial effect of the storage of such compounds has been recently questioned (Munns 1993), there is now interesting evidence from work with transgenic tobacco that at least mannitol functions in stress protection (Tarczynski et al. 1993).

Tobacco normally does not accumulate mannitol. By introduction of a bacterial gene that encodes mannitol-1-phosphate dehydrogenase mannitol-containng transformants were produced which stored up to 100 mM assuming a cytosolic location. Exposing wild-type and transgenic plants to 250 mM NaCl resulted in marked differences in growth after 30 days. Only the transformants were able to produce new roots and flowers, and fresh weight increase compared to controls without NaCl was much less affected. Thus this is unequivocal evidence that mannitol increased the ability of these plants to endure high salinity.

In this context, it is interesting to mention that quite a considerable number of mangrove species accumulate mannitol (Popp et al. 1990), whereas for herbaceous halophytes, so far only *Plantago* species are known to store sorbitol.

Also cyclic polyols (cyclitols) are much more frequent in mangroves than in herbaceous halophytes (Popp et al. 1993). Nevertheless, work with *Mesembryan-*

themum crystallinum, a facultative annual halophyte, has greatly promoted our understanding of the regulation of the synthesis of these compounds (Thomas et al. 1992; Vernon et al. 1993). These authors were able to show that the gene which encodes a myo-inositol-O-methyl-transferase (catalyzing the first step in the synthesis of pinitol from myo-inositol to ononitol) is induced by salt treatment (400 mM NaCl).

For the same plant, we also have evidence that pinitol is, in fact, located in the cytoplasm. According to Paul and Cockburn (1989), cytosol and chloroplast contained all the pinitol, while none was detected in the vacuole. This seems feasible for a plant which on a tissue-water basis contains overall 10 mol m^{-3} pinitol. For mangroves, where values of more than 200 mol m^{-3} can be found (Popp et al. 1993), the distribution in the cell compartments might be different. However, compartmental analysis is still a tricky business, and not easy to apply to the xeromorphic mangrove leaves.

5. Outlook

In 1983 Stewart and Ahmad ended their review on adaptation to salinity with the statement that it "is necessary to elucidate the molecular mechanisms underlying ... the regulation of salt accumulation, the compartmentation of salt and intracellular osmotic adjustment". Eleven years later, we see the first results in this direction. They look very promising, and will certainly also have a positive effect on breeding programs for salt-resistant varieties. From an ecophysiologist's point of view, it should be kept in mind that the halophytes so far studied in detail are a minute part of those which exist around the world (Aronson, cited in Flowers et al. 1986). In this richness of species there might still be some features hidden which we have not yet recognized as being useful in salinity adaptation.

References

Ahmad I, Wainwright SJ (1977) New Phytol 79:605–612. **Albert R** (1982) In: Kinzel H (ed): Pflanzenökologie und Mineralstoffwechsel. Ulmer, Stuttgart.

Biebl R, Kinzel H (1965) Österr Bot Z 112:56–93. **Blits KC**, Gallagher JL (1990) Plant Cell Environ 13:419–425. **Bradley PM**, Morris JT (1991) J Exp Bot 42:1525–1532. **Breckle SW**, Freitas H, Reimann C (1990) Plant Cell Environ 13:871–873.

Cheeseman JM (1988) Plant Physiol 87:547–550. **Clough BF**, Andrews TJ, Cowan IR (1982) In: Clough BF, Mangrove ecosystems in Australia. Structure, function and management. Australian National University Press, Canberra.

Drennan PM, Berjak P, Pammenter NW (1992) S Afr J Bot 58:486–490.

Flowers TJ, Hajibagheri MA, Clipson NJW (1986) Rev Biol 61:313–337. **Flowers TJ**, Flowers SA, Hajibagheri MA, Yeo AR (1990) New Phytol 114:675–684. **Freitas H**, Breckle SW (1992) Flora 187:283–297.

Glenn EP, Watson MC, O'Leary JW, Axelson RD (1992) Plant Cell Environ 15:711–718. **Gorham J** (1987) Plant Cell Environ 10:191–196.

Hassidim M, Braun Y, Lerner HR, Reinhold L (1990) Plant Physiol 94:1795–1801. **Hedrich R**, Schröder IJ (1989) Annu Rev Plant Physiol 40:519–569.

Ish-Shalom-Gordon N, Dubinsky Z (1993) Pac Sci 47:51–58.

Jennings DH (1976) Biol Rev 51:453–486. **Jones HG** (1983) Plants and microclimate. Cambridge University Press, Cambridge.

Koyro HW, Stelzer R, Huchzermeyer B (1993) Bot Acta 106:110–119.

Larcher W (1983) Physiological plant ecology. Springer, Berlin Heidelberg New York. **Leach RP**, Wheealer KP, Flowers TJ, Yeo AR (1990) J Exp Bot 41:1089–1094. **Levitt J** (1980) Responses of plants to environmental stress. Academic Press, New York. **Liphschitz N**, Waisel Y (1982) In: Sen DW, Rajpurohit KS (eds) Contributions to the ecology of halophytes. W Junk, The Hague. **Lüttge U** (1993) R Bras Fisiol Veg 5: 217–224.

Maathuis FJM, Prins HBA (1990) Plant Physiol 92:23–28. **Maathuis FJM**, Flowers TJ, Yeo AR (1992) J Exp Bot 43:1219–1223. **Marcum KB**, Murdoch CL (1990) Ann Bot 66:1–7. **Marcum KB**, Murdoch CL (1992) New Phytol 120:281–288. **Martinez U**, Läuchli A (1993) Planta 190:519–524. **Mennen H**, Jacoby B, Marschner H (1990) J Plant Physiol 137:180–183. **Morris C** (1992) Academic Press dictionary of science and technology. Academic Press, San Diego. **Munns R** (1993) Plant Cell Environ 16:15–24. **Munns R**, Greenway H, Kirst Go (1983) In: Lange OL, Nobel PS, Osmond CB, Ziegler H (eds) Physiological plant ecology III. Encyclopedia of plant physiology NS. Springer, Berlin Heidelberg New York.

Osmond CB (1980) In: Osmond CB, Bjorkman O, Anderson DJ (eds) Physiological processes in plant ecology. Towards a synthesis with Atriplex. Ecological Studies 36. Springer, Berlin Heidelberg New York.

Paul MJ, Cockburn W (1989) J Exp Bot 40:1093–1098. **Polania J** (1990) Anatomische und physiologische Anpassungen von Mangroven. PhD Thesis, University Vienna. **Popp M**, Polania J (1989) Ann Sci For 46:842s–844s. **Popp M**, Feldl C, Polania J, Aswathappa N (1990) In: Werner D, Müller P (eds) Fast growing trees and nitrogen fixing trees. Fischer, Stuttgart. **Popp M**, Polania J, Weiper M (1993) In: Lieth H, Al Masoom A, Towards the rational use of high salinity tolerant plants, vol I. Kluwer, Dordrecht.

Rada F, Goldstein G, Orozco A, Montilla M, Zabala O, Azocar A (1989) Aust J Plant Physiol 16:477–486. **Reimann C** (1992) J Exp Bot 43:503–510. **Reimann C**, Breckle SW (1993) Plant Cell Environ 16:323–328. **Rengel Z** (1992) Plant Cell Environ 15:625–632. **Rozema J** (1976) Flora 165:197–209. **Rozema J**, Buys E, Otte ML, Broekman RA, Ernst WHO (1991) Z Pflanzenernaehr Bodenkd 154:307–313.

Schachtman DP, Munns R, Whitecross MI (1991a) Crop Sci 31:992–997. **Schachtman DP**, Tyerman SD, Terry BR (1991b) Plant Physiol 97:598–605. **Shachar-Hill B** (1990) Proc R Soc Lond B 241:159–163. **Shennan C**, Hunt R, Macrobbie EAC (1987) Plant Cell Environ 10:59–65. **Spickett CM**, Smirnoff N, Ratcliffe RG (1993) Plant Physiol 102:629–638. **Staal M**, Maathuis FJM, Elzenga JTM, Overbeek JHM, Prins HBA (1991) Physiol Plant 82:179–184. **Stewart GR**, Ahmad I (1983) In: Robbs DA, Pierpoint WS, Metals and micronutrients. Academic Press, London.

Taleisnik EL (1989) Ann Bot 63:221–228. **Tarczynski MC**, Jensen RG, Bohnert HJ (1993) Science 259:508–510. **Thomas JC**, McElwain EF, Bohnert HJ (1992) Plant Physiol 100:416–423. **Thomson WW**, Faraday CD, Oross JW (1988) In: Baker DA, Hall JL (eds) Solute transport in plant cells and tissue. Longman, New York.

Ungar IA (1991) Ecophysiology of vascular halophytes. CRC Press, Boca Raton.

Vernon DM, Ostrem JA, Bohnert HJ (1993) Plant Cell Environ 16:437–444.

Waisel Y (1991) Physiol Plant 83:506–510. **Walter H**, Steiner M (1936) Z Bot 30:65–193. **Wang XY**, Suhayda CG, Redmann RE (1992) Can J Bot 70:1123–1130. **Warming E** (1909) Oecology of plants. Clarendon Press, Oxford. **Warwick NWM**, Halloran GM (1991) New Phytol 119:161–168. **Werner A**, Stelzer R (1990) Plant Cell Environ 13: 243–255. **Winter E** (1982) Aust J Plant Physiol 9:227–237.

Yeo AR (1983) Physiol Plant 58:214–222. **Youngman AL**, Heckathorn SA (1992) Funct Ecol 6:686–692.

Fr. Prof. Dr. M. Popp
Universität Münster
Institut f. Angewandte Botanik
Hindenburgplatz 55
48143 Münster, Germany

Present address:
Universität Wien
Institut für Pfhanzenphysiologie
Althanstraße 14
A-1090 Wien, Austria

F. Special Topics

I. Mycorrhizae: Ectomycorrhiza and Ectendomycorrhiza

By Reinhard Agerer

As new class of "mycorrhizal" interactions the "lycoperdioid mycothallus interaction" (Schmid and Oberwinkler 1993) is proposed and studied in detail. It is characterized by aseptate, intracellularly regularly coiled, partly swollen hyphae, forming vesicles but no arbuscles and is formed within the gametophyte of *Lycopodium clavatum*. Intercellular hyphae may branch, resembling the hyphal behaviour of a Hartig net.

1. Ectomycorrhiza

a) Symbiotic Organisms and Morphology/Anatomy of the Symbiotic Organs

Egli et al. (1993) pointed out that microscopical characterization of ectomycorrhizae can reveal more types than the exclusive application of macroscopic features. Moreover, they stated that, in addition, methods of biochemistry and molecular biology are necessary to differentiate ectomycorrhizae more precisely and convincingly.

α) Detailed Descriptions of Ectomycorrhizae, Including Keys

Two additional deliveries of the *Colour Atlas of Ectomycorrhizae* have been published, altogether now seven deliveries are available; roughly 190 species of ectomycorrhizae can now be determined (Agerer 1987–1993). Waller and Agerer (1993) described ectomycorrhizae of *Tricholoma acerbum* on *Fagus sylvatica* and of *Dermocybe cinnamomeolutea* on *Salix* and provided keys for all ectomycorrhizae described to date of either genera. Haug and Pritsch (1992) provided descriptions and keys for several identified and binomially named, unidentified ectomycorrhizae on *Picea abies*.

Progress in Botany, Vol. 56

β) Comprehensive Descriptions of Selected Ectomycorrhizae

Ectomycorrhizae of *Laccaria amethystina* and of *Russula ochroleuca* both on *Betula pendula* are comprehensively described by Cuvelier (1991), and compared to already existing descriptions of ectomycorrhizae of these fungi which, however, were formed on different tree genera. Tree-dependent and fungus-dependent characteristics could be shown. Pillukat and Agerer (1992) compared ectomycorrhizae of *Russula ochroleuca* on eight different tree genera and found that characteristics caused by the fungus are independent of the trees; morphology and features of the Hartig net were variable with respect to hosts. *Cortinarius odorifer* ectomycorrhizae were described from 42 sites on *Picea abies, Abies alba,* and *Pinus sylvestris*. The fungus-specific features of the ectomycorrhizae were revealed as being independent of site and host (Egli 1992). Agerer et al. (1993) described the ectomycorrhizae of *Gyrodon lividus* on *Alnus incana,* Waller et al. (1993a) those of *Scleroderma citrinum* on *Pinus sylvestris* and *Betula pendula,* Agerer (1992b) of *Boletopsis leucomelaena* on *Picea abies*, Agerer (1992c) of *Phellodon niger* on *Picea abies*. Ectomycorrhizae of *Amanita muscaria, Laccaria laccata, Rhizopogon luteolus, Scleroderma citrinum, Suillus brevipes*, and *Thelephora terrestris* all on *Pinus patula* were characterized by Mohan et al. (1993a, b, c). Ectomycorrhizae of *Pisolithus tinctorius* on *Picea abies* are characterized by Weiss (1992), those of *Hydnellum peckii* on *Picea abies* by Agerer (1993a).

Brunner et al. (1992) compared synthesized ectomycorrhizae of three different fungi (*Tricholoma vaccinum, Scleroderma citrinum, Boletus luridus*) with naturally occurring ectomycorrhizal types of a spruce site, and came to the conclusion that identification of the fungal partners of natural symbioses by means of in vitro-synthesis with potential ectomycorrhizal fungi was not possible. Brunner et al. (1991) compared natural and synthesized ectomycorrhizae of *Hebeloma crustuliniforme* on *Picea abies* and found only differences very likely due to developmental stages. Brunner (1993) characterized the ectomycorrhizae of *Hebeloma crustuliniforme* on *Picea abies*, naturally grown as well as synthesized including three different methods. Only a few characters, such as thickness of the mantle and the development of the Hartig net, were variable due to the synthesis method; rhizomorphs could be found only in naturally grown ectomycorrhizae.

Massicotte et al. (1992) described in detail an unidentified tuberculate ectomycorrhiza on *Pseudotsuga menziesii*, and found associated bacteria and crystals apparently containing calcium-oxalate.

γ) Unidentified Ectomycorrhizae Named Binomially

Waller et al. (1993b) described *Piceirhiza oleiferans*. Agerer (1993b) compiled all binomially named ectomycorrhizae published in 1991. Haug and Pritsch (1992) described from *Picea abies* many new species of ectomycorrhizae.

δ) Verification of Ectomycorrhizal Nature of Fungi, Including Short Descriptions

Spores of *Helvella lacunosa, Russula brevipes*, and *Lycoperdon perlatum* were successfully used to establish ectomycorrhizae on *Pinus patula* and *P. radiata* (Martines-Amores and Quintos 1991). *Tomentella crinalis* ectomycorrhizae were synthesized on *Pinus sylvestris* seedlings using mycelium obtained by germination of spores (Kõljalg

1992). Bougher et al. (1993) described a new *Hydnangium* species (*H. sublamellatum*) and included in the description a short characterization of synthesized ectomycorrhizae.

ε) Unusual "Ectomycorrhizae"

McGee and Furby (1992) described a mycorrhiza on *Ceratopetalum apetalum* which formed only a mantle consisting of a single hyphal layer; a Hartig net was lacking. Buscot (1992a) reports on ectomycorrhizae of *Morchella esculenta* with *Picea abies*, with infrequent mantle hyphae, and with intercellular hyphae not forming a typical Hartig net, root cells soon became necrotic. Fortas and Chevalier (1992) synthesized ectomycorrhizae in phosphorus-rich substrates between *Terfezia* and *Tirmania* species and *Helianthemum guttatum* which lacked a sheath, whereas in phosphorus-deficient substrates, ectendomycorrhizae, also without a sheath, were formed. Agerer and Waller (1993) described the mycorrhizae of *Entoloma saepium* on *Rosa* and *Prunus* with a very thick hyphal mantle, with intracellular hyphae, and a zone where root cells were digested; a few hyphae, however, seem to grow intercellularly.

ζ) Strain Variability of Fungi Regarding Ectomycorrhiza Formation

Jacobson and Miller (1992) compared allopatric populations of *Suillus granulatus* with respect to functional compatibility (different items) with three different pine species. The isolates from the five-needle pine *Pinus strobus* were most compatible and most specific to this species in comparison to the five-needle pine *P. wallichiana* and the two-needle pine *Pinus densiflora. S. granulatus* isolations from the two latter species were not host specific. Thoen et al. (1990) demonstrated that the ability of *Pisolithus tinctorius* to form ectomycorrhizae with *Casuarina* and *Allocasuarina* varied isolate-specifically. Even ectendomycorrhizal development could be shown with two species of *Allocasuarina*. A wide range of intraspecific and intrastrain-variation of different physiological properties with suggested influences on mycorrhiza formation and efficiency are demonstrated by Debaud et al. (1990) for *Hebeloma cylindrosporum*. It is indicated by the authors that chromosomal genetics can be used to improve the fungal strains. Lamhamedi et al. (1992) found differences in reconstituted dikaryons with respect to water stress of ectomycorrhizal seedlings. Secreted substances of *Eucalyptus globulus* roots are suggested to influence compatibility/incompatibility of different *Pisolithus tinctorius* strains (Lei et al. 1990a).

η) Verification of Ectomycorrhizal Nature of Some Selected Plants

Blaschke (1991) verified the ectomycorrhizal habit of *Dryas octopetala*, *Helianthemum nummularium* ssp. *grandiflorum* and *Polygonum viviparum* on a rendzina soil in the Alps; *P. viviparum* and *H. nummularium* were colonized also by VAM fungi. In Australia, new ectomycorrhiza forming species were shown (Bellgard 1991) in the genera *Platysace* (Apiaceae), *Amperaea* (Euphorbiaceae), *Dillwynia, Pultanaea* (Fabaceae), *Dampiera* (Goodeniaceae), *Acacia* (Mimosaceae), and *Angophora, Baeckea, Eucalyptus, Kunzea, Leptospermum* (Myrtaceae). In *Dillwynia, Pultanaea, Dampiera* and *Acacia*, VA-mycorrhizae were present as well. On herbaceous and shrubby species of a Jarrah forest, Brundrett and Abbott (1991) detected dual colonization with VA- and ectomycorrhizal fungi on *Allocasuarina* (Casuarinaceae), *Gompholobium, Mirbelia, Oxylobium* (Fabaceae), *Agonis, Eucalyptus, Melaleuca, Pericalymma* (Myrtaceae), and *Cryptandra, Trymalium* (Rhamnaceae); no species with exclusively ectomycorrhizal colonization was found. Koske et al. (1992), studying Hawaiian flora, could prove ectomycorrhizae only in *Coccoloba uvifera* (Polygonaceae). Väre et al. (1992) found ectomycorrhizae on *Pedicularis dasyantha* in middle-northern Arctic zone. Dighton and Coleman (1992) described ectomycorrhizae on *Rhododendron maximum.*

θ) Influence of Tree Clones on Ectomycorrhiza Formation

Dixon and Johnson (1992) inoculated *Quercus rubra* seedlings of different strain families with *Suillus luteus* and obtained significantly deviating amounts of ectomycorrhizal colonization due to seedling family in the first year.

b) Ontogeny and Ultrastructure

A very detailed study on the ontogeny of synthesized *Hebeloma crustuliniforme* ectomycorrhizae on *Picea abies*, using TEM and SEM, is published by Brunner and Scheidegger (1992). Freeze-fractured ectomycorrhizae of *Hebeloma crustuliniforme* and *Tricholoma vaccinum*, both on *Picea abies* are described by Scheidegger and Brunner (1993); the authors focus especially on the Hartig net. The fungal penetration between root cells took place only after an already differentiated mantle was formed around the root. Tam and Griffiths (1993a) investigated the ontogeny of *Pisolithus tinctorius* ectomycorrhizae on *Quercus myrsinaefolia.* Downes et al. (1992) studied aging ectomycorrhizae of *Paxillus involutus* and *Tylospora fibrillosa* on *Picea sitchensis* in detail, and calculated that most of the mycorrhizae became nonfunctional after 85 days. Paris et al. (1993) observed that the hyphae of *Cenococcum geophilum* become simplified when changing from saprophytic to symbiotic stage. The cell wall alteration is suggested to be due to a decrease in the amount of melanin or of melanin plus cystine-containing proteins. Pargney (1990) did not detect chitin in the matrix material between the hyphae and the plant cell wall in the Hartig net zone of *Tuber melanosporum* ectomycorrhizae on *Corylus avellana.* Dexheimer and Pargney

(1991, 1992) compared the ultrastructure of the interfaces of ecto-, ectendo-, and endomycorrhizae. Pargney and Gourp (1991) focused on ultrastructural features of the interface of *Pinus pinaster* ectomycorrhizae of *Hebeloma cylindrosporum* and *Paxillus involutus*. Martin and Hilbert (1991) review morphological, biochemical and molecular changes during ectomycorrhiza development and include also genetical aspects. Timonen et al. (1993) were able to identify cytoskeletal components in pine ectomycorrhizae. Orlovich and Ashford (1993) studied mycelium of *Pisolithus tinctorius* and obtained evidence that the frequently reported polyphosphate granules are an artifact of specimen preparation.

With the aid of ultrastructural research of septal pores of ectomycorrhiza-forming ascomycetes Berndt and Oberwinkler (1992) suggested that members of Tuberales or Pezizales formed these ectomycorrhizae.

c) Protein Pattern, Pigments

Dupre and Chevalier (1991) found a specific protein pattern of mycelium and ectomycorrhizae of *Tuber melanosporum* on *Corylus avellana* and other *Tuber* species; the pattern of the mycelium was richer. The authors regard their results as a first step in designing a new rapid identification method for *Tuber* species.

d) Nucleic Acid Researches

The amount of ribosomal DNA variation in *Cenococcum geophilum* indicates either an extremely heterogeneous species or a fungal complex representing a broader taxonomic rank than presently considered (Lobuglio et al. 1991).

e) Physiology

α) Germination of Spores

It was possible to germinate *Tricholoma robustum* spores by application of gluconic acid (Iwase 1992). Phosphate and mineral nutrient concentrations used for raising *Pinus radiata* seedlings depressed both spore germination and germ tube growth of *Rhizopogon luteolus* (Theodorou 1993).

β) Substances Assumed to be Important for Formation of Ectomycorrhizae

Hormones. The ability of *Hebeloma hiemale* to produce IAA from exogenous tryptophan is negatively influenced by high concentrations of glucose, phosphate, and nitrogen, and it is suggested that the formation of ectomycorrhizae is likewise influenced by these concentrations via IAA production (Gay et al. 1990). Strzelczyk et al. (1992) investigated the effect of pH on plant growth regulators production in mycelia of ectomycorrhizal fungi. Contrary to current theory, Wallander et al. (1992) found a lower IAA content in mycorrhizal roots than in non-mycorrhizal roots, and higher contents during high-nitrogen nutrition. Kraigher et al. (1993) found *Thelephora terrestris* strains which showed different levels of the same cytokinins. The strain which produced higher amounts of cytokinins in pure culture, also influenced positively the cytokinin contents of the needles of *Picea abies* when ectomycorrhizal with this strain in comparison to the other, although the differences were smaller than in pure culture exudates.

Phenolics. Tam and Griffiths (1993b) detected histochemically polyphenols, in some Fagaceae roots and after invasion by compatible ectomycorrhizal fungi, condensed tannins disappeared from cells of the root cap and the epidermal layer but hyphae were prevented from colonizing the cortex presumably due to suberin barriers.

γ) Recognition (cf. also Sect. I.f.α)

Sun and Fries (1992) found that mycelium of several possibly pine-compatible ectomycorrhizal fungi react positively with growth rates on pine-root exudates; but the pine specific *Suillus luteus* did not, although several strains were tested. Secreted substances of *Eucalyptus globulus* roots are suggested to influence compatibility/incompatibility of different *Pisolithus tinctorius* strains (Lei et al. 1990a).

It is stressed that recognition and compatible interaction between *Pisolithus tinctorius* and *Eucalyptus*- and *Pinus*-species, respectively, are indicated by a typical fibrillar bridge (Lapeyrie 1990; Lei et al. 1990b; Gourp and Pargney 1991) and by the appearance of fungal acid phosphatase activity (Lapeyrie 1990; Gourp and Pargney 1991), which can be detected only on and in the fungus (Gourp and Pargney 1991). Giollant et al. (1993) purified a glycine and especially methionine-rich lectin from the spruce-specific *Lactarius deterrimus* fruitbodies, which recognized binding sites on spruce roothairs; a lectin isolated formerly from the *Pinus*-specific *Lactarius deliciosus* did not bind on spruce roots. Lapeyrie and Mendgen (1993) pointed out that the hyphal wall-bound lectins of *Pisolithus tinctorius* are coated by nonreactive compounds which can be removed by protease or laminariase. The lectins are rather stable and do not depend on the age of the colony or the nutrient medium, but they are species- and strain-specifically variable. After inoculation of *Eucalyptus* roots, hyphae of *P. tinctorius* did not change their characteristics dramatically, except that some additional affinity to concanavalin A was observed in hyphal cross sections.

Simoneau et al. (1993) presented data that seven novel polypeptides were detected 4 days after inoculation of *Betula pendula* seedlings with a compatible strain of *Paxillus involutus*, three of them being already present in 2-day-old ectomycorrhizae. Ectomycorrhizins, new synthesized proteins which accumulate during formation of ectomycorrhizae, can be found already in early ontogeny stages, even when hyphae of *Pisolithus tinctorius* did not directly contact the root of *Eucalyptus* seedlings. An incompatible strain did not initiate ectomycorrhizin synthesis. Ectomycorrhizins could be shown as a result of differential protein biosynthesis rather than posttranslational modification of the polypeptides (Hilbert et al. 1991). They could not be detected in *Picea abies* seedlings ectomycorrhizal with *Amanita muscaria* using electrophoresis on a microscale (Guttenberger and Hampp 1992). The authors pointed out that conventional analyses, which use extracts of a large number of root tips, are not suitable for differentiating between, on the one hand, composition of the media and/or different developmental stages and, on the other, symbiosis-specific changes. The earlier reported "ectomycorrhizins" are regarded as either specific for *Eucalyptus-Pisolithus* symbiosis or simply as artifacts.

δ) Enzymes

Acid phosphatase activity with two pH optima (pH 3.5 and 5.0) were found in *Tuber albidum* ectomycorrhizae on *Pinus pinea*. The phosphatase activity was approximately 40% above the nonmycorrhizal controls (Pasqualini et al. 1990, 1992). Kieliszewska-Rokicka (1992) found stimulated activity of acid phosphatase on *Pinus sylvestris* seedlings due to higher levels of N in the soil and it was higher in seedlings ectomycorrhizal with *Paxillus involutus* than in nonmycorrhizal roots; moreover, the pine roots had greater acid phosphatase activity at reduced than at higher P concentrations. No differences could be found in the glutamine synthetase activity between *Pinus sylvestris* seedlings either ectomycorrhizal with *Piloderma croceum* or with *Paxillus involutus*, although differences were obvious in pure culture of these fungi (Sarjala 1993). The results of Rabbani et al. (1992) suggest a repression of fungal aspartate aminotransferase in ectomycorrhizae of *Eucalyptus globulus* and *Pisolithus tinctorius*. Chalot et al. (1991) pointed out a major role of glutamate dehydrogenase and glutamine synthetase in nitrogen assimilation in the *Picea abies-Hebeloma* sp. symbiosis and do not suggest any significant role for glutamate synthetase in the synthesis of glutamine. In addition, a NH_4^+ pool in the host plant is suggested. Salzer and Hager (1993) proved two isoforms of invertase on *Picea abies* cells, an ionically and a tightly wall-bound isoform. Since the activity of the invertase depends on the actual pH (highest activity at pH 4.5) and the fructose concentration in the mycorrhizal interface, it is suggested that the fungus can actively influence the activity of the plant invertase by acidification of the cell wall and by fructose uptake.

ε) Carbon Nutrition of Ectomycorrhizae

Cairney (1992) suggests that rapid acropetal translocation of carbon occurs along part of the rhizomorph length by a turgor-driven mass flow of solution in the apoplast of empty "vessel" hyphae. A substrate pH-dependent carbon translocation into the root system of mycorrhizal *Pinus sylvestris* seedlings could be observed only at the beginning of an experiment; in substrate of pH 3.5 it was faster than in substrate of pH 5.2. Rather quickly the differences disappeared and the supplied carbon amounted to 5% in the mycelium in either soil pH (Erland et al. 1991). Salzer and Hager (1991) obtained evidence that apoplastic invertase activity of the host *Picea abies* is a precondition for the utilization of sucrose by the mycorrhizal fungi *Amanita muscaria* and *Hebeloma crustuliniforme* during the nutritional interaction of the symbiotic partners. Rieger et al. (1992) showed that in ectomycorrhizae of *Amanita muscaria* on *Picea abies* seedlings, a longitudinal distinction of sucrose pools had its lowest levels in the middle parts, which represent areas of most intense symbiotic interactions; the total amount of the sugars sucrose, glucose, fructose, and mannose was about 30% higher in nonmycorrhizal roots compared to mycorrhizae. In *Tylospora fibrillosa* ectomycorrhizae on *Picea sitchensis* translocation of ^{14}C-labelled photosynthate was, in the same root system, reduced in older compared with young mycorrhizae (Cairney and Alexander 1992b). The soluble sugars arabitol, glucose, and mannitol were detected in significantly lower quantities in older mycorrhizae, whilst fructose and trehalose content was similar to that of young mycorrhizae. A few hours after hyphal contact between *Picea abies* roots and colonizing mycelium of *Pisolithus tinctorius*, the pool of soluble carbohydrates of the inoculum (trehalose, arabitol) declined and was almost zero after 2 weeks (Ineichen and Wiemken 1992).

ζ) Growth Responses of Plants

Positive growth responses of different tree species by several ectomycorrhizal fungi was proven repeatedly (Browning and Whitney 1992b; Chen and Pei 1992; Gong et al. 1992; Jones et al. 1991; Lee 1992; MacFall and Slack 1991; Martinez-Amores and Quintos 1991; Meng et al. 1991; Pirazzi 1992; Zhao et al. 1992; Zhou et al. 1991) in containers and/or greenhouses and/or in the field, but it also varied with the applied fungal species (Lee 1992). After outplanting, seedling survival was elevated (MacFall and Slack 1991); growth promotion, however, was left unchanged (MacFall and Slack 1991) or was declined after a series of years (Lee 1992). Initial growth depressions of seedlings are also known (Browning and Whitney 1992b; Svenson et al. 1991). Colpaert et al. (1992) demonstrated an important negative correlation between the extent of the fungal development (abundant versus infrequent external hyphae) and the growth of the host plants. In most cases, this is attributed to an energy drain by the fungus. A considerable amount of N is retained by the fungus, resulting in slowing down of growth. Dosskey et al. (1992) found that seedlings of *Pseudotsuga menziesii* ectomycorrhizal with *Rhizopogon vinicolor* had the same percentual distribution of biomass between roots, stems, and needles, but only 56% of the total length of roots compared to nonmycorrhizal seedlings; no significant alteration was caused by *Laccaria laccata*. Wu et al. (1993) found in outplanted nursery-grown seedlings of *Pinus resinosa* after 3–4-year growth a negative correlation between numbers of ectomycorrhizae per gram of dry root to changes in all seedling characteristics (shoot weight, root weight, total height, basal diameter, and shoot : root ratio), except of leaf water potential.

η) Phosphate Nutrition

Cairney (1992) suggests that basipetal nutrient translocation in ectomycorrhizal rhizomorphs seems likely to be confined to the symplast of living hyphae, driven down concentration gradients towards sinks by mixing resulting from cytoplasmic streaming. Early stages of mycorrhizal infections of *Thelephora terrestris* are particularly effective in supplying P to willow, and the rapid initial uptake of P was due to higher P inflow rates in mycorrhizal plants rather than to nonmycorrhizal root systems; it is suggested that the highest benefit (phosphate uptake) to C costs (carbohydrate used by the fungus) is in this association at the earliest stages of growth and infection (Jones et al. 1991). The authors stress the importance of investigating benefit/cost relations at different times during plant development. Cairney and Alexander (1992a) also obtained evidence that P uptake of older ectomycorrhizae of *Tylospora fibrillosa* on *Picea sitchensis* was reduced compared to younger ones. Rousseau et al. (1992) found a positive correlation between fungal biomass of mycorrhizae of *Pisolithus tinctorius* on *Pinus taeda* and total seedling P content. At the same time, they showed a close correlation between fungal dry weight and glucosamine content. Phosphate starvation of *Pinus sylvestris* seedlings ectomycorrhizal with different fungi increased production of extramatrical mycelium (Wallander and Nylund 1992). Browning and Whitney (1992a) found evidence that inoculation of *Picea mariana* seedlings with *Laccaria bicolor* will produce larger seedlings, but only where low P fertilizer is applied, preferably three times weekly; *Pinus banksiana* may also benefit from artificial inoculation, and especially in a higher fertility growing medium. It is evidenced that *Hebeloma* ectomycorrhizae on *Pinus resinosa* strongly mediate the supply of phosphate to the tree through the production of polyP, even at growth-limiting levels of soil P, and regulates compartmentalization of P in the mycorrhizal roots (Macfall et al. 1992). The results of Pacheco and Cambraia (1992) obtained on *Pinus caribaea* seedlings ectomycorrhizal with *Pisolithus tinctorius* suggest that increased P uptake by mycorrhizal plants was due to an increase in the number of absorption sites per unit of weight of root and mainly to an increase in the affinity of these sites to phosphorus. Looking for hints whether ectomycorrhizal fungi can change the values of $^{15/14}N$ or $^{13/12}C$ in the plants, Handley et al. (1993) found no fractioning with respect to these isotopes in *Eucalyptus* seedlings ectomycorrhizal with *Hydnangium carneum*.

θ) Nitrogen Nutrition (cf. also Sect. I.f.ε)

Wallander and Nylund (1991) presented data that the development of *Laccaria bicolor* ectomycorrhizae on *Pinus sylvestris* reduced sugar concentrations in both roots and shoots. Differences between ammonium and nitrate in reducing mycorrhizal development (ammonium strongly reduced ergosterol contents as a measure of mycorrhization) was due to differences in uptake alone. Finlay et al. (1992) clearly showed that *Pinus contorta* seedlings ectomycorrhizae with different fungi can differ in growth response due to access to bovine serum albumin for nitrogen nutrition. Plassard et al. (1991) review the role of nitrate, ammonium, ion-balance, and carbohydrate in assimilation of nitrogen by ectomycorrhizae. *Pseudotsuga menziesii* seedlings ectomycorrhizal with *Laccaria bicolor* and grown in hydroculture evidenced that, in general, very low and limiting external nutrient conditions resulted in rapid and good ectomycorrhiza formation, with the mycorrhizal plants having the same or a lower fresh weight than those of the nonmycorrhizal plants. While under higher nutrient conditions mycorrhiza formation was slower and less abundant, and the mycorrhizal plants had a larger total fresh weight than nonmycorrhizal plants (Kamminga-Van Wijk et al. 1992). In a translocation experiment between, on the one hand, *Alnus glutinosa* seedlings nodulated with *Frankia* and ectomycorrhizal with *Paxillus involutus* and, on the other, *Pinus contorta* ectomycorrhizal with *P. involutus*, Arnebrant et al. (1993) could show that between 5 and 15% of the recovered ^{15}N was found in the pine seedlings. Newman (1991) concluded that there is a tendency for the ratio of N : P to be higher in ectomycorrhizal plant species compared to VA-mycorrhizal species.

ι) Micronutrients

Roots of *Pinus elliotii* mycorrhizal with *Pisolithus tinctorius* were less able to use iron of the streptomycete siderophore Ferrioxamine B than nonmycorrhizal roots; the same is true of an unpurified siderophore of *P. tinctorius* culture; a negative feedback was suggested due to higher iron concentrations in mycorrhizal roots (Leyval and Reid 1991). Starvation of magnesium had no effect on production of extramatrical mycelium of different ectomycorrhizae on *Pinus sylvestris* (Wallander and Nylund 1992). Species specific differences are found in ectomycorrhizal fungi in their ability to provide nutrients to *Eucalyptus* and *Pinus* seedlings; with *Pisolithus tinctorius* being more efficient than *Thelephora terrestris* (Dixon and Hiol-Hiol 1992a). Jones et al. (1992) identified whewellite, the monohydrate form of calcium oxalate, on mantle hyphae of an unidentified *Larix* ectomycorrhiza. Behrmann and Heyser (1992) found evidence for an apoplastic pathway for water and ions at least in the ectomycorrhizae of *Suillus bovinus* on *Pinus sylvestris*; the mobility of charged molecules, however, is suggested to be limited in the fungal matrix. *Pinus rigida* seedlings ectomycorrhizal with *Pisolithus tinctorius* maintained under P-limitation foliar K concentrations and prevented the accumulation of Na (Cumming 1993); further, in contrast to noninoculated seedlings, mycorrhizal *Pinus rigida* were able to use $AlPO_4$, but both were unable to use inositol hexaphosphate. Dixon et al. (1993) provided evidence that colonization of *Pinus taeda* seedlings with *Laccaria laccata* and *Pisolithus tinctorius*, respectively, the dry weight of the seedlings was significantly reduced by 80 mM NaCl after 14 weeks.

κ) Water

Lehto (1992a) could show species-dependent differences in abundance of ectomycorrhiza due to water content of the substrate. It is suggested that breeding of ectomycorrhizal fungi could constitute a new tool for improving reforestation success in arid and semiarid zones (Lamhamedi et al. 1992). *Pisolithus tinctorius* ectomycorrhizae on *Eucalyptus camaldulensis* appeared to be more efficient than those of *T. terrestris* in assisting seedlings to maintain gas exchange and photosynthesis under limited soil moisture conditions (Dixon and Hiol-Hiol 1992b). Drought-acclimated *Pinus taeda* seedlings ectomycorrhizal with *Pisolithus tinctorius* had in comparison to noninoculated ones the lowest predawn fascicle water potential, conserved water with lowest bulk fascicle diffusive conductance and maintained low conductance and transpiration during recovery; this enhanced drought acclimation was independent of plant size and fascicle nutrient content (Svenson et al. 1991). *Picea sitchensis* seedlings ectomycorrhizal with *Paxillus involutus* had in well-watered but nutrient-deficient conditions a higher stomatal conductance, net photosynthesis rate, and water potential than nonmycorrhizal plants; an improved drought resistance of the mycorrhizal seedlings was probably due to their better P and K nutrition and their more extensive root systems with mycelial strands (Lehto 1992b); but under conditions of adequate nutrition, there was no effect of mycorrhizal infection on stomatal conductance or net photosynthetic rate in either well-watered or droughted seedlings (Lehto 1992c). Drought-stressed *Leucaena leucocephala* seedlings either ectomycorrhizal with *Boletus suillus* or nonmycorrhizal, and drought-stressed ectomycorrhizal (with *B. suillus) Acacia auriculiformis* seedlings tolerated lower xylem pressure potentials and larger water losses than drought-stressed ectomycorrhizal (with *B. suillus*) and nonmycorrhizal *Albizzia lebbeck* and *Gliricidia sepium* seedlings; these latter plants avoided drought by maintaining higher xylem pressure potentials and leaf relative water content (Osonubi et al. 1991). Reconstituted dikaryons of *Pisolithus* sp. were used to test water stress of *Pinus pinaster* seedlings: certain dikaryons were more sensitive to water stress and showed less mycorrhiza formation under water stress. Transpiration rate, stomatal conductance, and xylem water potential varied among mycorrhizal treatments during the water stress and recovery periods.

f) Ecology

α) Host Specifity (cf. also Sect. I.e.γ)

Kasuya et al. (1992) in a laboratory experiment tested the abundance of ectomycorrhiza formation between six different fungi and five different *Pinus* species and one subspecies. The degree of colonization varied considerably. *Pisolithus* was shown to form most abundant mycorrhizae on all pines, *Cenococcum geophilum* formed only view mycorrhizae exclusively on *Pinus caribaea* ssp. *hondurensis*. Though the ectomycorrhiza formation began at different times, this is not reflected in numbers after a 120-day experiment. Erland and Finlay (1992) found that incubation time of *Pinus sylvestris* seedlings for maximal colonization with *Piloderma croceum, Paxillus involutus*, and a pink ectomycorrhizal isolate corresponded to the relative growth rates of the fungi in pure culture. Jacobson and Miller (1992) found different degrees of host specifity for allopatric populations of *Suillus granulatus* and three pine species. *Chan* and *Griffiths* (1991) expect that *Eucalyptus microcorys* and *E. torelliana* could be succesfully introduced into Hong Kong in the future, because they proved these *Eucalyptus* species as being able to form ectomycorrhizae with local forest basidiomycetes.

β) Ecological Laboratory Research

Ba et al. (1991) found, using *Afzelia africana* seedlings as baits in sieved soils, successional formation sequence of four mycorrhizal types. The time of colonization correlated positively to the vegetative propagules in the soil of the two studied types. Successions in species composition of ectomycorrhizae are recorded (Edwards and Kelly 1992). Waterlogging experiments with different *Salix* species and strains showed that waterlogged plants had no ectomycorrhizae, control plants did (Good et al. 1992). Miller et al. (1992) studied, using *Alnus rubra* and *Pseudotsuga menziesii* seedlings, the potential for mycorrhizal formation and *Frankia* nodulation in soils from six differently managed sites. Feces of Australian marsupials contained viable spores frequently of hypogeous ectomycorrhizal fungi. The spores of *Mesophellia pachytrix* tested after intestinal passage, were able to cause ectomycorrhizae on *Eucalyptus* seedlings, in contrast to spores taken from fruitbodies. It is suggested that animals contribute to reestablish mycorrhizal colonies in regeneration sites (Claridge et al. 1992). Uninoculated *Pinus elliotii* seedlings visualized a "blue shift" in far-red reflectance in comparison to seedlings ectomycorrhizal with *Pisolithus tinctorius* (Cibula and Carter 1992).

γ) Researches in Natural Habitats

Although the soil core method is very useful to compare different ecosystems with respect to root demography, Fogel (1991) points out that direct observation using mini-rhizotrons and rhizotrons, in conjunction with radiolabeling experiments, appears to have the greatest potential for improving our understanding of interaction among roots and soil biota. Meier (1991) breaks a lance to include not only numbers of ectomycorrhizae in researches of plants under stress; rather qualitative than quantitative researches are necessary, as it can be expected that numbers of ectomycorrhizae can be kept unchanged, while alterations of species composition perhaps owing to different effectiveness can occur. The author presented data based on changes in morphologies.

Read (1990) discusses evidences for the importance of soil-mycorrhiza interactions and guild structures in determining the composition of the plant community. Faber (1991) divides fungivorous soil animals into three different "leagues", the "league" of euedaphic fungus engulfers is characterized by feeding on mycorrhizal hyphae or saprophytic fungi growing in the rhizosphere. They are assumed to affect plant nutrient uptake and plant growth, and to stimulate microbial colonization of the rhizosphere.

All *Pseudotsuga menziesii* seedlings were found to be associated and integrated in hyphal mats in a 60–75-year-old stand (Griffiths et al. 1991), a nursery function by the overstory trees is suggested. Entry et al. (1992) found evidence that the mat-forming ectomycorrhizal fungus *Hysterangium setchelii* is able to concentrate nutrients in hyphal tissue that are important to the growth and productivity of *Pseudotsuga menziesii* forests. Erland and Söderström (1991) provide evidence that mycelial connections to the mature host plants (*Pinus sylvestris*) may significantly alter the ability of different fungi to colonize host plant roots in competition with each other compared to when the fungi colonize from propagules in the soil. In addition, they showed that species composition of ectomycorrhizae is altered due to application of lime or wood ash; *Piloderma croceum*, e.g., was significantly more abundant in limed than in ash-treated soil. Calcifuge plants could be urged to be calcitolerant after inoculation with calciphilous fungi; some fungi are able to immobilize calcium as amorphous calcium oxalate in intracellular vesicles (Lapeyrie 1990).

Naturally grown *Shorea leprosula* seedlings showed ectomycorrhizae before they had true leaves, but seedlings isolated from contact with the roots of mature *Shorea* trees remained uninfected in the field for up to 10 months (Alexander et al. 1992). Morgan et al. (1992) found evidence that in an oligotrophic site, spruces in a mixed stand differ in contrast to a pure spruce stand in ability to access soil organic-N, with growth promotions as a consequence, due to differential rooting patterns and different mycorrhizal associations of the participating tree species; a "nurse" effect is assumed. A greater diversity of ectomycorrhizae was found in mixed than in pure *Picea sitchensis* stands (Heslin et al. 1992). Surveying a young (2–15-month-old) *Eucalyptus* plantation in Brazil, Richter and Bruhn (1993) reported on ordered changes and succession of ectomycorrhizal types on *Pinus resinosa* after outplanting in a recently cleared hardwood site; common nursery types decreased and new mycorrhizal types became abundant, the total number of types increased from 5 to 22 in the fourth year. Data suggested that the fruitbody succession of ectomycorrhizal fungi of a young *Pinus sylvestris*/*Picea abies* stand is more advanced in plots with low levels of available phosphate (Shaw et al. 1992a). Natarajan et al. (1992) reported on a (qualitative) correlation of the occurrence of fruitbodies and ectomycorrhizae. De Bellei et al. (1992) produced evidence that the overall trend of mycorrhizal colonization in an *Eucalyptus viminalis* plantation showed an early decline for VA-mycorrhizae and a later increase of ectomycorrhizae

δ) Coexistence with Other Organisms

Saprophytic and Parasitic Fungi. The role of ectomycorrhizal fungi in plant disease control caused by pathogenic soil fungi is reviewed by Jalali and Jalali (1991) and some explanations are discussed. The authors emphasize the importance of investigating the processes of interactions under field and not only under artificial laboratory conditions. Concluding from changes in hyphal characteristics of the pathogen, ectomycorrhizae of *Laccaria laccata* and *Paxillus involutus* on *Picea abies* increased the endogenous plant resistance against the infection by *Cylindrocarpon destructans*, although *C. destructans* invaded the plant cortical cells in the case of *L. laccata* (Buscot et al. 1992).

Ectomycorrhizal Fungi. Mamoun and Olivier (1993a) investigated competition phenomena and found that high irrigation favored the competitiveness of *T. melanosporum* against *T. brumale*, and an additional inoculation with *T. melanosporum* spores proved to promote the development of the introduced inoculated seedlings. *T. melanosporum* mycorrhizae on seedlings normally declined without inoculation with *T. melanosporum* or *T. brumale* spores; they retained their number even after inoculation with *T. brumale* spores. It is pointed out that the result of competition is dependent on the native soil microflora. Ectomycorrhizal competitors of *T. melanosporum* were found to be hampered by fluorescent pseudomonads (Mamoun and Olivier 1992). In a further competition experiment, Mamoun and Olivier (1993b) found that, although high water content is unfavorable for the development of *Tuber melanosporum* in the absence of other *Tuber* species, *T. brumale* stimulates the development of *T. melanosporum* under these conditions by protecting the roots from other ectomycorrhizal symbionts. Parlade and Alvarez (1993) found in a competition experiment that *Laccaria bicolor* is a very aggressive root colonizer of *Pseudotsuga menziesii*, and retained its levels of colonization constant at increasing inoculum rates of *Pisolithus tinctorius*; it could even replace *Rhizopogon subareolatus*. Agerer (1992b) proved that the ectomycorrhiza forming *Boletopsis leucomelaena* can produce haustoria within a yet unidentified ectomycorrhiza.

Bacteria. Mycorrhization helper bacteria (MHB) isolated from *Laccaria laccata* fruitbodies or ectomycorrhizae elevated numbers of mycorrhizae after inoculation (Duponnois and Garbaye 1991a; Garbaye et al. 1992). The most efficient bacterium species increased mycorrhization to different degrees – in comparison to controls without bacteria – depending on nursery type, and on inoculation doses of 10^6 living cells per m^2 (Duponnois and Garbaye 1991a). Garbaye and Duponnois (1992) found MHB of *Laccaria laccata* to be fungus-specific. One principle of their influence on stimulation of ectomycorrhiza formation and fungus-specifity is elucidated as direct stimulation of fungal growth due to releasing volatile substances by the bacteria. In a synthesis experiment with *Picea abies*, *Cantharellus cibarius*, and *Pseudomonas fluorescens* isolated from *C. cibarius* fruitbodies, Danell et al. (1993) could show that, although the bacteria are motile and were incorporated within the fruitbody, they were not able to penetrate the hyphal mantle of the ectomycorrhizae; they are found only on the outer mantle surface. In tuberculate ectomycorrhizae of *Pseudotsuga menziesii*, a spore-forming, nitrogen-fixing *Bacillus* species was found (Li et al. 1992), and a close nutritional relationship between the bacteria and the mycorrhizae could be shown. Leyval and Berthelin (1993) proved in the rhizosphere of *Fagus sylvatica* ectomycorrhizal with *Laccaria laccata* increased amounts of sugar and amino acids, and organic acids differed from those released from nonmycorrhizal beeches. In pines, also ectomycorrhizal with *L. laccata*, however, all compounds decreased. Dual inoculations with *Agrobacterium radiobacter* (beech) or *Agrobacterium* sp. (pine) mycorrhizal inoculation increased, and virtually no organic acids were detected.

Animals. Fitter and Sanders (1992) reviewed interactions between mycorrhizae and soil fauna.

Plants. Nilsson et al. (1993) reported that even a low concentration of *Empetrum hermaphroditum* leaf extract impaired nitrogen uptake in both *Pinus sylvestris* seedlings ectomycorrhizal with *Paxillus involutus* and nonmycorrhizal plants.

ε) Influence by Man

Fertilization with Lime or Nitrogen (cf. also Sect. I.e. θ). Liming of a *Pinus resinosa* forest did not appear to affect diversity of ectomycorrhizal morphotypes; however, lime did increase the relative frequency of certain morphotypes. Seasonal and liming-related differences in phosphatase activities were observed in individual morphotypes, but these were minor compared to differences existing among morphotypes (Antibus and Linkins 1992). Fertilization of low-productivity *Pinus sylvestris* stands with nitrogen of different compounds (ammonium nitrate, urea, balanced nutrients solution, or as solid fertilizer) and quantities, as well as application in different intervals (continuously daily for 15 years, once yearly for 15 years, once 13 years ago) caused different reactions regarding mycorrhization level (Arnebrant and Söderström 1992). In the nonfertilized plots, 70% of root tips were ectomycorrhizal and the corresponding figure for over 15 years continuously fertilized plots was 55%, and with no reaction when fertilized 13 years ago. One unidentified type decreased in all plots significantly, except in the urea-treated plot. Turner et al. (1993), with *Shorea* seedlings obtained preliminary evidence that the seedlings may only be responsive to NP(K)-fertilization when grown at very low nutrient availabilities, and that ectomycorrhizal colonization may be of great importance to seedling growth under such conditions. Erland (1992) showed results of liming on pine ectomycorrhiza and found changes in the relative distribution of mycorrhizal types, without resulting in a total decrease in colonization.

Pesticides. The application of pesticides, especially of fungicides, was discussed by Jalali and Jalali (1991); effects on mycorrhizae and on fruitbody production are apparent. Stimulating or depressing effects of methyl bromide on mycorrhization of *Quercus robur* with *Laccaria laccata* depended upon the initial microflora in the peat substrate (Garbaye et al. 1992).

Pollution – Acidification. Jortikka et al. (1991) reported that after a 4-year application of acid rain of pH 3 the formation of pine ectomycorrhizae diminished, but not significantly in comparison to not irrigated and to water control. Acid treatment (pH 3.8, *Pinus taeda* seedlings) had no effect on number or percent of mycorrhizae, and responses of two morphotypes to soil Mg treatments were probably due to differences in the soil environment rather than a result of changes in above-ground processes (Edwards and Kelly 1992). Walker and McLaughlin (1991) found that, although ectomycorrhizal development was greatest on *Pinus taeda* seedlings inoculated with *Pisolithus tinctorius* and treated with increasing acidity (pH 4.8, 4.2 or 3.6) of simulated acid rain, there was no consistent effect of increasing acidity on ectomycorrhizal development of *Quercus alba.* Of the variables examined, the data suggested the primary effect of acid precipitation on *Q. alba* is on shoot diameter, whereas its primary effect on *Pinus taeda* seedlings is on ectomycorrhizal development. Using acidified (pH 4.3; 3.6) rain throughfall to irrigate a mixed-aged *Pinus taeda* and *P. palustris* stand, Esher et al. (1992) found a significant (pH 3.6) or nearly significant (pH 4.3) decrease of percent and number of ectomycorrhizae. Total number of morphotypes decreased with increasing rainfall acidity in a O_3-sensitive *Pinus taeda* seedling family, but was not altered in a different family (Qiu et al. 1993).

Pollution – Effects of Ozone (and SO_2). Twice ambient ozone treatment of *Pinus taeda* seedlings exhibited smaller percentages of total ectomycorrhiza short roots, changes of morphotype composition were not consistent (Edwards and Kelly 1992). In an ozone-sensitive *Pinus taeda* seedling family the number of ectomycorrhiza morphotypes per cm long root increased with increasing O_3 concentrations. Elevated O_3 concentrations caused a shift in some individual morphotypes in the sensitive and in the insensitive family (Qiu et al. 1993). In fumigation experiments with ozone and SO_2 in young stands of *Pinus sylvestris/Picea abies* or of young *Pinus nigra*, no effects could be found in the mycorrhizal community regarding ectomycorrhizae or fruitbodies (Shaw et al. 1992a, 1992b). Bonello et al. (1993) showed that ozone exposure of *Pinus sylvestris* seedlings (200 nl l^{-1}, 8 h per day, for 28 days) increased disease incidence against *Heterobasidion annosum* significantly, but ectomycorrhizal colonization with *Hebeloma crustuliniforme* completely prevented this negative effect.

Effects of Nitrogen. Exposure of *Pinus sylvestris* seedlings mycorrhizal with *Pisolithus tinctorius* to 20–30 ppb NO_x had no effect on seedling growth, mycorrhizal infection, or total nitrogen concentrations, but resulted in a significant uptake of NO_x and measurable change in amino-acid composition (Nasholm et al. 1991).

Ammonium treatment affected *Pinus sylvestris* seedlings ectomycorrhizal with either *Paxillus involutus* or *Suillus bovinus* more positively and the mycorrhizae more negatively than nitrate. The effects on seedlings inoculated with *P. involutus* were more pronounced (Termorshuizen and Ket 1991). No extramatrical mycelium was formed on ectomycorrhizae of different fungi on *Pinus sylvestris* if the nitrogen content of the medium was raised above 200 mg l^{-1}; excess of nitrogen resulted in lowered P and particularly Mg concentration in the needles, regardless of the mycorrhizal status of the seedlings (Wallander and Nylund

1992). Holopainen and Heinonen-Tanski (1993), using an N fertilization experiment with different N species and concentrations on *Pinus sylvestris* seedlings, found evidence that numbers of mycorrhizal rootlets decreased with increasing concentrations of nitrate, but the ammonium N and urea formaldehyde treatments did not significantly affect the total number of mycorrhizae counted against root length. Gundersen (1991) discusses and defines the concept of "nitrogen saturation" and describes the interactions of effects of excess nitrogen, e.g., on loss of ectomycorrhiza, and concluded that these may destabilize the forest ecosystem and contribute to forest decline; estimates of long-term critical N loads for forest ecosystems of 2–20 kg N ha^{-1} per year, depending on biomass utilization, are derived from simple mass balance considerations. Denitrification is supposed to be an insignificant process in the N cycling of most forests. Termorshuizen (1993) pointed out for a 6–7-year-old *Pinus sylvestris* stand that carpophore formation is apparently more sensitive to nitrogen fertilizer than the mycorrhiza formation.

Ultrastructural researches of two species of ectomycorrhizae revealed that first changes occurred in the sheath tissue as a formation of dark accumulations in fungal vacuoles, decrease in numbers of cytoplasmic storage bodies, swelling of mitochondria, and dark staining and disintegration of the cytoplasm. Increased severity of the changes was correlated with increasing N application level, being most severe in the ammonium acetate treatment.

Greenhouse Effect. Shafer and Schoeneberger (1991) expect an impact of elevated concentrations of CO_2, and CO, on mycorrhizal interactions and suggest an increased dependence of the plant on nutrient supply from the mycorrhizal fungi. A role of CH_4, and chlorofluorocarbons is expected via temperature and precipitation effect. Stulen and den Hertog (1993) reviewed the root growth and functioning under atmospheric CO_2 enrichment, and concluded that the available information on CO_2 enrichment and mycorrhizae suggest that carbon allocation to the roots might be increased, but more experiments are needed.

Effects of Toxic Metals. A protective effect against cadmium toxicity in *Pinus sylvestris* seedlings by ectomycorrhizal fungi was observed by Colpaert and Van Assche (1993). The authors showed that fungi which produce a large mycelial biomass had the greatest effect in overcoming toxicity. Gadd (1993) reviews mycorrhizal interactions with toxic metals. *Pinus sylvestris* seedlings ectomycorrhizal with *Paxillus involutus* and grown in a heavily polluted forest showed two different granules within the hyphal cells: the first, similar in appearance to polyphosphate granules, was characterized by high contents of phosphorus accompanied by sulfur, calcium, and often aluminum; the second, more common, was granules containing more nitrogen, sulfur and cadmium, while phosphorus was much lower; it is suggested that the symbiotic fungus possibly can detoxify heavy metal (Turnau et al. 1993).

Pinus rigida seedlings ectomycorrhizal with *Pisolithus tinctorius* and grown on primarily NO_3^- based nutrient solutions, maintained, in contrast to nonmycorrhizal seedlings, growth and foliar P levels under Al exposure, suggesting that Al-induced P limitation due to coprecipitation of Al and P was a critical factor in nonmycorrhizal seedlings (Cumming and Weinstein 1991). Cumming (1991) stressed that in seedlings of *Pinus rigida*, ectomycorrhizal with *Pisolithus tinctorius*, foliar N concentrations, root nitrate reductase (NR) activity, and Al inhibition of NR activity were highly dependent on the proportion of NO_3^- of the nutrient solution. The fungus reduced both root and foliar NR activity in comparison to nonmycorrhizal seedlings, suggesting that NO_3^- uptake and translocation to foliage was reduced by the fungus. Preferential use of NH_4^+ by ectomycorrhizal roots may thus result in reduced movement of Al into root tissue and amelioration of Al toxicity. Aluminum was found in the mantle and in cell walls of hyphae of the Hartig net; together with lead, aluminum was also detected in the endodermis of ectomycorrhizae on *Picea abies* (Eeckhaoudt et al. 1992). Hentschel et al. (1993) concluded that mycorrhizal colonization of *Picea abies* seedlings by *Paxillus involutus* modifies the phytotoxic effects of Al in Norway spruce seedlings. However, differences in physiological responses to Al between mycorrhizal and nonmycorrhizal seedlings may be largely reduced in the long term as a result of impaired mineral nutrient uptake.

Ectomycorrhizal fungi vary considerably in their protection of *Pinus sylvestris* seedlings against zinc toxicity, and zinc retention was dependent on density of extramatrical mycelium (Colpaert and Van Assche 1992); however, one *Thelephora terrestris* strain increased zinc concentrations in the plants.

Other Toxic Substances. Tosh et al. (1993b) tested seedlings of *Betula pendula*, ectomycorrhizal with either *Laccaria proxima, Hebeloma crustuliniforme*, or *Paxillus involutus*, regarding their tolerance to different landfill leachate acids and found, in comparison to nonmycorrhizal seedlings, mostly an increased seedling tolerance to the acids.

Damage to Trees. Hilber (1991) reported on his investigations on revitalization of *Picea abies* affected by forest decline with either application of *Pisolithus tinctorius* spore suspensions or *Fagus sylvatica* seedlings inoculated with the same fungus. Some positive tendencies could be shown.

Afforestation, Clear-Cut, Application. Duponnois and Garbaye (1991b) review the available techniques for controlled synthesis of the Douglas-fir – *Laccaria laccata* ectomycorrhizal symbiosis with respect to the experimental purpose. A dual inoculum with *Laccaria laccata* mycelium and mycorrhization helper bacteria (MHB) included in alginate beads appears to be a valuable technique for increasing the efficiency of ectomycorrhizal inoculation of planting stocks in forest nurseries (Duponnois and Garbaye 1991a). *Betula pendula* seedlings ectomycorrhizal with *Paxillus involutus* lost their ectomycorrhizae when planted in refuse; if mycorrhizal or nonmycorrhizal seedlings were planted in topsoil as cover material of the refuse, mycorrhizae proliferated through the refuse, thus indicating a reduction in the toxicity towards the end of the 24-month experiment (Tosh et al. 1993a). Soil erosion (2–4 cm) by heavy storms after a wildfire resulted in a reduced inoculum potential for ectomycorrhizal formation on *Pseudotsuga menziesii* (Amaranthus and Trappe 1993). Helm and Carling (1993a) studied the impact of soil transfer and P fertilization on growth of outplanted *Populus balsamifera* cuttings and concluded that neither soil transfer alone nor P fertilization alone increased plantlet height, only the combination of both resulted in a significant promotion; a soil transfer from a mature forest was more effective than from a forest of a successionally intermediate site (Helm and Carling 1993b). Roth and Berch (1992) report on a high percentage persistence of *Thelephora terrestris* ectomycorrhizae on *Pseudotsuga menziesii* and *Tsuga heterophylla* 1 year after outplanting of nursery seedlings into the field. Browning and Whitney (1992b) reported on site and species-dependent persistence of ectomycorrhizae on *Picea mariana* and *Pinus banksiana* seedlings 2 years after outplanting. Species specific differences of persistence are also indicated by Buschena et al. (1992). Two briefly shown main types of ectomycorrhizae on bare root *Picea abies* seedlings showed different metabolic activity after a cold storage period (Al Abras et al. 1991). The authors suggested that for controlled nursery inoculation, mycorrhizal fungi could also be selected on their ability to resist lifting and storage stress.

g) Methods

Freeze-fractioning for low-temperature scanning electron microscopy was used by Scheidegger and Brunner (1993) for the study of ectomycorrhizae. Thomson et al. (1993) concluded that reisolating ectomycorrhizal fungi from mycorrhizas and fruitbodies can increase the colonization capacity and effectiveness of isolates which have been grown on agar media for extended periods. Henrion et al. (1992) describe and apply a combination of PCR (Polymerase Chain Reaction) and RFLPs (Restriction Fragment Length Polymorphism, received by endonucleases) to detect defined isolates of ectomycorrhizal fungi within a complex ecosystem. Gardes and Bruns (1993) designed two basidiomycete-specific PCR primers for the internal transcribed spacer region in the nuclear ribosomal repeat unit. Zambonelli et al. (1993) used enzyme-linked immunosorbent assay (ELISA) to detect ectomycorrhizae of *Tuber albidum*. They were able to distinguish *T. aestivum* ectomycorrhizae from several *Tuber* species and other ectomycorrhizae. Wingler et al. (1993) developed two sensitive enzymatic microassays to determine mannitol in mycelium and ectomycorrhizae. In testing the ergosterol method for estimation of fungal biomass, Antibus and Sinsabaugh (1993) found evidence that the range of values obtained with fungal isolates from field collected ectomycorrhizae agrees well with reported values for other mycorrhizal and decomposer fungi. A mycocosm is described which allows investigations of carbon allocation to roots under different environmental stresses (Andersen and Rygiewicz 1991).

h) Reviews

The role of mycorrhizae in dominant ecosystems is highlighted by Read (1991a, 1991b) and it is concluded that the nutrient dynamics of terrestrial ecosystems can be understood only in terms of activities of their mycorrhizal fungi. Pankow et al. (1991) came to the conclusion that the main ecological significance of mycorrhizae concerns the protective rather than the productive phase of plant individual as well as of an ecosystem. Perry et al. (1992) review and discuss the role of mycorrhizal fungi in mixed-species forests, positive feedback, redundancy, and stability. Trappe and Luoma (1992) point out the role of mycorrhizae in ecosystems. Molina et al. (1992b) reviewed specifity phenomena in mycorrhizal symbioses and community-ecological consequences and practical implications. The authors provide extended lists on host-specifity of fungi of several mycorrhizal classes. Gadd (1993) reviews interactions of fungi with toxic metals. Newton (1992) came to the conclusion that the phenomenon of succession of ectomycorrhizal fungi can be explained simply by the relative ability of different fungi to colonize and spread from different sources of inoculum, i.e., forming strands, or colonizing root systems by secondary infection.

Dighton (1991) reviews the possibilities of direct nutrient cycling by ectomycorrhizae. Garbaye (1991) compiles results of biological interactions in the mycorrhizosphere. Possible interactions of ectomycorrhizae with other soil organisms, and with the soil itself, are discussed by Plaits and Laundrymen (1991). They included positive and adverse effects of bacteria, and their role in solubilizing phosphate as well as the ectomycorrhizal influence on the population densities of bacteria. Fungal pathogens, protozoa, and soil invertebrates are included in their overview. The authors come to the conclusion that it is the right time for the development of "inoculum packages" (including mycorrhizal fungi and other soil organisms) that are compatible with host and soil and that can benefit plants in ways not possible by chemical additives. Berthelin et al. (1991) review the knowledge of chemical weathering of soil minerals of plant roots, mycorrhizae, and soil bacteria. The mechanisms for the increase in the uptake of P by mycorrhizae and the sources of soil P from mycorrhizal and nonmycorrhizal plants are critically reviewed (Bolan 1991). Torrey (1992) asks the question and tries to find answers if plant productivity can be increased by inoculation of tree roots with soil microorganisms. One part focuses on ectomycorrhizae. Dighton and Jansen (1991) reviewed the influence of atmospheric pollution on ectomycorrhizae and on their fruitbodies with respect to acid rain, SO_2, O_3, and nitrogen immission. Also including possible impacts of climate changes.

Allen (1992) edited the book *Mycorrhizal functioning* with some contributions covering ectomycorrhizae or including them in some aspects. Koske and Gemma (1992) report on fungal reactions to plants prior to mycorrhizal formation, Anderson (1992) on the influence of the plant-root on mycorrhizal formation. Bonfante-Fasolo and Scannerini (1992) highlight the cellular basis of plant-fungal interchanges in mycorrhizal associations, Read (1992) the structure and function of ectomycorrhizal mycelium. Finlay and Söderström (1992) discuss the role of mycorrhiza in carbon flow to the soil, Azcon-Aguilar and Barea (1992) interactions between mycorrhizal fungi and other rhizosphere organisms. Deacon and Fleming (1992) stress interactions of ectomycorrhizal fungi, Miller and Allen (1992) nutrient translocation and interactions between plants. Fitter and Sanders (1992) review interactions of mycorrhizae with the soil fauna, Molina et al. (1992b) specify phenomena in mycorrhizal symbiosis. Bledsoe (1992) points out the implication of physiological ecology for field applications.

The contributions of the 3rd European Conference on Mycorrhizae are published by Read et al. (1992); important topics have been covered: Söderström (1992) pointed out the necessity to investigate the amounts of carbon exported to the soil ecosystem via ectomycorrhizae with respect to ecological effects of energy input into the soil. Agerer (1992a) showed some modes of physical interactions of ectomycorrhizal rhizomorphs with soil and ectomycorrhizal mantle. Finlay (1992) outlined some central problems of uptake and translocation of nutrients by ectomycorrhizal fungal mycelia. Griffiths and Caldwell (1992) reviewed the impact of mycorrhizal mat communities in forest soils. Molina et al. (1992a) highlighted the ecological role of specifity phenomena in ectomycorrhizal plant communities. Erland (1992) showed results of liming on pine ectomycorrhiza. Le Tacon et al. (1992) stressed on the variation in field response of forest trees to nursery ectomycorrhizal inoculation. Dahlberg (1992) showed that somatic incompatibility of ectomycorrhizal fungi could be a tool to reveal spatiotemporal mycelial structures. Brand (1992) reported on mixed associations of fungi in ectomycorrhizal roots. Amaranthus (1992) discussed the fate and impact of mycorrhizae in forest disturbance and regeneration. Shaw et al. (1992b) reported on studies regarding effects of SO_2 and O_3 on mycorrhizae of *Pinus sylvestris*; no effects could be found, but it is suggested that low phosphate availability accelerated fruitbody succession. Last et al. (1992) discuss possible reasons for the occurrence of successions in ectomycorrhizal fruitbody communities. The influence of mycorrhizal succession on morel biology is discussed by Buscot (1992b). Ryan and Alexander (1992) concentrated on the phenomenon of better growth of *Picea sitchensis* in mixed-tree stands in heathland comparing to pure. Smits (1992) reported on ectomycorrhizal studies on Dipterocarpaceae in Indonesia. Different aspects of physiological ecology are covered: especially phosphate nutrition (Tinker et al. 1992), nitrogen distribution (Martin et al. 1992), uptake of cations (Kottke 1992), and water relations and soil drought (Guehl et al. 1992).

2. Ectendomycorrhiza

Ectendomycorrhizal development could be shown on two species of *Allocasuarina* with two strains of the usually ectomycorrhizal species *Pisolithus tinctorius* (Thoen et al. 1990). Pachlewski et al. (1992) isolated an ascomycetous fungal symbiont of *Pinus sylvestris* nursery seedlings, with uniform morphological, physiological and mycorrhizal features, and evidenced a permanent occurrence of N_2-fixing *Bacillus polymyxa* in the ectendomycorrhizosphere. Fortas and Chevalier (1992) synthesized in phosphorus-deficient substrates ectendomycorrhizae without a sheath between *Terfezia* and *Tirmania* species and *Helianthemum guttatum*, whereas in phosphorus-rich substrates ectomycorrhizae, also without a sheath, were formed. Different indole compounds could be isolated and identified from a culture of an ectendomycorrhizal fungus of pine; synthesis of indole compounds was possible (i.e., IAA) without tryptophan in the medium, but was enhanced by addition of this amino acid (Rudawska et al. 1992).

3. Some Highlights of This Report Period

Regarding recognition and ectomycorrhiza formation it became more evident that secreted substances not only may attract fungi but even may determine compatibility. It could also be shown that compatibility may be influenced by soil bacteria. Specific lectins, isolated from an ectomycorrhizal fungus compatible to *Picea*, and lacking in an incompatible one, focus on an important field of research, which should be more intensely covered. Ectomycorrhizins seem to be important hints toward a successful establishment of an ectomycorrhiza, although their existence is not undisputed. A far-reaching realization is that, fungi being under species differentiation, can have different preferences regarding hosts. This very likely indicates an important role of coevolution between ectomycorrhizal fungi and trees.

Considerable progress has been made in investigation of ectomycorrhizal plant nutrition due to realization of nutritional soil influences on the establishment of ectomycorrhizal emanating mycelium. Actually, phosphate starvation was shown to lead to a more pronounced formation of extramatrical mycelium; excess of nitrogen can have the detrimental effect. Even improved detoxification of Cd and Zn could be shown to be connected to a higher amount of extramatrical mycelium. Improved drought resistance can be caused by extensive mycelial strands. However, an additional carbon drain is also proven for the formation of extramatrical mycelium. This is an important step to directing ecophysiological research towards uptake sites and away from exclusive consideration of changes in numbers of ectomycorrhizae and/or alterations of species composition.

Important is also the ascertainment that nitrogen, especially nitrate, and phosphate excess may influence negatively the ability to produce IAA from exogenous (bacterial?) tryptophan and thus perhaps disturbs also the formation of ectomycorrhizae. Further, it could be shown that the fungi obviously play an important part in elevation of cytokinins levels in the plants.

The knowledge on the function of carbohydrate nutrition of ectomycorrhizal fungi by roots became more substantial by detecting pH as well as fructose concentration-dependent invertases in the host cell walls.

Concluding from an ecological standpoint, most informative are the results that there can be competition even between closely related ectomycorrhizal species within a root system, and being influenced by soil conditions and native soil microflora; and in addition, these species can protect the roots against other ectomycorrhizal competitors. Secondly, evidence is provided that better growth of mixed, compared to pure stands, is mediated by the activity of ectomycorrhizae. Both topics represent wide fields for future investigations on ectomycorrhizal ecology.

References

Agerer R (1987–1993) Colour Atlas of Ectomycorrhizae. 1st–7th delivery. Einhorn. Schwäbisch Gmünd. **Agerer R** (1992a) In: Read DJ, Lewis DH, Fitter AH et al. (eds) Mycorrhizas in ecosystems, CAB Int, Wallingford pp 84–90. **Agerer R** (1992b) Nova Hedwigia 55:501–518. **Agerer R** (1992c) Mycorrhiza 2:47–52. **Agerer R** (1993a) Mycologia 85:75–83. **Agerer R** (1993b) Mycorrhiza 2:183. **Agerer R**, Waller K (1993) Mycorrhiza 3:145–154. **Agerer R**, Waller K, Treu R (1993) Z Mykol 59:131–140. **Al Abras K**, Le Tacon F, Lapeyrie F (1991) Ann Sci For 48:489–495. **Alexander I**, Ahmad N, Lee Su See (1992) Philos Trans R Soc Lond B 335:379–388. **Allen MJ** (ed) (1992) Mycorrhizal functioning, an integrative plant-fungal process. Chapman and Hall, New York. **Amaranthus MP** (1992) In: Read DJ, Lewis DH, Fitter AH et al. (eds) Mycorrhizas in ecosystems. CAB Int, Wallingford pp 202–207. **Amaranthus MP**, Trappe JM (1993) Plant Soil 150:41–49. **Andersen CP**, Rygiewicz PT (1991) Environ Pollut

73:217–244. **Anderson AJ** (1992) In: Allen MJ (ed) Mycorrhizal functioning, an integrative plant-fungal process. Chapman and Hall, London, pp 37–64. **Antibus RA**, Linkins AE (1992) Soil Biol Biochem 24:479–487. **Antibus RK**, Sinsabaugh RL (1993) Mycorrhiza 3: 137–144. **Arnebrant K**, Söderström B (1992) For Ecol Mange 53:77–89. **Arnebrant K**, Ek H, Finlay R et al. (1993) New Phytol 124:231–242. **Azcon-Aguilar C**, Barea JM (1992) In: Allen MJ (ed) Mycorrhizal functioning, an integrative plant-fungal process. Chapman and Hall, London, pp 163–198.

Ba AM, Garbaye J, Dexheimer J (1991) Can J Bot 69:2442–2447. **Behrmann P**, Heyser W (1992) Bot Acta 105:427–434. **de Bellei MM**, Garbaye J, Gil M (1992) For Ecol Manage 54:205–213. **Bellgard SE** (1991) Aust J Bot 39:357–364. **Berndt R**, Oberwinkler F (1992) Mycologia 84:360–366. **Berthelin J**, Leyval C, Laheurte F et al. (1991) In: Atkinson D (ed) Plant root growth: An ecological perspective. Spec Publ Ser Brit Mycol Soc 10:187–200. **Blaschke H** (1991) Arct Alp Res 23:444–450. **Bledsoe CS** (1992) In: Allen MJ (ed) Mycorrhizal functioning, an integrative plant-fungal process. Chapman and Hall, London, pp 424–438. **Bolan NS** (1991) Plant Soil 134:189–207. **Bonello P**, Heller W, Sandermann H (1993) New Phytol 124:653–663. **Bonfante-Fasolo P**, Scannerini S (1992) In: Allen MJ (ed) Mycorrhizal functioning, an integrative plant-fungal process. Chapman and Hall, London, pp 65–101. **Bougher NL**, Tommerup IC, Malajczuk N (1993) Mycol Res 97:613–619. **Brand F** (1992) In: Read DJ, Lewis DH, Fitter AH et al. (eds) Mycorrhizas in ecosystems, CAB Int, Wallingford, pp 142–147. **Browning MHR**, Whitney RD (1992a) Can J For Res 22:1263–1270. **Browning MHR**, Whitney RD (1992b) Can J For Res 22:1974–1982. **Brundrett MC**, Abbott LK (1991) Aust J Bot 39:445–457. **Brunner I**, Amiet R, Schneider B (1991) Mycol Res 95:1407–1413. **Brunner I**, Amiet R, Zollinger M et al. (1992) Mycorrhiza 2:89–96. **Brunner I**, Scheidegger C (1992) New Phytol 120:359–369. **Brunner I** (1993) Water Air Soil Pollut 68:231–240. **Buschena CA**, Doudrick RL, Anderson NA (1992) Can J For Res 22:1883–1887. **Buscot F** (1992a) J Plant Physiol 141:12–17. **Buscot F** (1992b) In: Read DJ, Lewis DH, Fitter AH et al. (eds) Mycorrhizas in ecosystems. CAB Int, Wallingford, pp 220–224. **Buscot F**, Weber G, Oberwinkler F (1992) Trees 6:83–90.

Cairney JWG (1992) Mycol Res 96:135–141. **Cairney JWG**, Alexander IJ (1992a) New Phytol 122:159–164. **Cairney JWG**, Alexander IJ (1992b) New Phytol 122:153–158. **Chalot M**, Stewart GR, Brun A et al. (1991) New Phytol 119:541–550. **Chan WK**, Griffiths DA (1991) For Ecol Manage 43:15–24. **Chen L**, Pei Z (1992) For Res 5:65–70. **Cibula WG**, Carter GA (1992) Int J Remote Sens 13:925–932. **Claridge AW**, Tanton MT, Seebeck JH et al. (1992) Aust J Ecol 17:207–217. **Colpaert JV**, Van Assche JA (1992) Plant Soil 143:201–211. **Colpaert JV**, Van Assche JA (1993) New Phytol 123:325–333. **Colpaert JV**, Van Assche JA, Luijtens K (1992) New Phytol 120:127–135. **Cumming JR** (1991) Can J Bot 68:2653–2659. **Cumming JR** (1993) Tree Physiol 13:173–187. **Cumming JR**, Weinstein LH (1991) Can J Bot 68:2644–2652. **Cuvelier J-J** (1991) Belg J Bot 124:195–203.

Dahlberg A (1992) In: Read DJ, Lewis DH, Fitter AH et al.(eds) Mycorrhizas in ecosystems. CAB Int, Wallingford, pp 135–141. **Danell E**, Alstrom S, Ternstrom A (1993) Mycol Res 97:1148–1152. **Deacon JW**, Fleming LV (1992) In: Allen MJ (ed) Mycorrhizal functioning, an integrative plant-fungal process. Chapman and Hall, London, pp 249–300. **Debaud JC**, Gay G, Bruchet G (1990) Endocytobiology IV, INRA, Paris pp 107–110. **Dexheimer J**, Pargney J-C (1991) Experientia 47:312–321. **Dighton J** (1991) Experientia 47:362–369. **Dighton J**, Coleman C (1992) Mycorrhiza 1:175–184. **Dighton J**, Jansen AE (1991) Environ Pollut 73:179–204. **Dixon RK**, Hiol-Hiol F (1992a) Commun Soil Sci Plant Anal 23:1387–1396. **Dixon RA**, Hiol-Hiol F (1992b) Plant Soil 147:143–149. **Dixon RK**, Johnson PS (1992) J Arboric 18:266–272. **Dixon RK**, Rao MV, Garg VK (1993) Mycorrhiza 3:63–68. **Dosskey MG**, Linderman RG, Boersma L (1992) Plant Soil 142:147–150. **Downes GM**, Alexander IJ, Cairney JWG (1992) New Phytol 122:141–

152. **Duponnois R**, Garbaye J (1991a) Plant Soil 138:169–176. **Duponnois R**, Garbaye J (1991b) Ann Sci For 48:641–650. **Dupre C**, Chevalier G (1991) Crypt Mycol 12: 243–250.

Edwards GS, Kelly JM (1992) Environ Pollut 76:71–77. **Eeckhaoudt S**, Vandeputte D, Van Praag H et al. (1992) Tree Physiol 10:209–215. **Egli S** (1992) Mitt Eidgenoss Forsch Wald Schnee Landsch 67:317–412. **Egli S**, Amiet R, Zollinger M et al. (1993) Trees 7:123–129. **Entry JA**, Rose CL, Cromack K (1992) Soil Biol Biochem 24:447–453. **Erland S** (1992) In: Read DJ, Lewis DH, Fitter AH et al. (eds) Mycorrhizas in ecosystems. CAB Int, Wallingford, pp 113–118. **Erland S**, Finlay R (1992) Mycol Res 96:270–272. **Erland S**, Söderström B (1991) Scand J For Res 6:519–526. **Erland S**, Finlay R, Söderström B (1991) New Phytol 119:235–242. **Esher RJ**, Marx DH, Ursic SJ et al. (1992) Water Air Soil Pollut 61:269–278.

Faber JH (1991) Oikos 62:110–117. **Finlay R** (1992) In: Read DJ, Lewis DH, Fitter AH et al. (eds) Mycorrhizas in ecosystems. CAB Int, Wallingford, pp 91–97. **Finlay R**, Söderström B (1992) In: Allen MJ (ed) Mycorrhizal functioning, an integrative plant-fungal process. Chapman and Hall, London, pp 134–160. **Finlay RD**, Frostegard A, Sonnerfeldt A-M (1992) New Phytol 120:105–113. **Fitter AH**, Sanders IR (1992) In: Allen MJ (ed) Mycorrhizal functioning, an integrative plant-fungal process. Chapman and Hall, London, pp 333–354. **Fogel R** (1991) In: Atkinson D (ed) Plant root-growth: an ecological perspective. Spec Publ Ser Br Mycol Soc 10:89–102. **Fortas Z**, Chevalier G (1992) Can J Bot 70:2453–2460.

Gadd GM (1993) New Phytol 124:25–60. **Garbaye J** (1991) Experientia 47:370–375. **Garbaye J**, Duponnois R (1992) Symbiosis 14:335–344. **Garbaye J**, Churin J-L, Duponnois R (1992) Biol Fertil Soils 13:55–57. **Gardes M**, Bruns TD (1993) Mol Ecol 2:113–118. **Gay G**, Durand N, Debaud JC (1990) Endocytobioloy IV, INRA, Paris, pp 111–114. **Giollant M**, Guillot J, Damez M et al. (1993) Plant Physiol 101:513–522. **Gong M**, Chen Y, Wang F (1992) For Res 5:639–645. **Good JEG**, Winder JD, Sellers E et al. (1992) Proc R Soc Edinb 98B:21–48. **Gourp V**, Pargney J-C (1991) Crypt Mycol 12:293–304. **Griffiths RP**, Caldwell BA (1992) In: Read DJ, Lewis DH, Fitter AH, et al. (eds) Mycorrhizas in ecosystems. CAB Int, Wallingford, pp 98–105. **Griffiths PR**, Castellano MA, Caldwell BA (1991) Plant Soil 134:255–259. **Guehl J-M**, Garbaye J, Wartinger A (1992) In: Read DJ, Lewis DH, Fitter AH et al. (eds) Mycorrhizas in ecosystems. CAB Int Wallingford, pp 323–332. **Gundersen P** (1991) For Ecol Manage 44:15–28. **Guttenberger M**, Hampp R (1992) Planta 188:129–136.

Handley LL, Daft MJ, Wilson J et al. (1993) Plant Cell Environ 16:375–382. **Haug I**, Pritsch K (1992) Ectomycorrhizal types of spruce (Picea abies (L.) Karst.) in the Black Forest: a microscopical atlas. Kernforschungszentrum, Karlsruhe. **Helm DJ**, Carling DE (1993a) Mycorrhiza 3:97–106. **Helm DJ**, Carling DE (1993b) Mycorrhiza 3:107–114. **Henrion B**, LeTacon F, Martin F (1992) New Phytol 122:289–298. **Hentschel E**, Godbold DL, Marschner P et al. (1993) Tree Physiol 12:379–390. **Heslin MC**, Blasius D, McElhinney C et al. (1992) Eur J For Pathol 22:46–57. **Hilber O** (1991) In: Maher MJ (ed) Science and cultivation of edible fungi Balkema, Rotterdam, pp 625–633. **Hilbert J–L**, Costa G, Martin F (1991) Plant Physiol 97:97–984. **Holopainen T**, Heinonen-Tanski H (1993) Can J For Res 23:362–372.

Ineichen K, Wiemken V (1992) Mycorrhiza 2:1–7. **Iwase K** (1992) Can J Bot 70: 1234–1238.

Jacobson KM, Miller OK (1992) Can J Bot 70:26–31. **Jalali BL**, Jalali I (1991) In: Arora DK, Rai B, Mukerji KG et al. (eds) Handbook of Applied Mycology 1. Marcel Dekker, New York, pp 131–154. **Jones D**, McHardy W, Wilson MJ et al. (1992) Micron Micros

Acta 23:315–317. **Jones MD**, Durall DM, Tinker PB (1991) New Phytol 119:99–106. **Jortikka S**, Ratio-Lehtimäki A, Oleander M et al. (1991) Leonine Tektite 95:31–33.

Kamminga-Van Wijk C, Prins HBA, Kuiper PJC (1992) Acta Bot Neerl 41:481–495. **Kasuya MCM**, Muchovej RMC, Bellei MM et al. (1992) For Ecol Manage 47:127–134. **Kieliszewska-Rokicka B** (1992) Acta Soc Bot Pol 61:253–264. **Köljalg U** (1992) Mycol Res 96:215–220. **Koske RE**, Gemma JN (1992) In: Allen MJ (ed) Mycorrhizal functioning, an integrative plant-fungal process. Chapman and Hall, London, pp 3–36. **Koske RE**, Gemma JN, Flynn T (1992) Am J Bot 79:853–862. **Kottke I** (1992) In: Read DJ, Lewis DH, Fitter AH et al. (eds) Mycorrhizas in ecosystems. CAB Int, Wallingford, pp 316–322. **Kraigher H**, Strnad M, Hanke DE et al. (1993) Forstwiss Centralbl 112: 107–111.

Lamhamedi MS, Bernier PY, Fortin JA (1992) Tree Physiol 10:153–167. **Lapeyrie F** (1990) In: Werner D, Müller P (eds): Fast growing trees and nitrogen fixing trees. Fischer, Stuttgart, pp 312–317. **Lapeyrie F**, Mendgen K (1993) Mycol Res 97:603–609. **Last FT**, Natarajan K, Mohan V et al. (1992) In: Read DJ, Lewis DH, Fitter AH et al. (eds) Mycorrhizas in ecosystems. CAB Int, Wallingford, pp 214–219. **Lee KJ** (1992) Korean For Soc 81: 156–163. **Lehto T** (1992a) Scand J For Res 7:177–182. **Lehto T** (1992b) New Phytol 122:661–668. **Lehto T** (1992c) New Phytol 122:669–673. **Lei J**, Piche Y, Ding H et al. (1990a) Endocytobiology IV, INRA, Paris, pp 119–122. **Lei J**, Ding H, Lapeyrie F et al. (1990b) Endocytobiology IV, INRA, Paris, pp 123–126. **Le Tacon F**, Alvarez IF, Bouchard D et al. (1992) In: Read DJ, Lewis DH, Fitter AH et al. (eds) Mycorrhizas in ecosystems. CAB Int, Wallingford, pp 119–134. **Leyval C**, Berthelin J (1993) Biol Fertil Soils 15:259–267. **Leyval C**, Reid CPP (1991) New Phytol 119:93–98. **Li CY**, Massicotte HB, Moore LVA (1992) Plant Soil 140:35–40. **Lobuglio KF**, Rogers SO, Wang CJK (1991) Can J Bot 69:2331–2343.

MacFall JS, Slack SA (1991) Can J For Res 21:1459–1465. **Macfall JS**, Slack SA, Wehrill S (1992) Plant Physiol 100:713–717. **Mamoun M**, Olivier JM (1992) Plant Soil 139:265–274. **Mamoun M**, Olivier JM (1993a) Plant Soil 149:219–225. **Mamoun M**, Olivier JM (1993b) Plant Soil 149:211–218. **Martin F**, Chalot M, Brun A et al. (1992) In: Read DJ, Lewis DH, Fitter AH et al. (eds) Mycorrhizas in ecosystems. CAB Int, Wallingford, pp 311–315. **Martin FM**, Hilbert J-L (1991) Experientia 47:321–331. **Martinez-Amores EMV**, Quintos M (1991) New For 4:237–246. **Massicotte HB**, Melville LH, Li CY et al. (1992) Trees 6:137–146. **McGee PA**, Furby JH (1992) Aust J Bot 40:291–304. **Meier S** (1991) Environ Pollut 73:205–216. **Meng F**, Shao J, Zhao Y et al. (1991) For Res 4:528–532. **Miller SL**, Allen EB (1992) In: Allen MJ (ed) Mycorrhizal functioning, an integrative plant-fungal process. Chapman and Hall, London, pp 301–332. **Miller SL**, Koo CD, Molina R (1992) Mycorrhiza 2:53–61. **Mohan V**, Natarajan K, Ingleby K (1993a) Mycorrhiza 3:39–42. **Mohan V**, Natarajan K, Ingleby K (1993b) Mycorrhiza 3:43–49. **Mohan V**, Natarajan K, Ingleby K (1993c) Mycorrhiza 3:51–56. **Molina R**, Massicotte HB, Trappe JM (1992a) In: Read DJ, Lewis DH, Fitter AH et al. (eds) Mycorrhizas in ecosystems. CAB Int, Wallingford, pp 106–112. **Molina R**, Massicotte H, Trappe J (1992b) In: Allen MJ (ed) Mycorrhizal functioning: an integrative plant-fungal process. Chapman and Hall, London, pp 357–423. **Morgan JL**, Campbell JM, Malcolm DC (1992) In: Cannell MGR, Malcolm DC, Robertson PA (eds) The ecology of mixed-species stands of trees. Spec Publ Ser Br Ecol Soc 11. Blackwell, London, pp 65–85.

Nasholm T, Högberg P, Edfast A-B (1991) New Pnytol 119:83–92. **Natarajan K**, Mohan V, Ingleby K (1992) Soil Biol Biochem 24:279–280. **Newman EJ** (1991) In: Atkinson D (ed) Plant root growth: an ecological perspective. Spec Publ Br Ecol Soc 10: 437–451. **Newton AC** (1992) Mycorrhiza 2:75–79. **Nilsson M-C**, Högberg P, Zackrisson O et al. (1993) Can J Bot 71:620–628.

Orlovich DA, Ashford AE (1993) Protoplasma 173:91–102. **Osonubi O**, Mulongoy K, Awotoye OO et al. (1991) Plant Soil 136:141–143.

Pacheco S, Cambraia J (1992) Rev Microbiol 23:260–263. **Pachlewski R**, Kermen J, Chrusciak E et al. (1992) Acta Mycol 27:49–61. **Pankow W**, Boller T, Wiemken A (1991) Experientia 47:391–394. **Pargney J-C** (1990) Can J Bot 68:2722–2728. **Pargney JC**, Gourp V (1991) Phytomorphology 41:161–173. **Paris F**, Dexheimer J, Lapeyrie F (1993) Arch Microbiol 159:526–529. **Parlade J**, Alvarez IF (1993) Mycorrhiza 3:93–96. **Pasqualini S**, Panara F, Antonielli M (1990) Giorn Bot Ital 124:198–199. **Pasqualini S**, Panara F, Antonielli M (1990) Giron Bot Ital 124:198–199. **Pasqualini S**, Panara F, Antonielli M (1992) Can J Bot 70:1377–1383. **Perry DA**, Bell T, Amaranthus MP (1992) In: Cannell MPG, Malcolm DC, Robertson PA (eds) The ecology of mixed-species stands of trees. Spec Pbl Ser Br Ecol Soc 11 Blackwell, 151–179, London. **Pillukat A**, Agerer R (1992) Z Mykol 58:211–242. **Pirazzi R** (1992) Micol Ital 21:27–36. **Plaits TC**, Laundrymen RG (1991) In: Arora DK, Rai B, Mukerji KG et al. (eds) Handbook of Applied Mycology 1. Marcel Dekker, New York, pp 77–129. **Plassard C**, Scheromm P, Mousain D et al. (1991) Experientia 47:340–349.

Qui Z, Chappelka AH, Somers GL et al. (1993) Environ Exp Bot 33:423–431.

Rabbani S, Chalot M, Botton B et al. (1992) Bull Acad Soc Lorraines Sci 31:131–139. **Read DJ** (1990) Endocytobiology IV. INRA, Paris, pp 99–106. **Read DJ** (1991a) In: Hawksworth DL (ed) Frontiers in Mycology. CAB Int, pp 101–130. **Read DJ** (1991b) Experientia 47:376–391. **Read DJ** (1992) In: Allen MJ (ed) Mycorrhizal functioning, an integrative plant-fungal process. Chapman and Hall, London, pp 102–133. **Read DJ**, Lewis DH, Fitter AH et al. (1992) Mycorrhizas in ecosystems. CAB Int, Wallingford. **Richter DL**, Bruhn JN (1993) Soil Biol Biochem 25:355–369. **Rieger A**, Guttenberger M, Hampp R (1992) Z Naturforsch 47c:201–204. **Roth AL**, Berch SM (1992) Can J For Res 22:1646–1655. **Rousseau JVD**, Reid CPP, English RJ (1992) Soil Biol Biochem 24:183–184. **Rudawska M**, Bernillon J, Gay G (1992) Mycorrhiza 2:17–23. **Ryan EA**, Alexander IJ (1992) In: Read DJ, Lewis DH, Fitter AH, Alexander IJ (eds) Mycorrhizas in ecosystems. CAB Int, Wallingford, pp 237–245.

Salzer P, Hager A (1991) Bot Acta 104:439–445. **Salzer P**, Hager A (1993) Physiol Plant 88:52–59. **Sarjala T** (1993) Tree Physiol 12:93–100. **Scheidegger C**, Brunner I (1993) New Phytol 123:123–132. **Schmid E**, Oberwinkler F (1993) New Phytol 124:69–81. **Shafer SR**, Schoeneberger MM (1991) Environ Pollut 73:163–177. **Shaw PJA**, Dighton J, Poskitt J et al. (1992a) Mycol Res 96:785–791. **Shaw PJA**, Dighton J, Poskitt J (1992b) In: Read DJ, Lewis DH, Fitter AH et al. (eds) Mycorrhizas in ecosystems. CAB Int, Wallingford, pp 208–213. **Simoneau P**, Viemont JD, Moreau JC et al. (1993) New Phytol 124:495–504. **Smits WTM** (1992) In: Read DJ, Lewis DH, Fitter AH et al. (eds) Mycorrhizas in ecosystems. CAB Int, Wallingford, pp 283–292. **Söderström B** (1992) In: Read DJ, Lewis DH, Fitter AH, Alexander IJ (eds) Mycorrhizas in ecosystems. CAB Int, Wallingford, pp 77–83. **Strzelczyk E**, Pokojska A, Kampert M (1992) Symbiosis 14:201–215. **Stulen I**, den Hertog J (1993) Vegetation 104/105:99–115. **Sun Y-P**, Fries N (1992) Mycorrhiza 1:63–69. **Svenson SE**, Davies FT, Meier CE (1991) Hortscience 26:1406–1409.

Tam PCF, Griffiths DA (1993a) Mycorrhiza 2:125–131. **Tam PCF**, Griffiths DA (1993b) Mycorrhiza 3:165–170. **Termorshuizen AJ** (1993) For Ecol Mange 57:179–189. **Termorshuizen AJ**, Ket PC (1991) Eur J For Pathol 21:404–413. **Theodorou C** (1993) Soil Biol Biochem 25:647–649. **Thoen D**, Sougoufara B, Dommergues Y (1990) Can J Bot 68:2537–2542. **Thomson BD**, Malajczuk N, Grove TS et al. (1993) Mycol Res 97:839–844. **Timonen S**, Finlay RD, Söderström B et al. (1993) New Phytol 124:83–92. **Tinker**

PB, Jones MD, Durall DM (1992) In: Read DJ, Lewis DH, Fitter AH et al. (eds) Mycorrhizas in ecosystems. CAB Int, Wallingford, pp 303–310. **Torrey JG** (1992) Can J For Res 22:1815–1823. **Tosh JE**, Senior E, Smith JE et al. (1993a) Lett Appl Microbiol 16:187–191. **Tosh JE**, Senior E, Smith JE et al. (1993b) Mycol Res 97:129–135. **Trappe JM**, Luoma DL (1992) In: Caroll GC, Wicklow DT (eds) The fungal community: its organization and role in the ecosystem, 2nd edn. Mycological series 9: Dekker, New York, pp 17–27. **Turnau K**, Kottke I, Oberwinkler F (1993) Bot Acta 106:213–219. **Turner IM**, Brown ND, Newton AC (1993) For Ecol Manage 57:329–337.

Väre H, Vestberg M, Eurola S (1992) Mycorrhiza 1:93–104.

Walker RF, McLaughlin SB (1991) For Ecol Mange 46:123–133. **Wallander H**, Nylund J-E (1991) New Phytol 119:405–411. **Wallander H**, Nylund J-E (1992) New Phytol 120:495–503. **Wallander H**, Nylund J-E, Sundberg B (1992) Mycorrhiza 1:91–92. **Waller K**, Agerer R (1993) Sendtnera 1:23–38. **Waller K**, Raidl S, Agerer R (1993a) Z Mykol 59:141–153. **Waller K**, Agerer R, Brand F et al. (1993b) Sendtnera 1:11–22. **Weiss M** (1992) Cryptogam Bot 2:337–344. **Wingler A**, Guttenberger M, Hampp R (1993) Mycorrhiza 3:69–73. **Wu Y**, Gale MR, Cattelino PJ et al. (1993) Can J For Res 23:810–815.

Zambonelli A, Giunchedi L, Poggo Pollini C (1993) Symbiosis 15:71–76. **Zhao Z**, Wang T, Guo X (1992) Acta Microbiol Sin 32:227–232. **Zhou Y**, Liu C, Qi Y et al. (1991) For Res 4:528–532.

Professor Dr. Reinhard Agerer
Institut für Systematische Botanik
der Universität München
Menzingerstr. 67
80638 München, Germany

II. Symbioses: Mycorrhizae

By Fritz Schönbeck and Ingo Raschen

1. VA-Mycorrhizae

a) Taxonomy of Endophytes

Taxonomy has been one of the major problems in mycorrhizal research ever since the discovery of endomycorrhizal fungi. Identification of individual species is a prerequisite for taxonomy; however, owing to the fact that hyphae of different fungi are often virtually indistinguishable from one another, traditional means of taxonomy still rely mostly on morphological characters of resting spores in soil, such as spore size, shape, wall structure and composition, color, and ornamentation. Unfortunately, some of these characters vary not only with the developmental stages of spores as these are formed, but are also instable after fixation, resulting in difficulties producing viable specimens for future reference.

Spain (1992) points out that the ephemeral nature of germination shields reported for several *Acaulospora* and *Entrophospora* species is probably due to the mountants commonly used, such as polyvenyl-lactic acid-glycerine and lactophenol. These rather delicate structures can be seen in plane view and may hence be employed for taxonomic purposes if spores are mounted in water. In order to gain more detailed information on the ultra structures of spore walls of glomalean fungi the use of electron microscopy (EM) is suggested. Among nine methods of fixation of material for EM studies none was found suitable for spores of *Acaulospora*, *Gigaspora*, and *Scutellospora*, while one method proved efficient for those of *Glomus* species (Maia et al. 1993). EM investigations on *Glomus globiferum* and *G. fuegianum* provided some detailed information on spore ontogeny (Wu and Sylvia 1993) and spore wall ultrastructure (Yao et al. 1992), respectively. A rather different approach to identification is the use of N/C ratios in spores. Li et al. (1992) report that both the percentages and ratios of N and C in spores and sporocarps differed between genera and individual species and that the differences they observed were independent of host plants or other environmental factors. It is therefore suggested that these features may provide an additional indicator for classification and identification of VA-mycorrhizal fungi.

Progress in Botany, Vol. 56

General aspects of classification of VA-mycorrhizal fungi are reviewed by a number of authors. Wilcox (1991) provides a global review on classification of all mycorrhizal and pseudomycorrhizal systems currently known, while Simon et al. (1993) give sound genetic evidence for the classification in use. To investigate the origin and to clarify the phylogenetic relationships of VA-mycorrhizal fungi, ribosomal DNA genes from 12 species were sequenced. Phylogenetic analyses confirmed the existence of three families within arbuscular fungi on the basis of morphological characters. Approximal dates for the divergence of major branches on the phylogenetic tree were obtained and include an estimate for the origin of VAM-like fungi of 353–462 million years ago, which is consistent with the hypothesis that VA-mycorrhizal fungi were instrumental in the colonization of land by ancient plants. Morton (1993) discusses the status of glomalean systematics, identifying problem areas and proposing some solutions. Furthermore, he presents his conclusions within a hierarchical framework to pinpoint levels where processes generating biotic and ecological diversity interact, and where they are independent of each other, finally considering the implications of systematic relationships for mycorrhizal studies.

b) Physiology and Effects on Plants

In the past, much research has been carried out on different mycorrhizal associations in terms of effects mediated by increased phosphorus levels and improved nutrition in mycorrhizal systems. Results obtained were often of descriptive nature, while there were comparatively few reports on underlying mechanisms of plant/fungus interactions. Fortunately, this has started to change in recent years and more information is becoming available on mycorrhizal physiology and effects other than those obviously phosphorus-mediated.

Employing new techniques, more detailed research has been carried out on the development of different arbuscular mycorrhizal fungi. Scanning electron microscopy of *Glomus mosseae* revealed rather detailed information on the ultrastructure of the peridium during development. The presence of fine radial canals and lignituber-like ingrowths in both de-peridiate spore wall and shed peridium, compared with other sporocarpic genera, may indicate similarities in the processes of peridium formation among these fungi, even though more research is needed (Meier and Charvat 1992).

Observation of germination and subsequent hyphal growth of *Scutellospora gregaria* spores isolated from marine sand dunes led to the conclusion that there is a clear difference in fungal morphogenesis between the extramatrical and intramatrical phase when fungal development is under host control (Bianciotto et al. 1989). Further in vitro experiments employing root-inducing T-DNA-transformed carrot roots and *Gigaspora margarita* revealed that ATPase activity in fungal plasmalemma was essential for hyphal growth and directly related to the presence of stimuli from root exudates and volatiles (Lei et al. 1991). Subsequent studies showed that hyphal growth of VAM fungi is promoted in vitro in the presence of 2% CO_2 and flavonols, while all flavones, flavonones, and isoflavones tested were generally inhibitory. Quercetin (10 μM) prolonged hyphal growth of germinated spores of *Gigaspora margarita* from 10 to 42 days and also stimulated growth of *Glomus etunicatum*. Glycosides of quercetin, rutin and quercitrin were found not to be stimulatory (Becard et al. 1992). These authors support that there are rather specifically controlled interactions between mycorrhizal fungi and their hosts. Types and specificities of mycorrhizal associations are outlined by Bonfante (1991) comparing infection processes of mycorrhizal and pathogenic fungi. Considering plant defence reactions observed, she draws some conclusions on differentiation of haustoria and arbuscules and develops a theory of nonaggressive strategies employed by VAM-fungi.

DNA analysis showed that nuclei of *G. mosseae* and *G. versiforme* were well developed in young expanding hyphae, whereas in those of older hyphae, especially of collapsing arbuscules, chromatin appeared condensed and nuclei were pyknotic with blebs in the envelopes and conspicuous spaces between envelopes and nucleoplasm. Since nuclear degeneration was reflected by decreased immunogold labeling, this is suggested as a parameter for evaluation of viability and provides further evidence of the ephemeral life of arbuscules (Balestrini et al. 1992). DNA contents of *G. versiforme* and *Gigaspora margarita* differed significantly with 0.25–0.27 and 0.74–0.77 pg DNA per nucleus, respectively, and there were no differences in genome size at different stages in the life cycle of either species (Bianciotto and Bonfante, 1992). That there are host/fungus interactions controlled at the DNA level was shown by Garcia-Garrido et al. (1993), who also demonstrated the appearance of eight new polypeptides in onion roots colonized by *Glomus mosseae* and the disappearance of seven host polypeptides upon establishment of the VA-mycorrhiza. In tobacco roots, the pathogenesis-related protein PR-b_1 was found to be present in roots infested with a pathogen, but absent in the association with *Glomus mosseae*. Furthermore, indirect immunogold labeling showed that the respective antigen was strictly limited to host material in the interfacial matrix directly surrounding living intracellular fungal arbuscules in parenchymal cortical cells (Gianinazzi-Pearson et al. 1992). Yet more evidence for rather specific interactions comes from a report by Dumas-Gaudot et al. (1992). They report on the induction of two specific acidic chitinases with an estimated molecular size of about 33 kDa after infection of tobacco roots with *G. versiforme*, *G. intraradices*, or *G. fasciculatum*, while other known isoforms were induced by both, VA-mycorrhizal fungi and the root pathogen *Chalara elegans*. Palma et al. (1993) conclude from their experiments on *Trifolium patense* and *G. mosseae* that there are two isoforms of superoxide dismutase specifically induced by the VA-mycorrhizal fungus. A more complex hypothesis on the role of endochitinase, β-1,3-endoglucanase and chalcone isomerase on the one hand and P concentration in soil on the other hand in the control of mycorrhizal infection of *Phaseolus vulgaris* by *G. intraradices* is propounded by Lambais and Mehdi (1993). They suggest that there is a general suppression of the plant defence response during VA-mycorrhizal development, and discuss possible mechanisms of regulation and functions of endochitinase and β-1,3-endoglucanase in VA-mycorrhiza. However, that endochitinase activity alone does not interfere with VA-mycorrhizal development was demonstrated on transgenic tobacco plants expressing active chitinase constitutively without any influence on VA-mycorrhiza (Vierheilig et al. 1993). A change of endoglucanase activity at the logarithmic stage of VAM development in lettuce roots and a subsequent decline was evident in the lettuce/*G. fasciculatum* system. Results presented suggest that endoglucanases are rather likely to be involved in the process of colonization (Garcia-Garrido et al. 1992). In planta histochemical staining for fungal alkaline phosphatase activity revealed that this enzyme was induced by colonization of host roots. However, while the enzyme was induced in the *Platanus acerifolia*/*Glomus* sp., it could not be detected in myc$^-$ and nod$^-$ mutants of peas (Tisserant et al. 1993). These results suggest that complex signaling between host and fungal symbiont is required during the establishment of a functional symbiosis, similar to that of nitrogen-fixing symbioses between legumes and *Rhizobia*.

That these regulatory signals do not necessarily have to originate from the organ directly involved in the interactions, but may arise from the plant as a whole, was shown by Gianinazzi-Pearson and Gianinazzi (1992), carrying out stem grafts between mycorrhizal and non-mycorrhizal legumes. They discovered that self-grafts of three pea varieties affected neither development nor arbuscule formation of *G. intraradices* and *G. mosseae*, while lupin scions greatly reduced colonization and completely inhibited arbuscule formation of both fungi in grafted pea root stocks. Colonization of plant roots by VA-mycorrhizal fungi often also depends on external environmental factors, such as the concentration of available P in soil. McArthur and Knowles (1992) investigated in the response of potato to available P in terms of increased resistance to VA-mycorrhiza at different P levels. They conclude that *G. fasciculatum* alters enzyme activities in the phenolic pathway of plants, resulting in reduced ethylene production and defence reactions of the root. Raising external P supplies partly restored different enzyme activities and thus ethylene production, and may in this way increase the root's resistance to VAM infection.

That VA-mycorrhiza are part of, and have many beneficial effects on, the vast majority of vascular plants has been known for a long time. Since literature on effects of improved P status and plant nutrition has been reviewed rather extensively in the past the so-called P- and nutritional effects will be omitted here, even though both still constitute major areas in VA-mycorrhizal research today. However, it is perhaps worth mentioning here that VAM fungi were demonstrated to transport considerable amounts of P and N from soil zones several centimeters from the root (George et al. 1992) and may increase ammonium assimilation, glutamine production, and xylem translocation. Unlike ectomycorrhizal fungi, they do not appear to alter the pathway of ammonium assimilation in roots (Cliquet and Stewart 1993). Furthermore, Frey and Schüepp (1993) were able to demonstrate interplant transfer of ^{15}N between *Trifolium alexandrinum* inoculated with *G. intraradices* and maize plants or apple seedlings, and Hamel and Smith (1992) concluded that N transport was mainly by means of direct transfer through hyphae. Intercropping soybean and maize, they found that the amount of ^{15}N transferred was considerable between plants grown in the same row, but negligible between different rows. That VA-mycorrhizal fungi also mediate the transport of carbohydrates from plant to plant is reported by Martins (1992, 1993) using ^{14}C-labeled CO_2. He claims that 46% of the total amount of labeled ^{14}C transferred from mycorrhizal *Lolium perenne* to *Plantago lanceolata* associated with the same mycorrhizal fungus was due to direct fungal transfer via mycorrhizal hyphae, 15% was due to indirect transfer mediated by mycorrhizal mycelium, and 39% to indirect transfer not mediated by the mycorrhizal mycelial net.

As pointed out above, there is in most cases a fundamental physiological difference between mycorrhizal and nonmycorrhizal plant roots. However, physiological effects may also be observed in shoots. Many cultural stresses may be alleviated by inoculation with VA-mycorrhizal fungi, the management of which in agroecosystems should be studied in more detail, since these organisms are of vital importance to sustainable agriculture (Johnson and Pfleger 1992). Augé and Duan (1991) propose that VA-mycorrhizal colonization has pronounced effects on stomatal control when draught stress is applied to *Rosa hybrida* cv. Samantha colonized by *G. intraradices*. Similar effects are also observed for *Acacia albida* and *A. nilotica* inoculated with *G. clarum* (Osonubi et al. 1992) and for cowpea and soybean inoculated with *G. intraradices* and *G. mosseae*, respectively. It is concluded that mycorrhizal systems either scavenge water more effectively or exhibit a different root-to-shoot communication (Augé et al. 1992) and thus lead to improved draught tolerance of mycorrhizal plants.

Looking at all the information on VA-mycorrhizal physiology that has become available within the last 2 or 3 years, it is evident that we are approaching a detailed understanding of the overall principles of the physiology of this mutual symbiosis, most complex in organization. Moreover, the above results clearly leave much scope for, and justify the enthusiasm in VA-mycorrhizal research.

c) Ecology

There is a considerable species diversity in mycorrhizal fungi in all common soils, often including all three main types of mycorrhizal communities, ericoid, ecto- and vesicular-arbuscular mycorrhiza. The species compositions of communities with reference to field and microcosm studies are discussed by Read (1992). It is suggested that there is a close correlation between the biochemical potential shown by mycorrhizal fungi and the chemical complexity of the nutritional resources in the soils these inhabit. That nutritional factors are not the only ones determining the species composition of a VA-mycorrhizal community is shown by Kühn (1991) and Kühn et al. (1991), comparing species compositions of two fallow agricultural sites in a wet and a dry habitat. Examining 43 species of flowering plants in both habitats, 40 and 29 were found to be mycorrhizal in the dry and wet habitat, respectively. Providing a complete list of host plants examined and VA-mycorrhizal fungi found they also discuss the importance of nonstained mycorrhizal root sections for isolation of efficient VAM isolates. Kahn (1993) reports on the occurrence of VAM in aquatic trees of New South Wales, and concludes that even within the roots of one tree VA-mycorrhiza are only found in the terrestrial root sections least exposed to flooding, while aquatic root sections were always found to lack mycorrhizal infections. Thus, it is obvious that mycorrhizal propagules frequently found in aquatic sediments must result from surface runoff. These results are consistent with other reports suggesting that there generally is a negative correlation between waterlogging and a low redox potential and VA-mycorrhizal abundance.

Superimposed on the general mycorrhizal frequency of any particular soil there are also seasonal variations in population dynamics. Investigations of 15 host species in eroded soils from southern Spain revealed that mycorrhizal development was closely related to host physiology with maximum spore numbers occurring during fruit-bearing periods of the plants investigated, remaining high during autumn, falling to a minimum in winter, and increasing again in spring. Root infections were at a maximum as plants flowered and decreased to a minimum in summer (López-Sánchez and Honrubia 1992). Furthermore, Sanders and Fitter (1992a, 1992b) reported a more or less constant pattern of population dynamics for some host plants over 2 years, while other hosts on the same site showed considerable variations within 1 and between the 2 years of investigation. Results are discussed in relation to timing of P influx and demands of different hosts, concluding that the mycorrhizal associations studied are still most likely to be mutualistic. The total number of mycorrhizal propagules in agricultural soils in north Germany was subject to seasonal dynamics similar to those reported above, but was also found to vary with applications of fertilizers, fungicides, and crops grown. Yet there appear to be characteristic species compositions for each soil investigated (Land et al. 1993). An et al. (1993) also found similar seasonal variations in mycorrhizal population dynamics in tobacco and tall fescue and, employing tetrazolium blue viability staining of mycorrhizal roots, Bentivenga and Hetrick (1992) were able to show that VA-mycorrhizal activity very closely resembled host activities of cold and warm season grasses, the dependence on which was most pronounced at the temperature that favored growth. Surprisingly, colonization was relatively high in both groups of grasses late in the growing season, suggesting that mycorrhizal fungi may proliferate internally and even be parasitic at this time, and that cold season grasses may receive a greater benefit from the symbiotic association, since these plants would still be actively growing under relatively cool temperature regimes. Seasonal population dynamics of *G. intraradices* closely related to physiological age and root activities of mung bean and cashew were observed by Haugen and Smith (1992). Comparing spore germination at a variety of temperatures, they established that it was most rapid and peaked after 3 weeks at 30 °C in soil. Furthermore, early rapid infection led to greater plant growth of mung beans and cashew under these conditions, even though timing and extent of the effects differed between host plants. However, it must be stressed that VA-mycorrhizal fungi – even different isolates considered the same species – are individualistic in their responses to edaphic factors (Johnson et al. 1991). Therefore, a high diversity of different VAM fungi capable of responding to varying environmental conditions and stresses may contribute to high productivities of soils, as demonstrated in crop rotation experiments by Ellis et al. (1992). In general, one might conclude that mycorrhizal symbioses may still be mutualistic, even where nutrient supply and demand do not necessarily coincide, since plants may store some of the nutrients taken up early during the symbiosis.

VA-mycorrhizal fungi play an important role in natural successions in ecosystems. The role of VAM fungi in competitive interactions among native bunchgrasses and alien weeds in Oregon, USA is reviewed by Goodwin (1992), providing a background for discussion of their role in inter- and intraspecific competition in sagebrush (*Artemisia tridentata*) grass systems, and Rincón et al. (1993) present results suggesting that these fungi may also be of importance for the establishment, growth, water relations, and nutrition of cacti (e.g., *Pachycereus pecten-aboriginum*) in the arid tropics. Nelson and Allen (1993) report that plants grown in a particular soil alter the species composition of VA-mycorrhizal fungi and vice versa. From experiments on grass land restoration, they report three different *Glomus* sp. in natural grass land where *Stipa pulchra* is predominant. After the introduction of *Avena barbata*, the same three species were found, but at different relative abundancies. Furthermore, mycorrhizal fungi in the composition found favored *A. barbata*, conferring improved competitiveness to the plant. Yet the original fungal species composition returned relatively fast as the original plant species was reintroduced. They conclude therefore that inoculation with mycorrhizal fungi alone will not suffice, and the usual practices for control of weed competition need to be employed for re-establishing *S. pulchra*.

The presence of fungal propagules is a prerequisite for successful establishment of VA-mycorrhiza. There are, however, different fungal structures capable of initiating new infections. This may be vital for spreading of the mycelium and re-infecting plant roots after mechanical soil disturbance. Investigating on VA-mycorrhizal distribution of propagules in forest soils and an annual pasture using *Trifolium subterranum* as bait plant, Jasper et al. (1991) demonstrate that disturbance of a forest soil has rather drastic effects on recolonization, while the annual pasture investigated was comparatively little affected. They attribute the effects observed to much higher propagule numbers in a pasture, mainly consisting of mycorrhizal root residues. Comparing these residues of mycorrhizal roots with hyphae and spores for their abilities to start new infections, hyphae were found to be least effective, spores may initiate new infections after 28, and infected root debris in less than 14 days, thus constituting the most important inoculum which is commonly found within the upper 15 cm of the top soil (Bellgard 1992, 1993a,b).

Similar results are reported for five nurseries in Ethiopia and Somalia. The use of soil from densely vegetated areas, mainly top soil, is implemented in nursery practices to provide seedlings grown with indigenous mycorrhiza and rhizobia (Michelsen 1992). Also, Vörös and Szegi (1991) found that application of top soil to mine spoils in Yugoslavia resulted in good revegetation and mycorrhizal abundance after 7 and 14 years, while VA-mycorrhiza were generally absent in noninoculated sites. Investigating disturbed and revegetated sites from La Gran Sabana, Venezuela, Cuenca and Lovera (1992) found that the use of native mycorrhizal inoculum provided for revegetation with common mycorrhizal plants, whereas in noninoculated areas nonmycotrophic Amaranthaceae predominated.

That ecology and interactions of these fungi and the effects which have to be considered when assessing ecological impacts in terms of resulting plant fitness are rather complex in nature is demonstrated by Bellei et al. (1992) and Koide and Lu (1992), respectively. Showing that *Eucaplyptus viminalis* associates with a variety of different fungi during plant development, the authors of the former article point out that there are many woody plants engaging in both ecto- and VA-mycorrhiza during plant development. They discuss the implications of some ecological factors in dynamics of the dual mycorrhizal system in relation to their findings. The authors of the latter article suggest that mycorrhizal infection may influence plant fitness by increasing offspring vigor and offspring reproductive success is addition to previously reported increases in maternal fecundity.

VA-mycorrhizal fungi are a major factor in natural soil fertility. However, where and how these organisms fit into the overall hierarchy of components involved is rather difficult to establish. While the theory of soil conservation and the hierarchical model, as well as the role of VA-mycorrhizal fungi within this model are reviewed by Miller and Jastrow (1992), citing several case studies from field research, including tillage practices of prairie restoration, and discussing further research directions, Christensen and Jakobsen (1993) point out that interactions between VA-mycorrhizal fungi and soil microbes may be of considerable importance. Experimenting on the rhizosphere of cucumber they show that VA-mycorrhizal fungi may decrease total bacterial biomass in the rhizosphere and may also result in altered spatial patterns of growth. It is suggested that VAM may compete for root-derived carbohydrates and organic matter, resulting in increased spatial bacterial viabilities and species diversities in soil. In general, VA-mycorrhizal plants seem better adapted to compete with saprophytic soil microflora for common nutrients, such as N and P. Krishnaraj and Sreenivasa (1992) found that certain rock phosphate-dissolving bacteria were much more abundant in the vicinity of *Capsicum annuum* roots inoculated with *Glomus fasciculatum* and *G. macrocarpum* than in noninoculated systems. Pronounced synergistic effects in terms of growth responses were observed for *G. fasciculatum* and *Pseudomonas striata* with increased levels of rock phosphates (Sreenivasa and Krishnaraj 1992).

d) Practical Applications

Since the importance of VA-mycorrhiza and their role in sustainable agricultural production has been recognized much research has been devoted to inoculum production and effectiveness in different host/mycorrhiza combinations. Host dependence on the VA-mycorrhizal symbiosis varies with soil type and phosphorus availability, and effectiveness also depends on the host plant involved. Thus, Hao et al. (1991) carried out extensive experiments on mycorrhizal development in relation to host plants and soil types, providing information on the efficiency of VAM fungi and the prospect of their practical application in some soils in China. With particular reference to the practical agricultural conditions met in Zaïre 19 commonly cropped species were assessed for mycorrhizal dependency and effects of supplemental inoculation. The results presented clearly show that added endomycorrhizal inoculum improves root infection and plant growth significantly in mycotrophic species (Khasa et al. 1992).

One of the main problems, however, is the availability of inoculum in appropriate quantities and qualities. While adding top soil with sufficient mycorrhizal propagules often constitutes the simplest solution, it may not be readily available, so that inoculum has to be produced commercially. The uses of different culturing techniques, such as root hair cultures, hydroponic cultures in expanded clay particles, aeroponic and aero-hydroponic systems in different areas of plant research, such as cell cultures and micropropagation, also including VA-mycorrhizal research, is reviewed by Weathers and Zobel (1992). Sulla (1991) provides a review of systems especially adapted for VAM inoculum production. The Beltsville method for soilless production of VA-mycorrhizal inoculum (Millner and Kitt 1992) relies on sand cultures automatically watered and fertilized. Futhermore, the pH of the substrate may be adjusted to meet specific culture requirements with the result that many different fungi may be cultured and produce sufficient amounts of spores, exceeding those commonly found in soil systems on a variety of hosts. This method is suitable where large quantities of spores are required, either as inoculum or for research.

Spores may, however, not always be the inoculum source favored. Expanded clay particles provide an excellent carrier material, as shown by Grunewaldt-Stöcker and Dehne (1989). To assess the quality of expanded clay particles as inoculum carrier fluorescence microscopy was employed scrutinizing particle surfaces for the presence of mycorrhizal structures. Even though these were found in more than 90 % of the particles studied, rather low frequencies of chlamydospores were observed, regardless of the fungus cultured. Both the chlamydospores found and hyphae showed no differences to those found in soil. Subsequent bioassays confirmed good infectivities, even at high dilutions, suggesting that chlamydospores are of minor importance while mycelial fragments well protected within the particles cavities may play a more important role as source of inoculum than previously thought. Dried mycorrhizal plant roots comprising several naturally occurring VA-mycorrhizal fungi, predominantly *Glomus intraradices*, are also reported to be suitable as mass inoculum with the advantage of being low in weight and small in volume (Nemec 1992). Furthermore, Singh (1992) reports that fungal development and sporulation are enhanced by the presence of *Azospirillum brasilense*, while von Alten et al. (1993) present similar results for *Bacillus mycoides*. Champawat (1990) concludes that dual inoculation with *Gigaspora calospora* and *Rhizobium* resulted in synergistic effects and better sporulation. Thus the use of additional microorganisms may improve inoculum production and establishment of VA-mycorrhizal fungi under field conditions. In a review on mycorrhiza preparations, Myers (1993) concludes that commercial products currently available can be of financial and ecological benefit mainly in nutritionally poor soils.

There is accumulating evidence that VA-mycorrhizal fungi may also be applied to control plant disease beyond the mere improvement of plant health by means of enhanced nutrient acquisition. Research on VAM/pathogen interactions, with emphasis on mechanisms of altered plant resistance to pathogens, is reviewed by Sharma et al. (1992). They propose a working model to explain VAM/disease interaction under varied environmental conditions. Several other reports concern results suggesting that VA-mycorrhizal fungi constitute a viable means of controlling fungal diseases, rendering plants more resistant or controlling root nematodes effectively. Under greenhouse conditions, *Sclerotium rolfsii* on chilli and *S. rolfsii* Sacc. on wheat are controlled to acceptable degrees by simultaneous application of *Glomus fasciculatum* and *G. macrocarpum* both in single and dual inoculation (Sreenivasa et al. 1992; Harlapur et al. 1990, respectively), a negative correlation between mycorrhizal infection levels and *Trichosporium vesiculosum* was observed in casuarina (Gunjal and Patil 1992) and that *Glomus* spp. may also control *Pythium ultimum* effectively in cucumber grown in vermiculite is shown by Rosendahl et al. (1992). Although infection levels of *Rhizoctonia solani* were not reduced by *G. leptotichum*, *Acaulospora morrowae*, and *Gigaspora margarita*, tomato plants infested are still shown to be much more resistant to the disease than nonmycorrhizal controls (Cassiolata and De Melo 1991). Another interesting report comes from Guo et al. (1992), who observe that an *Acremonium* endophyte of tall fescue caused considerable reductions of VAM infection levels and spore formation, even when inoculated on pearl millet. Most impressive results controlling root knot nematodes (*Meloidogyne* spp.) were achieved on *Phaseolus vulgaris* (Osman et al. 1991), yellow and red pitaya (Palacino and Leguizamon 1991), tobacco (Krishna and Prasad 1991), and *Piper nigrum* (L.) (Sivaprasad et al. 1992). Furthermore, Francl (1993) points out that phosphate fertilizer alone is unable to duplicate the effects of mycorrhizae on plant vigor in soils infested with nematodes. In *Actinidia deliciosa*, inoculation with *Glomus etunicatum* resulted in improved resistance towards *M. javanica* without affecting the rate of reproduction of the nematode (Verdejo et al. 1990).

The results presented clearly indicate that VA-mycorrhizal fungi can be applied successfully to improve plant nutrition and health, and to control certain plant diseases, some of which are not readily controllable otherwise. Under practical conditions there are, however, often several problems to be solved and combined application of pesticides with VA-mycorrhiza may be inevitable. Thus further research was needed to assess both the effects of pesticides on VA-mycorrhizal fungi in soil and the influence of these fungi on the fate of a pesticide, its metabolism and metabolic products.

Nelson and Khan (1992) demonstrated that hyphal systems of *Glomus* spp. were generally capable of removing the herbicide atrazine from soil. They conclude that VAM fungi exert significant influence on the fate of atrazine in agricultural systems. Herbicides may also lead to inimical conditions for VA-mycorrhizal fungi in soil. Tebuthiuron, for example, may have detrimental effects on mycorrhizal fungi, depending on the host they are engaged with (Allen and West 1993). The use of cytokinins and thidiazuron stimulates mycorrhizal development in tomatoes and increases infection rates and yields by about 35 and 28%, respectively (Nasr 1993).

Among several fungicides investigated, mepronil, furalaxyl, and thiram may affect VA-mycorrhizal fungi adversely in a dosage-related manner, while the use of benomyl (Perrin and Plenchette 1993) and chlorothalonil (Habte et al. 1992) must generally be avoided in combination with mycorrhizal endophytes. Furthermore, different fungal isolates may differ in their sensitivities to fungicides. An isolate of *G. aggregatum* was shown to be extremely sensitive to thiram, while the indigenous population was not affected at all at the concentration applied (Habte and Manjunath 1992).

2. Ericales-Mycorrhizae

Taxonomy and identification have always been areas where new molecular techniques have been applied to yield reliable results. According to r-RNA analysis, there is strong evidence to suggest that *Scytalidium vaccinii* and *Hymenoscyphus ericae* must be considered anamorph and teleomorph states of a single taxon, even though there was some variation among the two isolates. This finding was further supported by morphological studies (Egger and Sigler 1993). While much research has been devoted to VA-mycorrhizal systems, comparatively few reports have been released concerning ericoid mycorrhiza. It has been known for some time that there are fungal species belonging to several different fungal orders involved in ericoid mycorrhiza. However, Duckett and Read (1991) report that it is possible to differentiate between asco- and basidiomycetous structures in host tissue. They used the lipophilic fluorochrome 3,3′dihexyloxacarbocyanine iodine [$DiO_6(3)$] at a concentration of 0.1–0.5 μg/ml to selectively stain for ascomycetous hyphae in ericaceous roots and in rhizoids of five families of liverworts. The same dye applied at a concentration of 50 μg/ml, however, also stained for bisidiomycetous structures, while VA-mycorrhizal structures did not stain at all. In contrast to other fluorescent probes recognizing fungal wall components, the dye applied here is a vital stain of fungal cytoplasm. Looking at host/fungus interactions, Berta et al. (1992) report that *H. ericae* caused nuclear hypertony in cells of infected hair roots of *Calluna vulgaris* and *Vaccinium myrtillus*. The hypertony was not observed in nonmycorrhizal plants, and is explained by a lower degree of chromatin condensation, rather than by polyploidation. Furthermore, Lemoine et al. (1992) and Straker et al. (1992) describe an acid invertase and an acid phosphatase of fungal origin, the activity of which is inhibited upon contact with host or nonhost root. They discuss possible implications of these fungal enzymes in ericoid endomycorrhizal infection processes, together with mechanisms that may regulate enzyme activities.

3. Orchidaceae-Mycorrhizae

Mycorrhizal associations in the terrestrial and aerial roots of 20 tropical orchid species and hybrids were examined by Goh et al. (1992). Apart from aerial roots, plants were generally found to harbor extensive mycorrhizal fungi. Thus, it is concluded that non-mycorrhizal aerial root explants could be used for micropropagation. In vitro, most isolates of anastomosis groups of *Rhizoctonia solani* (including some binucleate anastomosis groups) originating from nonorchid sources induced the symbiotic germination of *Spiranthes sinensis* var. *amoena*, whereas no germination occurred when seeds were inoculated with *Sebacina vermifera* (*R. globularis*) and *Waitea circinata* (*R. zeae* and *R. oryzae*) (Masuhara et al. 1993). To investigate beneficial effects of mycorrhizal fungi on *Cattleya* hybrids, mycorrhizal and nonmycorrhizal plantlets were grown on split dishes containing different amounts of starch as carbon source and different nutrient minerals on one side and water agar on the other. Results suggest that there is little evidence of nutrient translocation by the mycorrhizal fungus from the nutrient to the non-nutrient side in the absence of carbohydrates. Mycorrhizal infection is discussed as a mechanism for heterocyclic carbon assimilation. In advanced leafy orchids of *Cattleya*, external carbon resulted in increased root growh, decreased shoot/root ratio and sometimes yellowishgreen plantlets (Beyrle and Smith 1993). Seasonal and climatic regulation of plant establishment in *Dactylorhiza majalis* was investigated by Rasmussen and Rasmussen (1992) carrying out experiments under different controlled environmental conditions. Plants were germinated in vitro with the compatible fungus *Tulasnella calospora*, and results suggest that seeds normally germinate in autumn when saprophytic nutrition of the mycelium is at an optimum. Requirement of a cold season for further seedling differentiation may be restrictive for the distribution of this species. Investigating on the effect of nitrogen concentration in the association of *D. incarnata* (L.) Soó and *Rhizoctonia* sp., it was found that increased N concentrations may result in a change from mycorrhizal to parasitic behavior of the fungus, while the carbon source did not influence fungal behavior. Further research showed that phytohormone levels remained low under asymbiotic conditions, while the presence of the fungus led to a large increase in phytohormone concentration, correlating with growth stimulation in mycorrhizal protocorms. It is suggested that phytohormones are derived mainly from the mycorrhizal fungus (Beyrle et al. 1991).

References

Allen EB, West NE (1993) Mycorrhiza 3(2):75–78. **von Alten H**, Lindemann A, Schönbeck F (1993) Mycorrhiza 2(4):167–173. **An ZQ**, Guo BZ, Hendrix JW (1993) Soil Biol Biochem 25(7):813–817. **Augé RM**, Duan XR (1991) Plant Phys 97(2):821–824. **Augé RM**, Stodola AJW, Brown MS, Bethlenfalvay GJ (1992) New Phytol 120(1): 117–125.

Balestrini R, Bianciotto V, Bonfante-Fasolo P (1992) Mycorrhiza 1(3):105–112. **Bécard G**, Douds DD, Pfeffer PE (1992) Appl Environ Microbi 58(3):821–825. **Bellei M**, Garbaye J, Gil M (1992) For Ecol Manage 54(1–4):205–213. **Bellgard SE** (1992) Mycorrhiza 1(4):147–152, Bellgard SE (1993a) Mycorrhiza 3(1):19–24; Bellgard SE (1993b) Mycorrhiza 3(1):25–29. **Bentivenga SP**, Hetrick BAD (1992) Can J Bot 70(8): 1596–1602. **Berta G**, Sgorbati S, Giaccone P, Gianinazzi-Pearson V (1992) Protoplasma 170(3–4):160–165. **Beyrle HF**, Smith SE (1993) Mycorrhiza 3(2):57–62. **Beyrle HF**, Penningsfield F, Hock B (1991) New Phytol 117(4):665–672. **Bianciotto V**, Bonfante P (1992) Myc Res 96(12):1071–1076. **Bianciotto V**, Palazzo D, Bonfante-Fasolo P (1989) Allionia 29:17–24. **Bonfante P** (1991) Petria 1(3):157–169.

Cassiolata AMR, De Melo IS (1991) Summa Phytopathol 17(3–4):195–200. **Champawat RS** (1990) Folia Micorbiol 35(3):236–239. **Christensen H**, Jakobsen I (1993) Biol Fertil soils 15(4):253–258. **Cliquet JB**, Stewart GR (1993) Plant Physoil 101(3):865–871. **Cuenca G**, Lovera M (1992) Can J Bot 70(1):73–79.

De Duckett JG, Read DJ (1991) New Phytol 118(2):259–272. **Dumas-Gaudot E**, Furlan V, Grenier J, Asselin A (1992) Mycorrhiza 1(3):133–136.

Egger KN, Sigler L (1993) Mycologia 85(2):219–230. **Ellis JR**, Roder W, Mason SC (1992) Soil Sci Soc 56(3):789–749.

Francl LJ (1993) In: Khan MW (ed) Nematode interactions. Chapman and Hall, London, pp 206–216. **Frey B**, Schüepp H (1993) Soil Biol Biochem 25(6):651–658.

García-Garrido JM, Cabello MN, García-Romera I, Ocampo JA (1992) Soil Biol Biochem 24(10):955–959. **García-Garrido JM**, Toro N, Ocampo JA (1993) Mycorrhiza 2(4):175–177. **George E, Häussler KU**, Vetterlein D, **Gorgus E**, Marschner H (1992) Can J Bot 70(11)2130–2157. **Gianinazzi-Pearson V**, Gianinazzi S (1992) New Phytol 120(4):505–508. **Gianinazzi-Pearson V**, Tahiri-Alaoui A, Antoniw JF, Gianinazzi S, Dumas E (1992) Int J Endocytosis Cell Res 8(2–3):177–185. **Goh CJ**, Sim AA, Lim G (1992) Lindleyana 7(1):13–17. **Goodwin J** (1992) Northwest Sci 66(4):251–260. **Grunewaldt-Stöcker G**, Dehne HW (1989) Z Pflanzenkr Pflanzenschutz 96(6):615–626. **Gunjal SS**, Patil PL (1992) Agrofor Today 4(2):14–15. **Guo BZ**, Hendrix JW, An ZQ, Ferriss RS (1992) Mycologia 84(6)882–885.

Habte M, Manjunath A (1992) Mycorrhiza 2(1):25–31. **Habte M**, Aziz T, Yuen JE (1992) Plant Soil 140(2):263–268. **Hamel C**, Smith DL (1992) Soil Biol Biochem 24(5):499–501. **Hao WY**, Lin XG, Gu XX, Niu JQ (1991) Acta Pedol Sin 28(2): 129–131. **Harlapur SI**, Srikant Kulkarni, Hegde RK, Sreenivasa MN, Rajashekara E (1990) Karnataka J Agric Sci 3(3–4):203–207. **Haugen LM**, Smith SE (1992) Plant Soil 145(1):71–80.

Jasper DA, Abbott LK, Robson AD (1991) New Phytol 118(3):471–476. **Johnson NC**, Pfleger FL (1992) ASA Spec Publ 54:71–99. **Johnson NC**, Pfleger FL, Crookston RK, Simons SR, Copeland PJ (1991) New Phytol 117(4):657–663.

Koide RT, Lu X (1992) Oecologia 90(2):218–226. **Krishnaraj PU**, Sreenivasa MN (1992) Zentralbl Mikrobiol 147(1–2):131–133. **Krishna Prasad KS** (1991) Afro-Asian J Nematol 1(2):130–134. **Khan AG** (1993) Mycorrhiza 3(1):31–38. **Khasa P**, Furlan V, Fortin JA (1992) Trop Agric 69(3):279–283. **Kühn KD** (1991) Angew Bot 65(3–4):187–203. **Kühn KD**, Weber HC, Dehne HW, Gworgwor NA (1991) Angew Bot 65(3–4):169–185.

Lambais MR, Mehdi MC (1993) Mol Plant Microbe Interact 6(1):75–83. **Land S**, von Alten H Schönbeck F (1993) Mycorrhiza 2(4):157–166. **Lei J**, Bécaed G, Catford JG, Piché Y (1991) New Phytol 118(2):289–294. **Lemoine MC**, Gianinazzi-Pearson V, Gianinazzi S, Straker CJ (1992) Mycorrhiza 1(4):137–146. **Li HQ**, Wu GY, Wang HG, Sheng SS, Jin RC (1992) Microbiology 19(2):65–68. **López-Sánchez ME**, Honrubia M (1992) Mycorrhiza 2(1):33–39.

Maia LC, Kimbrough JW, Erdos G (1993) Mycologia 85(2):223–330. **Martins MA** (1992) Mycorrhiza 2(2):69–73. **Martins MA** (1993) Mycol Res 97(7):807–810. **Masuhara G**, Katsuya K, Yamaguchi K (1993) Mycol Res 97(6):746–752. **McArthur DAJ**, Knowles NR (1992) Plant Physiol 100(1):341–351. **Meier R**, Charvat I (1992) Am J Bot 79(8):928–936. **Michelsen A** (1992) For Ecol Manage 48(3–4):335–344. **Miller RM**, Jastrow JD (1992) ASA Spec Publ 54:29–44. **Millner PD**, Kitt DG (1992) Mycorrhiza 2(1):9–15. **Morton JB** (1993) Mycorrhiza 2(3):97–109. **Myers M** (1993) Horticulturist 2(1):8–12.

Nasr AA (1993) Mycorrhiza 2(4):179–182. **Nelson LL**, Allen EB (1993) Restor Ecol 1(1):40–50. **Nelson SD**, Khan SU (1992) Weed Sci 40(1):161–170. **Nemec S** (1992) Adv Hortic Sci 6(2):93–96.

Osman HA, Korayem AM, Ameen HH, Badr-Eldin SMS (1991) Beitr Trop Landwirtsch Veterinärmed 29(3)341–346. **Osonubi O**, Bakare ON, Mulongoy K (1992) Biol Fertil Soils 14(3):159–165.

Palacino JH, Leguizamon CJ (1991) Fitopatol Colomb 15(1):9–17. **Palma JM**, Longa MA, Río L A Del, Arines J (1993) Physiol Plant 87(1):77–83. **Perrin R**, Plenchette C (1993) Crop Protect 12(2):127–133.

Rasmussen H, Rasmussen FN (1992) Lindleyana 6(4):221–227. **Read DJ** (1992) In: Caroll GC, Wicklow DT (eds) The fungal community: its organization and role in the ecosystem. Marcel Dekker New York, pp 631–652. **Rincón E**, Huante P, Ramírez Y (1993) Mycorrhiza 3(2):79–81. **Rosendahl S**, Rosendahl CN, Thingstrup I (1992) Bull OILB/SROB 15(1):48–50.

Sanders IR, Fitter AH (1992a) New Phytol 120(4):517–524; Sanders IR, Fitter AH (1992b) New Phytol 120(4):525–533. **Sharma AK**, Johri BN, Gianinazzi S (1992) J Microbiol Biotechnol 8(6):559–563. **Simon L**, Bousquet J, Lévesque RC, Lalonde M (1993) Nature 363(6424):67–69. **Singh CS** (1992) Zentralbl Mikrobiol 147(7):447–453. **Sivaprasad R**, Jacob A, Sulochana KK, Visalaksyhy A, George B (1992) In: Proc 3rd Int Cong. on Plant Protection in the Tropics. Genting, Highlands, Malaysia, Ooi FAC, Lim GS, Teng PS (eds) College of Agriculture, Vellayani Trivandrum, Kerala, India. **Spain JL** (1992) Mycotaxon 43:331–339. **Sreenivasa MN**, Krishnaraj PU (1992) Zentralbl Mikrobiol 147(1–2):126–130. **Sreenivasa MN**, Nirmalnath PJ, Kulkarni S (1992) Zentralbl Mikrobiol 147(8):509–512. **Straker CJ**, Schnippenkoetter WH, Lemoine MC

(1992) Mycorrhiza 2(2):63–67. **Sulla SB** (1991) In: Prakash J, Pierik, RLM (eds) Horticulture – new technologies and applications. Kluwer Dordrecht, 49–53.

Tisserant B, Gianinazzi-Pearson V, Gianinazzi S, Gollotte A (1993) Mycol Res 97(2): 245–250.

Verdejo S, Calvet C, Pinochet J (1990) Bull Sanidad Vegetal, Plagas 16(3):619–624. **Vierheilig H**, Alt M, Neuhaus JM, Boller T, Wiemken A (1993) Mol Plant-Micorbe Interactions 6(2):261–264 Gier: **"von Alten" Vörös I**, Szegi J (1991) Agrokémia és Talatjan 40(3–4):431–440.

Weathers PJ, Zobel RW (1992) Biotechnol Adv 10(1):93–115. **Wilcox HE** (1991) Plant roots: the hidden half. Marcel Dekker, New York, pp 731–765. **Wu CG**. Sylvia DM (1993) Mycologia 85(2):317–322.

Yao YJ, Pegler DN, Young TWK (1992) Mycologist 6(3):132–137.

Prof. Dr. Fritz Schönbeck
Ingo Raschen, M.Sc. (Kent)
Institut für Pflanzenkrankheiten
und Pflanzenschutz
Herrenhäuser Straße 2
30419 Hannover, Germany

Subject Index

Springer-Verlag and the Environment

We at Springer-Verlag firmly believe that an international science publisher has a special obligation to the environment, and our corporate policies consistently reflect this conviction.

We also expect our business partners – paper mills, printers, packaging manufacturers, etc. – to commit themselves to using environmentally friendly materials and production processes.

The paper in this book is made from low- or no-chlorine pulp and is acid free, in conformance with international standards for paper permanency.

Printing: Mercedesdruck, Berlin
Binding: Buchbinderei Lüderitz & Bauer, Berlin